Sustainability and Resiliency of Building Materials and Structures

Sustainability and Resiliency of Building Materials and Structures

Editors

Shan Gao
Jingxuan Wang
Dewen Kong
Yong Liu

MDPI

Basel • Beijing • Wuhan • Barcelona • Belgrade • Novi Sad • Cluj • Manchester

Editors

Shan Gao
School of Civil Engineering
Harbin Institute
of Technology
Harbin
China

Jingxuan Wang
School of Civil Engineering
Lanzhou University
of Technology
Lanzhou
China

Dewen Kong
College of Civil Engineering
Guizhou University
Guiyang
China

Yong Liu
College of Civil Engineering
Hefei University
of Technology
Hefei
China

Editorial Office
MDPI AG
Grosspeteranlage 5
4052 Basel, Switzerland

This is a reprint of articles from the Special Issue published online in the open access journal *Buildings* (ISSN 2075-5309) (available at: www.mdpi.com/journal/buildings/special_issues/Resiliency_Materials).

For citation purposes, cite each article independently as indicated on the article page online and as indicated below:

Lastname, A.A.; Lastname, B.B. Article Title. *Journal Name* **Year**, *Volume Number*, Page Range.

ISBN 978-3-7258-2098-6 (Hbk)
ISBN 978-3-7258-2097-9 (PDF)
doi.org/10.3390/books978-3-7258-2097-9

Cover image courtesy of Shan Gao

Contents

About the Editors

Shan Gao

Dr. Shan Gao is an associate professor at the School of Civil Engineering, Harbin Institute of Technology, China. He obtained his PhD degree from Harbin Institute of Technology, China in July 2014. His expertise includes the analysis and testing of high-performance composite structures.

Jingxuan Wang

Dr. Wang Jing-Xuan is a professor at the School of Civil Engineering, Lanzhou University of Technology, China. He obtained his PhD degree from Lanzhou University of Technology, China in July 2015. His expertise includes the analysis and testing of building structures subjected to progressive collapse and the behaviour of concrete-filled, steel-tube structures under various loading conditions.

Dewen Kong

Dr. Kong Dewen is professor at the College of Civil Engineering, Guizhou University, China. He obtained his PhD degree from Harbin Institute of Technology, China in July 2015. His expertise includes the analysis and testing of steel–concrete composite structures, research and development of new civil engineering materials, and resource utilization of bulk industrial solid waste.

Yong Liu

Dr. Liu Yong is associate professor at the College of Civil Engineering, Hefei University of Technology, China. He obtained his PhD degree from Harbin Institute of Technology, China in January 2020. His expertise includes the analysis and testing of high performance steel–concrete composite structures under various loading conditions.

buildings

Article

Lateral Impact Response of Concrete-Filled Square Steel Tubes with Local Defects

Yulin Wang [1], Shan Gao [2,3,*], Youchun Xu [1] and Fangyi Li [2]

1 School of Civil Engineering, Xijing University, Xi'an 710123, China; wangyulin614@163.com (Y.W.); xuyouchun666@gmail.com (Y.X.)
2 School of Civil Engineering, Harbin Institute of Technology, Harbin 150090, China; lzm1637972346@163.com
3 Postdoctoral Station of Civil Engineering, Chongqing University, Chongqing 400045, China
* Correspondence: gaoshan@hit.edu.cn

Abstract: A numerical model of 84 high-strength concrete-filled square steel tubular columns (HSCFST) with local defects is developed using ABAQUS. The effects of parameters such as crack angle, crack length, impact surface and impact energy on the impact resistance of HSCFST columns are considered. The results show that under the effect of local corrosion, a model with horizontal cracks will show the phenomenon of crack closure when subjected to the front impact. The impact force platform value is mainly affected by the impact surface, followed by the crack angle, while the increase of the crack length mainly has a greater effect on the model of the rear impact. The impact resistance of the front impact model is better than that of the side and rear impact models. Increasing the crack length and decreasing the crack angle will increase the mid-span deflection of the model, and the mid-span deflection of the front impact model is smaller than that of the side and rear impact models. The energy absorption ratio of the model is proportional to the increase of the crack length and inversely proportional to the increase of the crack angle. Decreasing the crack angle will reduce the increase coefficient (R_d) of the dynamic flexural capacity of the model. A practical calculation method for the increased coefficient of the dynamic flexural capacity of HSCFST columns under local corrosion is proposed.

Keywords: localized corrosion; concrete-filled steel tube; impact; numerical simulation

Citation: Wang, Y.; Gao, S.; Xu, Y.; Li, F. Lateral Impact Response of Concrete-Filled Square Steel Tubes with Local Defects. *Buildings* **2022**, *12*, 996. https://doi.org/10.3390/buildings12070996

Academic Editors: Francisco López Almansa and Elena Ferretti

Received: 7 June 2022
Accepted: 11 July 2022
Published: 13 July 2022

1. Introduction

A concrete-filled steel tube (CFST) structure has the characteristics of high-bearing capacity, good plasticity and toughness. Meanwhile, a concrete-filled steel tube structure has good seismic performance and fire resistance, which is suitable for use in seismic zones as well as in structural engineering that requires fire resistance design [1,2]. Therefore, concrete-filled steel tube structures are widely used in all kinds of civil engineering structures, such as large-span space structures of super high-rise buildings, bridge piers, power towers, subway stations, etc.

Most of the actual engineering design only considers static loads such as constant load, live load and dynamic loads such as wind load and earthquake action, while ignoring the dynamic impact loads with very short action time and huge energy. The impact load will cause rapid deformation of structural members, which will lead to the collapse of the whole structure and the loss of load-bearing capacity in serious cases. Therefore, the lateral impact resistance of concrete-filled steel tube members is particularly important. Bambach et al. [3,4] conducted static and dynamic performance tests and numerical simulation studies on square concrete-filled steel tubular members under impact load, and the results indicated that slender concrete-filled steel tubular members were prone to early fracture, resulting in a decline in energy dissipation capacity. Hou et al. [5] obtained the correlation functions of the improved coefficients of dynamic flexural capacity of concrete-filled circular steel tubular members through regression analysis. The results of the parametric analysis showed that the yield strength of steel, the percentage of steel in the section, the diameter of the section and the impact velocity were the main parameters that affected the dynamic flexural strength of the section. Qu et al. [6] simplified the lateral impact model of concrete-filled circular steel tubes and found that when the section reaches the dynamic ultimate bending moment the angle could be calculated from the bending moment and impact energy. Yang [7] systematically studied the dynamic mechanical response of structural steel and high-strength concrete-filled square steel tubes. The strain rate effect model of S690 high-strength structural steel under three different working conditions was proposed. Cai et al. [8] numerically simulated the lateral impact of concrete-filled square steel tubes and proposed a practical calculation formula for the increased coefficient of section dynamic flexural capacity. Regarding hollow structural steel, Suzuki and Lignos [9] experimentally studied the collapse behavior of full-scale steel HSS under a symmetric lateral displacement loading protocol along with a near-collapse lateral displacement protocol, which represents the ratcheting behavior of steel columns in a special moment frame prior to collapse. Sediek et al. [10] concluded that the depth-to-thickness ratio and initial axial load ratio are the most influential parameters on the axial capacity of hollow structural steel columns under combined axial and lateral loading.

Concrete-filled steel tube structures in different service environments, in addition to the impact-resistant design of the components, should also pay special attention to the existence of easy corrosion of steel shortcomings. As CFST structures are exposed to the natural environment for a long time, it is almost inevitable that the outer steel tube will suffer from corrosion. The probability of local corrosion is much greater than uniform corrosion in construction engineering. When the concrete-filled steel tube member is locally corroded, the local corroded area of the steel tube cannot provide effective restraint to the concrete, which results in the deterioration of the overall performance of the member. Therefore, the effect of corrosive environment on the mechanical properties of concrete-filled steel tubular members is also a concern of many researchers. Gao et al. [11,12] studied the mechanical properties of concrete-filled circular steel tubular short columns under complex environments of freeze–thaw cycles, salt spray corrosion and freeze–thaw–salt spray interaction. The results showed that there was a linear decrease in the axial compression load capacity of the members with the action degree of the three environmental conditions. Han et al. [13] studied the mechanical properties of CFST under the combined action of medium and long-term loading and chloride salt corrosion. The studies showed that the stiffness, bearing capacity and ductility of CFST were significantly reduced by the overall corrosion of the steel tube. Ding et al. [14] studied the axial compression performance of CFST with notch and the results showed that the CFST bearing capacity with notch was significantly reduced. Chen et al. [15] studied the effect of acid rain environment on the mechanical properties of concrete-filled square steel tubes. The results showed that acid rain corrosion reduced not only the bearing capacity but also the energy dissipation capacity of the members.

In the service process of concrete-filled steel tube structures, the external impact on the structural members and the local corrosion on the steel tube occur almost simultaneously and cannot be avoided in the construction project. At present, there is no relevant research on the impact resistance of CFST members under local corrosion and the calculation method of the impact bearing capacity of CFST members based on local corrosion. Therefore, in this paper, ABAQUS software will be used to establish the lateral impact model of high-strength concrete-filled square steel tubular (HSCFST) columns with local corrosion. The corrosion crack of an HSCFST column will be simulated by setting local notches on the outer steel tube of the model. The influence of impact energy, impact surface and corrosion morphology on the impact resistance of the HSCFST column under local corrosion is observed. The design method of the lateral impact capacity of the HSCFST column under local corrosion is explored, which provides the basis for the improvement of the impact resistance design theory of concrete-filled steel tube structures regarding life-cycle service.

2. Numerical Simulation Program

2.1. Model Design

The finite element analysis (FEA) model design adopted the model parameters in the literature [7]. The cross-section size of the HSCFST model is D × D = 180 mm × 180 mm. The model has an overall length of 1700 mm and a steel tube thickness of 6mm. The rigid support is set 250 mm away from both ends of the model. In addition, the weight of the drop hammer is 424 kg and its impact height is 4 m, 8 m and 12 m. Based on the initial model, the local corrosion of the steel tube was simulated by setting local notch cracks with a width of 8 mm. Meanwhile, different crack lengths (l_c), crack angles (θ) and impact surfaces were designed to analyze the lateral impact response of the HSCFST model. The local corrosion parameters and model numbers used in this simulation are detailed in Table 1. The local corrosion locations and crack types of the model are detailed in Figures 1–3.

Table 1. Parameters of the HSCFST model.

Model Number	Impact Surface	Crack Length l_c/mm	Crack Angle θ/(°)	Impact Height H/m	Impact Mass m/kg	Impact Energy E_i/kJ
	F/R/S	180/135/90	H/O/V	4/8/12		
424-4/8/12	–	–	–	4/8/12	424	16.6/33.2/49.9
DH/DO/DV-F-424-4/8/12	F	1D	H/O/V	4/8/12	424	16.6/33.2/49.9
75DH/75DO/75DV-F-424-4/8/12	F	0.75D	H/O/V	4/8/12	424	16.6/33.2/49.9
RH/RO/RV-F-424-4/8/12	F	0.5D	H/O/V	4/8/12	424	16.6/33.2/49.9
DH/DO/DV-R-424-4/8/12	R	1D	H/O/V	4/8/12	424	16.6/33.2/49.9
75DH/75DO/75DV-R-424-4/8/12	R	0.75D	H/O/V	4/8/12	424	16.6/33.2/49.9
RH/RO/RV-R-424-4/8/12	R	0.5D	H/O/V	4/8/12	424	16.6/33.2/49.9
DH/DO/DV-S-424-4/8/12	S	1D	H/O/V	4/8/12	424	16.6/33.2/49.9
75DH/75DO/75DV-S-424-4/8/12	S	0.75D	H/O/V	4/8/12	424	16.6/33.2/49.9
RH/SO/SV-S-424-4/8/12	S	0.5D	H/O/V	4/8/12	424	16.6/33.2/49.9

Note: impact surface: F for front, R for rear, S for side; crack length: 1D = 180 mm, 0.75D = 135 mm, 0.5D = 90 mm; crack angle: H for 0°, O for 45°, V for 90°.

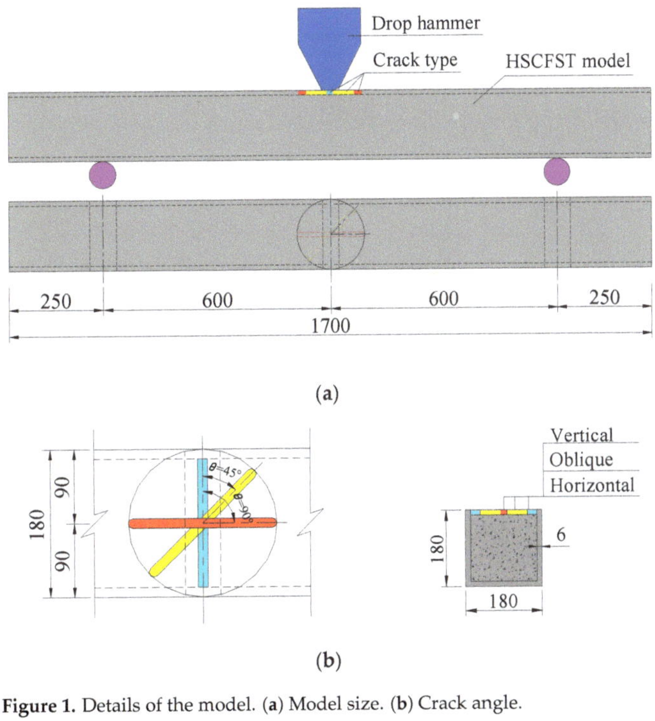

Figure 1. Details of the model. (**a**) Model size. (**b**) Crack angle.

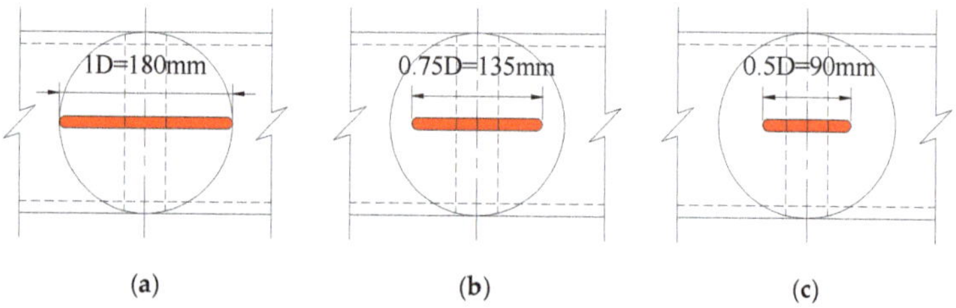

Figure 2. Crack type. (**a**) Type A: 1D. (**b**) Type B: 0.75D. (**c**) Type C: 0.5D.

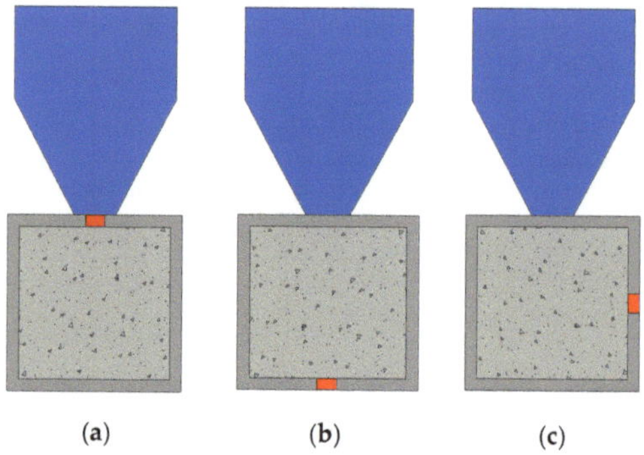

Figure 3. Impact surface (relative to the surface where the cracks are located). (**a**) Type A: front (F). (**b**) Type B: rear (R). (**c**) Type C: side (S).

2.2. Material Model

2.2.1. Material Model for Steel Tube

The structure or component will be in a high strain rate under dynamic conditions such as high-speed impact and explosion. Currently, the Cowper–Symonds (CS) model [16] and the Johnson–Cook (JC) model [17] are commonly used to predict the dynamic mechanical properties of metallic materials. The CS model is commonly used to characterize the dynamic increment coefficient of dynamic yield strength or ultimate strength, which is applied to the calculation of dynamical problems at medium strain rates. The JC model can well predict the relationship between the dynamic flow stress and plastic strain, strain rate and temperature of the material, and it is suitable for the calculation of high strain rate problems [18]. In this paper, the improved JC model (i.e., MJC model, as shown in Equation (1)) from the literature [7] was used as the dynamic mechanical constitutive model of S690 steel. The MJC model can well predict the dynamic mechanical properties of S690 steel at medium and high strain rates. The ideal elastic plastic model based on the Von Mises yield criterion was used to set the steel tube material properties in the finite element model. The material parameters of the steel tubes were adopted from S690 materiality test data in the literature [7], as shown in Table 2.

$$\sigma = (A + B\varepsilon_{\mathrm{p}}^{q})(1 + C\dot{\varepsilon}^{d}) \tag{1}$$

in which

$$\dot{\varepsilon}^{d} = \dot{\varepsilon}/\dot{\varepsilon}_{\mathrm{o}} \tag{2}$$

where σ is the plastic flow stress; ε_{p} is the plastic true strain; $\dot{\varepsilon}^{d}$ is the dimensionless strain rate parameter; $\dot{\varepsilon}$ is the strain rate; $\dot{\varepsilon}_{\mathrm{o}}$ is the quasi-static reference strain rate, i.e., 0.001 s^{-1}; A, B, C, q, and d are material constants. A is usually taken as the yield strength of the material and B and q are used to characterize the degree of strain reinforcement. C and d are the material parameter values of the model, which can be fitted using the dynamic increase coefficients corresponding to the strain rate of the material at each level.

Table 2. Material parameters of steel.

Density (kg/m^3)	fy (MPa)	fu (MPa)	Es (105MPa)	Poisson's Ratio	Elongation
7850	722	758	1.96	0.3	0.15

The material constant of S690 in this paper adopts the fitting results in reference [7], as shown in Table 3.

Table 3. Material constants of S690.

A (MPa)	B (MPa)	C	q	d
727	400	0.06	0.57	0.23

2.2.2. Material Model for Core Concrete

The static stress–strain relationship model and strain rate effect model were used to describe the dynamic mechanical properties of concrete. Since the compressive and tensile mechanical behaviors of concrete were inconsistent, different constitutive relations were used for the compressive and tensile properties of concrete in this simulation.

The compressive mechanical behavior of core concrete adopted the stress–strain relationship model given by Han [1], as shown in Equation (3).

$$y = \begin{cases} 2 \cdot x - x^2 \\ \frac{x}{\beta_o (x-1)^{\eta} + x} \end{cases} \quad (3)$$

in which

$$x = \frac{\varepsilon}{\varepsilon_o} \quad (4)$$

$$y = \frac{\sigma}{\sigma_o} \quad (5)$$

$$\sigma_o = f_c' \quad (6)$$

$$\xi = \frac{A_s f_y}{A_c f_{ck}} \quad (7)$$

$$\varepsilon_o = \varepsilon_c + 800 \times \xi^{0.2} \times 10^{-6} \quad (8)$$

$$\varepsilon_c = (1300 + 1.25 f_c') \times 10^{-6} \quad (9)$$

$$\eta = 1.6 + 1.5/x \quad (10)$$

$$\beta_o = \frac{(f_c')^{0.1}}{1.2\sqrt{1+\xi}} \quad (11)$$

where A_c is the concrete cross-sectional area; A_s is the steel tube cross-sectional area; f_y is the steel yield strength; f_{ck} is the standard value of concrete axial compressive strength; f_c' is the concrete cylindrical compressive strength.

For the tensile response of concrete, the tensile stress–strain relationship model for concrete from GB 50010-2010 [19] is used, as shown in Equation (12).

$$\sigma = (1 - d_t) E_c \varepsilon \quad (12)$$

in which

$$d_t = \begin{cases} 1 - \rho_t (1.2 - 0.2x^5) & x \le 1 \\ 1 - \frac{\rho_t}{\alpha_t (x-1)^{1.7} + x} & x > 1 \end{cases} \quad (13)$$

$$x = \frac{\varepsilon}{\varepsilon_{t,r}} \quad (14)$$

$$\rho_t = \frac{f_{t,r}}{E_c \varepsilon_{t,r}} \quad (15)$$

where α_t is the parameter value of the falling section of the uniaxial stress–strain curve of concrete, which can be calculated in accordance with $\alpha_t = 0.312 f_{t,r}^2$; $f_{t,r}$ is the representative value of the uniaxial tensile strength of concrete, which can be calculated by $f_{t,r} = 0.395 f_{cu}^{0.55}$; $\varepsilon_{t,r}$ is the peak tensile strain of concrete in uniaxial tension, which can be calculated by $\varepsilon_{t,r} = f_{t,r}^{0.54} \times 65 \times 10^{-6}$; E_c is the modulus of elasticity of concrete.

Concrete also showed obvious strain rate effects under dynamic load. In this paper, the empirical formula provided in the European Concrete Institute code [20] was used to calculate the coefficient of dynamic increase of concrete, as shown in Equations (16) and (17).

$$\text{Compression effect}: \begin{cases} f_{cd}/f_{cs} = (\dot{\varepsilon}/\dot{\varepsilon}_{co})^{1.026\alpha_s}, & |\dot{\varepsilon}| \le 30\ \text{s}^{-1} \\ f_{cd}/f_{cs} = \gamma_s(\dot{\varepsilon}/\dot{\varepsilon}_{co})^{1/3}, & |\dot{\varepsilon}| > 30\ \text{s}^{-1} \end{cases} \tag{16}$$

$$\text{Tension effect}: \begin{cases} f_{td}/f_{ts} = (\dot{\varepsilon}/\dot{\varepsilon}_{to})^{1.016\delta_s}, & \dot{\varepsilon} \le 30\ \text{s}^{-1} \\ f_{td}/f_{ts} = \beta_s(\dot{\varepsilon}/\dot{\varepsilon}_{to})^{1/3}, & \dot{\varepsilon} > 30\ \text{s}^{-1} \end{cases} \tag{17}$$

in which

$$\dot{\varepsilon}_{co} = -30 \times 10^{-6}\ \text{s}^{-1} \tag{18}$$

$$\dot{\varepsilon}_{to} = 3 \times 10^{-6}\ \text{s}^{-1} \tag{19}$$

$$\alpha_s = \frac{1}{5 + 9f_c'/f_{co}} \tag{20}$$

$$\delta_s = \frac{1}{10 + 6f_c'/f_{co}} \tag{21}$$

$$f_{co} = 10\ \text{MPa} \tag{22}$$

$$\log \gamma_s = 6.156\alpha_s - 2 \tag{23}$$

$$\log \beta_s = 7.112\delta_s - 2.33 \tag{24}$$

where f_{cd} is the dynamic compressive strength of concrete; f_{td} is the dynamic tensile strength of concrete; f_{cs} is the static compressive strength of concrete; f_{ts} is the static tensile strength of concrete.

The material setup of core concrete was described by the Concrete Damaged Plasticity (CDP model) of concrete built into the FEA. The mechanical parameters of concrete were based on the data obtained in the literature [7], where the elastic modulus is 50.8 GPa; density is 2450 kg/m^3; Poisson's ratio is 0.191. Other specific parameter settings of the model are shown in Table 4 [21].

Table 4. Material parameters of concrete.

e	α_f	K_c	$\mu/10^{-3}$	$\varphi/(°)$
0.1	1.16	0.6667	0.5	30

2.3. Modeling Techniques

The FEA model was primarily composed of the HSCFST model, drop hammer and support, as shown in Figure 4. The impact location was directly above the mid span of the HSCFST model. The drop hammer and support adopted discrete rigid body elements and set reference points to facilitate the application of boundary conditions and impact action. The reference points of the supports were set with fixed constraints. The drop hammer only allowed translational degrees of freedom in the direction of impact (vertically downward).

Hard contact was used in the normal direction of the contact surface in this numerical simulation. The tangential direction of the contact surface between the steel tube, drop hammer and support was set to frictionless. The Coulomb friction model was used to represent the relative sliding between the steel tube and the concrete and its corresponding friction coefficient was taken as 0.6 [22]. The steel tube and the core concrete were simulated by a C3D8R solid element in the ABAQUS library. Drop hammers and supports were simulated by an R3D4 discrete rigid body element from the ABAQUS library. The mesh size of the HSCFST model is 20 mm, which is 1/9 side length. In addition, the mesh size within 300 mm of the impact position of the drop hammer was encrypted to improve the calculation accuracy of the finite element model.

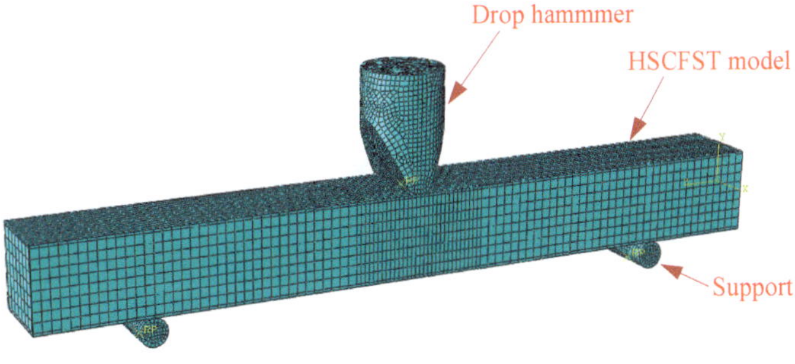

Figure 4. Overall view of FEA model.

Although the ABAQUS/Standard solver that requires a low computational resource is normally adopted in structural analysis, great efforts are required for the convergence of a static solver when complex contact phenomena exist in the model, not to mention model the impact loading. The ABAQUS/Explicit solver could overcome the difficulties of convergence problems associated with impact simulation and sophisticated contact. In this study, about 30–45 min was needed to run an impact simulation of the model by using the ABAQUS/Explicit solver.

2.4. Validation of Finite Element Model

In order to verify the correctness of the FEA model established in this paper, the above numerical simulation method was used to compare the specimen numbered HS3-100-1200-6 in reference [7]. The specimen has the same material strength, specimen size and boundary conditions as the finite element model established in this paper, except that the height of drop hammer impact is 3 m. The chemical composition of the high-strength steel is listed in Table 5. The mix proportion of C100 concrete is listed in Table 6.

Table 5. Chemical composition of steel (in mass %).

C	Si	Mn	S	P	N	Ni	Mo	Cr	Cu	V	Ti	Nb	CEV
0.14	0.29	1.26	0.0006	0.014	0.0018	0.02	0.11	0.2	0.01	0.003	0.023	0.015	0.42

Table 6. Mix proportion of C100 concrete.

Water Cement Ratio	Water kg/m^3	Cement kg/m^3	Sand kg/m^3	Gravel kg/m^3	Silica Fume kg/m^3	Superplasticizer kg/m^3
0.22	121	500	623	1156	50	10

Figure 5 shows the time history curve comparison of impact force F and displacement U of specimen HS3-100-1200-6. It can be seen that the finite element simulation results of the peak impact force, the maximum mid-span deformation and the platform value are in good agreement with the test data, which proves the reliability of the finite element model.

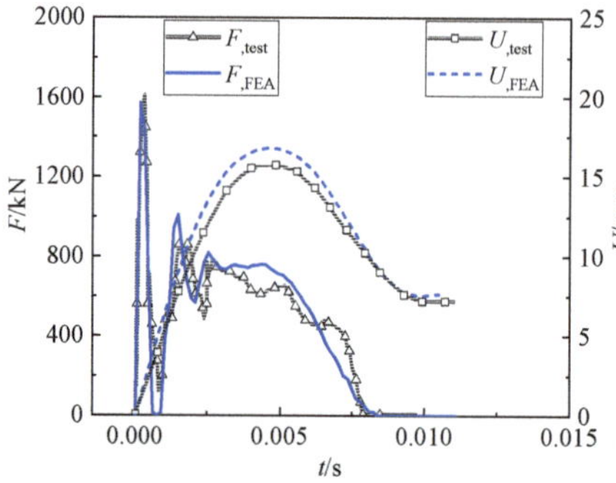

Figure 5. Time history curves of impact force and mid-span displacement of specimen HS3-100-1200-6.

3. Numerical Simulation Results

3.1. Failure Pattern of Crack-Free Model

The HSCFST model principally shows a symmetric overall bending failure on both sides under lateral load. As shown in Figures 6 and 7, the plastic hinge is concentrated in the mid-span drop hammer impact zone. The surface of the model is slightly depressed at the contact position with the drop hammer. Meanwhile, the local "drum" buckling phenomenon occurred on both sides of the mid span and near the impact position due to the extrusion of the drop hammer.

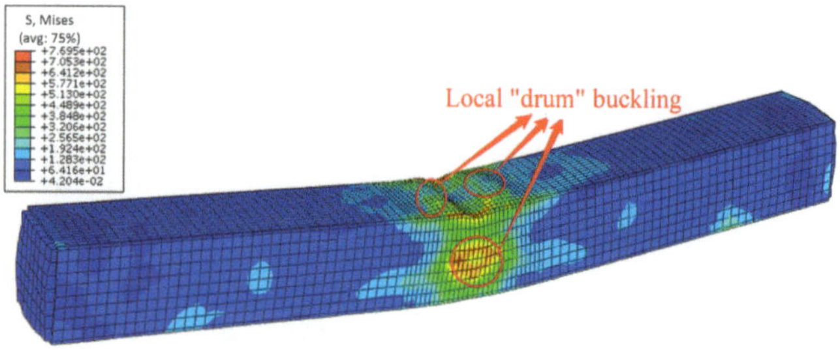

Figure 6. Typical overall bending failure.

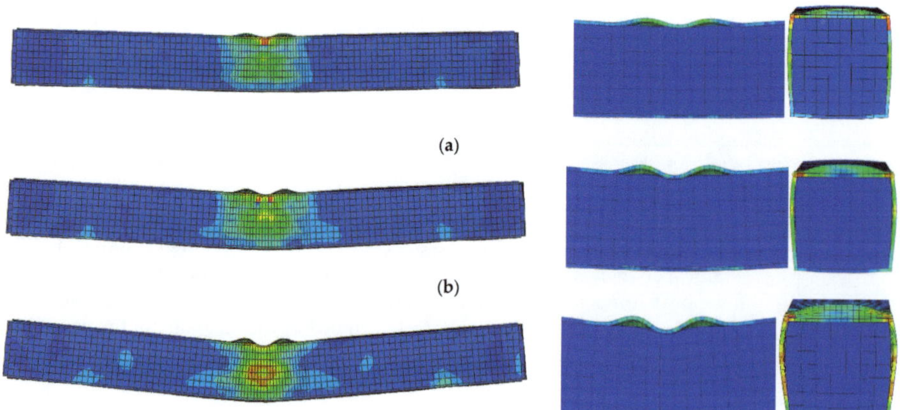

Figure 7. Local "drum" buckling diagram of crack-free model. (**a**) Model 424-4. (**b**) Model 424-8. (**c**) Model 424-12.

3.2. Failure Pattern of the Model under Local Corrosion

Local notch cracks generally increase the amplitude of overall bending and local "drum" buckling of the model. In addition, the change of crack angle and impact surface affects the failure pattern of the model. In order to better reflect the influence of each parameter on the HSCFST model, this paper adopts the failure pattern of the model with an impact height of 12 m for comparison.

3.2.1. Models with Horizontal Cracks ($\theta = 0°$)

Compared with the crack-free model, the front impact will reduce the amplitude of buckling near the impact point of the drop hammer and increase the amplitude of buckling on both sides of the model. The rear impact will reduce the amplitude of the buckling near the impact point of the drop hammer, while both sides of the model have a slight buckling near the top surface. The side impact model has a small amplitude of buckling near the impact point of the drop hammer and in the crack area. In addition, the crack width near the impact surface will decrease while the crack width far from the impact surface will increase when the model is side impacted. As the impact energy or the crack length increases, the horizontal crack will gradually shrink or crack close under front impact, as shown in Figure 8; crack width no longer decreases or crack closure generally occurs at the end of the impact platform segment. With the increase of impact energy, the time of crack closure is relatively earlier than the time of impact platform section end. Nevertheless, the model with crack width no longer decreasing or crack closure occurring earlier has lower impact platform values. Back impact will increase the width of cracks, while the increase of crack length and impact energy will lead to excessive local stress at both ends of cracks and yield failure phenomenon, as shown in Figure 9. In side impact, the increase of crack length and impact energy will lead to the yield failure of the steel tube at the bottom of the crack. In addition, the increase in crack length reduces the local buckling amplitude of the side impact model and the cracks eventually take on similar "triangular" shapes, as shown in Figure 10.

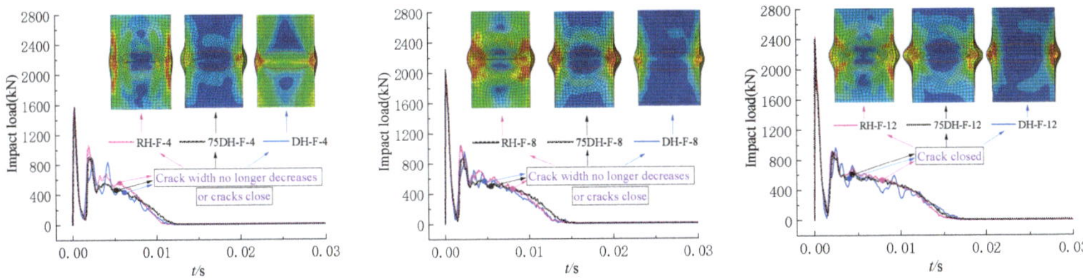

(a) (b) (c)

Figure 8. Horizontal crack closure time under front impact. (**a**) Impact height: 4 m. (**b**) Impact height: 8 m. (**c**) Impact height: 12 m.

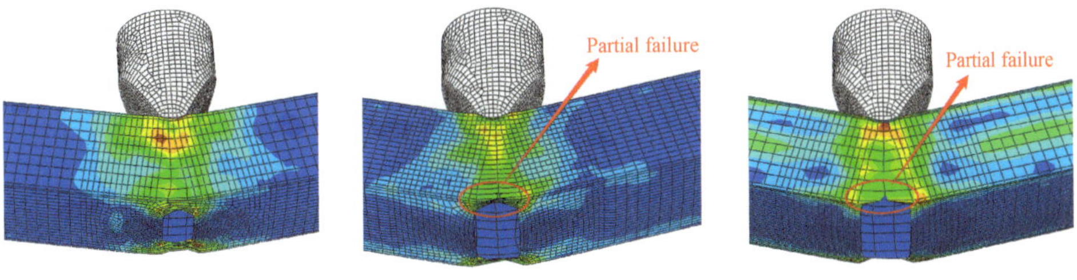

(a) (b) (c)

Figure 9. Failure pattern of a model with horizontal cracks under rear impact. (**a**) RH-424-R-12. (**b**) 75DH-424-R-12. (**c**) DH-424-R-12.

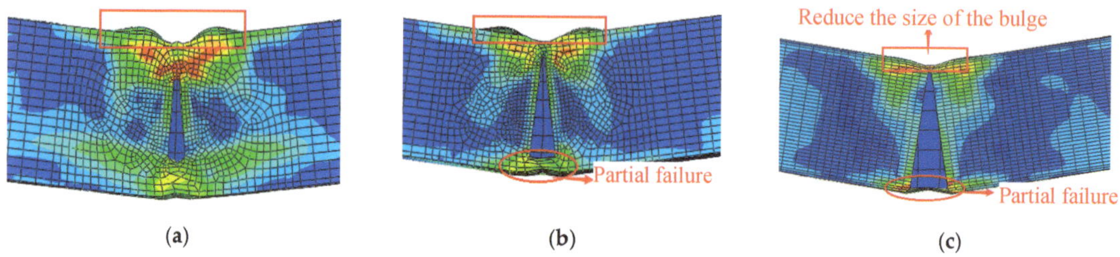

Figure 10. Failure pattern of a model with horizontal cracks under side impact. (**a**) RH-424-S-12. (**b**) 75DH-424-S-12. (**c**) DH-424-S-12.

3.2.2. Models with Vertical or Oblique Cracks ($\theta = 45^\circ, 90^\circ$)

The model with oblique and vertical cracks presents an obvious bending failure pattern. Compared with the crack-free model, the front impact will increase the local "drum" buckling amplitude of the model. For the model with oblique cracks, increasing the crack length will increase the amplitude and area of buckling on both sides of the mid span, while the amplitude of buckling near the impact point will decrease. The phenomenon of the vertical cracks model is the opposite to that of the oblique cracks model. In addition, the oblique crack width gradually decreases but does not close during impact, while the vertical crack width basically remains unchanged, as shown in Figure 11.

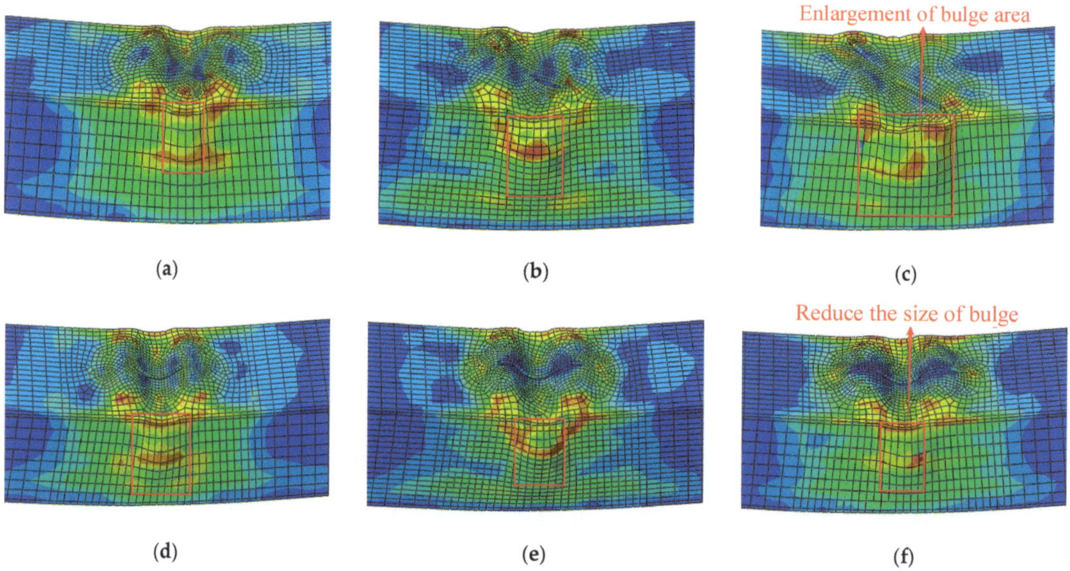

Figure 11. Failure patterns of models with oblique and vertical cracks under front impact. (**a**) RO-424-F-12. (**b**) 75DO-424-F-12. (**c**) DO-424-F-12. (**d**) RV-424-F-12. (**e**) 75DV-424-F-12. (**f**) DV-424-F-12.

When the model is subjected to rear impact, the model with oblique crack almost does not appear to buckle near the impact point, while the buckling amplitude of the model with vertical crack is not much different from that of the crack-free model. In addition, increasing the crack length will increase the width of oblique cracks and the local buckling area of the vertical crack model, as shown in Figure 12.

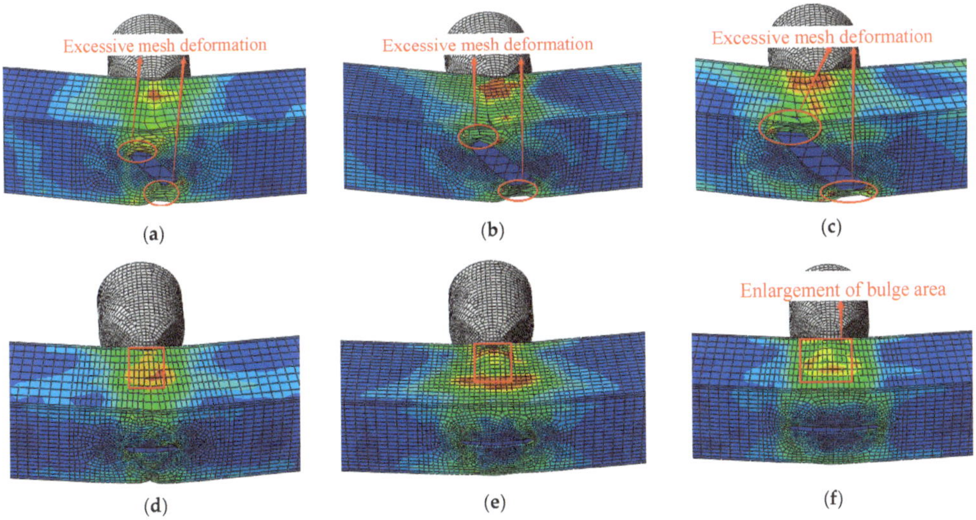

Figure 12. Failure patterns of models with oblique and vertical cracks under rear impact. (**a**) RO-424-R-12. (**b**) 75DO-424-R-12. (**c**) DO-424-R-12. (**d**) RV-424-R-12. (**e**) 75DV-424-R-12. (**f**) DV-424-R-12.

When the model is subjected to side impact, the buckling amplitude of the model with oblique cracks is lower than that of the crack-free model, while the buckling amplitude of the model with vertical cracks is increased. Increasing the crack length will enlarge the buckling area of the model on both sides of the mid span, especially the crack area near the impact surface, as shown in Figure 13.

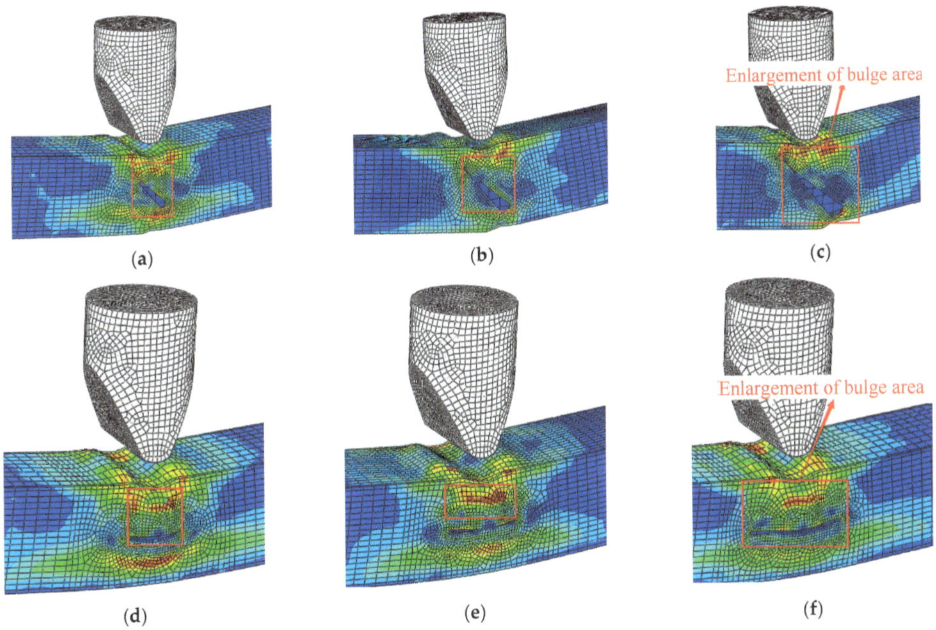

Figure 13. Failure patterns of models with oblique and vertical cracks under side impact. (**a**) RO-424-S-12. (**b**) 75DO-424-S-12. (**c**) DO-424-S-12. (**d**) RV-424-S-12. (**e**) 75DV-424-S-12. (**f**) DV-424-S-12.

3.3. *Impact Force versus Time Curves*

3.3.1. Comparison between Crack-Free Model and Local Corrosion Model

Figure 14 shows that raising the impact height has almost no effect on the impact platform value of the crack-free model, but it will significantly increase the peak segment of the impact time history curve and prolong the impact duration of the model. When the impact height is increased to 8 m and 12 m, the peak segment of the impact time history curve increases by 56% and 90%, respectively, while the impact duration increases by 16% and 39%, respectively.

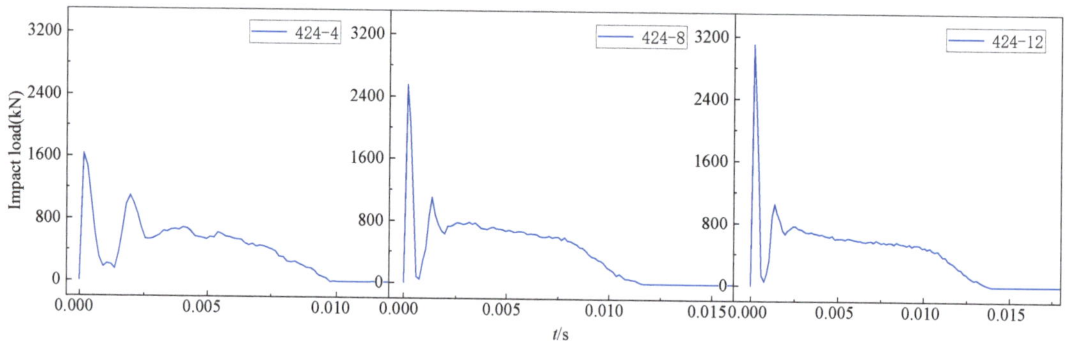

Figure 14. Time history curve of impact force for crack-free model.

Figure 15 takes the same impact energy (m = 424 kg, H = 4 m) as an example to analyze the influence of impact surface (F, R and S), crack angle (H, O and V) and crack length (0.5D, 0.75D and 1D) on the impact time history curve. Figure 15a shows that the three impact surfaces affect the impact duration to different degrees. The increase in impact duration for the models subjected to front, rear and side impacts are 12%, 16.4% and 6%, respectively. Figure 15b shows that the change of crack angle has a stable influence on impact duration. The influence of the crack angle on the impact duration of the model is about 11%. Figure 15c shows that the increase in crack length prolongs the impact duration of the model by about 12~30%. Meanwhile, the changes in impact surface, crack angle and crack length will reduce the impact platform value of the model to varying degrees. Among the three impact surfaces, front impact reduces the impact platform value by about 11%; rear impact reduces the impact platform value by about 26%; side impact reduces the impact platform value by less than 5%. Crack angles (H, O and V) reduce the impact platform values by 11%, 5% and 7%, respectively. In addition, when the crack length increases from 0.5D to 1D, the impact platform value reduction of the model is increased from 11% to 30%.

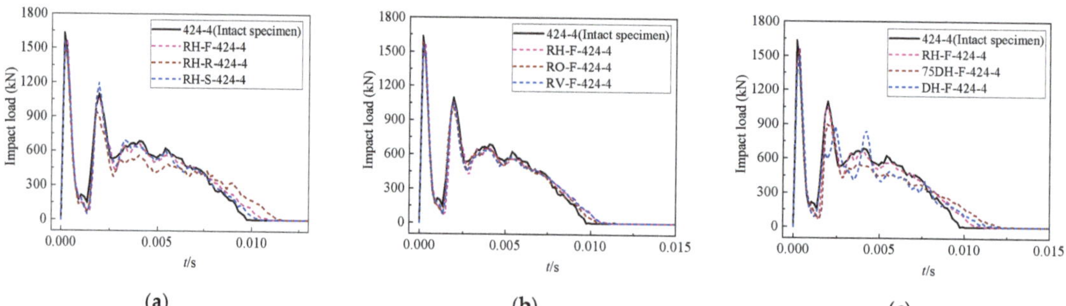

Figure 15. Comparison of each parameter with the crack-free model. (**a**) Change of impact surface. (**b**) Change of crack angle. (**c**) Change of crack length.

3.3.2. Comparison of Horizontal Crack Models

1. Comparison of peak segment of impact time history curve

As shown in Figure 16, the peak segment of the impact time history curve of the model with horizontal cracks is mainly influenced by the impact height. When the impact height is increased to 8 m and 12 m, the peak segment of the model impact time history curve will increase by 31% and 56%, respectively. Compared with the peak segment of the impact time history curve of the crack-free model, the peak segment of the impact time history curve of the horizontal crack model is basically unchanged when the impact height is 4 m. When the impact height is increased to 8–12 m, the peak segment of the impact time history curve will be reduced by about 21%.

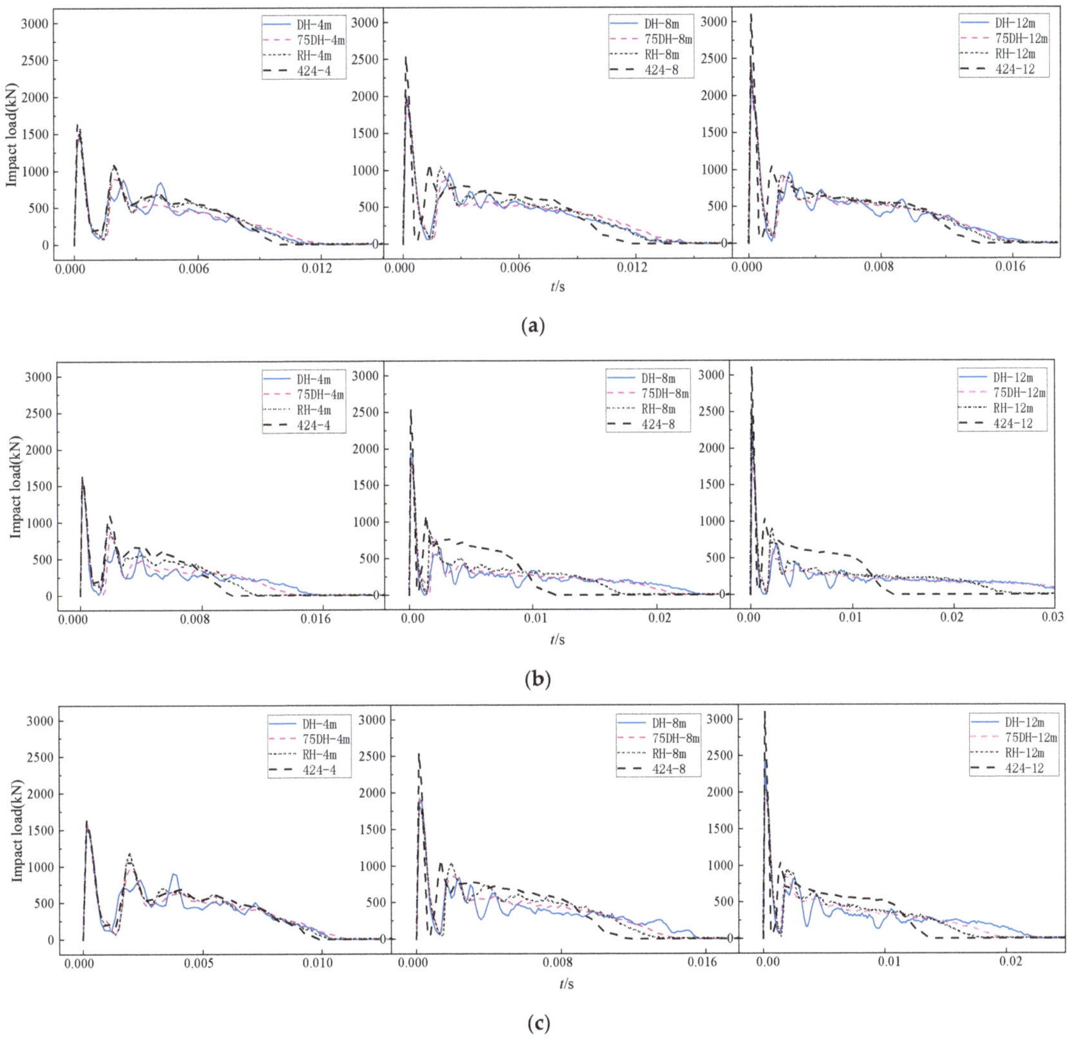

Figure 16. *Cont.*

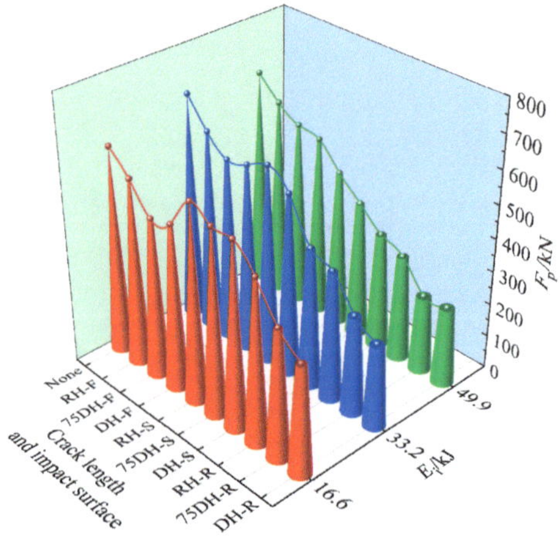

(d)

Figure 16. Comparison of impact time history curve and impact platform value of horizontal crack model. (**a**) Front impact. (**b**) Rear impact. (**c**) Side impact. (**d**) Impact platform value of horizontal crack model.

2. Comparison of impact duration

Figure 16a,c shows that crack length, impact surface and impact energy all affect the impact duration of the horizontal crack model. Overall, the impact surface has the greatest effect on the impact duration, while the crack length has the least effect on the impact duration. With the increase in impact energy, the impact duration of the front impact model is more stable in variation and its overall performance is better than that of the rear and side impact models.

3. Comparison of impact platform values

Figure 16d shows that the impact platform values show a downward trend with the change of impact surface (F, S and R). Increasing impact energy has a greater influence on the impact platform values of the side and rear impact models, while it has little influence on the impact platform values of the front impact model. Meanwhile, increasing the crack length will reduce the impact platform value of the model. Compared with the crack-free model, the impact platform value of the model with front impact decreases by about 17%. Under impact heights of 4 m, 8 m and 12 m, the impact platform values of the rear impact model decrease by 35%, 55% and 60% respectively, while the impact platform values of the side impact model decrease by 8%, 26% and 38% respectively. Figure 16a–d show that the impact platform value and impact duration of the model with horizontal cracks are relatively stable under the front impact load and it exhibits better impact resistance performance.

3.3.3. Comparison of Oblique Crack Models

1. Comparison of peak segment of impact time history curve

As shown in Figure 17a–c, the peak segment of the impact time history curve is mainly influenced by impact height. When the impact height is raised to 8 m and 12 m, the peak segments of the impact time history curves of the model increase by about 23% and 44%, respectively. Compared with the crack-free model, the peak segment of the impact time history curve of the model with oblique cracks is basically unaffected when the impact height is 4 m. However, when the impact height is increased to 8–12 m, the peak segment of the impact time history curve of the model decreases by about 24%.

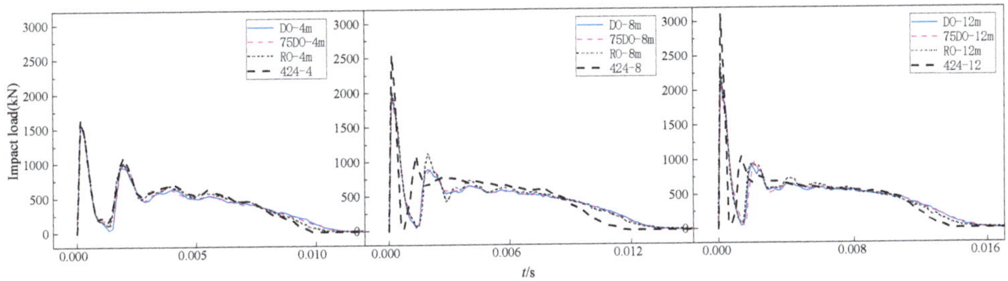

(a)

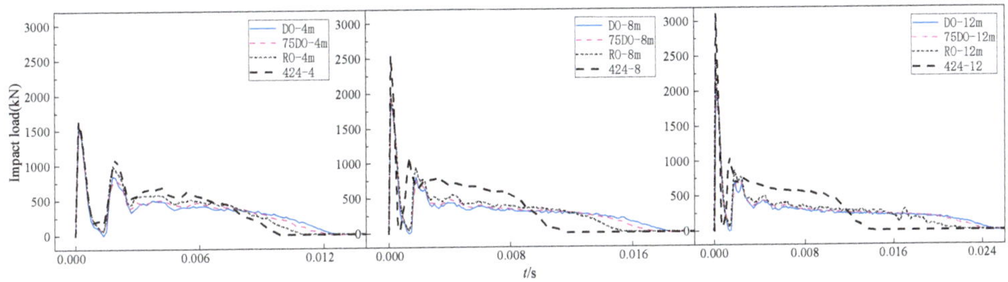

(b)

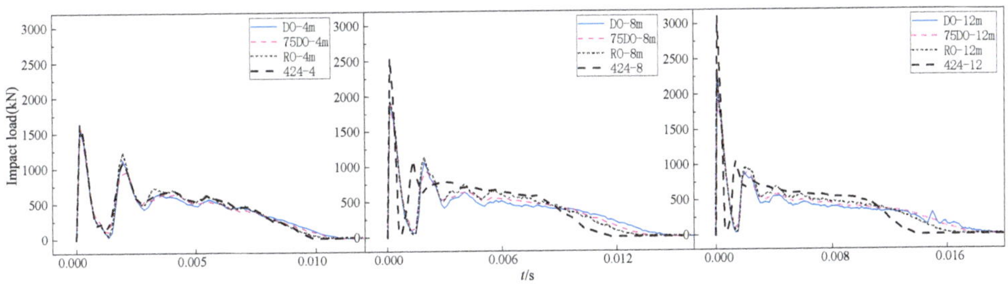

(c)

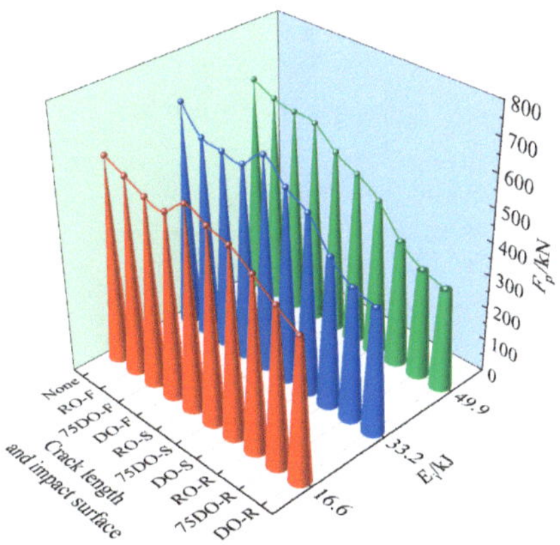

(d)

Figure 17. Comparison of impact time history curve and impact platform value of oblique crack model. (**a**) Front impact. (**b**) Rear impact. (**c**) Side impact. (**d**) Impact platform values of oblique crack model.

2. Comparison of impact duration

Figure 17a–c show that the impact duration of the oblique crack model is less affected by crack length than that of the horizontal crack model. Among the three impact surfaces, the rear impact significantly increases the impact duration of the model, while the front impact has the least effect on the model impact duration. In addition, the impact duration of the model is gradually prolonged with the increase of impact energy.

3. Comparison of impact platform values

Figure 17d shows that the impact platform values overall show a decreasing trend with the change of impact surface (F, S and R). Increasing impact energy will reduce the impact platform values of side and rear impact models, while the impact platform values of front impact models will increase. Meanwhile, increasing crack length will reduce the impact platform value of side and rear impact models, but it has basically no effect on the front impact model. Under impact heights of 4 m, 8 m and 12 m, the impact platform values of the front impact model decrease by about 10%, while the impact platform values of the rear impact model decrease by 24%, 42% and 49%, respectively. The impact platform values of the side impact model decrease by 6%, 16% and 22% at impact heights of 4 m, 8 m and 12 m, respectively. Compared with the horizontal cracked model, the crack length and impact energy have less effect on the oblique cracked model. In addition, the front impact model still shows good impact resistance in the three impact surfaces.

3.3.4. Comparison of Vertical Cracking Models

1. Comparison of peak segment of impact time history curve

As shown in Figure 18a–c, the peak segment of the impact time history curve is mainly affected by impact height. When the impact height is increased to 8 m and 12 m, the peak segment of the impact time history curve of the vertical cracks model increases by 24% and 45%, respectively. Compared with the crack-free model, the peak segment of the impact time history curve of the vertical crack model is basically unaffected when the impact height is 4 m. When the impact height is increased to 8–12 m, the peak segment of the impact time history curve of the model decreases by about 25%.

2. Comparison of impact duration

Figure 18a–c show that the crack length only has a great influence on the impact duration of the rear impact model. Overall, the impact duration of the vertical crack model is mainly affected by the impact energy followed by the crack length. The impact surface has little influence on the impact duration of the vertical crack model.

3. Comparison of impact platform values

Figure 18d shows that both the reduction of the impact energy and the increase of crack length will improve the impact platform value of the rear impact model. In general, the impact platform value of the vertical crack model is relatively stable as a whole, and its impact platform value varies in the range of 4% to 16%. The results show that the overall impact resistance of the vertical crack model is better than that of the model with horizontal and oblique cracks.

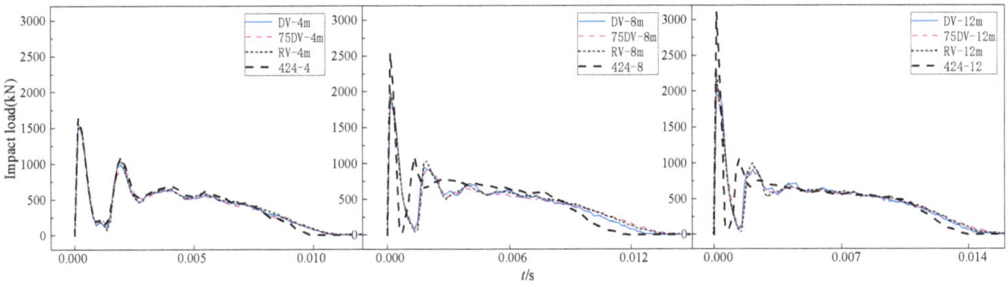

(a)

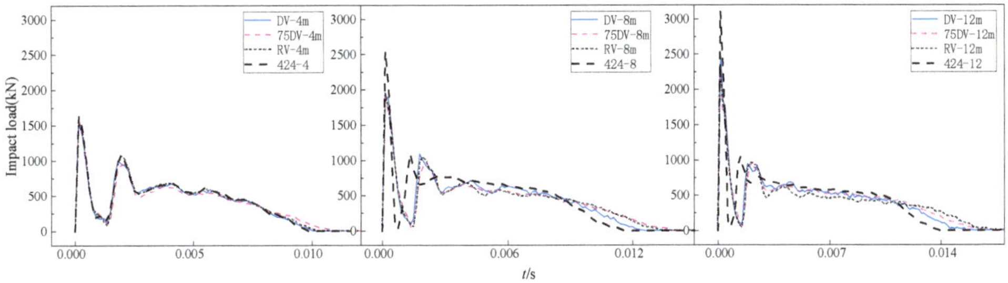

(b)

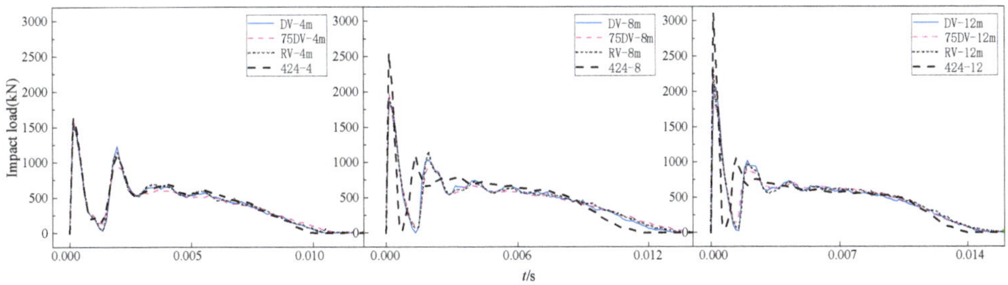

(c)

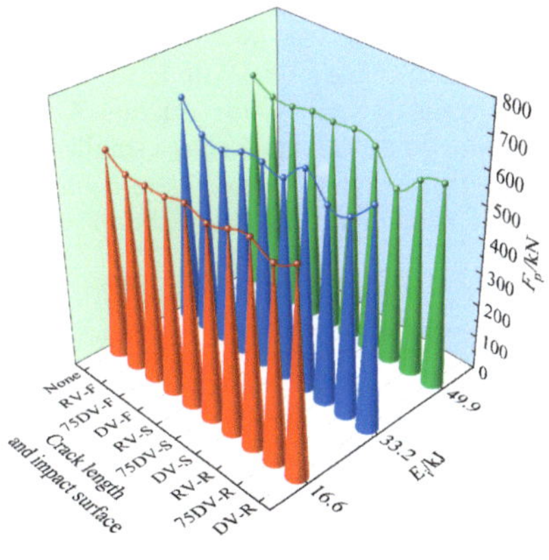

(d)

Figure 18. Comparison of impact time history curve and impact platform value of vertical cracking model. (**a**) Front impact. (**b**) Rear impact. (**c**) Side impact. (**d**) Impact platform values of vertical crack model.

3.4. Mid-Span Displacement versus Time Curves

The model starts to rebound after reaching the mid-span peak displacement, and its mid-span displacement reverses until the energy is completely absorbed. In this case, the mid-span final deflection of the model is the mid-span final displacement. The mid-span final displacement of the model is generally lower than its peak displacement. Figure 19 shows the influence of each parameter on mid-span peak displacement. Overall, the mid-span peak displacement of the model gradually increases with the increase of impact energy and crack length. Among the three impact surfaces, the mid-span peak displacement of the rear impact model is the largest, while the mid-span peak displacement of the front impact model is the smallest. Increasing the crack angle can effectively reduce the effect of impact surface and crack length on the mid-span peak displacement of the model. In addition, the crack length has the greatest effect on the rear impact model and the least effect on the front impact model.

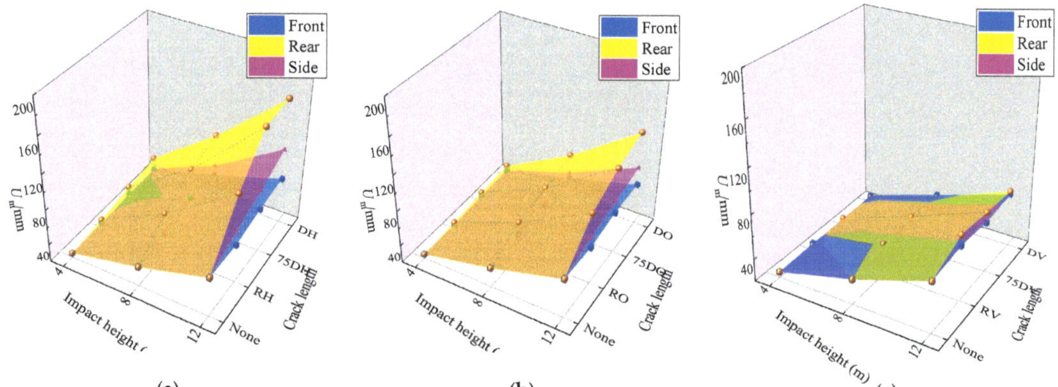

(a) (b) (c)

Figure 19. Comparison of peak displacement of models. (**a**) Horizontal crack. (**b**) Oblique crack. (**c**) Vertical crack.

Figure 20 shows that when the impact height of the crack-free model is increased to 8 m and 12 m, its mid-span peak displacement increases by 68% and 160%, respectively. The existence of cracks will deepen the influence of impact height on the mid-span peak displacement of the model. Meanwhile, increasing the impact height will reduce the rebound amplitude of the model with local corrosion, especially the model with horizontal cracks that has the most obvious reduction. Combining Figures 19 and 20, it can be seen that the model with vertical cracks has a smaller change in the mid-span peak displacement under the front impact and a larger rebound after reaching the mid-span peak displacement, which shows better impact resistance stability.

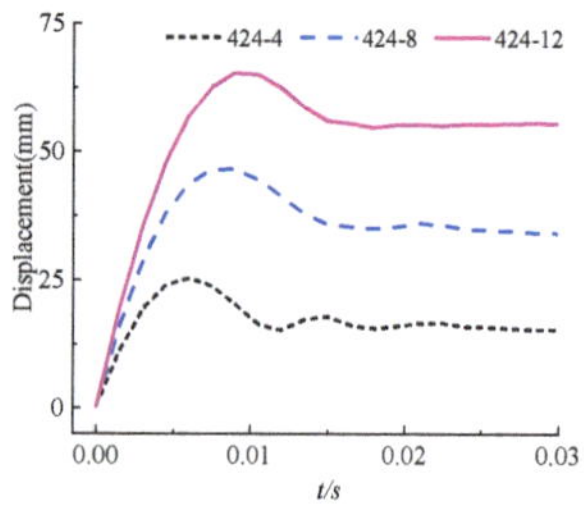

(**a**)

Figure 20. *Cont.*

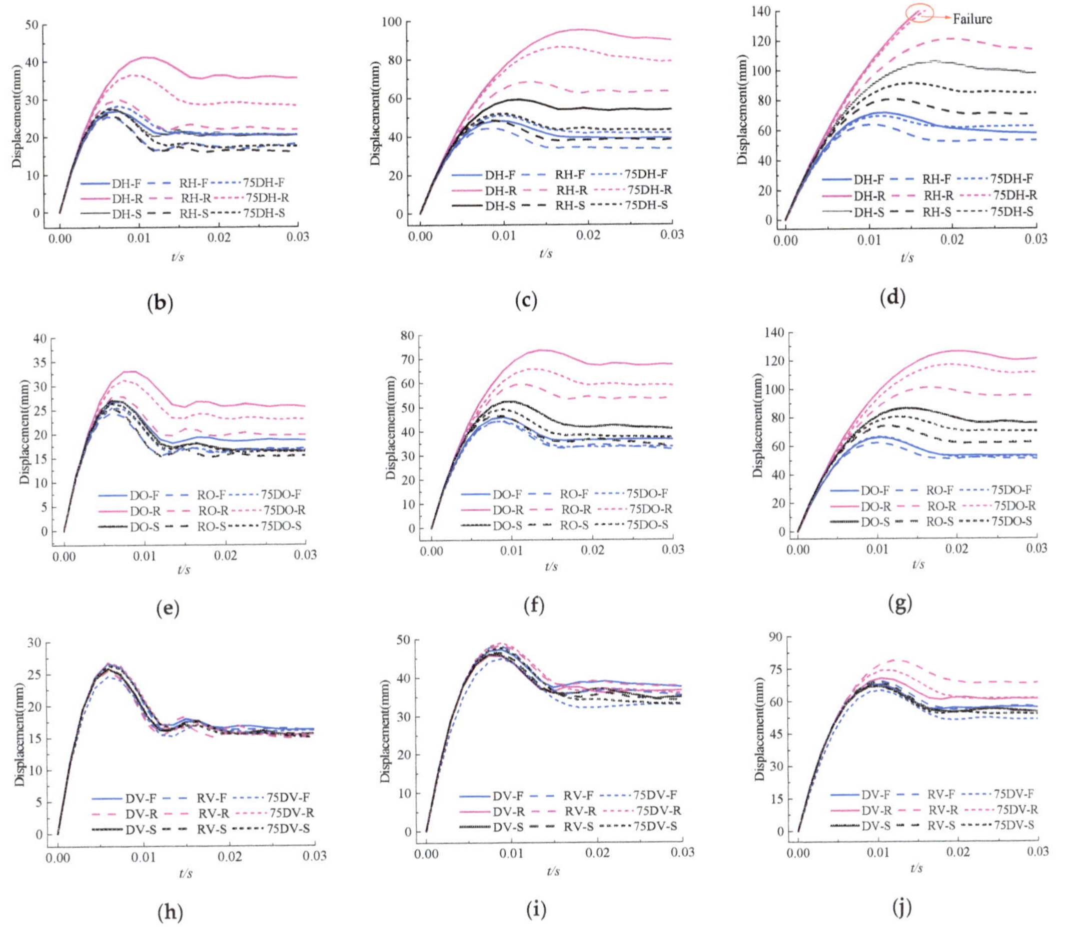

Figure 20. Mid-span displacement time history curve of the models. (**a**) 424-4/8/12 (**b**) H-424-4. (**c**) H-424-8. (**d**) H-424-12. (**e**) O-424-4. (**f**) O-424-8. (**g**) O-424-12. (**h**) V-424-4. (**i**) V-424-8. (**j**) V-424-12.

3.5. Comparison of Impact Platform Value and Mid-Span Peak Displacement

Figure 21 shows the influence of each parameter on the impact platform value and mid-span peak displacement of the model. As shown in the figure, the impact platform value is negatively correlated with the overall change trend of mid-span peak displacement. Increasing the crack length will decrease the impact platform value and increase the mid-span peak displacement. Among the three impact surfaces, the rear impact has the greatest influence on the impact platform value and mid-span peak displacement of the model, while the front impact has the least influence on the impact platform value and mid-span peak displacement of the model. The influence of crack length and impact surface on the model decreases with the increase of crack angle. Overall, the impact surface has the greatest influence on the impact platform value and mid-span peak displacement of the model, while increasing the crack angle can effectively reduce the influence of the impact surface.

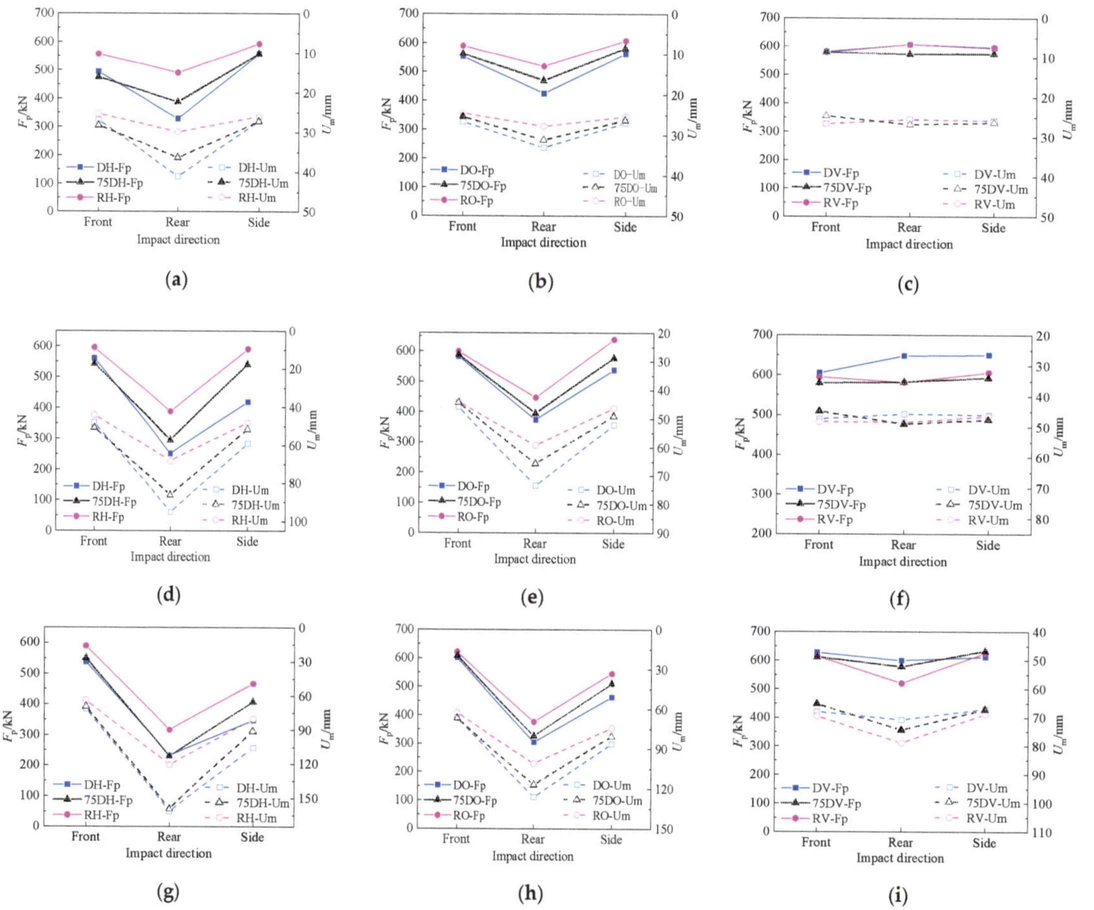

Figure 21. Comparison of impact platform value and mid-span peak displacement of the model. (**a**) H-424-4. (**b**) O-424-4. (**c**) V-424-4. (**d**) H-424-8. (**e**) O-424-8. (**f**) V-424-8. (**g**) H-424-12. (**h**) O-424-12. (**i**) V-424-12.

3.6. Energy Absorption Ratio of the Model

The energy loss caused by friction during impact is ignored in this paper. The total gravitational potential energy of the free-falling motion of the drop hammer is used as the impact energy of the whole system, i.e., $E_i = mgH$. The total impact energy of the drop hammer in the first impact process will be divided into three parts, namely the energy absorbed by the local denting and buckling, the energy absorbed by the overall bending deformation and the remaining kinetic energy of the drop hammer. Local corrosion will increase the overall bending and local buckling deformation of the model to a certain extent. Since the amplitude of the local deformation of the model is smaller than that of the whole bending deformation, most of the energy may be absorbed mainly by the whole bending deformation of the model [23].

The impact force mid-span displacement curve is obtained by combining the impact force and mid-span displacement time history curves of the model, thus quantifying the energy absorbed by the whole bending deformation of the model. A typical impact force mid-span displacement curve is shown in Figure 22. The energy value E_g absorbed by the whole bending deformation of each model can be obtained by mathematically integral calculation of the area enveloped by the impact force mid-span displacement curve of each model. Thus, the energy absorption ratio (EAR) of the whole deformation can be obtained, i.e., EAR = E_g/E_i. As shown in Figure 23a, the EAR of the crack-free model increased by 7.2% and 11% as impact height increased to 8 m and 12 m, respectively. Figure 23b–d show the influence of impact surface, crack length and crack angle on the EAR of the model. Overall, the EAR of the model is generally inversely proportional to the increase in crack angle and directly proportional to the increase in crack length. The front impact will reduce the EAR of the model, while the side and rear impacts will generally increase the EAR of the model. The energy absorption ratio of the rear impact model is the highest among the three impact surfaces. This indicates that the model is more favorable to absorbing impact energy when subjected to rear impact under local corrosion.

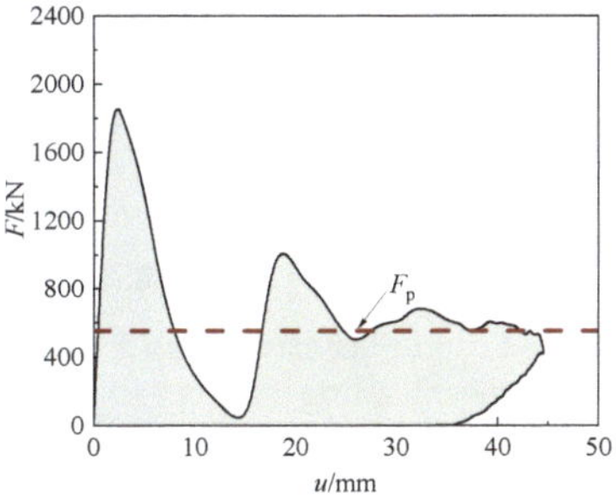

Figure 22. Typical impact force mid-span deformation curve (RV-F-424-8).

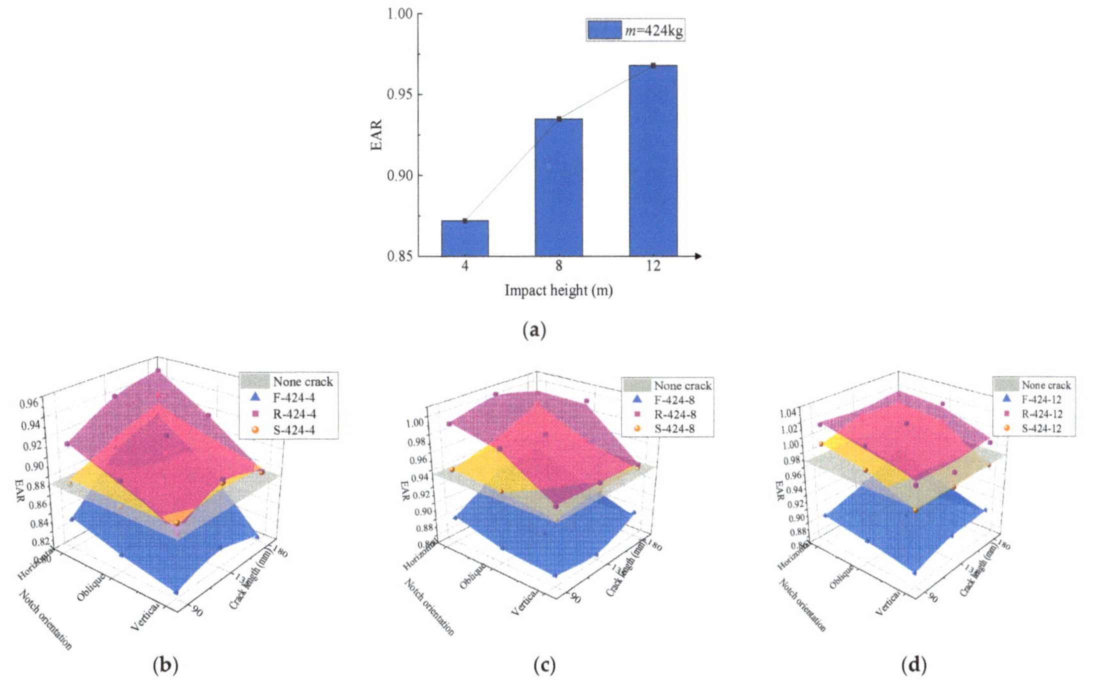

Figure 23. Comparison of energy absorption ratio. (**a**) 424-4/8/12. (**b**) Impact height: 4 m. (**c**) Impact height: 8 m. (**d**) Impact height: 12 m.

3.7. The Evaluation of Dynamic Flexural Strength

For the convenience of the analysis, the concept of increasing coefficient R_d of dynamic flexural strength of the model is defined as shown in Equation (25). According to the literature [1], the static ultimate bending moment M_{su} of CFST under bending failure is shown in Equation (26).

$$R_d = \frac{M_d}{M_{su}} \tag{25}$$

$$M_{su} = \gamma_m \cdot W_{scm} \cdot f_{scy} \tag{26}$$

in which

$$M_d = \frac{E_p \cdot L}{4U} \tag{27}$$

$$\gamma_m = 1.04 + 0.48\ln(\xi + 0.1) \tag{28}$$

$$\xi = \frac{A_s \cdot f_y}{A_c \cdot f_{ck}} \tag{29}$$

$$W_{scm} = B^3/6 \tag{30}$$

$$f_{scy} = (1.18 + 0.85\xi)f_c \tag{31}$$

where M_d is the simplified rigid-plasticity theory formula, which is the dynamic plastic moment of CFST under impact loading [8]; L is the effective length of the model; E_p and U are the plastic strain energy and mid-span final deflection obtained by FEA; M_{su} is the flexural strength value of the model under static loading; γ_m is the calculation coefficient of flexural strength; ξ is the degree of the confinement on the concrete fill provided by the steel tube; A_s and A_c are the cross-sectional areas of the steel tube and core concrete; f_y is the yield strength of steel; f_{ck} is the standard value of concrete compressive strength; W_{scm} is the modulus of flexural resistance of the model section; B is the cross-section side length of the model; f_{scy} is the index of axial compression strength bearing capacity of the model.

Figure 24a shows that the R_d of the crack-free model decreases with the increase of impact energy. It shows that the increase of flexural strength of the model decreases gradually. This is because when the impact height increases to 8 m and 12 m, the plastic strain energy E_p of the model increases only by 6% and 29.3%, while the mid-span final deflection increases by 99.1% and 264.8%, respectively. It can be seen that the mid-span final deflection values of the model basically increase exponentially. The relative value of the strength increase of the model is decreased due to the small difference of the strength increase caused by the strain rate effect of the material. Figure 24b–d show the influence of crack length and crack angle on R_d under different impact surfaces. Overall, the R_d value of the model can be reduced by decreasing the crack angle and increasing the crack length. With the increase of impact height, the overall variation amplitude of the rear impact model is more obvious than that of the front and side impact models. The R_d value of the side impact model is basically higher than that of the front and rear impact models when the impact height is 4 m. In addition, the maximum reduction in R_d value for the front impact model is 8% when the impact height is 12 m, while the maximum reduction in Rd value for the rear and side impact models is 62.2% and 38.5%, respectively. The results show that the decrease in crack length and the increase in crack angle are proportional to the increase of R_d. Under the same crack angle and crack length, the R_d reduction amplitude of the rear impact model is more obvious than that of the front impact and side impact models. The impact resistance of side impact models is better than that of front and rear impact models when impact energy is small. When the impact energy is large, the impact resistance of the front impact model is better than that of rear and side impact models.

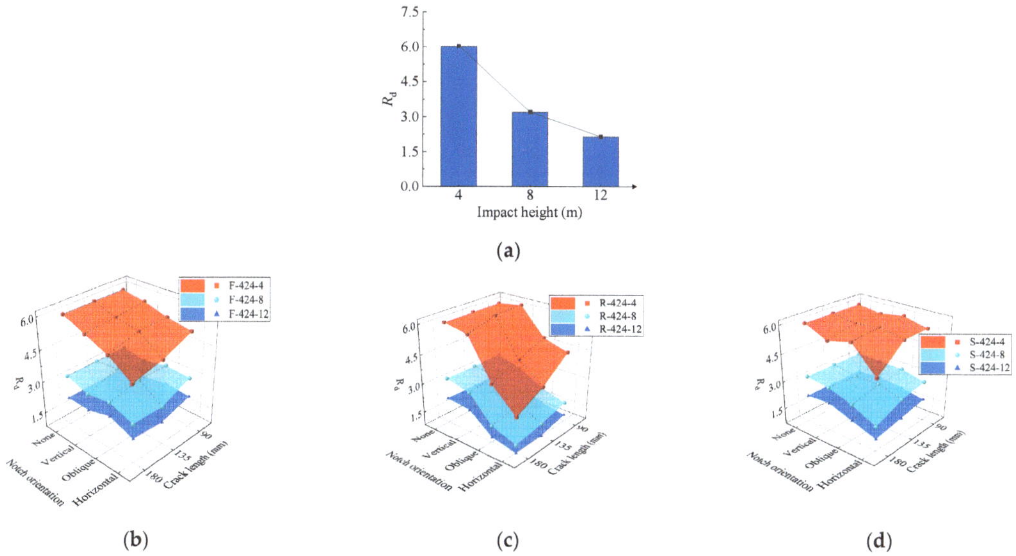

Figure 24. Comparison of the influencing factors of R_{d}. (**a**) 424-4/8/12. (**b**) Front impact. (**c**) Rear impact. (**d**) Side impact.

4. Practical Design Method for Dynamic Flexural Strength

Both the local corrosion type and the impact direction will affect the impact resistance of CFST. An important index to predict the impact resistance of CFST is the dynamic flexural strength improvement coefficient R_{d} under impact loading. In this paper, based on the parametric analysis results, the key factors affecting the improvement coefficient of dynamic flexural strength are selected, including crack length (l_{c}), crack angle (θ), impact surface (F, R and S) and impact energy (E_{i}). The practical design method (32–33) for the improvement coefficient R_{d} of the dynamic flexural strength of the concrete-filled square steel tube model section is obtained by regression analysis.

$$R_{\mathrm{d}} = 3.76682 \cdot f(n) \cdot f(E_i) \cdot f(\beta) \quad \theta = [0^{\mathrm{o}}, 90^{\mathrm{o}}) \tag{32}$$

in which

$$f(n) = -0.84476 + 2.26886n - 0.97719n^2$$

$$f(E_i) = 3.97012 - 0.12813E_i + 1.28 \times 10^{-3} {E_i}^2$$

$$f(\beta) = 1.72369 + 0.69479\beta - 1.19763\beta^2$$

$$R_{\mathrm{d}} = 2.05226 \cdot f(l_{\mathrm{c}}) \cdot f(E_i) \cdot f(\beta) \quad \theta = 90^{\mathrm{o}} \tag{33}$$

in which

$$f(l_{\mathrm{c}}) = 3.8859 + 0.03977l_{\mathrm{c}} - 1.58754 \times 10^{-4} {l_{\mathrm{c}}}^2$$

$$f(E_i) = 1.14824 - 0.03763E_i + 3.8461 \times 10^{-4} {E_i}^2$$

$$f(\beta) = 0.73013 + 0.06594\beta - 0.04551\beta^2$$

where n is the ratio of the cross-sectional area of the steel tube before and after local corrosion. The area after corrosion is the horizontal projection area and calculated according to $n = (A - t \cdot l_{\mathrm{c}} \cdot \cos\theta)/A$; A is the cross-sectional area of the steel tube; t is the wall thickness of the steel tube; β is the impact orientation factor, which is the angle formed by the impact direction of the drop hammer and the normal direction of the corrosion surface. The angle range 0–180° is normalized by taking β = 0 for front impact, β = 0.5 for side impact and $\beta = 1$ for rear impact. The specific application range of each parameter in the formula is shown in Table 7.

Table 7. Application scope of the practical formula of R_d.

l_c/mm	$\theta/(^o)$	β	E_i/kJ
90~180	0~90°	0, 0.5, 1	16.6~49.9

Figure 25a,b shows the comparison results of the improvement coefficient R_d of dynamic flexural strength by finite element simulation and proposed design method. The relative error between them is less than 10% basically, which shows that the calculation results of the proposed design method are in good agreement with the finite element simulation results. The proposed design method could be used to predict the dynamic plastic moment of CFST with local defects under impact loading.

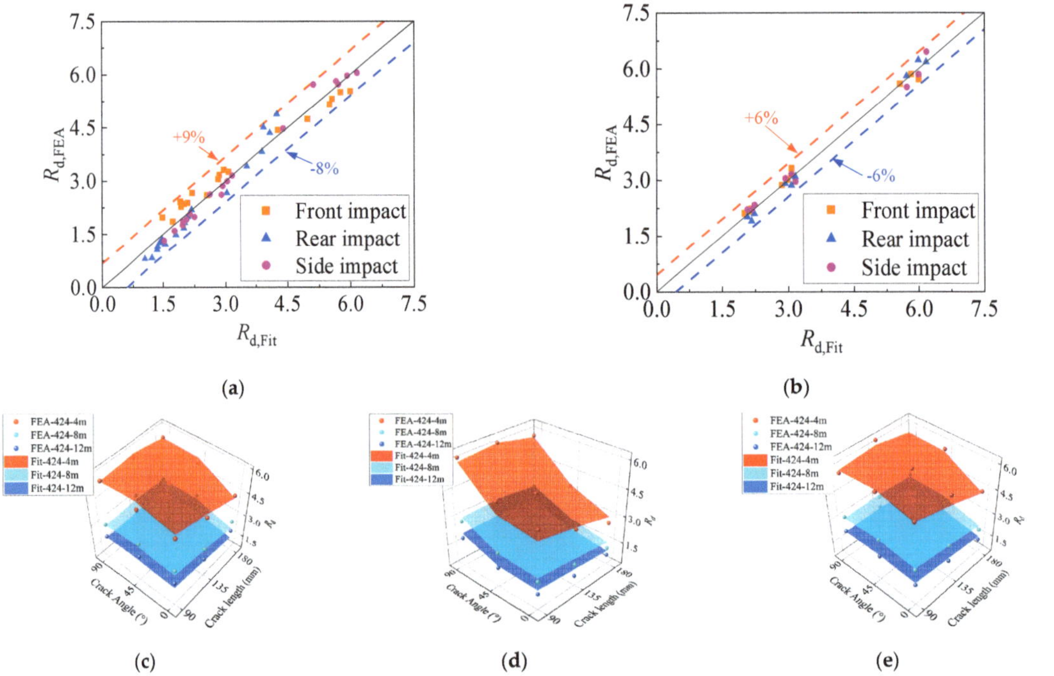

Figure 25. Comparison of simulated and fitted values of R_d. (**a**) $\theta = 0^o \sim 45^o$. (**b**) $\theta = 90^o$. (**c**) Front impact. (**d**) Rear impact. (**e**) Side impact.

5. Conclusions

In this paper, the impact resistance of locally corroded HSCFST members simulated by local notch cracks is numerically simulated. The influence of crack length, crack angle, impact surface and impact energy on the impact resistance of members is considered. The prediction method of impact resistance of HSCFST under the local corrosion is proposed. Some conclusions could be drawn as follows:

(1) The change in crack angle and crack length under the local corrosion will affect the amplitude and range of the local drum curvature. A model with horizontal cracks will show the phenomenon of crack closure when subjected to the front impact. The side impact will cause the horizontal crack to take on a "triangular" shape and will also cause the tensile failure of the steel tube at the bottom of the crack.

(2) The front impact and vertical cracks have little influence on the impact platform values of the model under local corrosion. Increasing the crack length and impact energy will reduce the impact platform value of the model. Among the three impact surfaces, the rear impact has the greatest influence on the impact platform value and impact duration of the model.

(3) The energy absorption ratio of the model under local corrosion is inversely proportional to the increase of crack angle and proportional to the increase of crack length in general. The side and rear impacts will promote the absorption of impact energy by the model and the energy absorption ratio of both is basically above 90%.

(4) The R_d of the model is generally proportional to the increase in crack angle and inversely proportional to the increase in crack length. The impact resistance of the side impact model is better than that of the front and rear impact models when the impact energy is small. The impact resistance of the front impact model is better than that of the side and rear impact models when the impact energy is larger.

(5) Based on the parametric analysis results, a practical design method for the improvement coefficient of dynamic flexural strength R_d of the HSCFST model under local corrosion under lateral impact is proposed and validated against the simulation results. The proposed design method could be used to predict the dynamic plastic moment of HSCFST with local defects under impact loading.

Author Contributions: Investigation and writing—original draft preparation, Y.W.; conceptualization and writing—review and editing, S.G.; formal analysis, Y.X.; validation F.L. All authors have read and agreed to the published version of the manuscript.

Funding: This research was funded by National Natural Science Foundation of China (No. 51908085), Natural Science Foundation of Chongqing (cstc2020jcyj-msxmX0010), Fundamental Research Funds for the Central Universities (AUGA5710010322, 2020CDJ-LHZZ-013) and The Youth Innovation Team of Shaanxi Universities (21JP138), which are gratefully acknowledged.

Conflicts of Interest: The authors declare no conflict of interest.

References

1. Han, L.H. *Concrete-Filled Steel Tubular Structures: Theory & Application*; Science Press: Beijing, China, 2016.
2. Zhong, S.T. *The Concrete-Filled Steel Tubular Structures*; Tsinghua University Press: Beijing, China, 2003.
3. Bambach, M.R.; Jama, H.; Zhao, X.L.; Grzebieta, R.H. Hollow and concrete-filled steel hollow sections under transverse impact loads. *Eng. Struct.* **2008**, *30*, 2859–2870. [CrossRef]
4. Bambach, M.R. Design of hollow and concrete-filled steel and stainless steel tubular columns for transverse impact loads. *Steel Constr.* **2011**, *49*, 1251–1260. [CrossRef]
5. Hou, C.C. Study on Performance of Circular Concrete-Filled Steel Tubular (CFST) Members under Low Velocity Transverse Impact. Master's Thesis, Tsinghua University, Beijing, China, May 2012.
6. Qu, H.; Li, G.; Sun, J.; Sozen, M.A. Numerical simulation analysis of circular concrete-filled steel tube specimen under lateral impact. *J. Archit. Civ. Eng.* **2010**, *27*, 89–96.
7. Yang, X.Q. Dynamic Constitutive Models of Structural Steels and Transverse Impact Resistance of High-Strength Concrete Filled Steel Tubes with Square Cross-Section. Ph.D. Thesis, Harbin Institute of Technology, Harbin, China, June 2020.
8. Cai, J.; Yu, Y.; Chen, Q.J.; Li, Y.N.; Ye, J.B. Parameter study on dynamic response of concrete filled square tube under lateral impact. *J. Cent. South Univ.* **2019**, *50*, 409–419.
9. Suzuki, Y.; Lignos, D.G. Collapse behavior of steel columns as part of steel frame buildings: Experiments and numerical models. In Proceedings of the 16th World Conference on Earthquake Engineering, Santiago, Chile, 9–13 January 2017.
10. Sediek, O.A.; Wu, T.-Y.; McCormick, J.; El-Tawil, S. Collapse Behavior of Hollow Structural Section Columns under Combined Axial and Lateral Loading. *J. Struct. Eng.* **2020**, *146*, 04020094. [CrossRef]
11. Gao, S.; Zhang, K.Y.; Li, J.Q.; Xu, Y.C.; Wang, Y.G. Experimental study on axial performance of circular CFST under freeze-thaw and corrosion complex environment. *J. Nat. Disasters* **2021**, *30*, 93–100.
12. Gao, S.; Peng, Z.; Guo, L.H. Experimental study on axial performance of concrete-filled steel tubular stub column under salt spray corrosion. *J. Build. Struct.* **2019**, *40*, 214–219.
13. Han, L.H.; Hou, C.C.; Wang, Q.L. Behavior of circular CFST stub columns under sustained load and chloride corrosion. *J. Constr. Steel Res.* **2014**, *103*, 23–36. [CrossRef]
14. Ding, F.X.; Fu, L.; Yu, Z.W. Behaviors of axially loaded square concrete-filled steel tube (CFST) Stub columns with notch in steel tube. *Thin-Walled Struct.* **2017**, *115*, 196–204. [CrossRef]

15. Chen, M.C.; Zhang, F.; Huang, H.; Wang, C. Study on seismic performance of concrete filled square steel tubes subjected to simulated acid rain attack. *J. China Railw. Soc.* **2018**, *40*, 106–114.
16. Cowper, G.R.; Symonds, P.S. *Strain-Hardening and Strain-Rate Effects in the Impact Loading of Cantilever Beams*; Brown University: Providence, RI, USA, 1957.
17. Johnson, G.R.; Cook, W.H. A constitutive model and data for metals subjected to large strains, high strain rates and high temperatures. In Proceedings of the 7th International Symposium on Ballistics, The Hague, The Netherlands, 19–21April 1983; Volume 21, pp. 541–547.
18. Zhu, M.; Liu, J.; Wang, Q.; Feng, X. Experimental research on square steel tubular columns filled with steel-reinforced self-consolidating high-strength concrete under axial load. *Eng. Struct.* **2010**, *32*, 2278–2286. [CrossRef]
19. China Building Industry Press. *Code for Design of Concrete Structures*; GB 50010-2010; China Building Industry Press: Beijing, China, 2010. (In Chinese)
20. Comite Euro-International du Beton. *CEB-FIP Model Code 1990*; Redwood Books: Trowbridge, UK, 1993.
21. Tao, Z.; Wang, Z.B.; Yu, Q. Finite element modelling of concrete-filled steel stub columns under axial compression. *J. Constr. Steel Res.* **2013**, *133*, 121–131. [CrossRef]
22. Han, L.H.; Yao, G.H.; Tao, Z. Performance of concrete-filled thin-walled steel tubes under pure torsion. *Thin-Walled Struct.* **2007**, *45*, 24–36. [CrossRef]
23. Wu, C.; Liu, J.L.; Wang, R.H. A study on the dynamic response of welded hollow spherical joints under impact loading. *Prog. Steel Build. Struct.* **2021**, *23*, 82–93.

buildings

MDPI

Article

Rational Use of Idealized Shear-Building Models to Approximate Actual Buildings

Zhanxuan Zuo [1,2], Yiting He [3,4] and Shuang Li [3,4,*]

1 Key Laboratory of Earthquake Engineering and Engineering Vibration, Institute of Engineering Mechanics, China Earthquake Administration, Harbin 150080, China; zuozhanxuan@sina.com
2 Key Laboratory of Earthquake Disaster Mitigation, Ministry of Emergency Management, Harbin 150080, China
3 Key Lab of Structures Dynamic Behavior and Control of the Ministry of Education, Harbin Institute of Technology, Harbin 150090, China; 18b933051@stu.hit.edu.cn
4 Key Lab of Smart Prevention and Mitigation of Civil Engineering Disasters of the Ministry of Industry and Information Technology, Harbin Institute of Technology, Harbin 150090, China
* Correspondence: shuangli@hit.edu.cn

Abstract: The paper aims to investigate the accuracies of idealization methods of the well-known shear-building models. Five idealization methods are adopted to idealize the structural story capacity curve within the range from zero to the deformation corresponding to the peak shear point. After the peak shear point, a skew branch followed by a constant branch are used to approximate the capacity curve. The five idealization methods are verified by using four reinforcement concrete (RC) frames with 3, 8, 12, and 18 stories. Results reveal that all the five idealization methods may cause remarkable errors in prediction of the period, displacements and accelerations of the actual buildings. The errors of the structural period by the five idealization methods are almost above 10–40%. The errors of the structural displacements and accelerations by the five idealization methods are almost above 30–90%. For all the five idealization methods, the prediction accuracy on displacement and acceleration will be dramatically increased if the comparison is only focused on the maximum value within all story rather than the maximum values of each story. The initial stiffness method provides the best predictions on periods of the actual buildings. The farthest point method provides better prediction than the other four idealization methods.

Keywords: shear-building model; idealization method; period of vibration; earthquake response; accuracy

Citation: Zuo, Z.; He, Y.; Li, S. Rational Use of Idealized Shear-Building Models to Approximate Actual Buildings. *Buildings* **2022**, *12*, 273. https://doi.org/10.3390/buildings12030273

Academic Editors: Jingxuan Wang, Shan Gao, Dewen Kong and Yong Liu

Received: 22 January 2022
Accepted: 21 February 2022
Published: 26 February 2022

1. Introduction

Currently, considering the wide availability of powerful computational tools and software, it is possible to use more complex models (e.g., the beam-column element models or solid element models) to perform analyses for obtaining the structural seismic responses. However, if providing structural responses only at the "story" level is the target, the use of simplified modeling assumptions such as the shear-building concept is necessary and convenience in these cases. Shear-building models are widely used to study the seismic response of multi-story buildings because of their simplicity and low computational cost [1]; especially, it facilitates performing comprehensive statistical analyses and parametric studies. Alongside the seismic response analyses on ordinary multi-story buildings, the idealizations of actual buildings to the shear-building model have been used in many research aspects, e.g., soil-structure-interaction analysis [2–4], pounding analysis [5], health monitoring and system identification [6–10], damper placements [11], isolated buildings [12,13], structural optimum design [14–17], pushover analysis [18,19], and city earthquake response analysis [20–23], etc. This type of model is also the basis of methods or regulations in some seismic design codes, e.g., the derivation of the vertical distribution of seismic forces for low and middle-height buildings [24–26], design requirements for nonbuilding structures [24], calculation models of isolated buildings [25], modification factors in nonlinear static procedure [27], etc.

The problems are as follows. (1) When the seismic response analysis is performed, how accurate are the results obtained by using shear-building models? Additionally, (2) if a structure is designed according to the regulations derived based on shear-building models, how does the capacity curve of the actual building need to be. The above problems are relevant to the idealization method of the shear-building model and its accuracy, which are the objective of this study. The main objective of this research is to investigate the accuracies of five different idealization methods and provide some advice on selection of proper idealization method for the shear-building model in dynamic time analysis.

2. Review of Idealization Methods of Shear-Building Models

In shear-building models, floor slabs are modeled as lumped masses and columns are modeled as elastic–plastic springs that only exhibit deformations in horizontal direction when subjected to lateral forces. The story lumped mass at the floor slab is set as the sum of masses between half-upper and half-lower of the story. The four-branch backbone curve shown in Figure 1 is adopted in this study to provide a representation of the non-linear behavior of the elastic–plastic springs. The story backbone curve, which has four key parameters including elastic stiffness K_e, post-yielding stiffness ratio α_1, degrading stiffness α_2, and residual shear V_r. In other words, the key points (D_y, V_y), (D_p, V_p), and (D_r, V_r) in the idealized curve need to be determined from the story capacity curve.

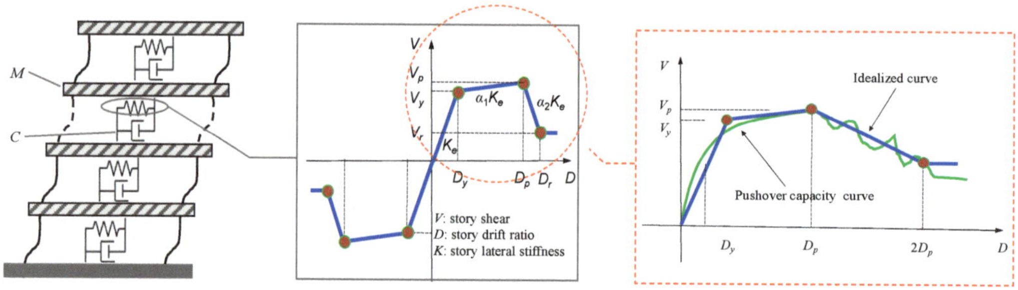

Figure 1. Shear-building model used to derive the design procedure.

Five idealization methods are verified in this study from previous studies. Table 1 shows the five idealization methods, and they are used to determine the elastic stiffness K_e and post-yielding stiffness ratio α_1 based on the story capacity curves. The point (D_p, V_p) in the idealized curve is defined as the point with maximum shear in the capacity curve in this study. In the references [18,27–32], some regulations are flexible to a practical analyst. In this study, what "has leveled off" (see Table 1) in the ATC-40 method means that the range starts from the point where tangent stiffness reduces to 20% of the initial stiffness and ends at the point (D_p, V_p), then the initial tangent stiffness is calculated using the average of three stiffness values adopted (0, 0) and the first, second, and third points in the capacity curve. Figure 2a–e provide the schematic diagram of these five methods.

Table 1. Idealization methods to determine the story backbone curve before the maximum shear point.

No.	Reference	Description of Idealization before Point (D_p, V_p)
1	ATC-40 [33]	1. $\alpha_1 K_e$ is determined to represent an average stiffness in the range in which the structural strength has leveled off; 2. K_e is determined by constructing a secant line passing through the point on capacity curve corresponding to a shear of $0.6V_y$. V_y is the intersection of the K_e and $\alpha_1 K_e$ lines; 3. The determination of V_y needs iteration to check the point where the K_e line crosses the capacity curve if it is equal to $0.6V_y$.
2	FEMA 356 [28] FEMA 440 [29] ASCE 41-13 [27]	This is the same as the ATC-40 method but the $\alpha_1 K_e$ is not pre-determined. K_e and $\alpha_1 K_e$ are defined based on the equal-area principle from 0 to the D_p. Iteration is needed in this procedure.
3	Park [30] ($0.75V_p$ method)	1. $\alpha_1 K_e$ is a leveled line passing the point (D_p, V_p); 2. K_e is a line passing the point on the capacity curve corresponding to a shear of $0.75V_p$.
4	Feng et al. [31] (farthest point method)	1. A line is drawn from 0 to the point (D_p, V_p), then the point that has the maximum perpendicular distance from the capacity curve to this line is found. The point is set as (D_y, V_y); 2. K_e and $\alpha_1 K_e$ can be determined by point (D_y, V_y).
5	Chopra and Goel [18] (initial stiffness method)	1. K_e is set as the initial tangent stiffness. 2. $\alpha_1 K_e$ is determined by equal-area method from 0 to the point (D_p, V_p).

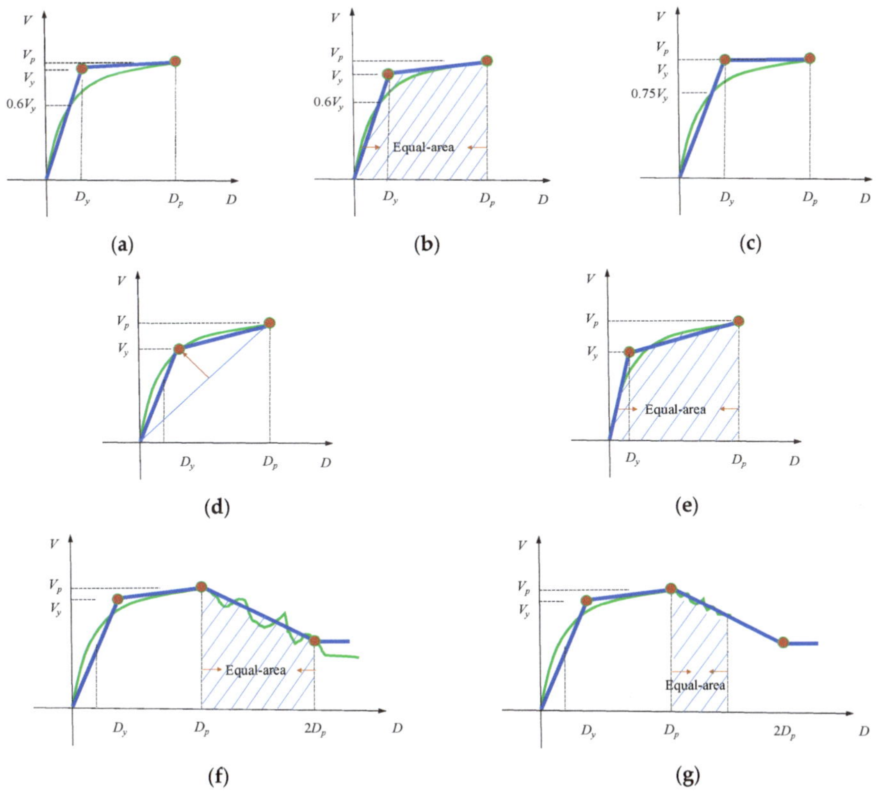

Figure 2. Schematic diagram of five methods for idealizing the story capacity curves. (**a**) ATC-40; (**b**) FEMA 356 (also FEMA 440, ASCE 41-13); (**c**) $0.75V_y$; (**d**) farthest point; (**e**) initial stiffness; (**f**) negative-stiffness region–Case 1; (**g**) negative-stiffness region–Case 2.

For the idealization of the story capacity curve after point (D_p, V_p), only ASCE 41-13 [27] and FEMA 440 [29] provide regulations. This segment is determined by the point (D_p, V_p) and the point at which the shear declines to 60% of the V_y. Actually, the above idealization method has some inconveniences in practical idealized processes because the unstable numerical analysis results are commonly obtained in the prediction of the story capacity curve in the negative-stiffness region, e.g., (1) the results in this segment may not be accurate because the calculation is in the negative stiffness region; and (2) the data may be lost and not reach to the point with shear declines to 60% of the V_y. Figure 2f,g provide a schematic diagram for idealization of the capacity curve after point (D_p, V_p). In this study, the following method is advised to idealize the curve: α_2K_e is a line passing the point (D_p, V_p) and crossing a vertical line with horizontal coordinate $2D_p$. α_2K_e is determined by having equal areas within the range of D_p–$2D_p$; if the data are not enough, the last point of the story capacity curve is used to determine α_2K_e, in which the equal-area principle is satisfied within D_p and the last point of the story capacity curve. Then, the segment will extend to the vertical line with horizontal coordinate $2D_p$. In this way, the point (D_r, V_r) is determined.

Figure 3 shows a flowchart of the paper in the following sections. Firstly buildings and ground motions used in this paper are selected in Section 3. In Section 4, the selected buildings are idealized to different shear-building models according to the five idealization methods, and the shear-building models are excited by ground motions. Then the structural responses (periods, displacements and accelerations) are compared to investigate the accuracy of different idealization methods. Finally, the conclusions are obtained and suggestions are provided to select proper idealization method for the buildings in dynamic time analysis.

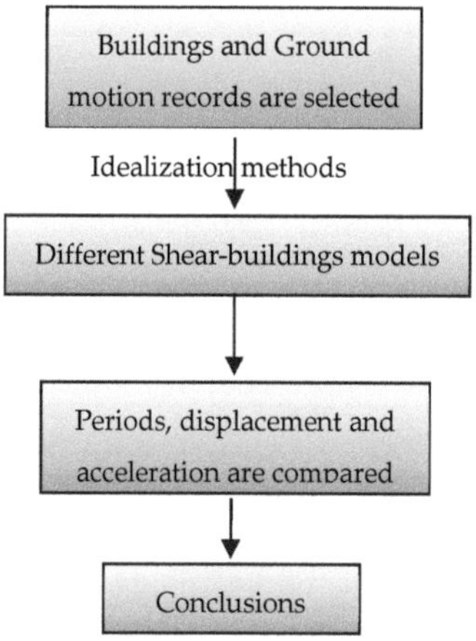

Figure 3. Flowchart of this paper.

3. Buildings, Selected Ground Motions and Modelling

Four RC frames with 3, 8, 12, and 18 stories are shown in Figure 4 (design information is in reference [32]). The architectural elevation of four buildings are similar. For each building, the height of the first story and the standard floor are 4.2 m and 3.3 m, respectively. In total, the height of 3-, 8-, 12-, 18-story buildings are 10.8 m, 27.3 m, 40.5 m, 60.3 m. The design earthquake intensities for the frames are PGA = 0.15 g (seismic hazard corresponds to 10% exceedance probability in 50 years at the building site). The site type is II and group is I in Chinese seismic design code [25]. A total of 80 ground motion records recommended in reference [32] are used in the analyses. The acceleration spectra of ground motions are shown in Figure 5. The ground motion records are selected from strong earthquakes, whose source magnitudes are mainly ranges from 5.0 to 8.0. The records from such strong earthquakes usually cause severe building damage. Three input ground motion intensities are considered with PGA = 0.056 g, 0.316 g, and 0.45 g (corresponds to frequent, rare and very rare earthquake intensities with 63% and 2~3% exceedance probabilities in 50 years, and 10^{-4} exceedance probability per year, respectively).

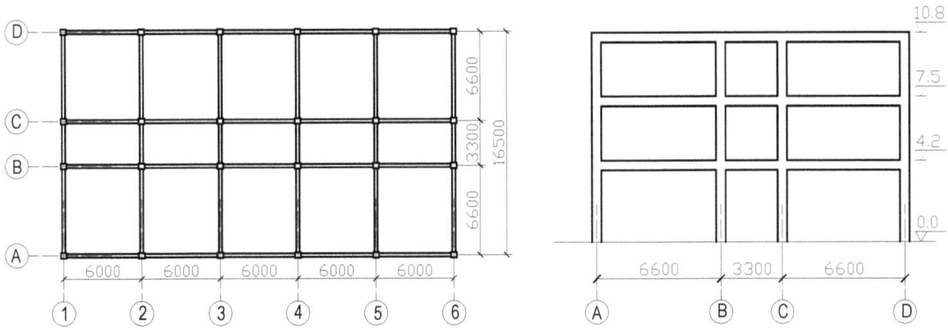

Figure 4. Elevation of the RC frame structures (Unit: mm) (3-story frame as an example and other frames having same configuration but different numbers of stories).

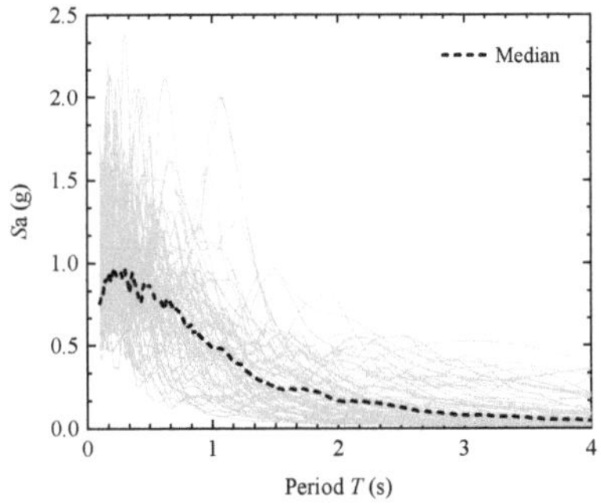

Figure 5. The acceleration spectra of ground motions (PGA = 0.45 g).

The well-known Idarc program [34] is used in the modelling of the frames to obtain the story capacity curves and structural responses in the later sections with the beam-column model. Idarc has been widely used by many researchers in inelastic structural damage analysis [35–37]. The OpenSees program is used to build the idealized shear-building models and obtain the structural responses. The "Hysteretic material" model in the OpenSees program [38] is used, in which the backbone curve parameters are determined with the methods in Table 1 and hysteretic rule parameters use the default values. The residual capacity point is set as ($2D_p$, $60\%V_y$). The damping ratio used in the analyses is 0.05.

4. Results

The actual story capacity curves should be obtained first, then the idealization methods are used to derive the shear-building models. The actual story capacity curves of structure can be obtained by two methods: (1) incremental dynamic analysis (IDA) method, through which the story capacity curves can be obtained by increasing the PGA of ground motion records as input to the structures, the maximum story shear and maximum story ratio are obtained during the time history analysis, thus producing story shear force versus inter-story drift ratio (IDR) curves; and (2) the pushover method, through which the story shear versus IDR curves can be obtained by monotonically increasing the lateral loads along the structural height until the prescribed displacement is reached.

Figure 6 shows the story capacity curves of the 3-, 8-, 12-, and 18-story frames by using IDA method. Every line in Figure 6 represents the capacity curve of each story of the frames. In order to compare the story capacity curves obtained by IDA and pushover methods. The actual story capacity curves of the 12-story frame obtained by the two methods are presented in Figure 7, using the 12-story frame as an example. The inverted triangular lateral load distribution is adopted in the pushover method.

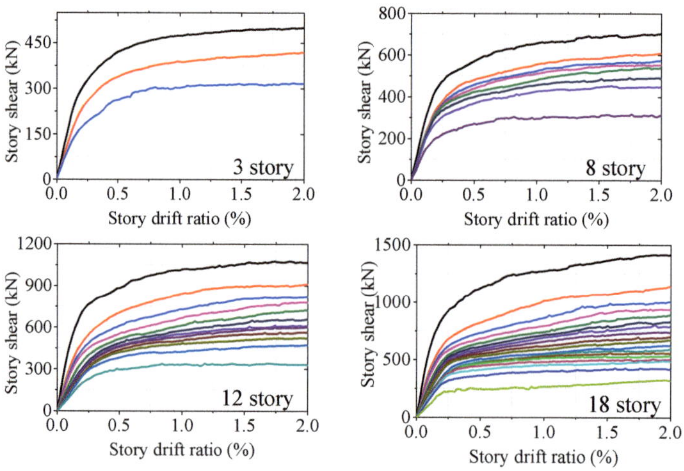

Figure 6. Actual story capacity curves by using IDA method.

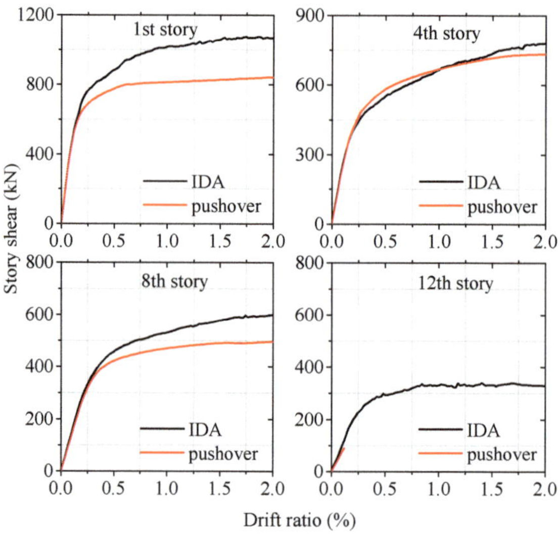

Figure 7. Comparison of actual story capacity curves by IDA and pushover methods (using a 12-story frame as an example).

As can be seen in Figure 7, the whole story capacity curves can be obtained using the IDA method. However, for the pushover method, the whole story capacity curves can be obtained in the low and middle stories (1st, 4th, and 8th stories), while the whole story capacity curve cannot be obtained in the upper story (12th story). The capacity curve of the 12th story does not increase with the increase of top displacement. This is because the damages mainly occur in the low or middle stories, the upper stories experience elastic unloading even when the top displacement is getting bigger. It means the capacity curve of the upper stories cannot reach the inelastic stage in the pushover procedure. A similar phenomenon can be also observed in the 3-, 8-, and 18-story frame structures. A similar observation was also obtained in the authors' previous study, regardless of the lateral load patterns used in pushover analysis, e.g., the uniform pattern, linear pattern, parabolic pattern, and modal adaptive pattern [39].

Note again that the pushover is a simplified method and lacks a rigorous theoretical foundation, the predefined lateral load distributed along the building height is different from the real situation in the earthquake, and this causes the result that some stories may experience an unloading phenomenon when the structure is severely damaged in the other stories. The whole story capacity curve cannot be obtained due to the unloading phenomenon for some stories in the pushover method. The existing non-seismically designed buildings are usually poorly detailed (i.e., the use of plain steel bars, the use of concrete with low compression strength, the short length of column lap splices, the inadequate anchorage of reinforcement, and the poorly confined beam-column joint regions), which substantially increase number of parameters related to structural weaknesses, and decisively affect the seismic response and structural integrity. The implementation of the pushover method in the case of existing non-seismically designed buildings is even more inaccurate. Compared with the pushover method, the dynamic analysis is regarded as a more accurate method [37,40–42]. Thus, the IDA method is adopted to obtain the story capacity curves in this study because it can get the whole story capacity curves of all stories.

In this section, the accuracies of the shear-building models by different idealization methods will be investigated. The structural dynamic property (period) and responses (displacement, i.e., the story drift ratio, and acceleration) calculated by the beam-column model in Idarc program and shear-building models using different idealization methods are compared. It is reasonable to assume that the results calculated by the beam-column model in Idarc program are more accurate. Thus, based on the results calculated by the beam-column model in the Idarc program, the accuracy and applicability of different idealization methods are discussed.

4.1. Comparison of Structural Period

The actual story capacity curves of the frames are obtained by the IDA method. Then the five idealization methods are adopted to provide idealization of the actual capacity curves, as shown in Figure 8. The $0.75V_p$ method greatly overestimates the yield point compared with the IDA curve. The $0.75V_p$ and farthest point methods underestimate the elastic stiffness.

The errors of the structural periods obtained by shear-building models using the five idealization methods are discussed. The error index used in the study is presented as

$$Error = (T_{\text{sm}} - T_{\text{Idarc}})/T_{\text{Idarc}} \quad (1)$$

where T_{Idarc} is the structural period obtained by beam-column model, and T_{sm} is the structural period obtained by shear-building models using different idealization methods.

Table 2 shows the error of the structural periods. For the 12-story frame, the period calculated using the $0.75V_p$ and farthest point methods overestimate the period due to the underestimation of the elastic stiffness, which can be seen in Figure 6. For the $0.75V_p$ method, the errors of the first and second periods are 35.39% and 34.43%, respectively. For the farthest point method, the errors of the first and second periods are 35.39% and 42.62%, respectively. The errors of the first and second periods from the ATC-40, FEMA 356 and initial stiffness methods are all below 10%.

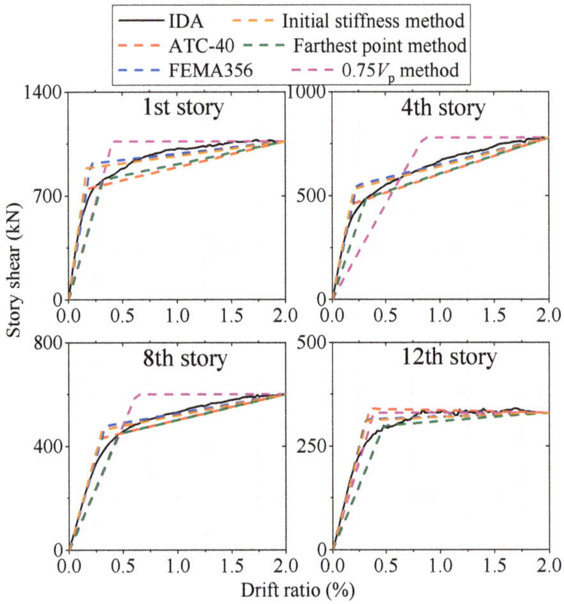

Figure 8. Idealization of the actual story capacity curves (using 12-story frame as an example).

Table 2. Errors of first and second periods calculated by using different models.

Frame	3-Story		8-Story		12-Story		18-Story		Average Error
Methods	T_1	T_2	T_1	T_2	T_1	T_2	T_1	T_2	
Idarc	0.54	0.16	1.19	0.38	1.78	0.61	2.73	0.93	
ATC-40	0.61 (12.96%)	0.23 (43.75%)	1.22 (2.52%)	0.42 (10.53%)	1.78 (0.00%)	0.65 (6.56%)	2.63 (−3.66%)	0.96 (3.23%)	(9.45%)
FEMA 356	0.61 (12.96%)	0.23 (43.75%)	1.21 (1.68%)	0.42 (10.53%)	1.79 (0.56%)	0.66 (8.20%)	2.68 (−1.83%)	0.98 (5.38%)	(10.15%)
$0.75V_p$	0.76 (40.74%)	0.28 (75.00%)	1.55 (30.25%)	0.54 (42.11%)	2.41 (35.39%)	0.82 (34.43%)	3.72 (36.26%)	1.23 (32.26%)	(40.81%)
Farthest point method	0.83 (53.70%)	0.32 (100.00%)	1.53 (28.57%)	0.55 (44.74%)	2.41 (35.39%)	0.87 (42.62%)	3.33 (21.98%)	1.21 (30.11%)	(44.64%)
Initial stiffness method	0.61 (12.96%)	0.21 (31.25%)	1.16 (−2.52%)	0.40 (5.26%)	1.77 (−0.56%)	0.66 (8.20%)	2.73 (0.73%)	0.93 (10.75%)	(8.26%)

Note: The numbers in the brackets are the error of periods.

Similar conclusions can be also observed in the 3-, 8-, and 18-story frames. The average errors of the $0.75V_p$ and farthest point method are 40.81% and 44.64%, respectively. The average errors of the ATC-40, FEMA 356 and initial stiffness methods are no more than 10.12%. In general, the $0.75V_p$ and farthest point method greatly overestimated the period, the average error of period of these two methods are more than 40%. The errors of the other three methods (the ATC-40, FEMA 356 and initial stiffness methods) are no more than 10.15%, which are in an acceptable range in engineering. The initial stiffness method provides the best prediction on periods of the actual buildings.

4.2. Comparison of Structural Responses

The structural responses are average values of the results obtained by the 80 ground motion records used in Section 3. The error index of structural response is presented as

$$Error = \mathrm{Abs}(RES_{\mathrm{sm}} - RES_{\mathrm{Idarc}})/RES_{\mathrm{Idarc}} \quad (2)$$

where RES_{Idarc} is the structural responses (inter-story drift ratio or accelerations) obtained by beam-column model, and RES_{sm} is the structural responses (inter-story drift ratio or accelerations) obtained by shear-building models using different idealization methods.

Figure 9a–c show the inter-story drift ratios (IDRs) obtained by Idarc program and the five shear-building models at different seismic intensities. Table 3 shows the error of IDRs obtained from the shear-building models compared with beam-column model in Idarc program. As shown in Table 3, the minimum error is 23.25%, and the maximum error reaches 109.8%. The average errors of IDRs are all above 30%. Among the five different idealization methods, the average errors of ATC-40 and initial stiffness methods are about 40%, while the average error of farthest point method reaches 64.14%.

Table 3. Errors of calculated inter-story drift ratios (average value of all the stories).

Error		ATC-40	FEMA356	$0.75V_p$	Farthest Point	Initial
0.45 g	3-story	23.25%	23.43%	35.92%	42.55%	25.06%
	8-story	40.82%	42.26%	41.95%	48.40%	48.51%
	12-story	46.60%	43.34%	47.98%	39.26%	43.32%
	18-story	61.44%	63.62%	54.96%	58.46%	56.87%
0.316 g	3-story	35.34%	26.70%	43.40%	46.26%	40.27%
	8-story	26.70%	33.41%	29.03%	28.55%	32.84%
	12-story	36.26%	34.11%	30.90%	30.69%	22.84%
	18-story	53.67%	49.56%	56.43%	56.02%	51.29%
0.056 g	3-story	52.12%	50.98%	95.47%	136.81%	28.00%
	8-story	26.73%	36.22%	62.00%	80.28%	34.55%
	12-story	53.59%	68.32%	109.80%	109.08%	50.63%
	18-story	52.06%	60.24%	94.44%	93.37%	58.47%
Average error		42.38%	44.35%	58.52%	64.14%	41.05%

Figure 9d shows the maximum IDRs of the structures obtained by different methods. The "maximum" here denotes that it firstly averages the IDR results of 80 ground motions of each story, and then the maximum value within all the stories is used. Table 4 shows the error of maximum IDRs obtained from the shear-building models compared with beam-column model. As can be seen form Table 4, the errors of the maximum IDRs are much smaller compared with those errors in Table 3 (the reason is because the comparison only focuses on the maximum values regardless of which story they occur in). The minimum error is 3.16% in Table 4. The average error of the ATC-40 method is about 30%, which is the most accurate among the five idealization methods.

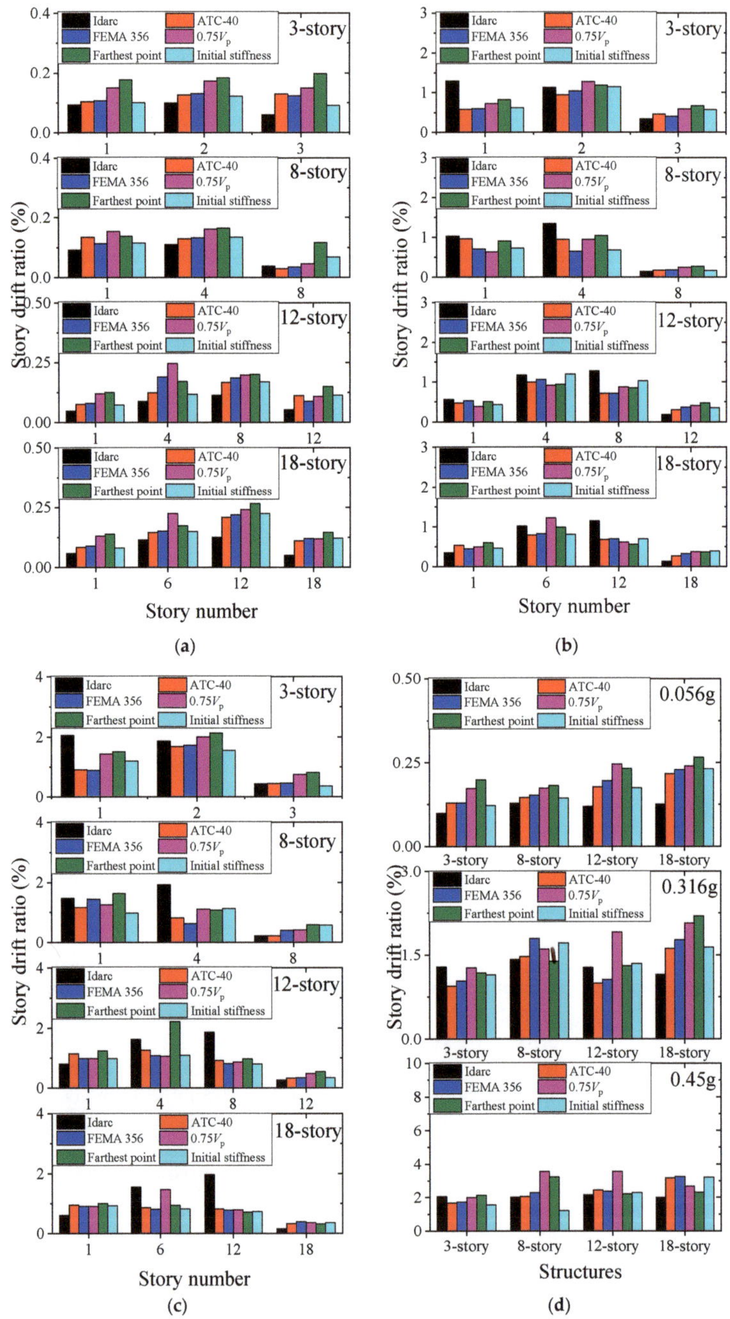

Figure 9. Story drift ratios caculated by beam-column model and five shear-building models. (**a**) PGA = 0.056 g; (**b**) PGA = 0.316 g; (**c**) PGA = 0.45 g; (**d**) PGA = 0.056 g, 0.316 g, and 0.45 g (maximum within all stories).

Table 4. Errors of calculated inter-story drift ratios (maximum value within all the stories).

Error		ATC-40	FEMA356	$0.75V_p$	Farthest Point	Initial
0.45 g	3-story	18.52%	16.11%	2.92%	3.16%	24.68%
	8-story	1.49%	12.71%	76.82%	60.41%	39.65%
	12-story	13.76%	9.52%	65.82%	3.00%	5.83%
	18-story	59.84%	63.13%	33.81%	14.93%	61.11%
0.316 g	3-story	26.68%	19.66%	1.19%	8.35%	11.23%
	8-story	3.22%	26.13%	12.51%	2.46%	20.46%
	12-story	21.98%	17.07%	49.51%	1.85%	5.38%
	18-story	39.76%	53.48%	79.58%	90.40%	42.13%
0.056 g	3-story	30.84%	31.59%	74.28%	100.43%	23.42%
	8-story	13.61%	18.80%	35.22%	41.03%	12.60%
	12-story	48.65%	64.02%	106.57%	95.67%	46.56%
	18-story	72.74%	82.50%	92.04%	112.71%	85.10%
Average error		29.26%	34.56%	52.52%	44.53%	31.51%

Figure 10 shows the accelerations obtained by beam-column model and the five idealized methods at different seismic intensities. Tables 5 and 6 show the error of accelerations obtained from the five idealized methods compared with beam-column model. As shown in Figure 10, all the five idealization methods cause remarkable errors in prediction of the accelerations compared with those of the beam-column model. The accuracy of the farthest point method is relatively more accurate than the other four methods. Comparing Tables 3 and 4, the errors of the acceleration estimation are higher than those of the IDRs.

Table 5. Errors of calculated story accelerations (average value of all the stories).

Error		ATC-40	FEMA356	$0.75V_p$	Farthest Point	Initial
0.45 g	3-story	43.93%	43.92%	47.86%	42.84%	39.95%
	8-story	72.65%	86.11%	91.02%	75.89%	93.60%
	12-story	96.72%	99.07%	109.65%	96.45%	100.26%
	18-story	147.50%	151.84%	159.08%	142.03%	151.72%
0.316 g	3-story	31.61%	36.37%	33.43%	33.43%	31.96%
	8-story	65.60%	74.58%	86.95%	75.88%	82.80%
	12-story	99.81%	116.21%	110.90%	95.01%	97.55%
	18-story	129.70%	134.95%	147.31%	131.70%	130.05%
0.056 g	3-story	22.34%	19.97%	14.93%	18.02%	17.81%
	8-story	60.03%	76.53%	56.31%	65.42%	90.12%
	12-story	116.12%	106.28%	98.74%	96.42%	119.68%
	18-story	145.84%	148.44%	120.21%	116.45%	150.51%
Average error		85.99%	91.19%	89.70%	82.46%	92.17%

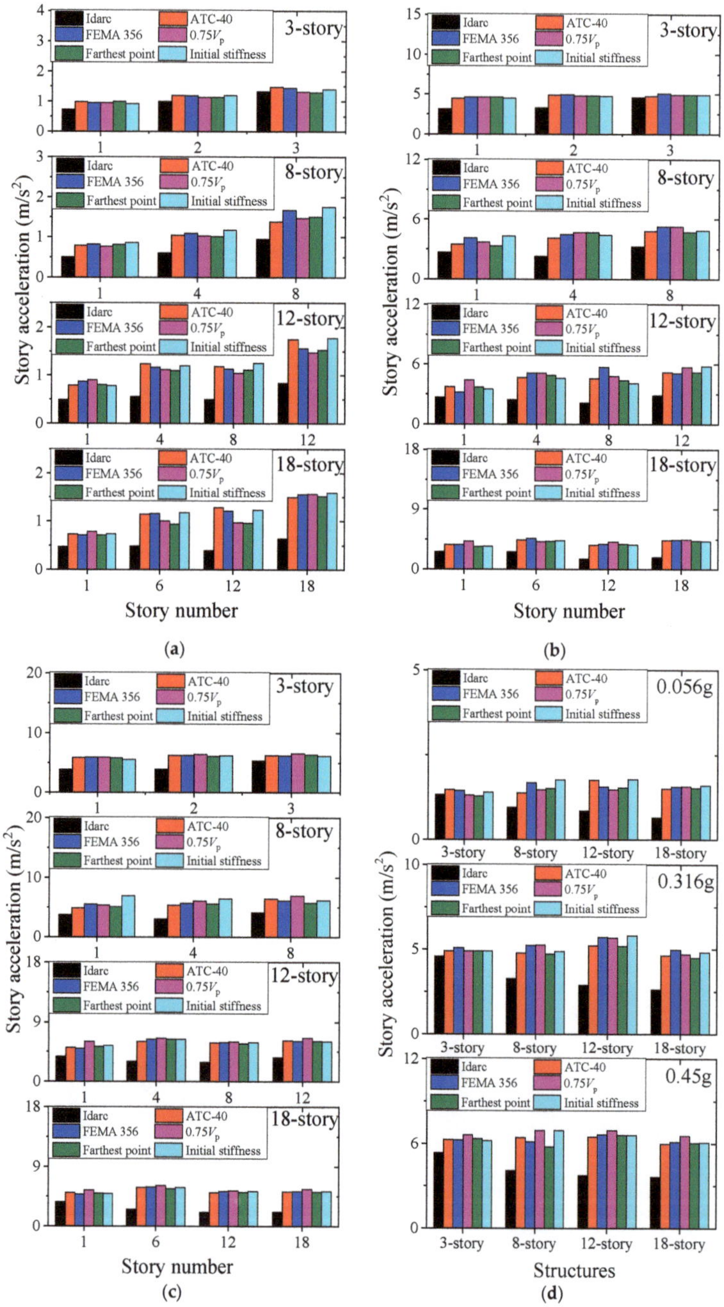

Figure 10. Story accelerations caculated by beam-column model and five shear-building models. (**a**) PGA = 0.056 g; (**b**) PGA = 0.316 g; (**c**) PGA = 0.45 g; (**d**) PGA = 0.056 g, 0.316 g, and 0.45 g (maximum within all stories).

Table 6. Errors of calculated story accelerations (maximum value within all the stories).

Error		ATC-40	FEMA356	$0.75V_p$	Farthest Point	Initial
0.45 g	3-story	17.25%	16.56%	23.41%	18.50%	15.77%
	8-story	57.37%	50.42%	70.05%	41.54%	69.66%
	12-story	71.87%	75.81%	84.08%	74.81%	74.67%
	18-story	63.77%	67.55%	78.99%	65.51%	66.09%
0.316 g	3-story	6.92%	10.89%	7.15%	7.15%	7.11%
	8-story	46.88%	60.71%	61.15%	44.76%	49.66%
	12-story	80.16%	97.10%	96.01%	78.94%	100.37%
	18-story	76.77%	89.90%	79.55%	72.03%	84.13%
0.056 g	3-story	11.62%	9.10%	0.63%	2.80%	5.90%
	8-story	45.25%	76.42%	54.76%	58.30%	85.31%
	12-story	107.53%	84.77%	75.12%	81.40%	109.88%
	18-story	130.86%	139.90%	141.15%	133.85%	145.86%
Average error		59.69%	64.93%	64.34%	56.63%	67.87%

5. Conclusions and Discussion

The study focuses on the accuracies of idealization methods of shear-building models. Five idealization methods are adopted to provide idealization of the actual story capacity curve. The main conclusions and discussions are summarized as follows:

(1) The story shear versus displacement curves need to be obtained first to idealize the shear-building model. The story shear versus displacement curves can be obtained by the pushover method. However, in some cases, not all story shear versus displacement curves can be obtained by the pushover method. This is caused by the elastic unloading phenomenon even when the top displacement is getting bigger in some stories of the structures. When the top displacement is monotonically increased in the pushover method, the damages usually mainly occur in some weak stories, while the other stories are still in the elastic stage, and will experience elastic unloading even if the top displacement is getting bigger. The story shear versus displacement curves can be obtained by using IDA method, which can avoid the limitation of the pushover method;

(2) The farthest point method and the $0.75V_p$ method greatly overestimated the period. The errors of periods by these two methods are more than 40%. The overestimation of the periods is due to the underestimation of the elastic stiffness in these two methods. The errors of the other three methods are in an acceptable range in engineering. The initial stiffness method provides the best prediction on periods of the actual buildings;

(3) For the structural responses, all the five methods provide better prediction on the maximum IDRs (or accelerations), i.e., without considering the maximum values occur at which story. For the structural maximum IDRs, the average errors of the ATC-40 and initial stiffness method are about 30%, which are smaller than the other three idealization methods. For the structural maximum accelerations, the average errors of all the methods are above 55%, and the farthest point method provides better prediction than the other four idealization methods;

(4) For the five idealization methods of shear-building models, which have different backbone curves but same hysteretic rule, the differences in the structural response prediction can be within about 20%. However, the differences between the results obtained by beam-column model and the shear-building models are much larger. This implies the hysteretic rule and negative-stiffness region in the backbone curve may also have obvious influence on the prediction results, because in this study the backbone curve before the negative-stiffness region is simplified based on the actual story capacity curve, whereas the hysteretic rule and negative-stiffness region are assumed in the calculation. Note that in practical analyses, the negative-stiffness

region is also usually assumed due to the story capacity curve in this region may lack of data or not be accurate.

Recent years, the city or regional earthquake response analysis are hot topics which need huge computational workload. It is impractical to adopt the beam-column models in dynamic time analysis in the city or regional earthquake response analysis. The modeling requires idealized shear-building models for building, because it is a feasible way to balance the accuracy and computational workload. Note that rational use of idealized shear-building models in such analyses is needed. The reason is that, in these studies, the analytical formulas in seismic design code and overstrength factors are usually used to determine the backbone curve before the negative-stiffness region, the backbone curve after the negative-stiffness region, and the hysteretic rule are all assumed. Therefore, the accuracy even may be lower than result of the present study because both the analytical formulas overstrength factors have errors to the actual structural story capacity. The other issue is many optimum seismic design methods are derived based on idealized shear-building models. Those methods are promising, but it is very important to guarantee that the designed structure has similar story behaviour (backbone curve and hysteretic rule) to the idealized shear-building model.

The main conclusions of this paper are obtained by analysis of the RC frames, and only periods, displacement and acceleration responses are selected to investigate the accuracy of the idealization methods. Whether the conclusions are applicable to other structural response parameters (i.e., energy dissipation and damage index) need to be investigated in future research.

Author Contributions: Conceptualization, S.L. and Z.Z.; methodology, S.L. and Z.Z.; software, Z.Z. and Y.H.; validation, Z.Z. and Y.H.; formal analysis, Z.Z. and Y.H.; investigation, Z.Z.; writing—original draft preparation, Z.Z., Y.H. and S.L.; writing—review and editing, Z.Z., Y.H. and S.L. All authors have read and agreed to the published version of the manuscript.

Funding: Institute of Engineering Mechanics, China Earthquake Administration: 2017B17; National Natural Science Foundation of China: U1939210, 41861134010; Natural Science Foundation of Heilongjiang: YQ2019E021, LH2019E097.

Data Availability Statement: The data presented in this study are available on request from the corresponding author.

Conflicts of Interest: The authors declare no conflict of interest.

References

1. Ganjavi, B.; Hajirasouliha, I.; Bolourchi, A. Optimum lateral load distribution for seismic design of nonlinear shear-buildings considering soil-structure interaction. *Soil Dyn. Earthq. Eng.* **2016**, *88*, 356–368. [CrossRef]
2. Abtahi, S.; Mahsuli, M.; Ghannad, M.A. Probabilistic evaluation of soil-structure interaction effects on strength demands of shear buildings. *J. Struct. Eng.* **2020**, *146*, 04019166. [CrossRef]
3. Tian, Y.; Sun, C.; Lu, X.; Huang, Y. Quantitative analysis of site-city interaction effects on regional seismic damage of buildings. *J. Earthq. Eng.* **2020**, 1–21. [CrossRef]
4. Asadi-Ghoozhdi, H.; Attarnejad, R. The effect of nonlinear soil-structure interaction on the ductility and strength demands of vertically irregular structures. *Int. J. Civ. Eng.* **2020**, *18*, 1209–1228. [CrossRef]
5. Naserkhaki, S.; Aziz, F.N.A.; Pourmohammad, H. Parametric study on earthquake induced pounding between adjacent buildings. *Struct. Eng. Mech.* **2012**, *43*, 503–526. [CrossRef]
6. Hoskere, V.; Park, J.-W.; Yoon, H.; Spencer, B.F., Jr. Vision-based modal survey of civil infrastructure using unmanned aerial vehicles. *J. Struct. Eng.* **2019**, *145*, 04019062. [CrossRef]
7. Ikeda, A.; Fujita, K.; Takewaki, I. Takewaki, Story-wise system identification of actual shear building using ambient vibration data and ARX model. *Earthq. Struct.* **2014**, *7*, 1093–1118. [CrossRef]
8. Valdez, J.M.; Alvarez-Icaza, L.; Sanchez-Sesma, F.J. Shear building stiffness estimation by wave traveling time analysis. *Struct. Control Health Monit.* **2018**, *25*, e2045. [CrossRef]
9. Yang, J.-H.; Lam, H.-F. An efficient adaptive sequential Monte Carlo method for Bayesian model updating and damage detection. *Struct. Control Health Monit.* **2018**, *25*, e2260. [CrossRef]
10. Yuen, K.-V.; Huang, K. Real-time substructural identification by boundary force modeling. *Struct. Control Health Monit.* **2018**, *25*, e2151. [CrossRef]

11. Rahbari, N.M.; Azar, B.F.; Talatahari, S.; Safari, H. Semi-active direct control method for seismic alleviation of structures using MR dampers. *Struct. Control Health Monit.* **2013**, *20*, 1021–1042. [CrossRef]
12. Alhan, C.; Sürmeli, M. Shear building representations of seismically isolated buildings. *Bull. Earthq. Eng.* **2011**, *9*, 1643–1671. [CrossRef]
13. Gu, X.; Yu, Y.; Li, Y.; Li, J.; Askari, M.; Samali, B. Experimental study of semi-active magnetorheological elastomer base isolation system using optimal neuro fuzzy logic control. *Mech. Syst. Signal Process.* **2019**, *119*, 380–398. [CrossRef]
14. Fujita, K.; Takewaki, I. Sustainable building design under uncertain structural-parameter environment in seismic-prone countries. *Sustain. Cities Soc.* **2011**, *1*, 142–151. [CrossRef]
15. Ganjavi, B.; Hao, H. Optimum lateral load pattern for seismic design of elastic shear-buildings incorporating soilstructure interaction effects. *Earthq. Eng. Struct. Dyn.* **2013**, *42*, 913–933. [CrossRef]
16. Li, S.; Yu, B.; Gao, M.; Zhai, C. Optimum seismic design of multi-story buildings for increasing collapse resistant capacity. *Soil Dyn. Earthq. Eng.* **2019**, *116*, 495–510. [CrossRef]
17. Lu, Y.; Hajirasouliha, I.; Marshall, A. Performance-based seismic design of flexible-base multi-storey buildings considering soil-structure interaction. *Eng. Struct.* **2016**, *108*, 90–103. [CrossRef]
18. Chopra, A.K.; Goel, R.K. A modal pushover analysis procedure for estimating seismic demands for buildings. *Earthq. Eng. Struct. Dyn.* **2002**, *31*, 561–582. [CrossRef]
19. Chopra, A.K.; Goel, R.K. Envelope-based pushover analysis procedure for the approximate seismic response analysis of buildings. *Earthq. Eng. Struct. Dyn.* **2014**, *43*, 77–96. [CrossRef]
20. Xu, B.; Wu, Z.; Yokoyama, K.; Harada, T.; Chen, G. A soft post-earthquake damage identification methodology using vibration time series. *Smart Mater. Struct.* **2005**, *14*, S116–S124. [CrossRef]
21. Xiong, C.; Lu, X.; Lin, X.; Xu, Z.; Ye, L. Parameter determination and damage assessment for tha-based regional seismic damage prediction of multi-story buildings. *J. Earthq. Eng.* **2017**, *21*, 461–485. [CrossRef]
22. Xiong, C.; Lu, X.; Guan, H.; Xu, Z. A nonlinear computational model for regional seismic simulation of tall buildings. *Bull. Earthq. Eng.* **2016**, *14*, 1047–1069. [CrossRef]
23. Lu, X.; Guan, H. *Earthquake Disaster Simulation of Civil Infrastructures: From Tall Buildings to Urban Areas*, 2nd ed.; Springer: Singapore, 2021; ISBN 978-981-15-9531-8. [CrossRef]
24. *Seismic Evaluation and Retrofit of Existing Buildings*; ASCE/SEI 7-16; American Society of Civil Engineers: Reston, VA, USA, 2016.
25. *Code for Seismic Design of Buildings*; GB50010-2010; Ministry of Housing and Urban-Rural Development of P. R. China: Beijing, China, 2010.
26. *Design of Structures for Earthquake Resistance. Part 1: General Rules, Seismic Action and Rules for Buildings*; Eurocode 8; European Committee for Standardization: Brussels, Belgium, 2004.
27. *Minimum Design Loads for Buildings and Other Structures*; ASCE/SEI 41-13; American Society of Civil Engineers: Reston, VA, USA, 2014.
28. *Prestandard and Commentary for the Seismic Rehabilitation of Buildings*; American Society of Civil Engineers: Reston, VA, USA, 2000.
29. *Improvement of Nonlinear Static Seismic Analysis Procedures*; FEMA 440; Applied Technology Council: Redwood City, CA, USA, 2005.
30. Park, K. State of the art report ductility evaluation from laboratory and analytical testing. In Proceedings of the Ninth World Conference on Earthquake Engineering, Tokyo, Japan, 2–6 August 1988.
31. Feng, P.; Cheng, S.; Bai, Y.; Ye, L. Mechanical behavior of concrete-filled square steel tube with FRP-confined concrete core subjected to axial compression. *Compos. Struct.* **2015**, *123*, 312–324. [CrossRef]
32. Sun, Y. Study on the Inelastic Displacement Estimation of Aseismatic Structures. Master's Thesis, Harbin Institute of Technology, Harbin, China, 2006.
33. *Seismic Evaluation and Retrofit of Concrete Buildings*; ATC-40; Applied Technology Council: Redwood City, CA, USA, 1996; Volume 1.
34. Valles, R.E.; Reinhorn, A.M.; Kunnath, S.K.; Li, C.; Madan, A. *Idarc 2D Version 4.0: A Program for the Inelastic Damage Analysis of Buildings*; Report No. NCEER-96-0010; National Center for Earthquake Engineering Research, State University of New York at Buffalo: New York, NY, USA, 1996.
35. Katsanos, E.I.; Sextos, A.G. Inelastic spectra to predict period elongation of structures under earthquake loading. *Earthq. Eng. Struct. Dyn.* **2015**, *44*, 1765–1782. [CrossRef]
36. Zhai, C.; Chang, Z.; Li, S.; Xie, L. Selection of the Most Unfavorable Real Ground Motions for Low- and Mid-rise RC Frame Structures. *J. Earthq. Eng.* **2013**, *17*, 1233–1251. [CrossRef]
37. Li, S.; He, Y.; Wei, Y. Truncation method of ground motion records based on the equivalence of structural maximum displacement responses. *J. Earthq. Eng.* **2021**, *5*, 1–22. [CrossRef]
38. Open System for Earthquake Engineering Simulation (OpenSees). 2016. Available online: https://opensees.berkeley.edu/wiki/index.php/Command_Manual (accessed on 20 February 2022).
39. Zhai, C.; Li, S.; Sun, Y. Inelastic displacement ratio spectra for reinforced concrete regular frame structures. *Eng. Mech.* **2009**, *9*, 88–94. [CrossRef]
40. Krawinkler, H.; Seneviratna, G. Pros and cons of a pushover analysis of seismic performance evaluation. *Eng. Struct.* **1998**, *20*, 452–464. [CrossRef]

41. Li, S.; Zuo, Z.; Zhai, C.; Xie, L. Comparison of static pushover and dynamic analyses using RC building shaking table experiment. *Eng. Struct.* **2017**, *136*, 430–440. [CrossRef]
42. Daei, A.; Poursha, M. On the accuracy of enhanced pushover procedures for seismic performance evaluation of code-conforming RC moment-resisting frame buildings subjected to pulse-like and non-pulse-like excitations. *Structures* **2021**, *32*, 929–945. [CrossRef]

buildings

MDPI

Article

Numerical Investigations of Progressive Collapse Behaviour of Multi-Storey Reinforced Concrete Frames

Qiao-Ling Fu [1], Liang Tan [2], Bin Long [3,*] and Shao-Bo Kang [3]

1 Chongqing Water Resources and Electric Engineering College, Chongqing 402160, China
2 China Nerin Engineering Co., Ltd., Nanchang 330103, China
3 School of Civil Engineering, Chongqing University, Chongqing 400045, China
* Correspondence: longbin@cqu.edu.cn

Abstract: This paper presents numerical simulations of multi-storey reinforced concrete frames under progressive collapse scenarios. Reinforced concrete frames with different storeys are modelled using DIANA. The load resistance and failure mode of frames are obtained from the numerical simulation. Variations in axial force and bending moment at the beam end are also determined and analysed to shed light on the force transfer mechanism. Numerical results show that the single-storey frame can develop compressive arch action at the initial loading stage and subsequent catenary action at large deformations. However, in multi-storey frames, only the first-storey beam develops compressive arch action and catenary action, whereas beams in other storeys show rather limited axial compression force. Based on numerical results, a design method is proposed for multi-storey frames to resist progressive collapse. Comparisons between numerical results and design methods suggest that the design method can evaluate the progressive collapse resistance of multi-storey frames with good accuracy.

Keywords: reinforced concrete frame; multiple storey; progressive collapse; numerical model; design method

Citation: Fu, Q.-L.; Tan, L.; Long, B.; Kang, S.-B. Numerical Investigations of Progressive Collapse Behaviour of Multi-Storey Reinforced Concrete Frames. *Buildings* **2023**, *13*, 533. https://doi.org/10.3390/buildings13020533

Academic Editors: Abdelhafid Khelidj, Shan Gao, Jingxuan Wang, Dewen Kong and Yong Liu

Received: 6 January 2023
Revised: 8 February 2023
Accepted: 13 February 2023
Published: 15 February 2023

1. Introduction

Since the disastrous failure of the World Trade Center in 2001, progressive collapse has become an important concern in the design of building structures. Different design methods have been proposed and incorporated in relevant guidelines [1,2].

In recent years, different types of experimental tests have been conducted by researchers from different countries [3–8], and the force transfer mechanism in beam–column sub-assemblages has been identified as compressive arch action and catenary action in beams with axial restraints. Numerical models were also developed for beam–column connections subjected to progressive collapse [9,10]. Parameters affecting the progressive collapse resistance of structures have been intensively investigated through experimental tests and numerical simulations [11,12]. Strengthening techniques have also been developed to mitigate the risk of progressive collapse [13]. This experimental study mainly focuses on the behaviour of beam–column sub-assemblages or single-storey frames. As for multi-storey frames, limited data are available due to the difficulties in conducting relevant tests and measuring resistance of different storeys [14–18]. Even though experimental tests were also conducted on whole structures [19] and design methods have been developed and incorporated in different codes [20], there is still a lack of quantitative methods for direct calculation of load capacity of structures against progressive collapse.

This paper describes numerical modelling of progressive collapse behaviour of multi-storey frames. In the numerical study, different frames are simulated using DIANA, and the force transfer mechanism at different storeys of multi-storey frames is identified. The contribution of each storey to total resistance of frames is also quantified using the numerical model, and finally, a design method is proposed based on the numerical results. The method fills the research gap in design methods for progressive collapse of multi-storey frames and can be directly used in the calculation of the load capacity under compressive arch action.

2. Numerical Modelling of Reinforced Concrete Frames

2.1. Development of Numerical Model

In this study, reinforced concrete frames with one supporting column removed are simplified as plane-stress models using DIANA [21], as shown in Figure 1. In the model, the plane-stress element CQ16M is used to define concrete beams, columns and footing, and the truss element is used for steel reinforcement. The stress–strain relationship of steel reinforcement is defined as piece-wise linear, as shown in Figure 2. In the constitutive model, hardening of reinforcement is considered, as it might affect the load capacity of reinforced concrete frames when fracture of reinforcement occurs. The specific mechanical properties of reinforcement are determined from tensile tests on reinforcement. As for concrete, the tension-softening stress–strain model developed by Hordijk is used for concrete in tension, and the compressive stress–strain model proposed by Maekawa is employed to model the behaviour of concrete in compression, as shown in Figure 3. In the concrete model, the confinement effect provided by stirrups in beams is not considered, as in general the spacing of stirrups is large and the confinement effect is weak.

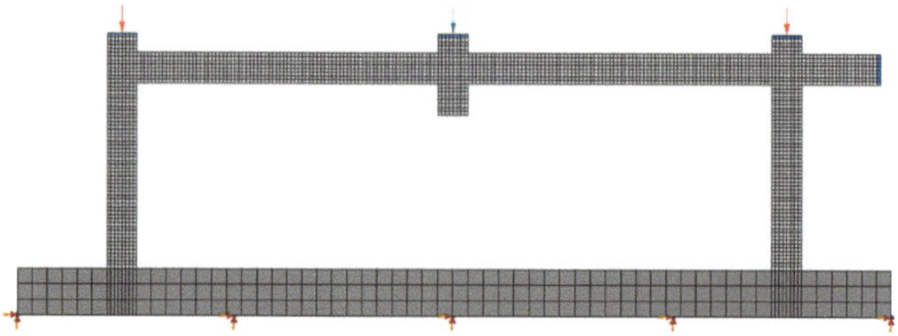

Figure 1. Numerical model for reinforced concrete frames. The arrows on top of the frame denote vertical loads, and those at the bottom represent vertical and horizontal restraints.

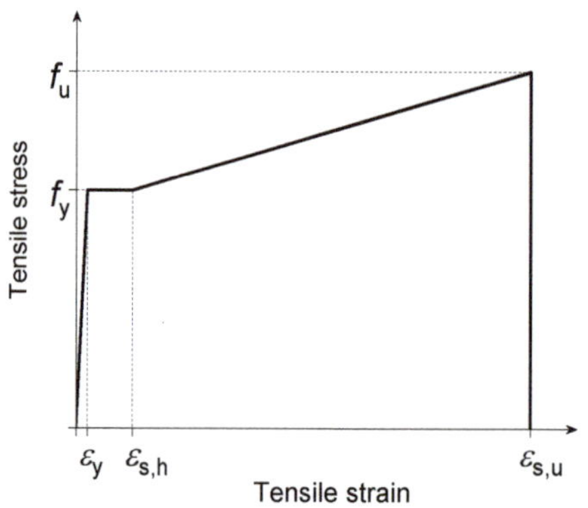

Figure 2. Constitutive models for steel reinforcement.

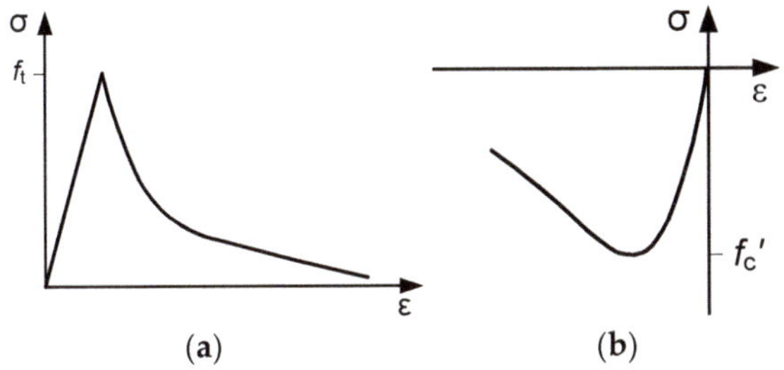

Figure 3. Constitutive models for concrete in tension and compression: (**a**) Hordijk model in tension; (**b**) Maekawa model in compression.

2.2. Simulation of Bond

In addition to the mechanical properties of steel reinforcement and concrete, the bond between reinforcement and concrete should also be defined in the numerical model. Previous studies show that the progressive collapse behaviour of reinforcement concrete frames can be affected by the bond–slip behaviour between steel reinforcement and concrete. DIANA provides two options to define the bond–slip behaviour of reinforcement, either by embedding reinforcement in concrete or by defining bond–slip damage models for reinforcement. When steel reinforcement is embedded in concrete, the normal and tangent stiffness moduli can be taken as 2000 N/mm^3 and 0.002 N/mm^3, respectively. In defining bond–slip models, the multi-linear bond–slip curve and cubic function by Dörr can be used. The bond stress in the models depends on the compressive or tensile strength of concrete.

Figure 4 shows the effect of bond models on the load–displacement curve of the single-storey frame tested by Tan [22]. It can be observed from the figure that by varying the bond–slip model between steel reinforcement and surrounding concrete, the peak load at the initial stage is insignificantly affected, whereas the ultimate load and associated displacement of the frame at the fracture of reinforcement is sensitive to the bond–slip model. The numerical result is in good agreement with the experimental result when the reinforcement is embedded in concrete. By contrast, the ultimate load and associated displacement of the frame are significantly overestimated when the multi-linear model or Dörr cubic function is defined in the numerical model. Therefore, embedded reinforcement is used in the subsequent numerical modelling.

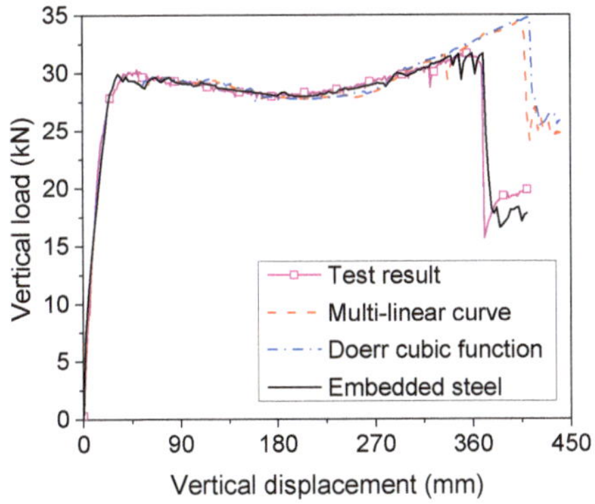

Figure 4. Effect of different bond models on load–displacement curve of frames.

2.3. Mesh Sensitivity Analysis

Mesh size in numerical models might affect the accuracy of the numerical result. To adopt the appropriate mesh size in the present study, a mesh sensitivity study is performed using numerical simulations, in which the single-storey reinforced concrete frame tested by

Tan [22] is used. Figure 5 shows the effect of mesh size on the load–displacement curve of the frame. It can be observed that when the mesh size is smaller than 30 mm, the overall load–displacement curve is not significantly affected in the numerical model, and the difference in the load capacity is within 1 kN. Therefore, the mesh size in the present study is selected as 30 mm to save the computational cost.

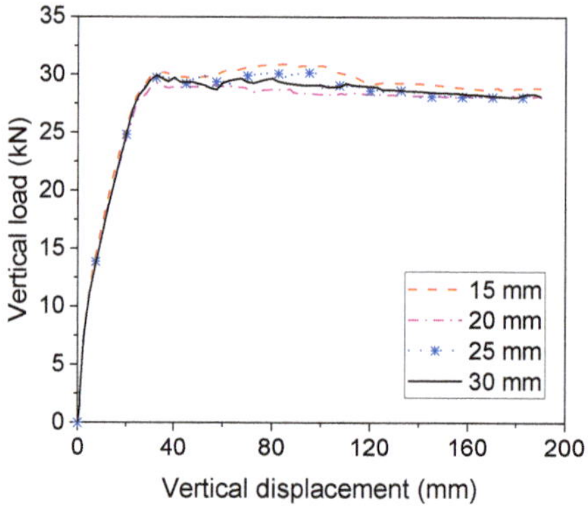

Figure 5. Effect of mesh size on load–displacement curve of frames.

3. Behaviour of Single-Storey Frame

In this section, the behaviour of a single-storey frame tested by Tan [22] is simulated under progressive collapse scenarios by using the proposed numerical model, as shown in Figure 6. Comparisons are made between experimental and numerical results at different levels to validate the accuracy of the numerical model. Moreover, the force transfer mechanism of the frame under progressive collapse is analysed in depth. More details of experimental tests can be found in [22].

Figure 6. A single-storey reinforced concrete frame under progressive collapse.

3.1. Load–Displacement Relationship

Figure 7 shows the comparisons of experimental and numerical results of the frame. In the figure, EXP represents experimental results and FEA denotes numerical results. It can be observed from Figure 7a that in general the experimental load–displacement curve is in good agreement with the numerical result from the initial loading to final failure. The vertical load can increase slowly as a result of the mobilisation of catenary action in beams. Besides the load–displacement curve, comparisons are also made among the variations of lateral displacement measured at the column top, as shown in Figure 7b. Note that D1 was the lateral displacement of the left column and D2 was the lateral displacement of the right column. It can be observed from the figure that the numerical lateral displacement is slightly larger than the experimental value, but the overall trends are close to each other. The lateral displacement of the left column exceeds that of the right column, indicating the potential failure of the left column.

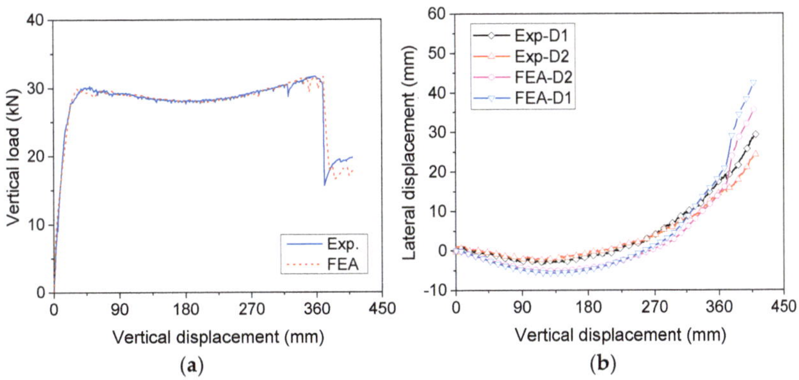

Figure 7. Comparisons between experimental and numerical results of single-storey frame: (**a**) load–displacement curve; (**b**) lateral displacement of side columns.

3.2. Failure Mode of Single-Storey Frame

Figure 8 shows the failure mode of the single-storey frame under progressive collapse scenarios. Note that the contour represents the failure mode of the frame at a vertical displacement of 400 mm. It is apparent that plastic hinges have formed at the beam end, and cracks even extend along the beam length at failure due to the combined effect of bending moment and tension force in the beam under catenary action. It is noteworthy that another hinge forms at the bottom end of the left column, indicating flexural failure of the column subjected to catenary action in the beam.

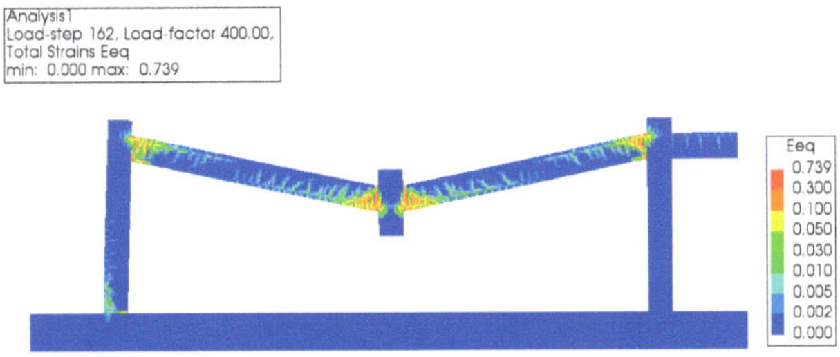

Figure 8. Numerical failure mode of single-storey frame.

3.3. Force Transfer Mechanism

To investigate the force transfer mechanism of the single-storey frame under progressive collapse scenarios, axial and shear forces acting at the beam end are extracted from the numerical model, as shown in Figure 9a. It can be observed from the figure that the axial force in the beam is compressive when the vertical displacement is less than 340 mm, suggesting the development of compressive arch action in the beam. The peak axial compression is 34 kN, attained at a vertical displacement of 132.5 mm. Once the vertical displacement exceeds 340 mm, axial tension develops in the beam, indicating the mobilisation of catenary action. The peak tension force at the failure of the beam is 13.7 kN. The shear force at the beam end decreases with the increasing axial tension in the beam. In addition to the beam force, the bending moment at the column base is also obtained in the numerical model, as shown in Figure 9b. It can be observed that when the vertical displacement is less than 275 mm, the bending moments at the left and right column bases are close to each other. However, at the failure of the frame, the bending moment at the left column base is roughly 50% greater than that at the right column base, which leads to flexural failure of the left column base (see Figure 6). Therefore, it can be concluded from the numerical simulation that the single-storey reinforced concrete frame can develop compressive arch action and catenary action to resist progressive collapse, even though flexural failure of one column hinders the full development of catenary action capacity.

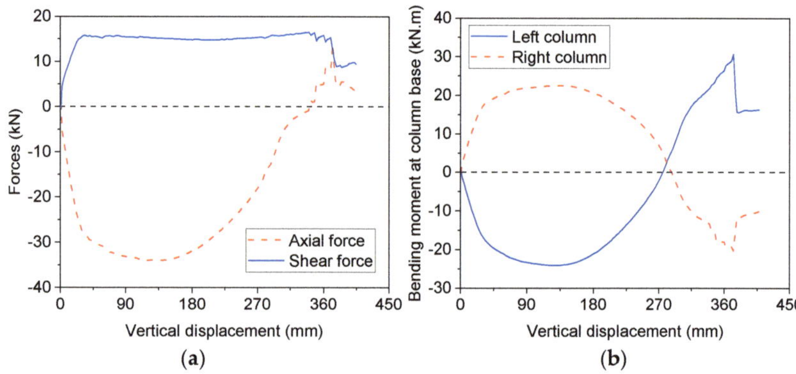

Figure 9. Comparisons between experimental and numerical results of single-storey frame: (**a**) axial and shear force; (**b**) bending moment.

4. Behaviour of Two-Storey Frame

Under progressive collapse scenarios, two-storey frames may behave differently due to the effect of Vierendeel action. Thus, numerical simulations were also conducted on two-storey frames tested by Shan et al. [23]. The frame consisted of four bays of beams and two storeys of columns. Details of the frame can be found in [23], including the geometric dimension and reinforcement details. During analyses, a displacement-control loading is applied to the middle column, and restraints are provided for the footing.

4.1. Global Behaviour of Frame

Comparisons are made between numerical and experimental results of the two-storey frame in terms of the load–displacement relationship and lateral deflections of columns at different storeys, as shown in Figure 10. It can be observed from Figure 10a that the numerical load–displacement curve agrees well with the experimental one, but the load capacity of the same at final failure is overestimated by the numerical model by roughly 28%. Figure 10b,c shows the comparison of numerical and experimental lateral deflections of columns at the first and second storeys. Note that the lateral displacement in the figure refers to the displacement measured at the left column. It can be observed that the lateral deflection of columns can also be obtained from the numerical model with good accuracy.

4.2. Failure Mode of Two-Storey Frame

Figure 11 shows the numerical failure mode of the two-storey frame under progressive collapse scenarios. The contour represents the failure mode of the frame at 360 mm vertical displacement. It can be observed that after the failure of a middle column, only the beams directly connecting to the column exhibit significant damage, whereas the beam and column adjacent to the affected spans only show limited cracks. Moreover, due to the presence of two columns on each side of the middle column, flexural failure of columns does not occur, even though catenary action develops in the bridging beam. In general, the crack pattern and failure mode of the frame agree well with experimental observations.

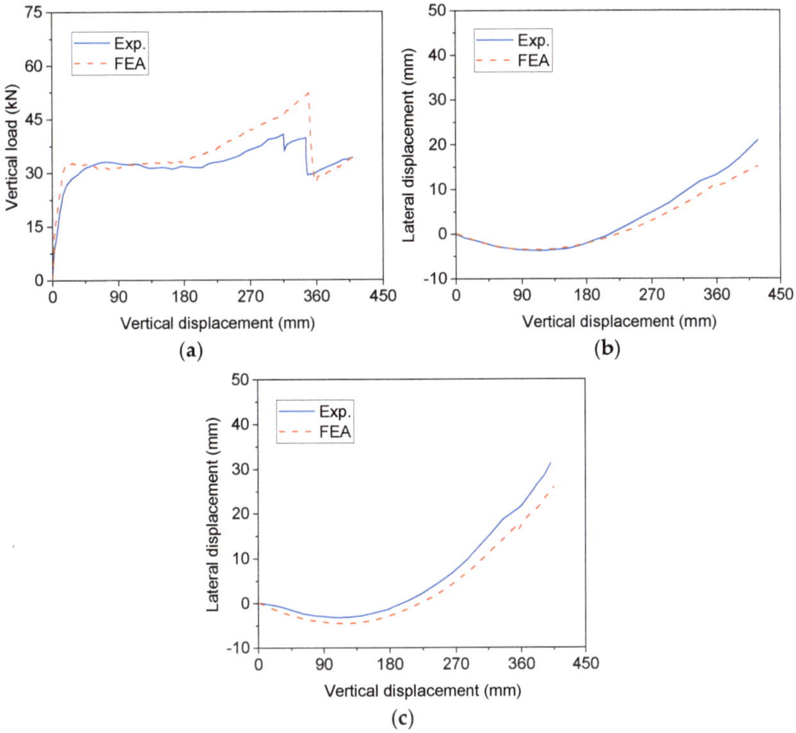

Figure 10. Comparisons between experimental and numerical results of two-storey frames: (**a**) load–displacement curve; (**b**) lateral displacement of first-storey column; (**c**) lateral displacement of second-storey column.

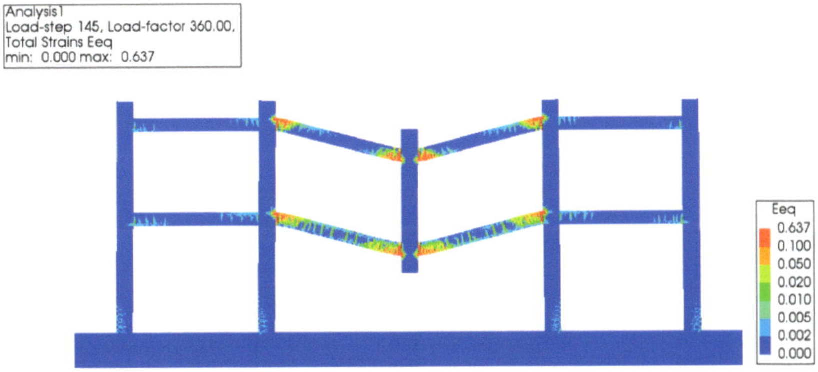

Figure 11. Numerical failure mode of two-storey frame.

4.3. Force Transfer Mechanisms

Figure 12 shows the variation of beam axial force, shear force and bending moment at different storeys. In the figure, IF and 2F denotes the beam at the first and second floors, respectively, and vertical displacement denotes the displacement of the removed column. Note that span BC refers to the beam on the left side of the damaged column. A similar definition is used in the subsequent section. It can be observed from Figure 12a that at the initial loading stage, only the first floor develops significant axial compression in the bridging beam, whereas the second storey only develops rather limited compression force. Moreover, with increasing vertical displacement, the axial compression in the first-storey beam is gradually shifted to tension, indicating the formation of catenary action in the beam, whereas the compression force in the second-storey beam increases slowly until final failure occurs. With the increase of the axial compression in the second-storey beam, the associated shear force at the beam end is also increased, as shown in Figure 12b. Different variations of bending moment are also obtained at the beam ends of different storeys, as shown in Figure 12c. At the first storey, the development of axial tension in the beam decreases the bending moment at the beam end. However, the beam-end bending moment is slightly increased due to the increasing axial compression in the beam at the second storey. Note that the shear force and bending moment are extracted at the left end of the beam. Therefore, when a two-storey reinforced concrete frame is considered in the progressive collapse design, the difference between beam behaviour at the first and second storey should be taken into account to accurately evaluate the progressive collapse resistance of the frame.

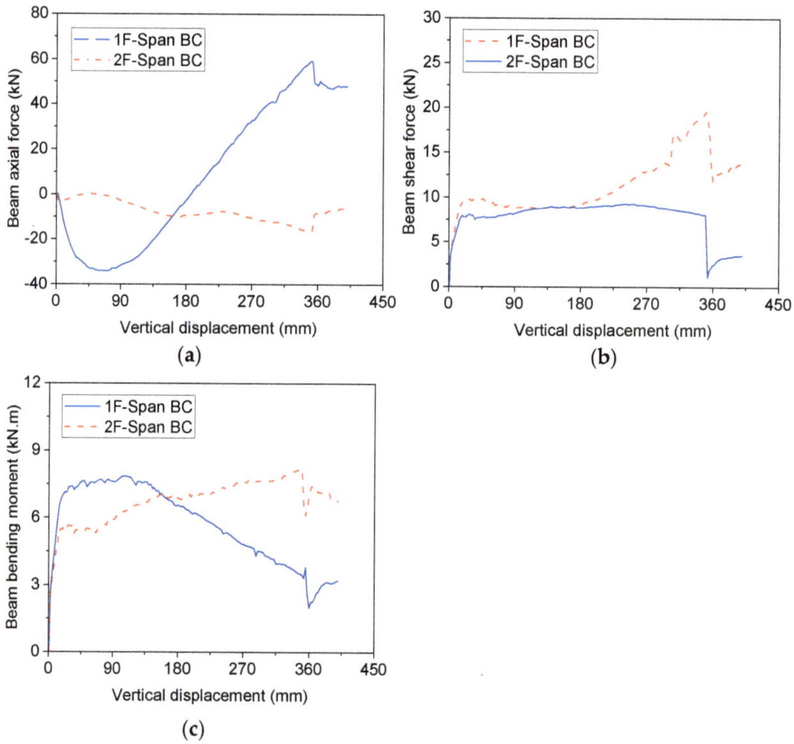

Figure 12. Behaviour of beams at different storeys of two-storey frame: (**a**) axial forces in beams; (**b**) shear forces in beams; (**c**) bending moments in beams.

5. Behaviour of Three-Storey Frame

Yi et al. [14] tested a four-bay three-storey frame under progressive collapse scenarios. In the frame, a middle column was assumed to have failed and the remaining structural members were tested to failure to determine the load resistance and failure mode of the frame. Details of the frame can be found in [14].

5.1. Global Behaviour of Frame

Figure 13 shows the behaviour of the three-storey frame subjected to progressive collapse. It can be observed from Figure 13a that the numerical load–displacement curve is in good agreement with the experimental one, with nearly the same load capacity under compressive arch action, but the ultimate load of the frame is slightly underestimated by 9% by the numerical model. The difference between the load–displacement curves mainly lies in the initial stiffness and the displacement at which the beam bottom reinforcement fractures. The difference in the initial stiffness might result from the different mechanical properties of concrete in the experimental tests and numerical simulations, such as the modulus of elasticity. In the numerical model, the bottom reinforcement fractures at a greater displacement that that in the experimental test, and thus the ultimate load capacity of the frame is overestimated. Comparisons are also made between numerical and experimental lateral displacements of columns at different storeys, as shown in Figure 13b–d. Note that the displacement refers to the value at the left side of the frame. It can be observed that the numerical model can also predict the lateral displacement of columns at different storeys. Similarly to single-storey and two-storey frames, the columns in the three-storey frame are pushed away from the middle column at the initial stage, namely, when the vertical displacement is less than one-beam depth. With increasing vertical displacement, the column is pulled towards the middle column, and the maximum lateral displacement can be up to 15 mm.

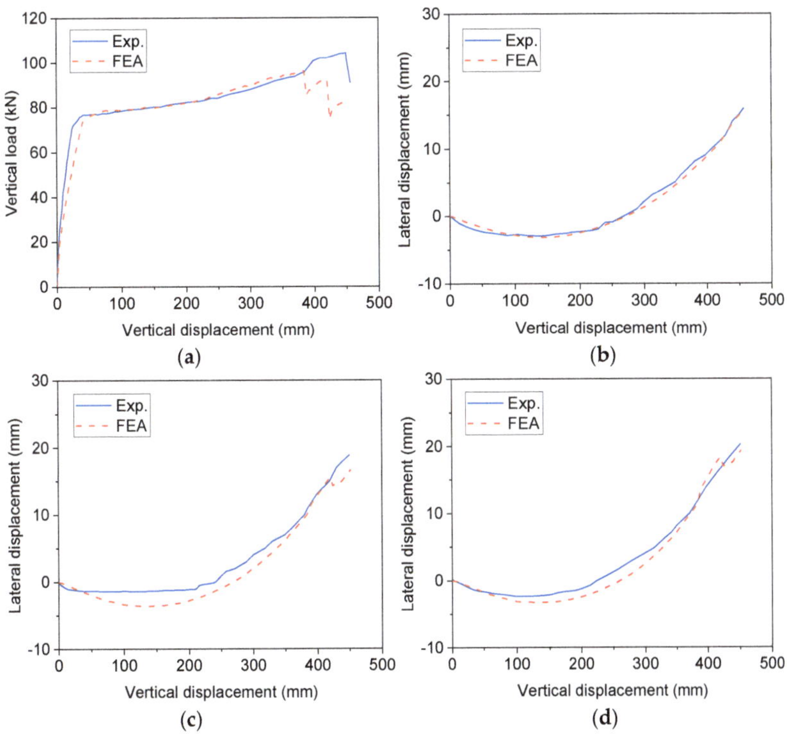

Figure 13. Comparisons between experimental and numerical results of three-storey frames: (**a**) load–displacement curve; (**b**) lateral displacement of first-storey column; (**c**) lateral displacement of second-storey column; (**d**) lateral displacement of third-storey column.

5.2. Failure Mode of Three-Storey Frame

Similarly to the two-storey frame, the three-storey frame develops cracks and flexural failure of beams above the damaged column, as shown in Figure 14. Note that the contour represents the failure mode of the frame at 460 mm vertical displacement. The beam at the first storey shows more severe damage than those at the second and third storeys, indicating that progressive collapse of the frame might occur at the first storey and propagate to the second and the third storey. As for the columns, only those at the first storey exhibit visible damage, whereas the columns at the second and third storeys remain intact. Thus, the numerical failure mode of the frame is in good agreement with the experimental results.

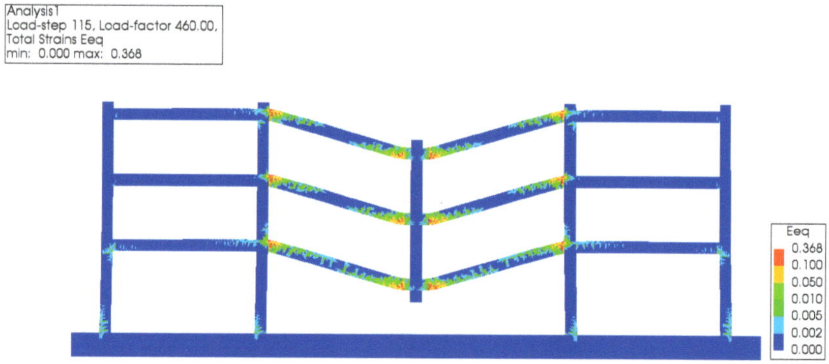

Figure 14. Numerical failure mode of three-storey frame.

5.3. Force Transfer Mechanisms

Figure 15a shows the variation of axial force of beams at different storeys. It can be observed that the beam at the first storey develops axial compression force when the vertical displacement is less than 200 mm, and then is gradually converted to tension force with increasing vertical displacement. By contrast, the beams at the second and third storeys only develop axial compression force during the whole loading process. Moreover, it should be pointed out that the beam axial compression at the second storey is rather limited and can be neglected. Therefore, it can be conceived that for the three-storey frame, catenary action is only mobilised at the first storey, whereas the beam at the third storey is always subjected to an axial compression force, which forms Vierendeel action along with the tension force at the first storey. For the beam at the second storey, its axial force can be neglected in design, and thus it can be treated as a pure flexural member.

Figure 15b shows the variation in shear force of beams at different storeys. At the initial stage, the shear force of beams at different storeys is close to one another, in spite of the differences in the beam axial force. However, when the vertical displacement exceeds 270 mm, the shear force of beams at the first storey starts increasing, indicating the increasing contribution of the first-storey beam to the global resistance of the frame. Unlike the shear force, the bending moment at the beam end shows a different variation with the vertical displacement, as shown in Figure 15c. Prior to the commencement of catenary action, the bending moment of beams at different storeys is rather close. Nonetheless, the bending moment of beams at the first storey decreases with increasing vertical displacement once catenary action develops in the beam.

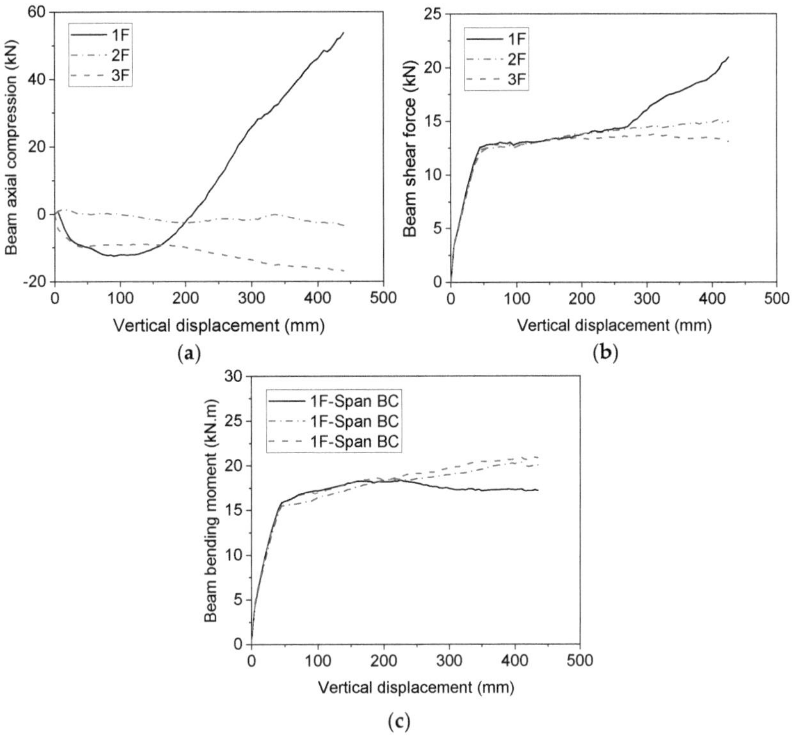

Figure 15. Behaviour of beams at different storeys of three-storey frame: (**a**) axial forces in beams; (**b**) shear forces in beams; (**c**) bending moments in beams.

6. Parametric Study

To explicitly investigate the influence of the number of storeys on the load resistance of frames under progressive collapse, a prototype planar frame is designed according to Chinese code GB50010-2010 [24], as shown in Figure 16. The frame consists of four spans of 2200 mm. The height of the ground storey is 1400 mm, and for other storeys the height is kept at 1200 mm. The cross-sections of beams and columns are 100 mm × 200 mm and 200 mm × 200 mm, respectively. Reinforcement details in the beam and column remain the same as those in the frame tested by Tan [22]. The number of storeys of the frame varies from one to five to investigate the influence on load resistance of frames by using numerical simulations. In the design of frames, only the vertical load is considered. The mechanical properties of steel reinforcement and concrete are also the same as those used by Tan [22]. More details of the frame can be found in [22].

Figure 17 shows the vertical load–displacement curves of frames with different storeys. In the figure, K1 through K5 represent frames with one through five storeys. It can be observed that by increasing the number of storeys, both the compressive arch action capacity and the catenary action capacity can be increased. Table 1 lists the load capacities of frames under compressive arch action and catenary action. By increasing one storey, the compressive arch action capacity of frames is increased by roughly 26 kN. Note the increase in the compressive arch action capacity is less than the load capacity of frame K1 with only one storey. Likewise, the catenary action capacity of the frame is increased by around 34 kN when it is increased by one more storey. The increase in the catenary action capacity of frames is much smaller than the catenary action capacity of K1. The foregoing behaviour results from the force transfer mechanism of the multi-storey frame. As concluded in the previous section, significant compressive arch action and subsequent catenary action only develop at the first storey of reinforced concrete frames. When the number of storeys is increased in multi-storey frames, the upper storey mainly sustains loads through flexural action, rather than compressive arch action and catenary action. Therefore, the increase in the load capacity of frames is smaller than the load capacity of the first-storey frame. With an increasing number of storeys, the ratio of catenary action capacity to compressive arch action capacity decreases.

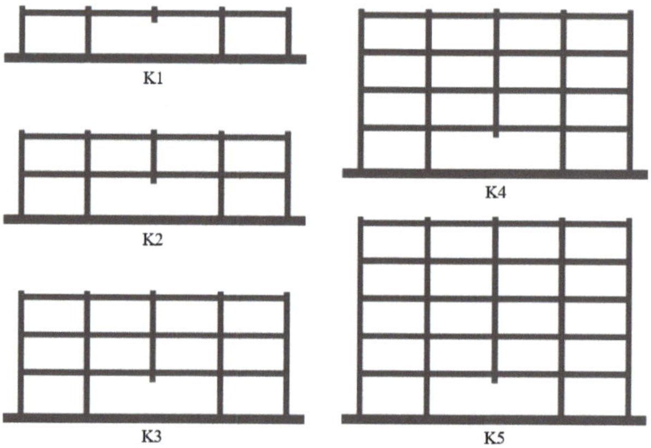

Figure 16. Multi-storey reinforced concrete frames under progressive collapse scenarios.

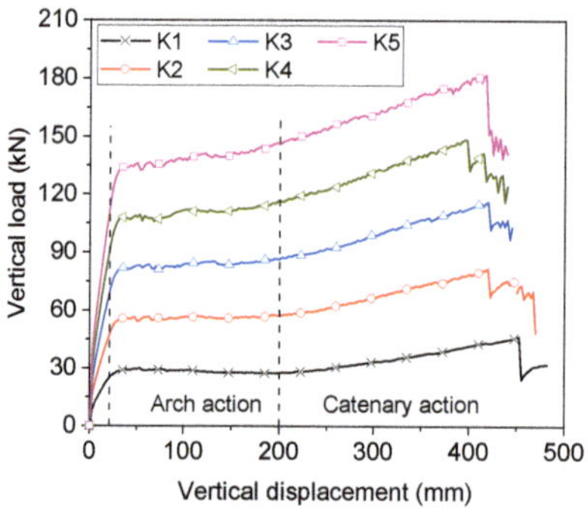

Figure 17. Load resistance of frames with different storeys.

Table 1. Load capacity of frames under progressive collapse.

Specimen	Compressive Arch Action Capacity P_{CA} (kN)	Compressive Action Capacity P_{CT} (kN)	$\Delta P = P_{CT} - P_{CA}$ (kN)	$\frac{\Delta P}{P_{CA}}$	Calculated Value P_{total} (kN)	
K1	30.0	46.5	16.5	0.55	30.0	1.00
K2	56.5	81.9	25.4	0.45	54.6	0.97
K3	83.1	116.2	33.1	0.40	79.2	0.95
K4	109.2	148.5	39.3	0.36	103.8	0.95
K5	136.3	182.0	45.7	0.34	128.4	0.94

Figure 18 shows the comparisons of beam axial forces at the first and top storeys. It can be observed from Figure 18a that at the initial loading stage, the beam axial force at the first storey of frames is rather close, whereas differences in the beam axial force of single-storey and multi-storey frames becomes increasingly significant when the vertical displacement is greater than 150 mm. In general, multi-storey frames can develop catenary action at a smaller vertical displacement than the single-storey frame. Moreover, the peak axial tension under catenary action is also larger in the multi-storey frame than that in the single-storey frame. Nevertheless, the beam axial force at the top storey remains compressive and limited differences exist in the axial force at the top storey, as shown in Figure 18b. Note that the axial force in beams of other storeys is rather limited during the whole loading process, and thus it is not included in the figure.

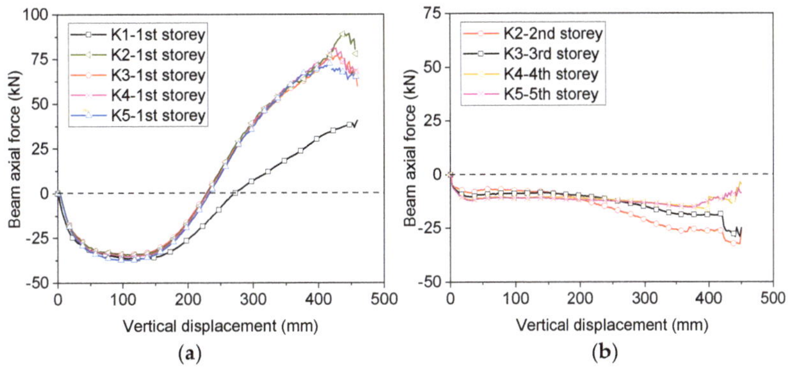

Figure 18. Variations in beam axial force at the first and top storeys: (**a**) beam axial force at the first storey; (**b**) beam axial force at the top storey.

7. Development of Design Method

In design of building structures, most standards only allow the use of compressive arch action capacity out of conservatism. Therefore, in accordance with the numerical results, a design equation is proposed to quantify the load capacity of multi-storey frames against progressive collapse, as expressed in Equation (1).

$$P_{total} = P_{CA}^{1} + \sum_{i=2}^{n} P_{FA}^{i} \tag{1}$$

where P_{CA}^{1} is the load capacity of the first-storey beam under compressive arch action and P_{FA}^{i} is the flexural capacity of beams at the upper storeys.

In the equation, the load capacity of beams under compressive arch action can be determined from Equation (2).

$$P_{CA}^{1} = \frac{2(M_1 + M_2 - N\delta)}{L} \tag{2}$$

where M_1 and M_2 are the positive and negative bending moments at beam ends, N is the axial compression of beams under compressive arch action, δ is the vertical deflection of beams at the compressive arch action capacity and can be determined in accordance with Lu's equation [25], and L is the clear span of beams.

In the calculation of M_1 and M_2, the axial compression in the beam should be considered, and thus the bending moments can be determined from the axial force-bending moment interaction diagram of beams. To simplify the calculation, the interaction diagram of axial force and bending moment can be assumed to be linear at the compressive arch action stage, and the bending moment at the beam end can be calculated from Equation (3).

$$M_1 = M_0 + \frac{N}{N_{cr}}(M_{cr} - M_0) \tag{3}$$

where M_0 is the moment capacity of beam sections in pure bending, N_{cr} is the axial compression of beams at balanced failure, and M_{cr} is the bending moment of beam sections at balanced failure.

In Equation (3), the axial compression force of beams can be calculated from the method derived by Kang and Wang [26]. A similar equation can also be used to calculated M_2.

The load capacity of beams at other storeys can be quantified from Equation (4). Note that in calculating the bending moment at beam ends, the axial force in the beam is neglected and the bending moment can be computed from pure flexural theory.

$$P^i_{FA} = \frac{2(M_1 + M_2)}{L} \tag{4}$$

Table 1 shows the calculated load capacity using Equation (1). Comparisons between P_{total} and P_{CA} suggest that the equation yields reasonably conservative estimations of the load capacity of multi-storey frames under compressive arch action. With the increasing number of storeys, the ratio of calculated to numerical results decreases slightly, as the axial compression force in the upper storey is neglected in the proposed equation.

8. Conclusions

This paper addresses the progressive collapse behaviour of multi-storey reinforced concrete frames. Numerical models are developed for different types of frames against progressive collapse, and the model is validated against test data. The variation in beam axial force, shear force and bending moment are extracted from the numerical model and analysed to gain insight into the force transfer mechanism of multi-storey frames under progressive collapse scenarios. The following conclusions can be drawn from the numerical study.

(1) A single-storey frame can develop compressive arch action and subsequent catenary action under progressive collapse scenarios, with axial compression force in the beam at the compressive arch action stage and axial tension force at the catenary action stage.
(2) Multi-storey reinforced concrete frames only develop compressive arch action and catenary action in the beam at the first storey, whereas beams at other storeys can develop limited axial compression force in the whole loading process.
(3) By increasing the number of storeys, the progressive collapse resistance of frames can be increased. The increase in the load resistance depends mainly on the flexural resistance of beams at upper storeys due to the limited axial compression in the beam, in particular the beams at the storeys between the first and top storeys.
(4) A design method is proposed based on numerical results of progressive collapse resistance of multi-storey frames. In the design method, only the compressive arch action capacity in the first-storey beam is considered, whereas flexural resistance of beams at other storeys is incorporated. Comparisons with numerical results suggest that the design method can evaluate the load resistance of multi-storey frames with reasonably good accuracy.

Author Contributions: Conceptualization, S.-B.K. and Q.-L.F.; methodology, L.T.; software, Q.-L.F. and L.T.; validation, L.T. and S.-B.K.; formal analysis, Q.-L.F. and S.-B.K.; investigation, Q.-L.F.; writing—original draft preparation, Q.-L.F. and L.T.; writing—review and editing, B.L. and S.-B.K.; supervision, S.-B.K.; project administration, B.L.; funding acquisition, Q.-L.F. All authors have read and agreed to the published version of the manuscript.

Funding: This research was funded by the Chongqing Commission of Education (KJQN202103802) and the Chongqing Water Resources and Electric Engineering College (K202214).

Data Availability Statement: The data presented in this study are available on request from the corresponding author.

Conflicts of Interest: The authors declare no conflict of interest.

References

1. DoD. *Design of Buildings to Resist Progressive Collapse;* Unified Facilities Criteria (UFC) 4-023-03; Department of Defence: Washington, DC, USA, 2013.
2. GSA. *General Services Administration Alternate Path Analysis & Design Guidelines for Progressive Collapse Resistance;* General Services Administration: Washington, DC, USA, 2013.
3. Su, Y.; Tian, Y.; Song, X. Progressive collapse resistance of axially-restrained frame beams. *ACI Struct. J.* **2009**, *106*, 600–607.
4. Yu, J.; Tan, K.H. Special Detailing Techniques to Improve Structural Resistance against Progressive Collapse. *J. Struct. Eng.* **2014**, *140*, 04013077. [CrossRef]
5. Yu, J.; Tan, K.H. Structural Behavior of Reinforced Concrete Frames Subjected to Progressive Collapse. *ACI Struct. J.* **2017**, *114*, 63–74. [CrossRef]
6. Kang, S.-B.; Tan, K.H. Behaviour of Precast Concrete Beam-Column Sub-assemblages Subject to Column Removal. *Eng. Struct.* **2015**, *93*, 85–96. [CrossRef]
7. Kang, S.-B.; Tan, K.H. Robustness Assessment of Exterior Precast Concrete Frames under Column Removal Scenarios. *J. Struct. Eng.* **2016**, *142*, 04016131. [CrossRef]
8. Alshaikh, I.M.H.; Bakar, B.H.A.; Alwesabi, E.A.H.; Akil, H.M. Experimental investigation of the progressive collapse of reinforced concrete structures: An overview. *Structures* **2020**, *25*, 881–900. [CrossRef]
9. Yu, J.; Tan, K.H. Structural Behavior of RC Beam-Column Subassemblages under a Middle Column Removal Scenario. *J. Struct. Eng.* **2013**, *139*, 233–250. [CrossRef]
10. Long, X.; Lee, C.K. Improved strut-and-tie method for 2D RC beam-column joints under monotonic loading. *Comput. Concr.* **2015**, *15*, 807–831. [CrossRef]
11. Long, X.; Lee, C.K. Modelling of two-dimensional reinforced concrete beam-column joints subjected to monotonic loading. *Adv. Struct. Eng.* **2015**, *18*, 1461–1474. [CrossRef]
12. Yu, J.; Tan, K.H. Experimental and Numerical Investigation on Progressive Collapse Resistance of Reinforced Concrete Beam-Column Sub-assemblages. *Eng. Struct.* **2013**, *55*, 90–106. [CrossRef]
13. Kiakojouri, F.; De Biagi, V.; Chiaia, B.; Sherdaii, M.R. Strengthening and retrofitting techniques to mitigate progressive collapse: A critical review and future research agenda. *Eng. Struct.* **2022**, *262*, 114274. [CrossRef]
14. Yi, W.J.; He, Q.F.; Xiao, Y.; Kunnath, S.K. Experimental study on progressive collapse-resistant behavior of reinforced concrete frame structures. *ACI Struct. J.* **2008**, *105*, 433–439.
15. Sagiroglu, S. *Analytical and Experimental Evaluation of Progressive Collapse Resistance of Reinforced Concrete Structures;* Northeastern University: Boston, MA, USA, 2012.
16. Qian, K.; Weng, Y.-H.; Fu, F.; Deng, X.-F. Numerical evaluation of the reliability of using single-story substructures to study progressive collapse behaviour of multi-story RC frames. *J. Build. Eng.* **2021**, *33*, 101636. [CrossRef]
17. Sagiroglu, S.; Sasani, M. Progressive Collapse-Resisting Mechanisms of Reinforced Concrete Structures and Effects of Initial Damage Locations. *J. Struct. Eng.* **2014**, *14*, 04013073. [CrossRef]
18. Yu, J.; Tian, J.-W. Is the Load Transfer Mechanism of Each Story in a Multi-Story Building the Same Subjected to Progressive Collapse. In Proceedings of the Structural Congress, Denver, CO, USA, 6–8 April 2017.
19. Xiao, Y.; Kunnath, S.K.; Li, F.W.; Zhao, Y.B.; Lew, H.S.; Bao, Y. Collapse Test of Three-Story Half-Scale Reinforced Concrete Frame Building. *ACI Struct. J.* **2015**, *112*, 429–438. [CrossRef]
20. Russell, J.M.; Sagaseta, J.; Cormie, D.; Jones, A.E.K. Historical review of prescriptive design rules for robustness after the collapse of Ronan Point. *Structures* **2019**, *20*, 365–373. [CrossRef]
21. DIANA. *DIANA Users' Manual Release 10.2: DIANA FEA bv;* DIANA: Memphis, TN, USA, 2017.
22. Tan, L. *Experimental Tests and Numerical Analysis on Progressive Collapse Behavior of Planar Reinforced Concrete Frames;* Chongqing University: Chongqing, China, 2022.
23. Shan, S.; Li, S.; Xu, S.; Xie, L. Experimental study on the progressive collapse performance of RC frames with infill walls. *Eng. Struct.* **2016**, *111*, 80–92. [CrossRef]
24. *GB50010-2010;* Ministry of Construction. Code for Design of Concrete Structures. China Architecture & Building Press: Beijing, China, 2010.
25. Lu, X.; Lin, K.; Li, C.; Li, Y. New analytical calculation models for compressive arch action in reinforced concrete structures. *Eng. Struct.* **2018**, *168*, 721–735. [CrossRef]
26. Kang, S.B.; Wang, S. Design Approach for Compressive Arch Action in Reinforced Concrete Beams. *ACI Struct. J.* **2020**, *117*, 197–206. [CrossRef]

buildings

MDPI

Article

Study on the Eccentric Compressive Performance of Steel Fibre Reinforced Coal Gangue Concrete Columns

Bin Cai *, Bingyang Bai, Wenfeng Duan *, Lin Wang and Shengda Wang

School of Civil Engineering, Jilin Jianzhu University, Changchun 130118, China
* Correspondence: caibin@jlju.edu.cn (B.C.); duanwenfeng@jlju.edu.cn (W.D.)

Abstract: Coal gangue is the waste created in the coal mining process and can be utilised as a coarse aggregate in construction projects to solve the environmental problems it causes. To study the mechanical properties of steel fibre reinforced coal gangue concrete (SFCGC) columns under eccentric compression, two natural aggregate concrete (NAC) columns and eight SFCGC columns were designed and fabricated for eccentric compression loading tests. The variables involved in the tests include gangue substitution rate (0%, 30%, 50%, 70%), steel fibre volume content (SFVC) (0%, 0.5%, 1%, 1.5%), and eccentricity (0.25, 0.5). The experimental work and theoretical analysis were used to investigate the failure mode, cracking, and ultimate bearing capacity of SFCGC columns. The effects of various variation parameters on the longitudinal strain, concrete strain, transverse displacement, crack width, and bearing capacity were analysed in detail. The digital image correlation (DIC) technique was used to compare with the conventional observation and to analyse the cracking trend of the specimens. The testing results revealed that the damage pattern of SFCGC columns under eccentric compression was similar to that of NAC columns. The eccentricity significantly affected the damage pattern (or bearing capacity) of SFCGC columns. The effect of the coal gangue replacement rate on the lateral displacement corresponding to the ultimate load can be neglected under the same eccentricity. The incorporation of steel fibres effectively inhibited the development of cracks in the columns, with an average increase in crack load and ultimate load of 7.36% and 17.1%. The equations were also established to determine the crack width and bearing capacity of the studied SFCGC columns, and the theoretical predictions agreed with the experimental results.

Keywords: steel fibre; coal gangue; column; eccentric compression; mechanical performance

Citation: Cai, B.; Bai, B.; Duan, W.; Wang, L.; Wang, S. Study on the Eccentric Compressive Performance of Steel Fibre Reinforced Coal Gangue Concrete Columns. *Buildings* **2023**, *13*, 1290. https://doi.org/10.3390/buildings13051290

Academic Editor: Andreas Lampropoulos

Received: 8 April 2023
Revised: 5 May 2023
Accepted: 13 May 2023
Published: 16 May 2023

1. Introduction

As the world's second-largest energy source and China's number one, coal's tremendous production has led to an impressive output of its derivative, coal gangue. According to statistics, China's production of raw coal in 2021 is as high as 4.13 billion tons, an increase of 5.9%, and the production of coal gangue is about 743 million tons, an increase of 5.84% [1,2]. The accumulation of coal gangue seriously affects human health, safety, and the ecological health of the environment [3]. Therefore, developing a comprehensive utilisation strategy for coal gangue and eliminating its negative environmental impact is urgent [4,5]. At the same time, concrete structures, the most widely used building structures, have had a significant negative impact on the environment due to their construction size with the development of urbanisation and industrialisation [6]. The main reason includes the consumption of natural resources (including aggregate, water, etc.) during construction [7]. The shortage of aggregates is the most significant, accounting for 75–80% of the volume of concrete [8]. Coarse aggregates are mainly obtained by blasting rock boulders and then crushing them to the required size, a process harmful to the environment as it destroys mountains, erodes the soil, and produces large amounts of carbon dioxide. As a result, coal gangue has entered the minds of researchers as one of the possible alternatives to coarse aggregates.

The replacement of crushed stone with coal gangue for concrete production can consume large amounts of coal gangue and reduce the use of crushed stone, which is beneficial to the development of eco-friendly construction materials and fulfils the purpose of protecting the natural environment [9,10]. With its stable structure and low reactivity, coal gangue must be activated by mechanical grinding and thermal and chemical activation methods for construction applications [11,12]. Sun and Li [13] explored the use of coal gangue concrete for gob-side entry retaining, which addressed both ecological and resource shortages, decreased the overall cost of filling materials and raised the economic efficiency of coal mining enterprises. Wang and Zhao [14] demonstrated that coal gangue could satisfy the compressive strength requirements of concrete aggregates by testing the basic physical and mechanical properties of coal gangue aggregates. Ma [15] showed that coal gangue concrete has good resistance to sulphate. As a coarse aggregate, coal gangue provides both higher compressive strength and durability in alkali-active gangue-slag concrete for chemically aggressive environments such as sulphate or chloride. Li [16] prepared high closed porosity foamed ceramics with good thermal insulation properties by using coal gangue. Qiu [17] et al. analysed the mechanical properties of coal gangue concrete under the freeze-thaw environment. They presented the evolutionary model and the intrinsic structure relationship of its freeze-thaw damage. Xiao et al. [18] discovered that when coal gangue was used as an aggregate, the concrete's compatibility in tensile strength continued to decrease as the amount of coal gangue increased. Liu et al. [19] suggested a prediction model for the modulus of elasticity of coal gangue concrete. The performance of coal gangue concrete significantly reduces when the coal gangue substitution rate is too large [20,21]. Zhu et al. [22] revealed that when the replacement rate of coal gangue exceeded 50%, it would result in a rapid decrease in the compressive strength of coal gangue concrete. Chen [23] illustrated that the 28-day compressive strength of C25 concrete with a 100% coal gangue replacement rate was 25.47% less than that of ordinary crushed stone concrete. Guan et al. [24] studied that the compressive strength of CGC was decreased by 21.2% when the replacement rate of coal gangue was 60%, while when the replacement rate of coal gangue was less than 40%, it had no significant effect on the strength of CGC.

As a structural material invented in 1874, steel fibres have been widely used in construction applications [25] for their ability to dramatically improve the mechanical properties of concrete, such as impact strength, toughness, flexural and tensile strength, ductility, and resistance to cracking and spalling [26]. In this case, however, there is still extensive research value in steel fibres [27,28]. Li and Qin [29] observed that the mechanical summation of the shaped steel fibres in the concrete resulted in enhanced adhesion within the member. When the volume content of steel fibres in concrete was 0–2%, the steel fibre-matrix adhesion properties increased with the increasing volume content of steel fibres [30]. In addition, steel fibres improve the protective layer-core zone, effectively confining the concrete to the core of the reinforced concrete specimens, which benefits both the strength and ductility of the reinforced concrete specimens [31,32].

The strength of the concrete produced using coal gangue as coarse aggregate is similar to that of ordinary concrete. Still, its durability is lower than ordinary concrete's and prone to brittle damage [33,34]. Liu et al. [35] found that gangue concrete columns eccentrically stressed were unable to prevent the expansion of cracks effectively and were prone to cracking and corrosion of reinforcement when conducting a study on gangue concrete columns. Luo et al. [36] revealed that the integration of steel fibres in the concrete with coal gangue as coarse aggregate is beneficial to its frost resistance. Wang et al. [37] discovered that the incorporation of steel fibres could significantly change the brittleness of coal gangue vitrified light aggregate concrete. Thus, adding steel fibres to coal gangue concrete can improve its properties and suitability for construction.

Digital Image Correlation (DIC) has the advantages of high accuracy, non-contact measurement, no special requirements for the environment, and the ability to give information on the deformation of a specimen in any area and in any direction during loading, and is therefore widely used in the fields of solid mechanics and applied physics, and is now gradually being applied to research on construction materials. The accuracy of DIC in testing the uniaxial compressive strain and crack opening displacement of specimens was verified by Dinh et al. [38] and Yu et al. [39] using strain gauges and displacement gauges, respectively. An objective of this test was to use DIC to monitor the development of cracks in concrete columns at different loading stages at mid-height and to extract data on the vertical displacement and mid-span deflection of the specimens and, in this way, compensate for possible errors in the contact measuring instruments.

Most of the current research on coal gangue remains in the field of physical and chemical properties of the material, while there is a lack of research on the mechanical performance testing of concrete specimens. People's knowledge of the performance of coal gangue concrete applied to actual buildings is relatively incomplete as compared to other mature materials. In particular, the study of SFCGC columns has not been reported. Therefore, this paper subjected ten SFCGC columns to eccentric compression tests with coal gangue replacement rate, steel fibre volume admixture, and eccentric distance as the main parameters. The performance of the specimens was studied and verified using current structural codes, proving the applicability of the current codes to SFCGC columns, which provides theoretical support for mitigating environmental problems through the promotion of coal gangue applications.

2. Materials and Methods

2.1. Raw Materials

JGJ51-2002 [40] is a test used to design the coal gangue concrete used in the test with design strength C30. SFCGC is formulated from coal gangue aggregate, silicate cement, shear wavy steel fibre (as shown in Figure 1, physical properties in Table 1), and tap water. P-O42.5 ordinary silicate cement was selected according to specification GB 175-2020 [41], and its chemical and physical properties are shown in Tables 2 and 3. The chemical composition of coal gangue is shown in Table 4.

Figure 1. Materials.

Table 1. Physical properties of steel fibres.

Steel Fibre	Length (mm)	Width (mm)	Thickness (mm)	Equivalent Diameter (mm)	Aspect Ratio	Density (g/cm^3)
CSF	38	1	0.35–0.5	0.76	50	7.8

Table 2. Chemical composition of cement.

Chemical Composition	SiO_2	Al_2O_3	MgO	Na_2O	CaO	Fe_2O_3	f-CaO	K_2O
Content (%)	22.4	4.6	3.00	0.4	64.2	3.5	1.3	0.6

Table 3. Physical properties of cement.

Cement	Conservation Times (Day)	Rupture Strength (Mpa)	Compressive Strength (Mpa)
P-O42.5	3	5.5	8.4
	28	8.4	51.6

Table 4. Chemical composition of coal gangue.

Chemical Composition	SiO_2	Al_2O_3	FeO	MgO	BaO	Na_2O	TiO_2	Fe_2O_3	P_2O_5	MnO	K_2O
Content (%)	36.83	36.83	5.83	5.00	5.66	2.83	1.77	3.51	0.39	0.15	1.65

2.2. Preparation of SFCGC Columns

Ten specimens were designed concerning specification GB50010-2010 [42], where the system variable parameters considered included two eccentricities (0.25, 0.5), four gangue replacement rates (0%, 30%, 50%, 70%), and four steel fibre volume contents (0%, 0.5%, 1%, 1.5%). The design parameters for the specimens are listed in Table 5.

Table 5. Specimen detail.

Specimens	V_c (%)	V_f (%)	e_0 (mm)	C (kg/m^3)	W (kg/m^3)	S (kg/m^3)	CA (kg/m^3) N	CA (kg/m^3) CG	Wr (kg/m^3)
S-0-1	0	1	50	503.7	201.5	623	1155.5	0	5.04
S-30-1	30	1	50	503.7	201.5	623	1048.8	360	5.04
S-50-1	50	1	50	503.7	201.5	623	749.6	598.5	5.04
L-0-1	0	1	100	503.7	201.5	623	1155.5	0	5.04
L-30-1	30	1	100	503.7	201.5	623	1048.8	598.5	5.04
L-50-1	50	1	100	503.7	201.5	623	749.6	598.5	5.04
L-70-1	70	1	100	503.7	201.5	623	450.3	840	5.04
L-50-0	50	0	100	503.7	201.5	623	749.6	598.5	5.04
L-50-0.5	50	0.5	100	503.7	201.5	623	749.6	598.5	5.04
L-50-1.5	50	1.5	100	503.7	201.5	623	749.6	598.5	5.04

Note: C = cement; W = water; S = sand; CA = coarse aggregate; N = natural; CG = coal gangue; Wr = water reducer; V_c = coal gangue replacement rate; V_f = steel fibre volume content.

The cross-section of all specimens is 200 mm × 150 mm, and the height is 1000 mm. All samples contained the same reinforcement: a longitudinal reinforcement of 16 mm in diameter with a yield strength of 417 MPa for HRB400 grade bars and hoop bars of 8 mm in diameter with a yield strength of 346 MPa for HPB300 bars. Figure 2 displays the detailed dimensions of the column specimens, the reinforcement configuration and the fabricated solid reinforcement cage, respectively. As shown in Figure 2b, the reinforcement was strictly assembled according to the set specifications while the wooden formwork was built, and finally, the assembled reinforcement cage was placed into the wooden formwork. Then, concrete was poured into the formwork and fully compacted with a vibrator to ensure uniform concrete dispersion. The concrete fabrication process is shown in Figure 3. While pouring the specimens, 150 mm × 150 mm cubic specimens were made from the same concrete batch. After three days, the specimens were de-moulded and immediately placed in a maintenance room at a temperature of 22 ± 2 °C and a relative humidity of 95 ± 5% for 28 days.

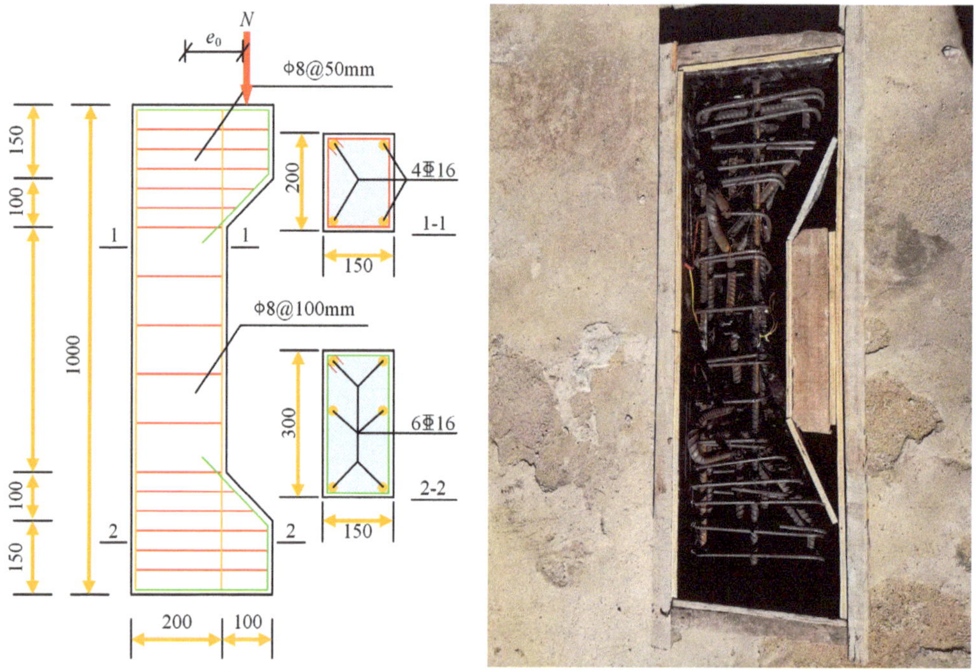

(a) (b)

Figure 2. Preparation of specimen: (**a**) Design of reinforcement (units: mm), (**b**) Steel bar cage.

Figure 3. The specimen fabrication process.

2.3. Test Preparation

2.3.1. Location of Measurement Points

In the test, three strain gauges were evenly distributed at the middle height of the compressive and tensile sides of the specimens to measure the concrete strain. Five strain gauges were evenly distributed at the middle height of the bottom surface of each specimen to verify whether the cross-sectional strain at the middle height of the specimen complied with the plane cross-section assumption. Five displacement meters are evenly placed along the vertical direction on the tensioned side of the bias column, with the outermost displacement meter 100 mm from the top and bottom. Two strain gauges are arranged in the middle of each longitudinal bar. Two strain gauges are arranged in the middle of each longitudinal bar. The layout of the displacement and strain gauges is depicted in Figure 4a.

Referring to the setup of Lin et al. [43], two one-way hinge supports were placed on the top and bottom of the specimen in contact with the press, as shown in Figure 4a.

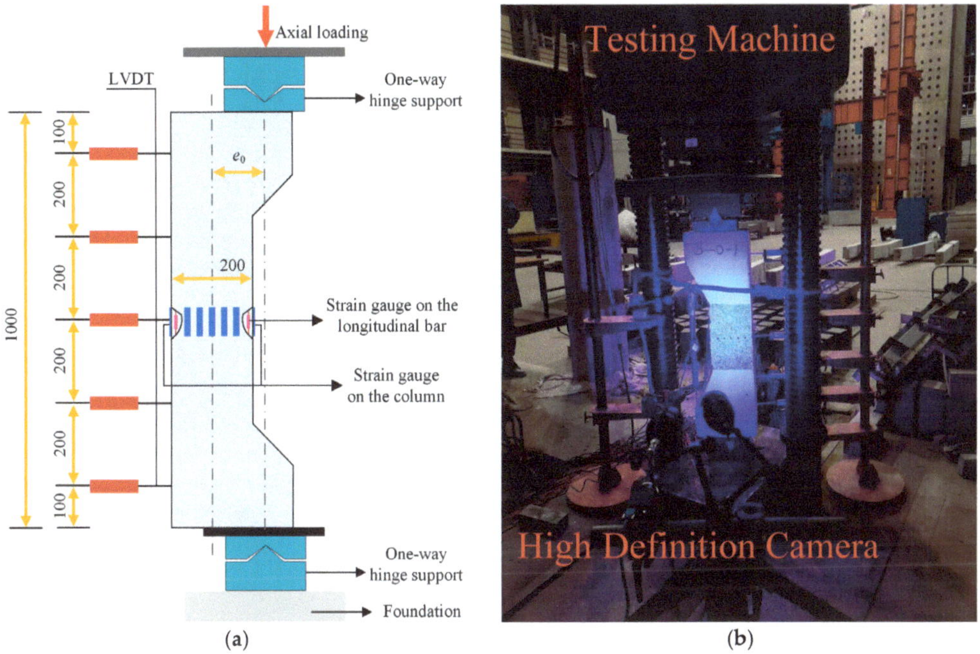

Figure 4. Test design: (**a**) measurement point layout, (**b**) DIC and testing machine.

2.3.2. Preparation for DIC

A non-contact optical measurement method based on digital image correlation (DIC) was applied to the other side of the specimen surface, as shown in Figure 4b. The scattered spots of global strain measurement are made on the specimen's front surface, and the diameter of the scattered spots is about 0.8 mm, covering the whole area in which the test is to be performed. A high-definition camera was used for image acquisition in this area, and the capture frequency was 1 s/time. By comparing the changes in the relative positions of these scattered spots during the loading process, the displacement field on the surface of the specimens can be calculated, and the strain distribution field can be further analysed. The strain field provided by the DIC can be verified and enhanced with the strain data gained from the strain gauge to uncover the details of deformation.

2.3.3. Test Loading Scheme

The compression tests of all the specimens in this research were conducted on the model NYL-500t hydraulic-type compression tester with a maximum load of 5000 kN. The arrangement and photos of the loading device are shown in Figure 4b. A load of 2 kN is preloaded on the specimens to check the correct operation of the strain gauges, displacement meters, loading apparatus, and data acquisition system. Then, reloading starts from 0 with the loading control first. During the initial loading phase, the loading hold time for each phase was 10 min. During this period, the development of concrete cracks was observed and recorded. As the load approaches the ultimate load, it is gradually loaded until it reaches the upper limit of the sample.

3. Experimental Results

3.1. Failure Modes

The crack morphology of the specimens under different eccentricity loading is shown in Figure 5. Figure 6 shows the crack development of specimens with different coal gangue replacement rates under large eccentric loading obtained by DIC system analysis. Each specimen is extracted with its nephogram at 0.2 Nu, 0.4 Nu, 0.6 Nu, and 0.8 Nu, respectively.

Specimens S-0-1, S-30-1, and S-50-1 were loaded under small eccentric loads. At 65% of the ultimate load, three–four transverse cracks appear in the tensile zone. At 30% of the ultimate load, the number of cracks gradually increases, and the length and width of the existing cracks develop slowly. At 80% of the ultimate load, longitudinal cracks appear in the compression zone. At 90% ultimate load, the transverse crack stretched to the neutral axis of the test column, the crack width became larger, and the transverse displacement kept increasing. When the ultimate load is reached, many longitudinal cracks appear in the compression zone, while the specimens emit a crisp cracking sound. The load is then rapidly reduced, and the specimens are destroyed.

Specimens L-0-1, L-30-1, L-50-1, L-70-1, L-50-0, L-50-0.5, and L-50-1.5 were loaded under a large eccentric load. At 20% ultimate load, transverse cracks appeared in the tensile zone of the test column, and an increase in the width and length of the transverse cracks accompanied the increase in load. With the continued loading of up to 90% of the ultimate load, longitudinal cracks appeared in the compression zone, and transverse cracks penetrated. The strain nephogram agrees well with the crack development under conventional observation, which verifies the reliability of DIC in examining crack development.

Notably, all specimens had no major crushing and spalling of the concrete in the compressed area, maintaining its integrity due to the binding of the steel fibres and the action of the bridge.

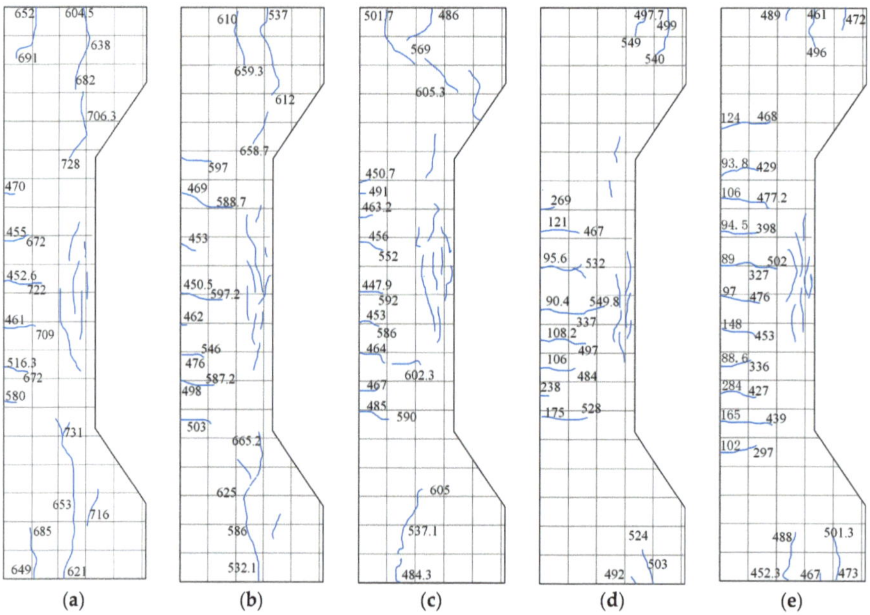

Figure 5. *Cont.*

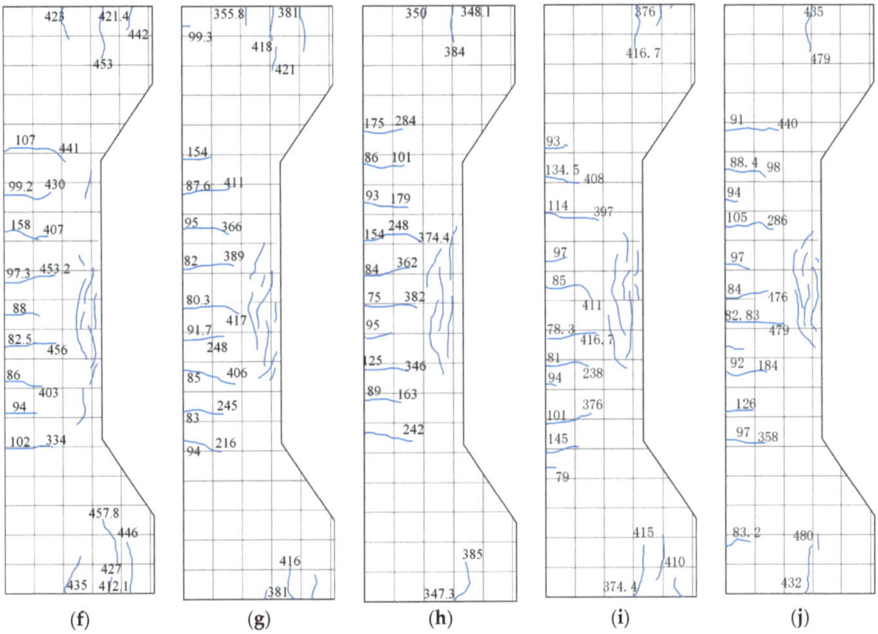

Figure 5. Specimen failure modes: (**a**) S-0-1, (**b**) S-30-1, (**c**) S-50-1, (**d**) L-0-1, (**e**) L-30-1, (**f**) L-50-1, (**g**) L-70-1, (**h**) L-50-0, (**i**) L-50-0.5, (**j**) L-50-1.5.

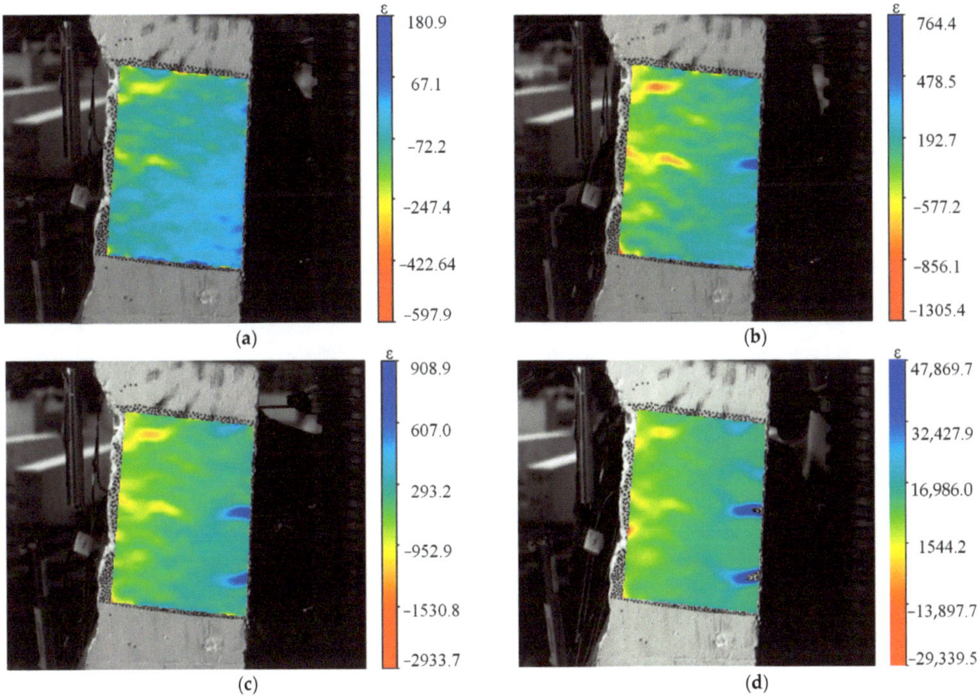

Figure 6. *Cont.*

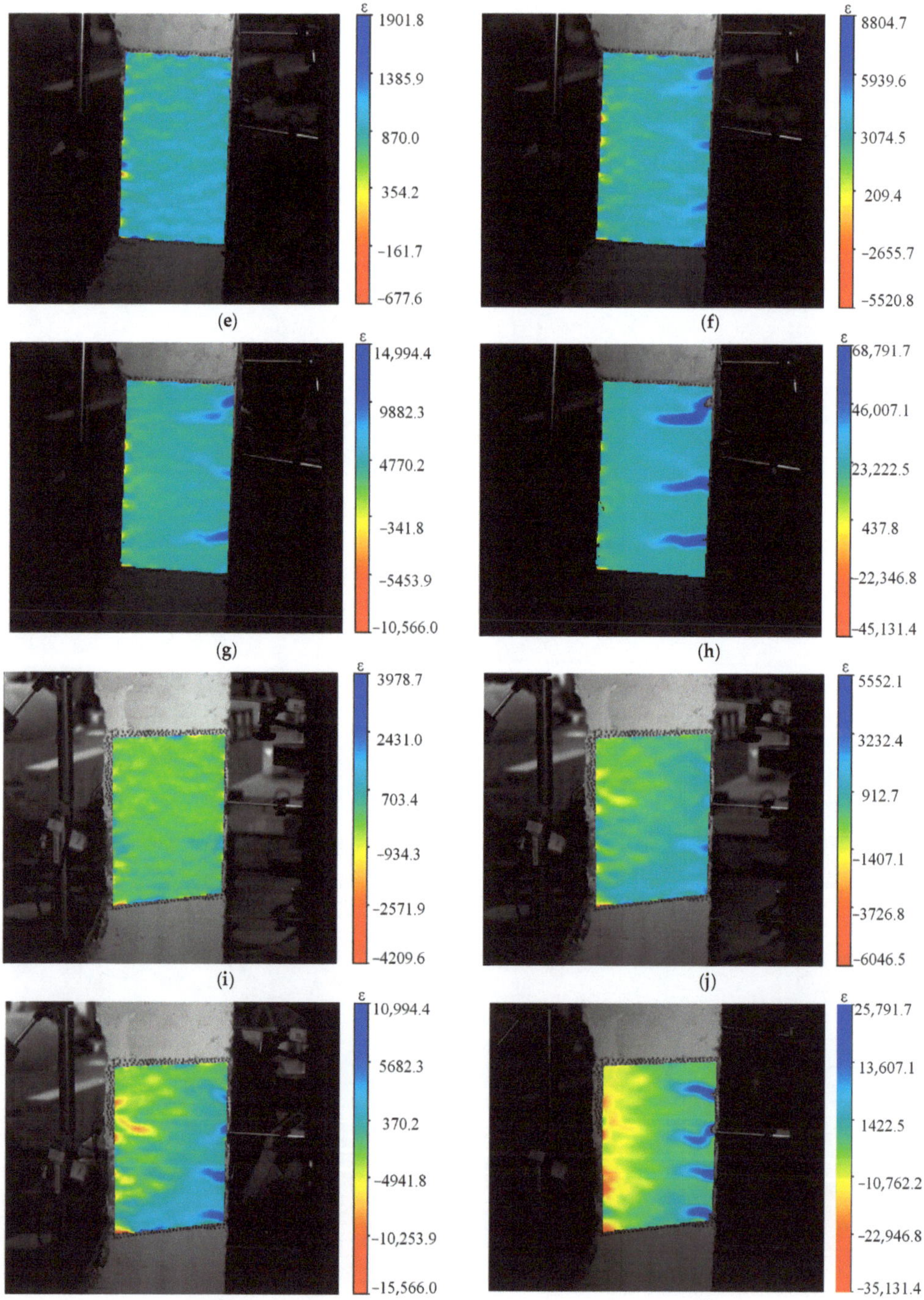

Figure 6. *Cont.*

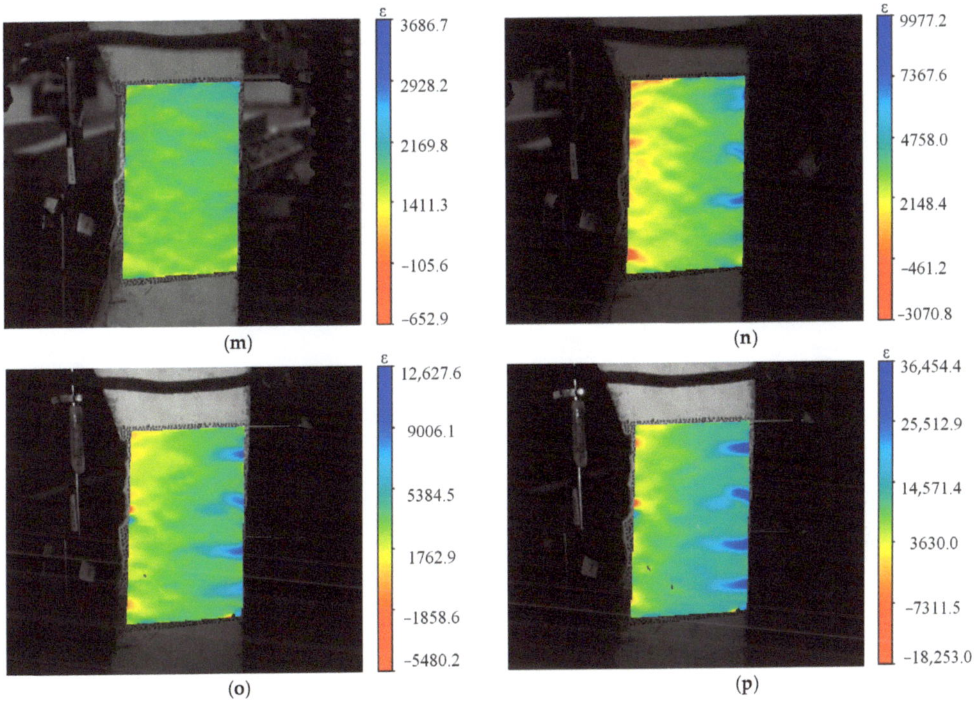

Figure 6. Nephogram of the strain process: (**a**) L-0-1 under 0.2 Nu, (**b**) L-0-1 under 0.4 Nu, (**c**) L-0-1 under 0.6 Nu, (**d**) L-0-1 under 0.8 Nu, (**e**) L-30-1 under 0.2 Nu, (**f**) L-30-1 under 0.4 Nu, (**g**) L-30-1 under 0.6 Nu, (**h**) L-30-1 under 0.8 Nu, (**i**) L-50-1 under 0.2 Nu, (**j**) L-50-1 under 0.4 Nu, (**k**) L-50-1 under 0.6 Nu, (**l**) L-50-1 under 0.8 Nu, (**m**) L-70-1 under 0.2 Nu, (**n**) L-70-1 under 0.4 Nu, (**o**) L-70-1 under 0.6 Nu, (**p**) L-70-1 under 0.8 Nu.

3.2. Concrete Strain of the Mid-Height Section

Figure 7 illustrates the strain distribution of concrete in the middle height section of the specimen at different loading stages. During the loading process, the average strain in the cross-section is almost linearly distributed, and the depth of the neutral axis gradually moves toward the compression edge as the applied load increases. This phenomenon confirms that the deformation of the SFCGC column section follows the plane cross-section assumption.

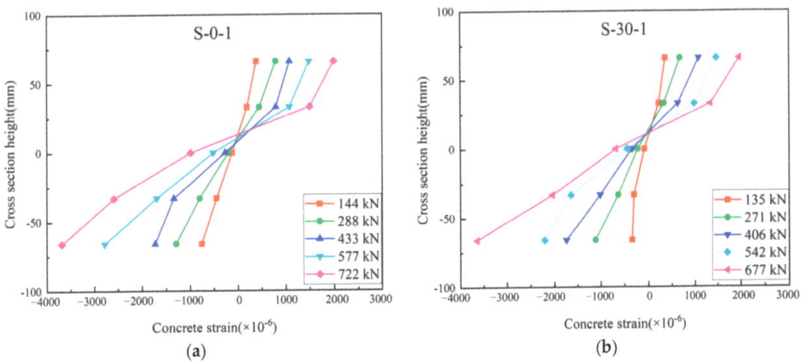

Figure 7. *Cont.*

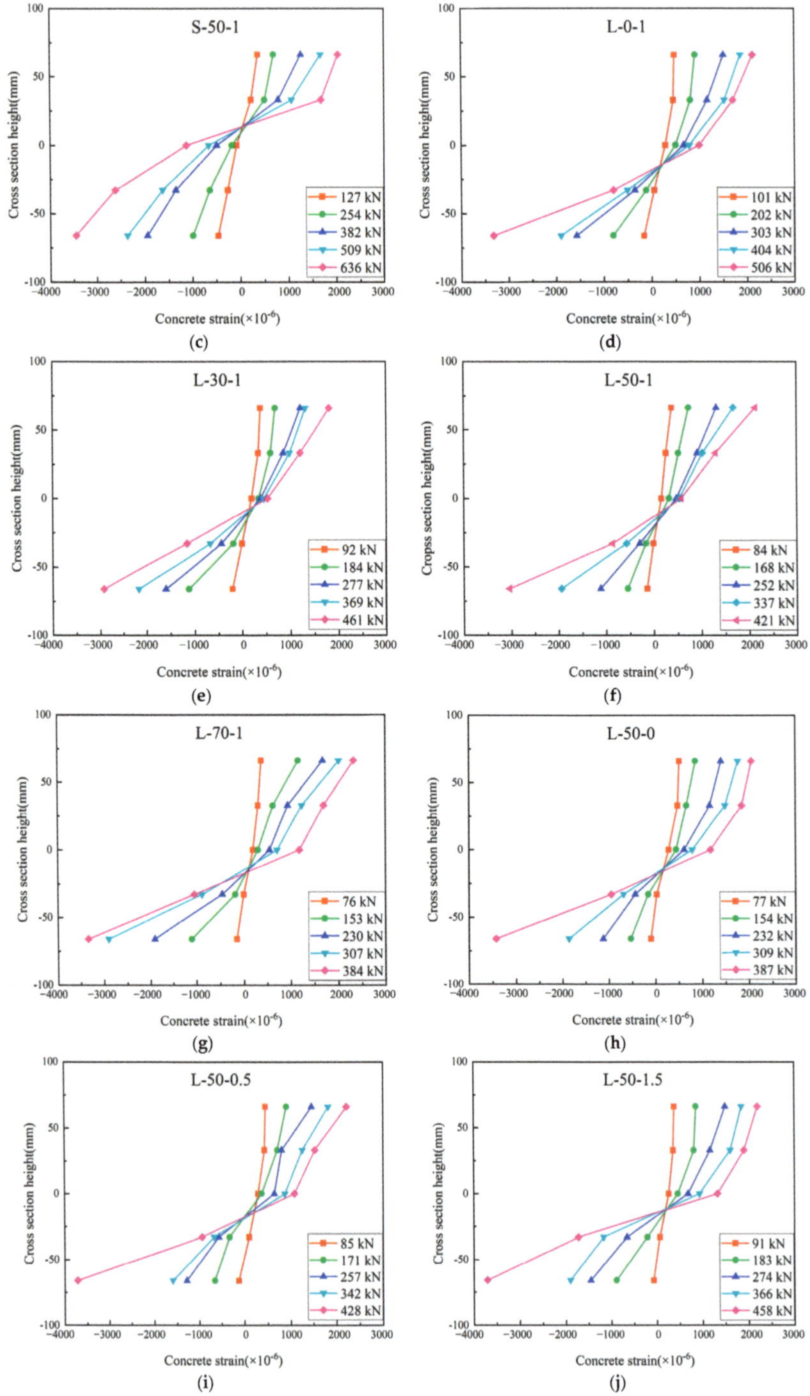

Figure 7. Strain distribution in mid-span sections of concrete: (**a**) S-0-1, (**b**) S-30-1, (**c**) S-50-1, (**d**) L-0-1, (**e**) L-30-1, (**f**) L-50-1, (**g**) L-70-1, (**h**) L-50-0, (**i**) L-50-0.5, (**j**) L-50-1.5.

3.3. Strains of the Concrete

Based on the experimental test results, the load-concrete strain curves of eccentrically compressed SFCGC columns with different conditions were plotted, as shown in Figure 8. Under the same residual conditions, the concrete deformation of the specimens was greater with an increase in the coal gangue replacement rate. At the same time, the increase in steel fibre admixture also led to the same results as the former one. This is because the coarse aggregate of the gangue has mechanical properties that increases deformation, while the steel fibres have a certain anti-cracking function, which improves the deformation capacity of the specimen, while this also corresponds to the conclusions reached in the paper [44] on the role of steel fibres.

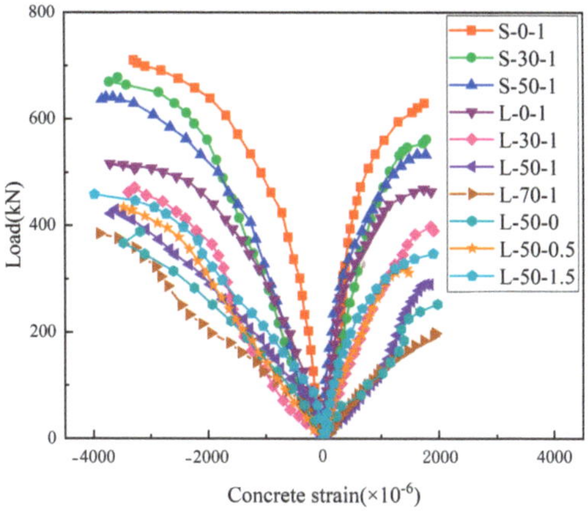

Figure 8. The load-strain curve of concrete.

3.4. Strains of the Longitudinal Steel Bars

Figure 9 provides the load-longitudinal strain curves of SFCGC columns under different conditions of eccentric compression. The Figure shows that the compressive reinforcement of the small eccentric compressive specimens has yielded when the load reaches the bearing capacity limit state. Moreover, its tensile reinforcement has not yielded yet. On the contrary, the compressive and tensile reinforcement of the large eccentric compressive specimen has yielded. This is consistent with the eccentric compressive damage characteristics.

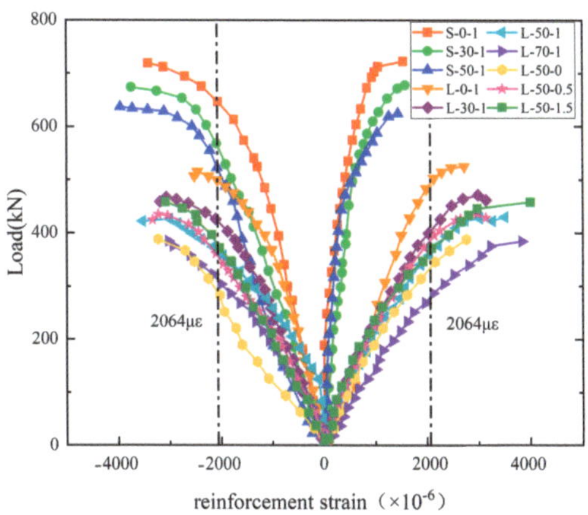

Figure 9. The load-reinforced steel strain curve.

The slope of the strain curve increases as the amount of steel fibres increases, and the strain on the reinforcement increases when the final load is reached, in agreement with the phenomenon obtained by reference [45], which is explained by the fact that the steel fibres present take up a portion of the tensile force when the concrete cracks, causing the stress on the reinforcement to decrease and the strain to increase at the same load level.

3.5. *Load-Displacement Analysis*

3.5.1. Load-Axial Displacement Analysis

The load-axial displacement curve of the specimens is shown in Figure 10. It can be seen that the axial displacement of the coal gangue concrete columns is proportional to the replacement rate of coal gangue. With an increase in gangue replacement rate, the maximum axial displacement of the member increases; this indicates that the gangue, as a coarse aggregate when its strength is low and easy to deform, characteristics directly affect the mechanical properties of the member not only reducing the ultimate bearing capacity of the columns but also increasing the axial deformation of the columns. From Figure 9, we can see that steel fibre has almost no effect on axial deformation, and the amount of coal gangue doping plays a dominant role in the axial deformation of the member.

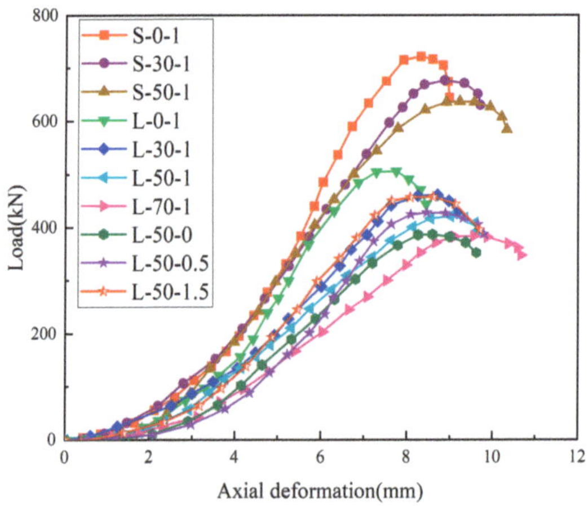

Figure 10. Axial displacement versus load curves for specimens.

3.5.2. Load-Lateral Displacement Analysis

Figure 11 presents the lateral deflection curves of each specimen under different loading stages. The lateral deflection curves of the specimens are symmetrically distributed along the middle of the columns and approximate sinusoidal curves. As expected, the maximum lateral displacement occurred in the middle of the test columns, and the deformation was shaped like an arch.

The deflection curves of the middle height of the specimen are shown in Figure 12. From the Figure 12, it can be seen that the load-lateral deflection curves of the eccentrically stressed columns can be summarised into three stages:

1. In the elastic stage, although small cracks appeared on the tensile side of the specimens, the stiffness of the specimens did not drop significantly due to the presence of reinforcement and steel fibres which inhibited the development of cracks, and the curves were straighter.
2. In the plastic stage, as the load increases, the reinforcement gradually yields, the cracks expand faster, the stiffness of the specimens decreases, their lateral deflection increases nonlinearly, the slope of the curves decreases, and the specimens enter the plastic stage.

3. In the descending stage, after reaching the ultimate load, the load decreases, and the lateral deflection increases.

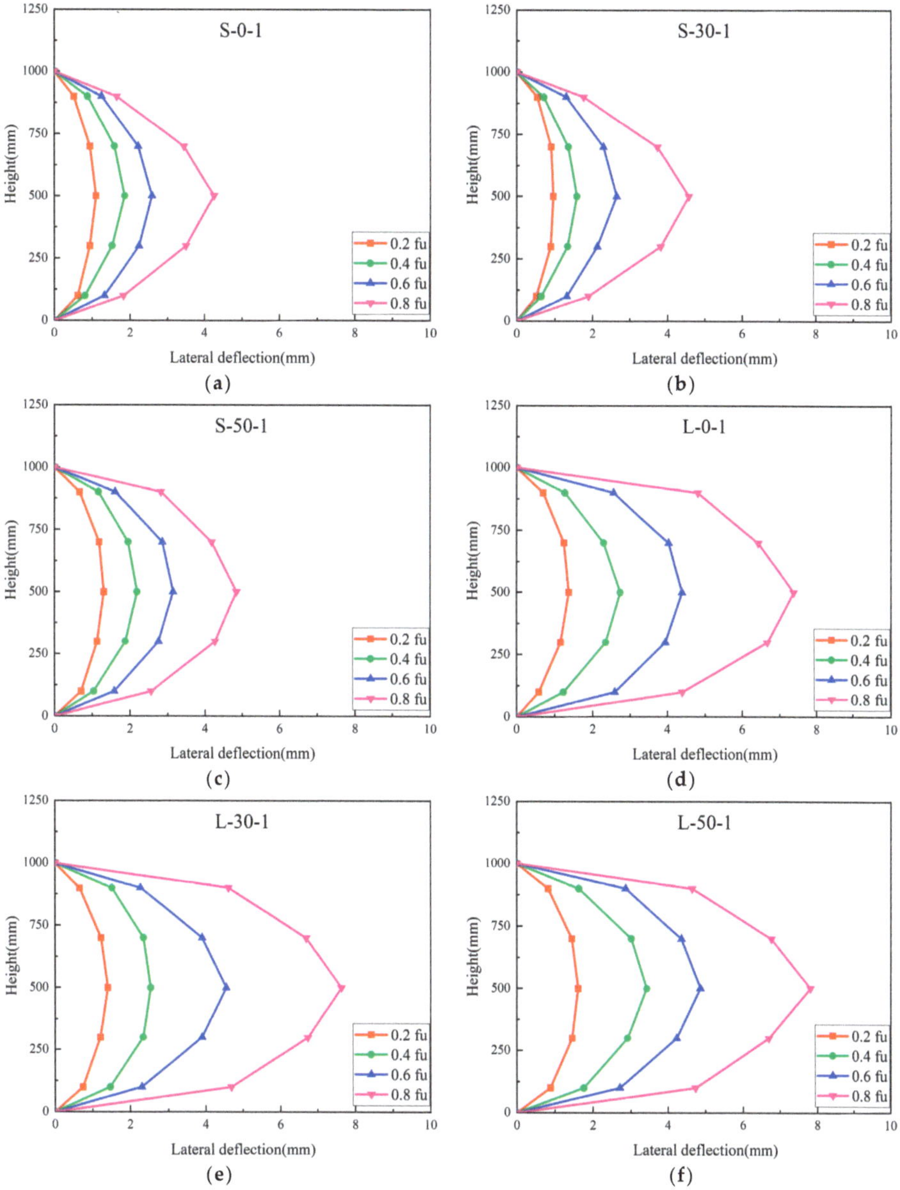

Figure 11. *Cont.*

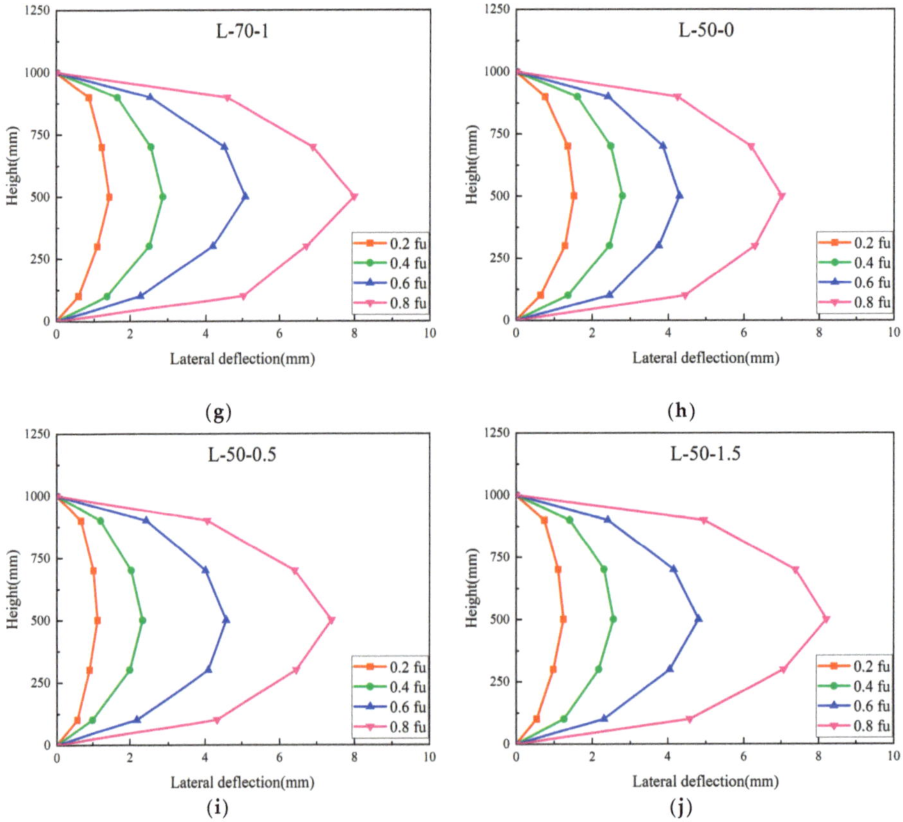

Figure 11. Lateral deflection curves: (**a**) S-0-1, (**b**) S-30-1, (**c**) S-50-1, (**d**) L-0-1, (**e**) L-30-1, (**f**) L-50-1, (**g**) L-70-1, (**h**) L-50-0, (**i**) L-50-0.5, (**j**) L-50-1.5.

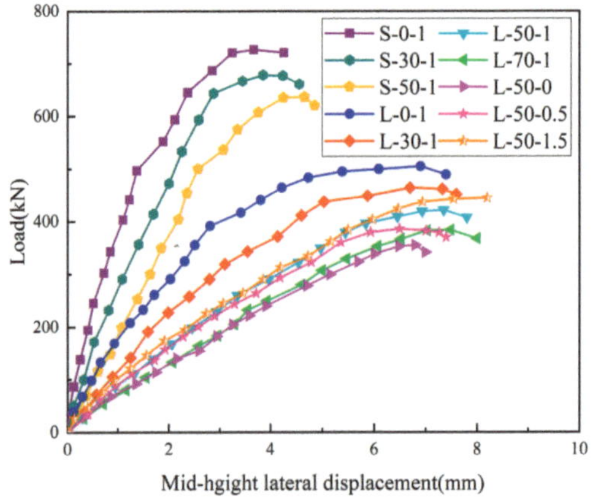

Figure 12. Load–lateral displacement curve.

As can be seen from Table 6, in the case of the same coal gangue replacement rate and steel fibre admixture, the bearing capacity of the specimens when damaged under the large eccentric load is 24%, 23%, and 24.3% lower than the bearing capacity when damaged under the small eccentric damage load, respectively. The above phenomenon is because the eccentrically pressurised columns receive axial load and bending moment joint action. It can also be seen from Figure 11 that the flexural stiffness of the specimens decreases as the eccentricity increases.

Table 6. Load-bearing capacity and ductility of specimens.

Specimens	f_{cu} (Mpa)	f_{ft} (Mpa)	N_{cr} (kN)			N_u (kN)		
			N_{exp}	N_{theo}	N_{exp}/N_{theo}	N_{exp}	N_{theo}	N_{exp}/N_{theo}
S-0-1	39.96	3.63	452.6	475.42	0.952	731	722.13	1.012
S-30-1	36.36	3.55	450.5	457.71	0.984	665.2	677.63	0.982
S-50-1	33.09	3.27	447.9	441.25	1.015	605.3	636.89	0.951
L-0-1	39.96	3.63	90.4	94.03	0.961	553	506.13	1.093
L-30-1	36.36	3.55	89	89.89	0.991	503.2	461.88	1.089
L-50-1	33.09	3.27	82.5	85.98	0.959	457.9	421.41	1.087
L-70-1	30.44	3.12	80.3	82.71	0.972	421.3	388.53	1.084
L-50-0	27.87	2.97	75	79.45	0.944	385.9	356.59	1.082
L-50-0.5	30.12	3.11	78.3	82.29	0.951	416.7	384.41	1.084
L-50-1.5	36.07	3.36	82.83	87.71	0.944	480.8	458.24	1.047

In addition, the lateral displacement of the specimens increased with the elevated doping of the steel fibres. This is due to the bridging effect of the steel fibres, which improves the deformation capacity of the specimens.

3.6. Effect of Coal Gangue Replacement Rate on the Bearing Capacity of Specimens

Figure 13 shows the effect of the coal gangue replacement rate on the bearing capacity of the specimens with steel fibre doping of 1% for different eccentricities. Figure 13 and Table 6 suggest that as the coal gangue replacement rate increases from 0% to 30% and 50%, the cracking load of small eccentric compressed concrete columns decreases by 3% and 7%, respectively, and its ultimate load decreases by 6% and 12%, respectively. Moreover, for large eccentric compressed concrete columns, when the coal gangue replacement rate was increased from 0% to 30%, 50%, and 70%, respectively, its cracking load was reduced by 4%, 8.5%, and 12%, and its ultimate load was reduced by 8.7%, 16.7%, and 23%, respectively. Therefore, the best replacement rate for the gangue concrete configured in this experiment is 30%. This is different from the 50% obtained in the literature [22], mainly because the gangue used was taken from different materials and a conservative value was taken in this paper.

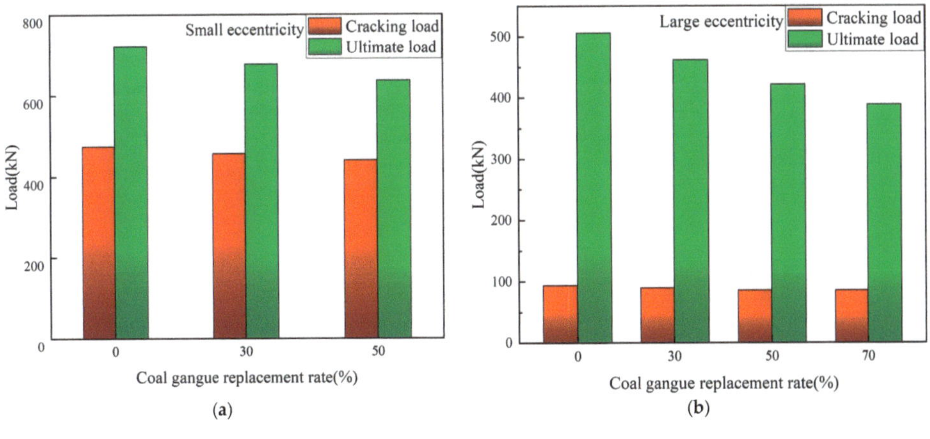

Figure 13. Effect of coal gangue replacement rate on column bearing capacity: (**a**) small eccentricity, (**b**) large eccentricity.

3.7. Effect of Volume Content of Steel Fibres on the Bearing Capacity of Specimens

The effect of different steel fibre volume content on the bearing capacity and lateral displacement of the specimens is shown in Figure 14. Under a large eccentric load, the bearing capacity of the columns rises, and the lateral displacement increases with an increase in steel fibre volume content. When the steel fibre admixture was increased from 0% to 0.5%, the cracking load of the specimens increased by 3.5% and the ultimate load by 7.1%. Similarly, when the volume content of the steel fibre was increased from 0.5% to 1%, the cracking load of the specimens was increased by 4.2%, and the ultimate load was increased by 9.3%. In contrast, when the steel fibre admixture was increased from 1% to 1.5%, the cracking load of the specimen was increased by only 2.1%, and the ultimate load was increased by only 5.4%. It can be clearly seen that when the amount of steel fibre doping exceeds 1%, there is a significant slowdown in the growth of the bearing capacity of the specimens. Considering the bearing performance and economical cost of the specimens, it is recommended that the volume content of steel fibres for SFCGC columns is about 1%.

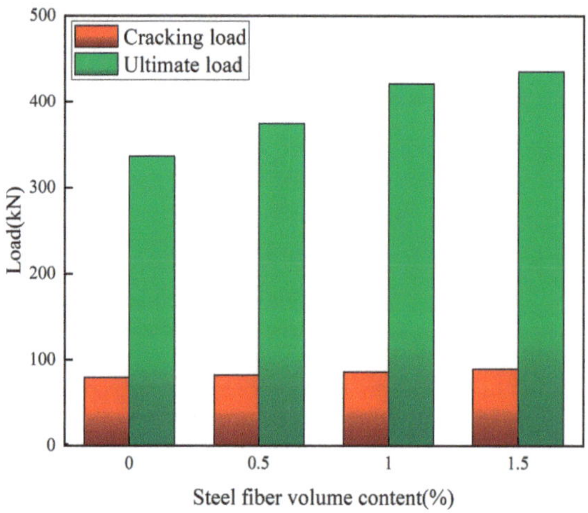

Figure 14. Effect of steel fibre volume content on column bearing capacity.

4. Discussion

4.1. Calculation of Cracking Load Capacity

Since steel fibres play a role in structural cracking resistance, the effect of steel fibres must be considered when calculating cracking loads in eccentric compressed SFCGC columns. The measured values of tensile strength and modulus of elasticity of SFCGC were used to calculate the cracking load, as shown in Table 6. The cracking load equation for SFCGC columns under short-term loading can be obtained from references [42,46,47]. The calculated results are compared with the experimental results in Table 6.

$$N_{cr} = \frac{\gamma_m \alpha_{ct} f_{ft} A_0 W_0}{e_0 A_0 - W_0} \tag{1}$$

$$A_0 = bh + \alpha_E As \tag{2}$$

$$W_0 = \frac{I_0}{h - y_0} \tag{3}$$

$$y_0 = \left(bh^2/2 + \alpha_E Ash_0\right)/(bh + \alpha_E As) \tag{4}$$

$$I_0 = by_0{}^3/3 + b(h - y_0)^3/3 + \alpha_E As(h_0 - y_0)^2 \tag{5}$$

$$\alpha_E = E_s/E_c \tag{6}$$

$$\rho = A_s/bh_0 \tag{7}$$

where N_{cr} is the concrete cracking load; γ_m is the plasticity coefficient of the resisting moment of the section, taken as 1.55; α_{ct} is the concrete tensile stress limiting factor, taken as 0.85; f_{ft} is the standard value of tensile strength of SFCGC, taking the actual measured value; A_0 is the converted cross-sectional area of the member; e_0 is eccentricity; W_0 is the elastic resisting moment of the converted section to the edge of the concrete tensile zone; y_0 is the distance from the axis of the centre of gravity of the converted section to the edge of the concrete compressive zone; I_0 is the moment of inertia of the section to its centre of gravity axis; ρ is the reinforcement ratio of longitudinal tensile reinforcement; E_s is the modulus of elasticity of longitudinal tensile reinforcement; E_c is the modulus of elasticity of concrete; A_s is the cross-sectional area of longitudinal tensile reinforcement.

The cracking load test values and the calculated values are shown in Table 6, their average ratio is 0.967, and the coefficient of variation is 0.0182; the predicted results are in good agreement with the test results.

4.2. Calculation of the Ultimate Bearing Capacity

The load-bearing capacity calculations for the large and small eccentric pressed columns are shown in Figure 15. According to the force and moment balance conditions and coordination conditions given in the literature [42,46], the ultimate bearing capacity equations for columns are shown below, where Equations (9)–(12) are the equations for the calculation of the small eccentricity of the specimens. Equations (13)–(15) are the formulae for the calculation of the large eccentricity of the specimens.

$$N_u = \alpha_1 f_c b x_c + f_y' A_s' - \sigma_s A_s \tag{8}$$

$$N_u e = \alpha f_c b x_c (h_0 - x_c/2) + f_y' A_s' (h_0 - a_s') \tag{9}$$

$$e = \eta_{ns} e_0 + h/2 - a_s \tag{10}$$

$$\sigma_s = E_s \varepsilon_{cu} (h_0/x_c - 1) \tag{11}$$

$$N_u = \alpha_1 f_c b x_c - f_{ft} b x_t + f_y' A_s' - f_y A_s \tag{12}$$

$$N_u e = \alpha_1 f_c b x_c (h_0 - x_c/2) + f_y' A_s' (h_0 - a_s') - f_{ft} b x_t (x_t/2 - a_s) \tag{13}$$

$$x_t = h - x_c/\beta \tag{14}$$

where N_u is the ultimate column load capacity; α, β is the concrete strength correlation coefficient, according to the specification GB50010-2010 [42] to take 0.91, 0.71; f_y' is the compression zone longitudinal reinforcement yield strength; f_y is the yield strength of the longitudinal reinforcement in tension; A_s' is the cross-sectional area of the compressive reinforcement; σ_s is the strength of the longitudinal tensile reinforcement; A_s is the cross-sectional area of the longitudinal tensile reinforcement; x_c is the height of the concrete compression zone; a_s' is the distance from the edge of the compressed zone to the point where the combined force of the compressed reinforcement acts; a_s is the distance from the edge of the tensile zone to the point where the combined force of the tensile reinforcement is applied; E_s is the longitudinal tensile reinforcement modulus of elasticity; ε_{cu} is the ultimate compressive strain of coal gangue concrete, taken as 0.0033; h_0 is the effective height of the section; f_{ft} is the tensile strength of the SFCGC; x_t is the height of the concrete in tension; η_{ns} is the second-order effect factor, which takes the value of one because second-order effects are not considered in this test.

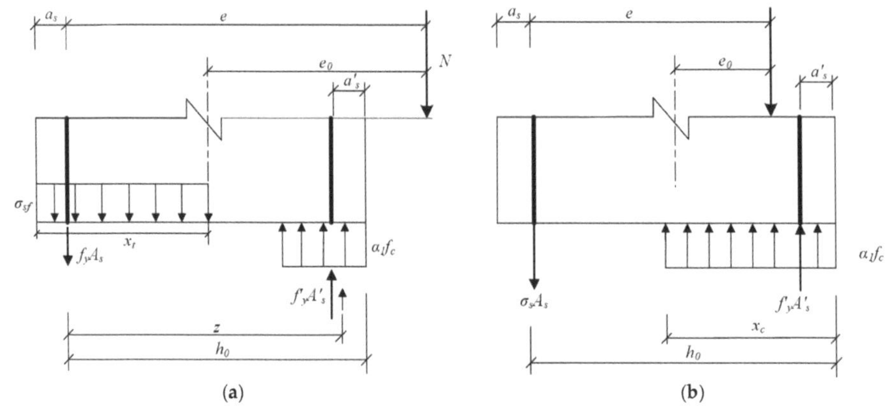

Figure 15. Equivalent force distribution in the positive section of the eccentrically pressed column: (**a**) Large eccentricity, (**b**) Small eccentricity.

Table 6 shows the experimental and calculated data with a mean ratio of 0.967 and a coefficient of variation of 0.0312, as expected.

4.3. Average Crack Spacing

Concerning the relevant provisions of the code CECS 2004 [46] and the relevant formulae of [48], the average crack spacing of SFCGC columns under eccentric compression was calculated using Equation (16); the average crack spacing of the specimens was calculated, and the test results are shown in Table 7.

$$l_{cr} = k_1 c_s + k_2 \frac{d_{eq}}{\rho_{te}\left(1 + \alpha_t \lambda_f\right)} \tag{15}$$

$$\rho_{te} = A_s / A_{te} \tag{16}$$

$$d_{eq} = \frac{\sum n d_i^2}{\sum n_i v_i d_i} \tag{17}$$

where l_{cr} is the average crack spacing of concrete; k_1 is the coefficient related to the protective layer of SFCGC for longitudinal tensile reinforcement, and k_2 is the coefficient related to the bonding performance between SFCGC and reinforcement; c_s is the distance from the outer edge of the longitudinal tensile reinforcement to the edge of the tensile zone; d_{eq} is the equivalent diameter of the longitudinal tensile reinforcement (mm); ρ_{te} is the effective reinforcement ratio of longitudinal tensile reinforcement calculated according to the effective tensile concrete cross-sectional area, when $\rho_{te} < 0.01$, take $\rho_{te} = 0.01$; α_t is the coefficient of bond enhancement due to the presence of steel fibres; A_{te} is the area of the effective tensioned concrete section; d_i, n_i, v_i denote the diameter, number of rods and relative bond characteristics of type i longitudinal tensile reinforcement, respectively, the relative bond characteristics of ribbed reinforcement are taken as 1.0.

Table 7. Average crack spacing of specimens.

Specimens	L_m		
	Tested	Calculated	Tested/Calculated
S-0-1	61.3	61.26	1.001
S-30-1	63.0	61.26	1.028
S-50-1	60.9	61.26	0.994
L-0-1	62.3	61.26	1.017
L-30-1	61.2	61.26	0.998
L-50-1	60.1	61.26	1.038
L-70-1	63.6	61.26	1.012
L-50-0	66.8	66.77	0.996
L-50-0.5	64.3	63.84	1.007
L-50-1.5	60.7	58.95	1.029

By fitting the mean crack spacing in this study, the results of k_1 = 1.86 and k_2 = 0.10 can be obtained. Combined with the Chinese code [42,46], the average crack spacing equation can be obtained as follow:

$$l_m = 1.86c_s + 0.1d_{eq}/\rho_{te} \tag{18}$$

4.4. Average Crack Width

The average crack width is equal to the difference between the average tensile elongation of the reinforcement and the concrete within the average crack spacing; due to the small tensile deformation of concrete, according to the literature [42,46,47], it is possible to introduce the coefficient of influence of tensile deformation of concrete α_c and the strain unevenness coefficient of reinforcement between cracks ψ, from which we can obtain the Equation for calculating the average crack width of ordinary concrete:

$$w_m = \alpha_c \psi \frac{\sigma_s}{E_s} l_m \tag{19}$$

$$\psi = 1.1 - 0.65\frac{f_{ft}}{\rho_{te}\sigma_s} \tag{20}$$

where w_m is the average crack width of concrete; α_c is the coefficient of influence of tensile deformation of concrete, which is taken as 0.77 according to the Chinese code [42,46]; ψ is the strain unevenness coefficient of longitudinal tensile reinforcement between cracks, which is taken as ψ = 0.2 when ψ < 0.2 and ψ =1 when ψ > 1; σ_s is the stress of longitudinal tensile reinforcement between cracks, which is calculated according to Equation (22).

In consideration of the effect of steel fibres on the mean crack width, the mean crack width equation was amended as follows:

$$w_m = 0.77\left(1 - \beta_w \lambda_f\right)\psi \frac{\sigma_s}{E_s} l_m \tag{21}$$

where β_w is the coefficient of influence of steel fibres on the average crack spacing; λ_f indicates the characteristic value of steel fibre content ($\lambda_f = \rho_f l_f / d_f$); ρ_f indicates the volume rate of steel fibres; l_f is the length of steel fibres; d_f is the diameter of steel fibres.

As shown in Figure 14, to obtain the equilibrium diagram of the internal forces between cracks, consider the influence of steel fibres in the tension zone, take the moment of the combined force point in the pressure zone to obtain the following formula, according to the equilibrium conditions to obtain the steel stress σ_s.

$$\sigma_s = \frac{N(e - z) - \sigma_{sf} b x_t (z - x_t/2 + a_s)}{A_s z} \tag{22}$$

$$z = \left[0.87 - 0.12\left(\frac{h_0}{e}\right)^2\right] h_0 \quad (23)$$

$$e = \eta_{ns} e_0 + h/2 - a_s \quad (24)$$

$$\sigma_{sf} = \alpha_t \lambda_f f_t \quad (25)$$

$$x_t = 0.5h \quad (26)$$

where e is the force loading point to the distance of the tension zone longitudinal reinforcement joint point; z is the distance of the tension zone longitudinal reinforcement joint point to the section concrete compression zone joint point; σ_{sf} is the crack between the steel fibre tensile stress; α_t is the steel fibre on the coal gangue concrete tensile strength coefficient, according to the test to take 0.37; f_t is the coal gangue concrete without considering the impact of steel fibre tensile strength, $f_{ft} = f_t(1 + \alpha_t \lambda_f)$; x_t is the height of the effective tensile zone of the section affected by steel fibres.

By fitting the measured values of the average crack width to this experiment, we obtain $\beta_w = 0.186$. The average crack width of SFCGC columns under eccentric compression is obtained according to Equation (21) as follows; the average crack width of the specimens is calculated and tested as shown in Table 8.

$$w_m = 0.77\left(1 - 0.186\lambda_f\right)\psi \frac{\sigma_s}{E_s} l_m \quad (27)$$

Table 8. Crack width of specimens.

Specimens	N/Nu	w_m			w_{max}		
		Tested	Calculated	Tested/Calculated	Tested	Calculated	Tested/Calculated
S-0-1	70%	0.016	0.0172	0.930	0.030	0.0286	1.050
	80%	0.021	0.0222	0.946	0.038	0.0368	1.032
	90%	0.027	0.0272	0.994	0.046	0.0451	1.020
S-30-1	70%	0.014	0.0159	0.878	0.028	0.0265	1.057
	80%	0.021	0.0206	1.018	0.035	0.0342	1.022
	90%	0.026	0.0253	1.028	0.041	0.0420	0.976
S-50-1	70%	0.015	0.0148	1.011	0.024	0.0246	0.975
	80%	0.021	0.0192	1.092	0.034	0.0319	1.065
	90%	0.024	0.0236	1.016	0.042	0.0392	1.071
L-0-1	30%	0.011	0.0095	1.158	0.016	0.0157	1.014
	50%	0.030	0.0276	1.087	0.050	0.0458	1.091
	70%	0.049	0.0457	1.072	0.077	0.0759	1.015
L-30-1	30%	0.009	0.0080	1.122	0.014	0.0133	1.051
	50%	0.024	0.0245	0.978	0.045	0.0407	1.104
	70%	0.041	0.0410	0.999	0.073	0.0682	1.071
L-50-1	30%	0.008	0.0067	1.195	0.013	0.0111	1.169
	50%	0.022	0.0218	1.011	0.040	0.0361	1.107
	70%	0.037	0.0368	1.004	0.064	0.0612	1.047
L-70-1	30%	0.006	0.0056	1.063	0.010	0.0094	1.067
	50%	0.021	0.0195	1.075	0.036	0.0324	1.110
	70%	0.036	0.0334	1.077	0.058	0.0555	1.045
L-50-0	30%	0.005	0.0057	0.869	0.011	0.0096	1.152
	50%	0.022	0.0205	1.072	0.033	0.0341	0.969
	70%	0.034	0.0353	0.964	0.060	0.0586	1.024
L-50-0.5	30%	0.007	0.0064	1.093	0.012	0.0106	1.129
	50%	0.022	0.0219	1.005	0.038	0.0363	1.046
	70%	0.041	0.0374	1.097	0.061	0.0621	0.983

Table 8. *Cont.*

Specimens	N/Nu	w_m			w_{max}		
		Tested	Calculated	Tested/ Calculated	Tested	Calculated	Tested/ Calculated
L-50-1.5	30%	0.008	0.0072	1.108	0.011	0.0120	0.918
	50%	0.023	0.0219	1.052	0.037	0.0363	1.019
	70%	0.038	0.0365	1.041	0.061	0.0606	1.007

From Table 8, the mean value of the ratio of the tested to the calculated mean crack width for the SFCGC columns is 1.035 with a coefficient of variation of 0.071, which indicates that the two values are close to each other. It is also clear from the test results that the average crack width decreases as the steel fibre dose increases. It further shows that the average crack width of SFCGC columns proposed in this test is in good agreement with the test results and has good calculation accuracy.

4.5. Maximum Crack Width

The maximum crack width w_{max} of the specimens under short-term loading can be obtained by multiplying the average crack width w_m by the enlargement factor α_s. Owing to the few specimens in this test, the average crack width expansion factor was taken as 1.66 according to the specification [46,48]. The formula is as follows:

$$w_{max} = 1.66 w_m \tag{28}$$

According to Table 8, the mean of the ratio between the tested and calculated values of the maximum crack width for the SFCGC columns is 1.047, and the coefficient of variation is 0.053, which gives a good agreement, indicating that the formula for calculating the maximum crack width for the NAC columns is applicable to the SFCGC columns.

5. Conclusions

Based on the experimental and theoretical study of the behaviour of SFCGC columns under eccentric compression loading, the following conclusions can be drawn:

(1) The whole process of loading the SFCGC column to the ultimate state is in accordance with the plane cross-section assumption. During the loading process, the lateral displacement curve of the SFCGC column is approximately sinusoidal and symmetrically distributed in the middle. Under the same eccentricity, the coal gangue replacement rate of the SFCGC columns has little effect on the lateral displacement corresponding to the ultimate load. The final damage mode of the SFCGC columns resembled that of the NAC columns, and their bearing capacity was greatly influenced by eccentricity.

(2) The increased crack resistance of SFCGC columns is directly associated with their increased tensile strength. By means of applying the corresponding tensile strength (f_{ft}) of SFCGC, the formula for NAC columns can be used for SFCGC columns.

(3) The cracking spacing decreased with an increase in steel fibres. The reduction of longitudinal reinforcement stress contributed to the decline of crack width. On the basis of the regression analysis of experiment results, the equations for computing the average crack spacing and average crack width of SFCGC columns were presented. For these results, there is excellent consistency with the experimental results.

(4) A salutary effect of steel fibres in improving the load-bearing capacity of SFCGC columns is shown. Considering the beneficial contribution of steel fibres in the tension zone, equations are presented to predict the axial load under load-bearing conditions. Acceptable prediction accuracy indicates the suitability of SFCGC for its structural applications.

(5) The increase in the coal gangue replacement rate will lead to a certain degree of reduction in the bearing capacity of the specimens. Compared with the replacement

rate of the 30% SFCGC column, when the gangue replacement rate is at 50% and 70%, the reduction of the ultimate bearing capacity of the specimens increases significantly. It is recommended that the coal gangue replacement rate for SFCGC columns is about 30%.

(6) Steel fibres can effectively compensate for the negative impact of gangue on the bearing capacity of SFCGC columns. As the volume content of steel fibres increases from 0% to 0.5% and 1%, the ultimate load capacity of the SFCGC increases by 7.1% and 9.3%, respectively. When the steel fibre content exceeds 1% by volume to 1.5%, the increase in ultimate load capacity of the SFCGC column begins to slow down and only increases by 5.4%. In view of the load-bearing performance and cost-effectiveness of the specimens, it is suggested that the volume content of steel fibres for SFCGC columns be about 1%.

Author Contributions: Conceptualization, B.C.; methodology, B.C.; test, B.B. and W.D.; validation, B.B. and S.W.; investigation, L.W. and S.W.; resources, B.C.; data curation, W.D. and L.W.; writing—original draft preparation, B.B.; writing—review and editing, B.B.; visualization, B.C.; supervision, B.C.; project administration, B.C.; funding acquisition. All authors have read and agreed to the published version of the manuscript.

Funding: This research was funded by Scientific research projects of the education department of Jilin Province, grant number JJKH20210279KJ, JJKH20200279KJ; China Scholarship Council, grant number 201805975002; and Jilin Provincial Science and Technology Development Plan Project, grant number 20220203082SF. The authors wish to acknowledge the sponsors. However, any opinions, findings, conclusions and recommendations presented in this paper are those of the authors and do not necessarily reflect the views of the sponsors.

Institutional Review Board Statement: Not applicable.

Informed Consent Statement: Not applicable.

Data Availability Statement: The data used to support the findings of this study are available from the authors upon request.

Conflicts of Interest: The authors declare no conflict of interest.

References

1. Chang, J.; Du, G.; Du, J.; Shi, X. Current Situation of the Comprehensive Utilization of Coal Gangue in China and the Related Problems and Recommendations. *China Environ. Prot. Ind.* **2022**, *8*, 13.
2. Wang, B.; Ma, Y.N.; Lee, X.Q.; Wu, P.; Liu, F.; Zhang, X.; Li, L.; Chen, M. Environmental-friendly coal gangue-biochar composites reclaiming phosphate from water as a slow-release fertilizer. *Sci. Total Environ.* **2020**, *758*, 143664. [CrossRef] [PubMed]
3. Gong, C.; Yan, J.; Liu, J.; Yu, H. Biology Migration and Distribution Characteristics of Trace Elements in Reconstructed Soil with Coal Gangue Filling. *Agric. Sci. Technol.* **2016**, *17*, 2167–2170.
4. Pone, J.D.N.; Hein, K.A.A.; Stracher, G.B.; Annegarn, H.J.; Finkleman, R.B.; Blake, D.R.; Mccormack, J.K.; Schroeder, P. The spontaneous combustion of coal and its by-products in the Witbank and Sasolburg coalfields of South Africa. *Int. J. Coal Geol.* **2007**, *72*, 124–140. [CrossRef]
5. Querol, X.; Zhuang, X.; Font, O.; Izquierdo, M.; Alastuey, A.; Castro, I.; Drooge, B.L.V.; Moreno, T.; Grimalt, J.O.; Elvira, J. Influence of soil cover on reducing the environmental impact of spontaneous coal combustion in coal waste gobs: Are view and new experimental data. *Int. J. Coal Geol.* **2011**, *85*, 2–22. [CrossRef]
6. Laurance, W.F.; Peletierjellema, A.; Geenen, B.; Koster, H.; Verweij, P.A.; Dijck, P.V.; Lovejoy, T.E.; Schleicher, J.; Kuijk, M.V. Reducing the global environmental impacts of rapid infrastructure expansion. *Curr. Biol.* **2015**, *25*, 259–262. [CrossRef]
7. Monteiro, P.J.M.; Miller, S.A.; Horvath, A. Towards sustainable concrete. *Nat. Mater.* **2017**, *16*, 698–699. [CrossRef]
8. Al-Oraimi, S.K.; Taha, R.; Hassan, H.F. The effect of the mineralogy of coarse aggregate on the mechanical properties of high-strength concrete. *Construct. Build. Mater.* **2006**, *20*, 499–503. [CrossRef]
9. Wu, J.; Bai, G.L.; Wang, P.; Liu, Y. Mechanical properties of a new type of block made from shale and coal gangue. *Constr. Build. Mater.* **2018**, *190*, 796–804. [CrossRef]
10. Qin, J.G.; Zhao, R.D.; Chen, T.J.; Zi, Z.Y.; Wu, J.H. Co-combustion of municipal solid waste and coal gangue in a circulating fluidized bed combustor. *Int. J. Coal. Sci. Technol.* **2019**, *6*, 218–224. [CrossRef]
11. Guo, Y.; Yan, K.; Cui, L.; Cheng, F. Improved extraction of alumina from coal gangue by surface mechanically grinding modification. *Powder Technol.* **2016**, *302*, 33–41. [CrossRef]

12. Li, Y.; Yao, Y.; Liu, X.M.; Sun, H.H.; Ni, W. Improvement on pozzolanic reactivity of coal gangue by integrated thermal and chemical activation. *Fuel* **2013**, *109*, 527–533. [CrossRef]
13. Sun, Y.X.; Li, X.D. Development and design of coal gangue concrete filling material. *Adv. Mater. Res.* **2011**, *295*, 1198–1201. [CrossRef]
14. Wang, Z.; Zhao, N. Influence of coal gangue aggregate grading on strength properties of concrete. *Wuhan Univ. J. Nat. Sci.* **2015**, *20*, 66–72. [CrossRef]
15. Ma, H.Q.; Zhu, H.G.; Wu, C.; Chen, H.Y.; Sun, J.W.; Liu, J.Y. Study on compressive strength and durability of alkali-activated coal gangue-slag concrete and its mechanism. *Powder Technol.* **2020**, *368*, 112–124. [CrossRef]
16. Li, X.; Pan, M.; Tao, M.; Liu, W.; Gao, Z.; Ma, C. Preparation of high closed porosity foamed ceramics from coal gangue waste for thermal insulation applications. *Ceram. Int.* **2022**, *48*, 37055–37063. [CrossRef]
17. Qiu, J.; Zhou, Y.; Vatin, N.I.; Guan, X.; Sultanov, S.; Khemarak, K. Damage constitutive model of coal gangue concrete under freeze-thaw cycles. *Constr. Build. Mater.* **2020**, *264*, 120720. [CrossRef]
18. Xiao, M.; Ju, F.; He, Z.Q. Research on shotcrete in mine using non-activated waste coal gangue aggregate. *J. Clean. Prod.* **2020**, *259*, 120810. [CrossRef]
19. Liu, H.; Xu, Q.; Wang, Q.; Zhang, Y. Prediction of the elastic modulus of concrete with spontaneous-combustion and rock coal gangue aggregates. *Structures* **2020**, *28*, 774–785. [CrossRef]
20. Ma, H.Q.; Yi, C.; Zhu, H.G.; Dong, Z.C.; Chen, H.Y.; Wang, J.X.; Li, D.Y. Compressive strength and durability of coal gangue aggregate concrete. *Mater. Rev.* **2018**, *32*, 2390–2395.
21. Gao, S.; Zhao, G.; Guo, L.; Zhou, L.; Yuan, K. Utilization of coal gangue as coarse aggregates in structural concrete. *Constr. Build. Mater.* **2021**, *268*, 121212. [CrossRef]
22. Zhu, H.; Yang, S.; Li, W.; Li, Z.; Fan, J.; Shen, Z. Study of mechanical properties and durability of alkali-activated coal gangue-slag concrete. *Materials* **2020**, *13*, 5576. [CrossRef] [PubMed]
23. Chen, B. Strength of coal gangue Concrete. *Ind. Constr.* **1994**, *7*, 29–32.
24. Guan, X.; Qiu, J.; Song, H.; Qin, Q.; Zhang, C. Stress–strain behaviour and acoustic emission characteristic of gangue concrete under axial compression in frost environment. *Constr. Build.* **2019**, *220*, 476–488.
25. Moghadam, M.J.; Ajalloeian, R.; Hajiannia, A. Preparation and application of alkaliactivated materials based on waste glass and coal gangue: A review. *Constr. Build. Mater.* **2019**, *221*, 84–98. [CrossRef]
26. Katzer, J.; Domski, J. Quality and mechanical properties of engineered steel fibres used as reinforcement for concrete. *Constr. Build. Mater.* **2012**, *34*, 243–248. [CrossRef]
27. Atiş, C.D.; Karahan, O. Properties of steel fibre reinforced fly ash concrete. *Constr. Build. Mater.* **2009**, *23*, 392–399. [CrossRef]
28. Wang, L.; Shen, N.; Zhang, M.; Fu, F.; Qian, K. Bond performance of Steel-CFRP bar reinforced coral concrete beams. *Constr. Build. Mater.* **2020**, *245*, 118456. [CrossRef]
29. Li, L.; Qin, H.G. Study of interface bonding performance of steel fibres within cement mortar. *J. Luoyang Univ.* **2002**, *17*, 82–85.
30. Cheng, J.; Liu, J.P.; Zhang, L.H. Testing and mechanism analysis of the fibre-matrix bonding properties of ultrahigh performance concrete. *China Concr. Cem. Prod.* **2016**, *5*, 62–66.
31. Campione, G.; La Mendola, L. Behavior in compression of lightweight fibre reinforced concrete confined with transverse steel reinforcement. *Cem. Concr. Compos.* **2004**, *26*, 645–656. [CrossRef]
32. Sharma, U.K.; Sheikh, S.A.; Bhargava, P. Tie-confined fibre-reinforced high-strength concrete short columns. *Mag. Concr. Res.* **2007**, *59*, 757–769. [CrossRef]
33. Li, Y.J.; Geng, X.; Zhang, X.; Yan, X. Experimental study on the durability of the concrete with coal gangue aggregate. *J. China Coal Soc.* **2013**, *38*, 1215–1219.
34. Zhang, J.X.; Chen, W.L.; Jin, S.S.; Chen, C.Z.; Yang, R.J. Investigation on durability of coal gangue aggregate concrete. *J. Beijing Univ. Technol.* **2011**, *37*, 115–125.
35. Liu, H.; Bai, G.; Gu, Y.; Yan, F. The influence of coal gangue coarse aggregate on the mechanical properties of concrete columns. *Case Stud. Constr. Mater.* **2022**, *17*, e01315. [CrossRef]
36. Luo, D.M.; Wang, Y.; Zhang, S.H.; Niu, D.T.; Song, Z.P. Frost resistance of coal gangue aggregate concrete modified by steel fibre and slag powder. *Appl. Sci.* **2020**, *10*, 3229. [CrossRef]
37. Wang, Y.; Qiu, J.; Deng, W.; Xing, J.; Liang, J. Factors affecting brittleness behavior of coal-gangue ceramsite lightweight aggregate concrete. *Front. Mater.* **2020**, *7*, 554718. [CrossRef]
38. Le, D.B.; Tran, S.D.; Dao, V.T.; Torero, J. Deformation capturing of concrete structures at elevated temperatures. *Procedia Eng.* **2017**, *210*, 613–621.
39. Yu, K.; Yu, J.; Lu, Z.; Chen, Q. Determination of the softening curve and fracture toughness of high-strength concrete exposed to high temperature. *Eng. Fract. Mech.* **2015**, *149*, 156–169. [CrossRef]
40. *JGJ 51-2002*; Technical Specification for Light Aggregate Concrete. China Architecture & Building Press: Beijing, China, 2002.
41. *GB 175-2020*; Common Portland Cement. China Standardization Administration Press: Beijing, China, 2020.
42. *GB 50010-2010*; Ministry of Housing and Urban-Rural Construction of the People's Republic of China (MHURC-PRC). China Building Industry Press: Beijing, China, 2010.
43. Lin, G.; Zeng, J.J.; Teng, J.G.; Li, L.J. Behavior of large-scale FRP-confined rectangular RC columns under eccentric compression. *Eng. Struct.* **2022**, *216*, 110759. [CrossRef]

44. Hadi, M.N. Behaviour of eccentric loading of FRP confined fibre steel reinforced concrete columns. *Constr. Build. Mater.* **2009**, *23*, 1102–1108. [CrossRef]
45. Qing, K.; Han, F.; Xing, L.; Wang, X. Experimental study on the flexural performance of BFRP reinforced steel fibre recycled concrete beams. *Fiber Reinf. Plast./Compos.* **2019**, *11*, 5.
46. *CECS 38:2004*; China Association for Engineering Construction Standardization. China Planning Press: Beijing, China, 2004.
47. Li, C.Y.; Ding, X.X.; Zhao, S.B.; Zhang, X.Y.; Li, X.K. Cracking resistance of reinforced SFRFLC superposed beams with partial ordinary concrete in compression zone. *Open Civ. Eng. J.* **2016**, *10*, 727–737. [CrossRef]
48. Zhao, S.B. *Design Principles of Concrete Structures*, 2nd ed.; Tongji University Press: Beijing, China, 2013.

buildings

MDPI

Article

A New Steel-Joint Precast Concrete Frame Structure: The Design, Key Construction Techniques, and Building Energy Efficiency

Xiaona Shi [1], Xian Rong [1,2], Lin Nan [1], Lida Wang [1,*] and Jianxin Zhang [1,2,*]

1 School of Civil and Transportation Engineering, Hebei University of Technology, Tianjin 300401, China
2 Hebei Civil Engineering Technology Research Center, Tianjin 300401, China
* Correspondence: 201711601010@stu.hebut.edu.cn (L.W.); zhangjianxin505@hebut.edu.cn (J.Z.); Tel.: +86-158-3225-5099 (L.W.); +86-136-1211-8433 (J.Z.)

Abstract: Assembled methods play a critical role in the construction of precast concrete structures. However, conventional dry-connections-like sleeve grouting joints in precast concrete structures lagged at a low construction and management efficiency with poor quality control. In this study, a novel steel joint for precast reinforced concrete beam-column components is proposed to improve constructability. New joints transform the assembled method from reinforced concrete members into a steel structure by setting a pre-embedded steel connector at both ends of reinforced concrete beams and columns, showing outstanding economic, durability, and fire resistance capabilities. The construction process, construction efficiency, economy, and energy consumption were discussed based on the material, structure, and construction hybrid characteristics. Numerical simulation and structural health monitoring methods are used to monitor and evaluate the deformation and stress state of the proposed system in the whole construction process, so as to optimize the construction scheme and ensure safe and orderly construction. The results reveal that the FEA-simulated values of key building components during construction are in good agreement with the actual monitoring values, which verifies the feasibility of the FEM models and provides a guarantee for construction safety; the construction period of the proposed assemble system is reduced by approximately 56% and 40%, compared with the conventional reinforced concrete frame structure and cast-in-place joints in the precast concrete frame structure, respectively. Meanwhile, the energy consumption of buildings decreases by 20%. This research provides a theoretical basis for the design, calculation, and application of assembled precast structural systems.

Keywords: steel joint; precast concrete structure; construction process; building energy efficiency

Citation: Shi, X.; Rong, X.; Nan, L.; Wang, L.; Zhang, J. A New Steel-Joint Precast Concrete Frame Structure: The Design, Key Construction Techniques, and Building Energy Efficiency. *Buildings* **2022**, *12*, 1974. https://doi.org/10.3390/buildings12111974

Academic Editors: Shan Gao, Jingxuan Wang, Dewen Kong and Yong Liu

Received: 20 October 2022
Accepted: 8 November 2022
Published: 14 November 2022

1. Introduction

As stated by the master architect Gropius in 1910, 'reinforced concrete buildings shall be prefabrication and factory'; architectural industrialization has been widely explored by researchers. Currently, the precast structure, the centerpiece of building industrialization, has the advantages of less wet work, better energy conservation, and faster construction than the conventional reinforced concrete structure [1–3]. It has been widely used, including in the Osaka Kitahama Apartment Building in Japan, the Chunli River Apartment Building in China, and Phoenix Library in the United States. However, according to numerous investigations of strong earthquakes worldwide, the core area of the beam-column joint of the precast structure can be severely damaged in the event of a disaster. The damage results in the collapse of buildings, endangering people's lives and property. The shortage of antiseismic, precast structures limits the development of industrialized construction [4–8]. Numerous experimental investigations [9–13] have revealed that the connection design of each precast component plays a key role in the seismic and mechanical performance of precast structure buildings. Accordingly, the connection parts should have sufficient stiffness, energy dissipation capacity, and ductility to ensure the transmission of the bending moment, shear force, and axial force between components [14–18]. Therefore, enhancing the mechanical properties of fabricated structures by changing the connection mode of precast components has become the focus of research in recent years.

Many types of precast concrete beam-column joints currently exist. Three examples include the following: (1) In slot-bond assembled concrete joints, the precast concrete components are spliced by way of sunk key and steel anchoring, and the concrete is poured at the connection. This type of precast joint has good energy dissipation capacity and ductility, but wet operations on construction sites are a common issue [19–21]. (2) In prestressed precast concrete joints, concrete members are connected with prestressed reinforcement. Many studies have found that, although this form of member connection has self-recovery ability and slow stiffness degradation, the prestressed reinforcement should be stretched during member assembly, making the construction complex [22–24]. (3) Bolted or welded precast concrete joints refer to the pre-embedding of steel members in the joint area. The concrete beams and columns are combined by bolting or welding connections between the steel components. Although this type of connection mode improves the construction efficiency, the joint performance is not significantly improved, compared to when using the cast-in-suit concrete structure [25–27]. The engineering applications of the above-mentioned precast joints are restricted, owing to the influence of factors, such as the post-pouring of concrete and the use of a complex joint structure [8,14,17,20,23–27]. Therefore, it is of practical significance to develop an assembly system with superior constructability and performance.

In this study, a new type of precast beam-column joint is proposed [28]. The joint has a simple steel connection mode, and steel members are only embedded in the connection of precast reinforced concrete members of the beam-column. During site construction, the members are assembled by a combination of bolting and welding methods. Buildings that rely on this form are called precast concrete structures with steel joints.

The connection form of the joints is the main factor affecting the production and assembly process of the prefabricated components. The load of the precast structure quickly rises during the construction process, then the overall stiffness, load, and stress of the structure are constantly changing. Taken together, it is highly significant to ensure safety during construction. Previous studies [29–34] in structural health monitoring are mostly focused on complex component design, construction monitoring, and monitoring systems, and they provide a volume of monitoring technologies [35]. However, this enormous effort has yielded only a small amount of construction monitoring and simulation research for the precast concrete structure; it is an ongoing research issue.

Construction monitoring is performed by means of monitoring and numerical simulation [35]. The former arranges monitoring devices on the construction site to carry out real-time monitoring of the structure status, so as to discover the construction risk in time; The latter pre-analyzes the state of the building structure during the construction process by means of numerical simulation and verifies the simulation results with the monitoring data of the construction site, thereby preventing the construction risk in advance. Due to the lack of theoretical verification, the accuracy of on-site monitoring data cannot be guaranteed, and real-time monitoring repair of the damaged part may require great effort. In response to some of the above shortcomings, this study applied both numerical simulation and structural health monitoring methods to discuss the whole process of construction, structural deformation, and stress monitoring and evaluation, so as to optimize the construction scheme and ensure the safety and orderly nature of construction.

This article intends to introduce this novel steel-joint precast concrete frame structure; analyze engineering examples of this type of structural system; and describe the advantages, component design, component production, and construction process. Meanwhile, a reasonable construction technology of steel-joint precast concrete structure and a scientific construction monitoring technology are proposed to guarantee the safety of the new system in the construction stage. Then, we will compare the conventional reinforced concrete structure and the precast concrete structure in terms of construction efficiency and building energy consumption. This research can provide a reference for future research on precast buildings and the application of the structural system itself.

2. Steel-Joint Precast Concrete Structure System

2.1. Advantages of Steel-Joint Precast Concrete Structure System

Precast concrete structures are usually connected by grouting sleeves and steel bar binding at the connection of the precast members. Moreover, the whole structure is formed by pouring concrete at the joint. This type of structural system requires additional process links, such as formwork, binding steel bar, concrete pouring, and maintenance, when assembling precast members, which significantly occupies both construction time and labor. Meanwhile, the concealment of the joint area in the concrete is not conducive to checking the connection quality of the members and the reinforcement and repair of the defects at the joint in the later stage.

The proposed structure improves upon the shortcomings of the precast concrete members' connectors. The main design purpose of the new joint is to use the steel structure connection technology to solve the difficult assembly problem of the conventional connection mode of the concrete precast members. However, the joint needs to be used in conjunction with the structure system. If the other selected components do not match the joint, the advantages of the joint or system cannot be utilized, and even the joint or system cannot be truly put into application. Therefore, this study designs a steel-joint precast concrete structure system based on the steel connection method.

The prefabricated steel-joint connection [28] refers to the reinforced concrete structure in which the precast beams and columns are embedded with steel components at the connection points, and the beam-column connection points are joined by the steel structure connection mode. The main purpose of the joint connection design is to solve the related problems of the difficult connection of the precast concrete structure with the steel structure connection mode. The precast beam in this project is embedded with H-shaped steel at the beam end, and the precast column connection node is embedded with a square steel pipe. The upper and lower horizontal plates, cross-type dividers, and extending connecting plates are built into the square steel pipe. The beam-column connection node is mainly connected by a combination of high-strength bolts and welds.

The novel structural system is applied to a precast office building in Lingshou Town, Shijiazhuang, Hebei Province, China. The building scene is shown in Figure 1.

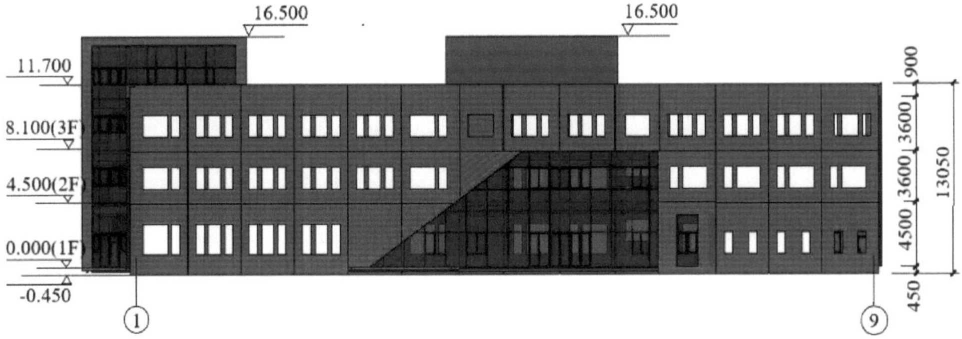

(a)

Figure 1. *Cont.*

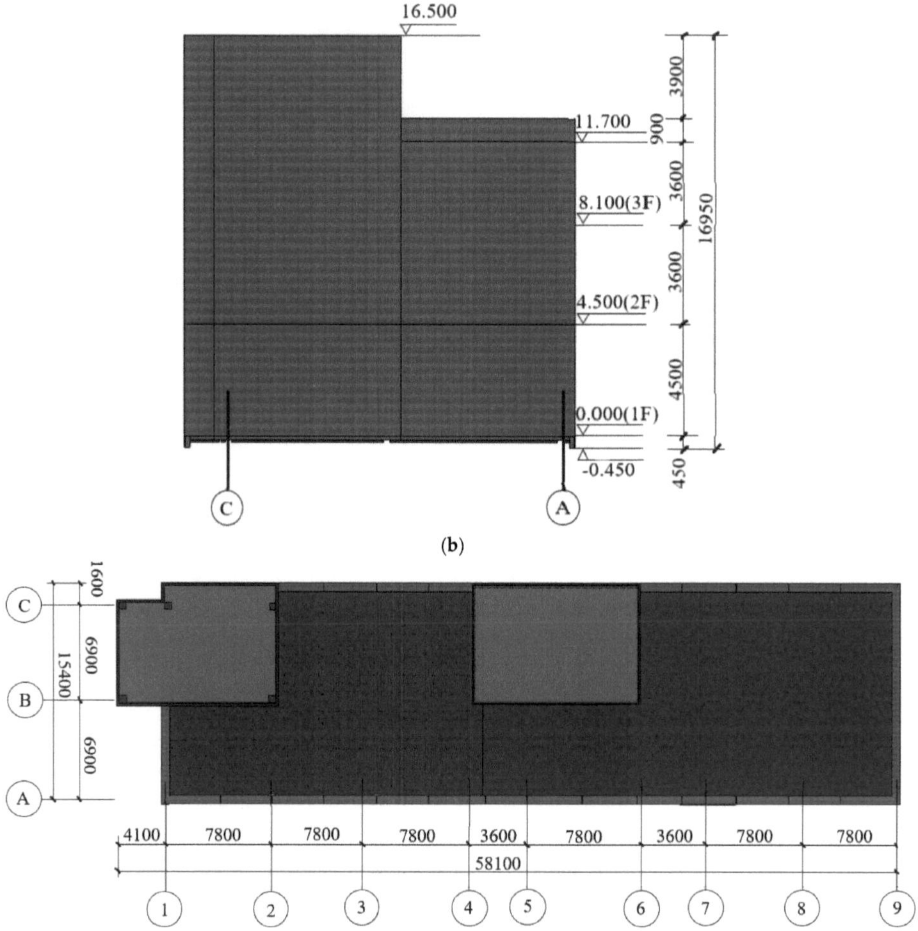

Figure 1. Real view of the building: (**a**) Front elevation; (**b**) Side elevation; (**c**) end face figure.

The project is a four-story frame structure building, which is partially three-story. The building area is 2821.2 m^2, the structure height is 16.5 m, and the building height from one to four layers are 4.5 m, 3.6 m, 3.6 m, and 4.8 m, respectively. The first floor is used as an exhibition hall, and the other floors are used as offices and meeting rooms. The building is divided into two structural monomers. The structural monomer from axis 1 to axis 4 is a steel-joint prefabricated concrete frame structure made of precast columns, precast beams, composite slabs, and precast stairs. The structural monomer from axis 5 to axis 9 is the casting reinforced concrete frame structure. The construction stage of axis 1~4 starts after the completion of the construction stage of axis 5~9, and the two construction stages do not affect each other.

The project building has the characteristics of low height and fewer layers, and the prefabricated columns are all non-spliced full-length prefabricated components with a cross-sectional area of 0.5 m × 0.5 m. The steel component material is Q345, the steel reinforcement strength grade used in the building is mainly HRB400, and the concrete strength grade used in the precast component and the concrete pouring of the building floor is C30. The volume, weight, and quantity of each precast component are shown in Table 1.

Table 1. Overview of precast components.

Component Type	Quantity/Pieces	Maximum Volume/m^3	Maximum Mass/Tons
Precast columns	14	3.225	9.11
Precast beams	103	1.64	4.095
Composite slabs	194	0.41	1.038
Precast stairs	6	1.44	3.6

This structural system is popular and practical for project construction. The construction of precast concrete structures with steel joints is shown in Figure 2. From the perspective of engineering applicability, the precast components connected by the steel structure are mature and reliable, which results in low requirements at the installation and management level of the construction personnel. From the perspective of engineering economy, the dry connection of precast components is fast and convenient, and concrete should not be poured after the connection of members. Therefore, the demand for manpower, construction time, and building materials is low in engineering construction. In addition, unlike when using integral steel components for steel structures, the project cost is reduced. From the perspective of engineering durability, the precast members are mainly reinforced concrete materials with excellent corrosion resistance and fire resistance. In summary, the precast concrete structure with steel joints is a form of building structure system that combines the merits of steel structure and reinforced concrete structure.

Figure 2. Construction drawing of steel-joint precast concrete structure building.

2.2. Design and Production of Precast Components

Reasonable modular design is an indispensable factor in determining whether the precast structure system building can be popularized. The modular design can realize the batch production of precast components and production with less specification, which can reduce the amortization cost of molds and reduce the engineering cost. The modular design is conducted in accordance with the requirements of building function, bearing capacity, number of building floors, building elevation, etc. For example, the number of steel-embedded parts of precast column members is selected according to the number of building floors. In addition, the beam span length is selected according to the building-use function. Moreover, steel members are buried in precast beams according to the location of the filling wall in the building to erect secondary beams, and so on. In addition, the precast concrete structure with steel joints can be applied to a variety of building types through a reasonable modular design, which can meet the different functions of the building and improve the engineering economy.

The production process of precast components is shown in Figure 3.

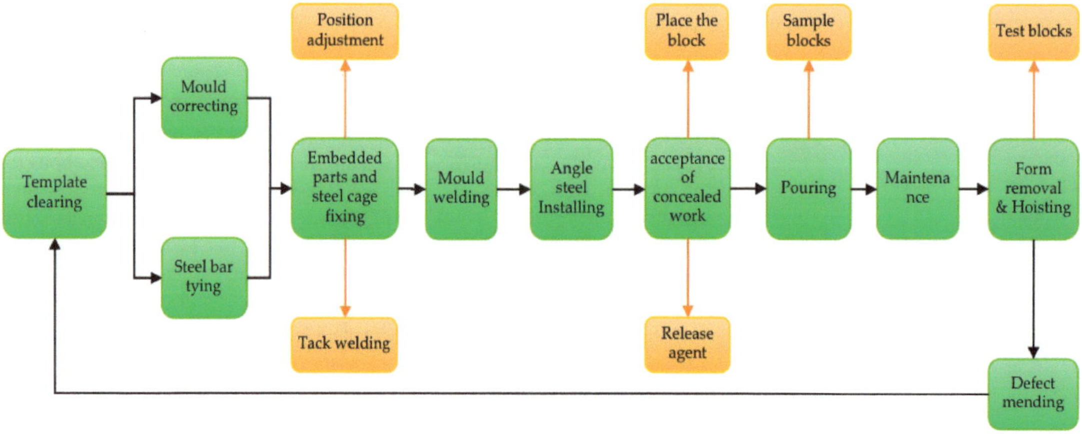

Figure 3. Production process flow chart of precast components.

Therefore, this study introduces the design and manufacture of precast members used in precast concrete structures with steel joints in accordance with building engineering examples. In steel-member precast concrete structure buildings, the precast columns are in the form of reinforced concrete, and the steel members are embedded only at the beam-column joints. The steel member is a square steel skeleton composed of a diaphragm, a cross-shaped partition, and an extension connecting plate with bolt holes, as shown in Figure 4.

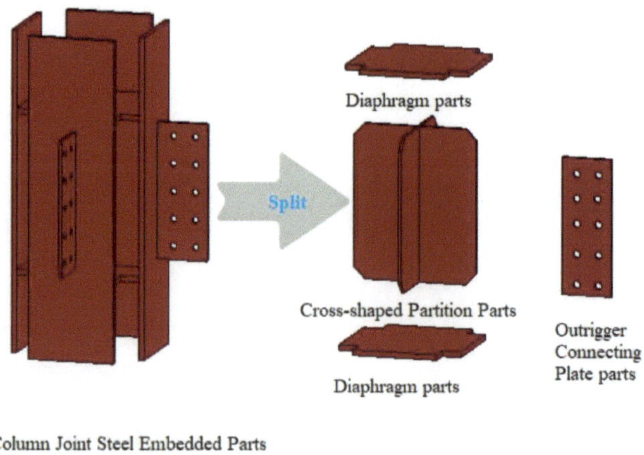

Figure 4. Steel-embedded parts in the precast column.

The all-steel mold is used in the production of the members. All the steel bars used in the precast columns are cut using industrial machinery and manually tied into steel frames. A diaphragm is located in the steel-embedded parts in the column members. Only the full-length longitudinal corner bars pass through the joint area of the precast column, and the steel skeleton composed of the central longitudinal steel bar and stirrups is divided into multiple sections by the joint. The multi-segment steel skeleton and the column nodal steel embedded parts are spliced outside the production mold, and the relative position between the rebar skeleton and the joint steel embedded parts is adjusted during the production of the column members. In addition, the longitudinal steel bar extending into the joint steel embedded parts and the vertical plate of the joint steel buried parts are welded on both sides. This requires that the weld length be no less than 10 times the longitudinal steel diameter of the column. Subsequently, the long corner bar is penetrated and tied. The connection between the steel reinforcement skeleton and the joint steel embedded parts is shown in Figure 5.

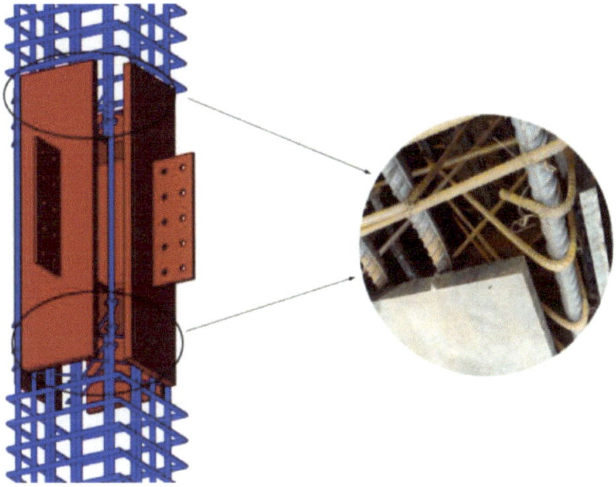

Figure 5. Connection diagram of reinforcing steel skeleton and joint steel embedded parts.

The crane hangs the precast column skeleton into the mold to pour concrete and completes the manufacture of the precast column components. The production process of the precast column is shown in Figure 6.

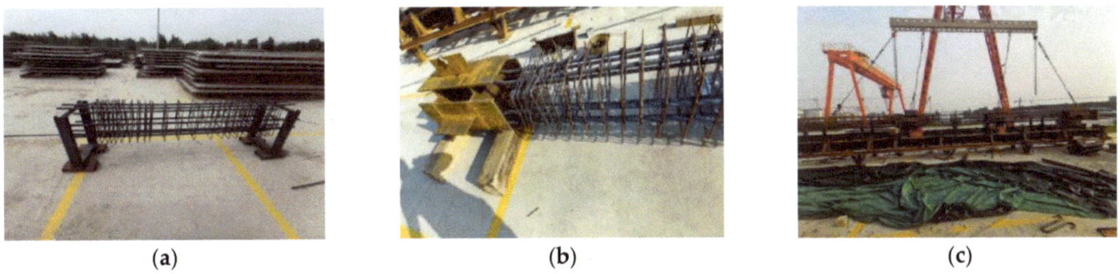

Figure 6. Production process diagram of the precast column: (**a**) binding of column reinforcement; (**b**) assembling steel joints and reinforcing steel skeleton; (**c**) column member is hoisted into the mold.

Because the system is mainly suitable for multi-story precast buildings, the general steel-joint precast concrete frame buildings use non-spliced precast column members to improve the assembly efficiency of the main frame of the building (Figure 7). The inability of non-spliced precast column members to properly resist lateral effects in construction can be effectively solved by using oblique supports to fix the precast columns in building construction.

Figure 7. Column unit.

In the project, the precast concrete structure system with steel joints adopts composite precast beam members, and the top of the beam protrudes part of the steel bar skeleton into the building floor to reduce the weight of the members and increase the height of the building floor. During construction, the precast beam laminated layer is poured together with the building floor slab. Simultaneously, H-shaped steel with web openings is embedded at both ends of the beam, and the H-shaped steel flange is welded with the longitudinal steel bar in the beam on both sides. The weld length should not be less than 10 times the longitudinal bar diameter of the beam. The precast beam members are shown in Figure 8.

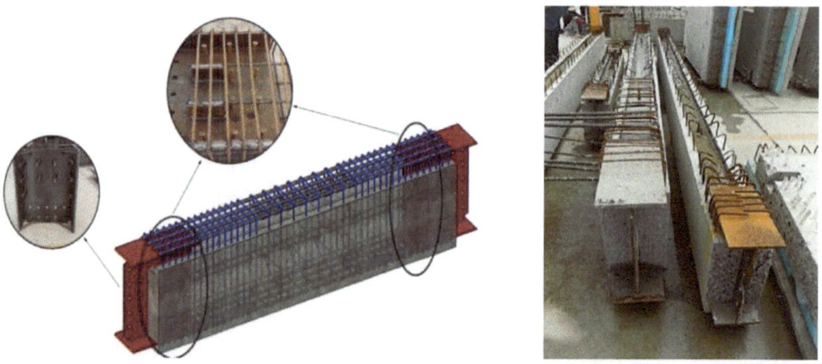

Figure 8. Beam unit.

The steel frame of the beam is first tied up in the production process of the precast beam. Next, the H-shaped steel embedded parts are inserted into the mold at both ends of the skeleton because tools in the mold can accurately locate the positions of the buried parts of the H-shaped steel and ensure the finished product of the precast beam members. It can meet the assembly accuracy requirements of the components.

The relative position of the steel bar skeleton and the H-section steel embedded parts is adjusted according to the requirements of the welding length of the H-section steel flange and the longitudinal steel bar in the beam. Owing to the small space in the mold, the welding operation cannot be completed. Therefore, the relative position of the reinforcing steel frame and the H-shaped steel embedded parts should be fixed by spot welding, then the mold is hoisted out as a whole for the welding operation. Finally, the beam precast components are hoisted into the mold to pour concrete to complete the fabrication of the precast beam components. The production process is shown in Figure 9.

(a)

(b) (c)

Figure 9. Production process diagram of precast beam: (**a**) binding of beam reinforcement; (**b**) fixed position; (**c**) welding connection.

2.3. Connection of Precast Column and Precast Beam

All beam-column precast members of the steel-joint precast concrete structures adopt the steel structure connection mode of bolting and welding. Specifically, the outrigger connecting plate of the square steel skeleton at the joint in the precast column and the web plate of the H-shaped steel at the beam end of the precast beam are threaded with high-strength bolts at the corresponding bolt holes. The upper and lower flanges of the H-shaped steel are welded and connected with the wall of the square steel skeleton. The connection is shown in Figure 10. During the construction of the steel-joint precast concrete structure, the installation of beam-column components is not limited by the construction of the floor slab. Simultaneously, no supporting formwork is needed, and concrete is poured at the joint connection parts. Therefore, during construction, precast columns can be integrally installed first. Second, precast beams and precast columns are spliced to form the main frame of the building, and then floor construction is conducted. The connection mode of this component has significantly reduced the manual requirements and construction period. The construction of the steel-joint precast concrete structure is shown in Figure 11.

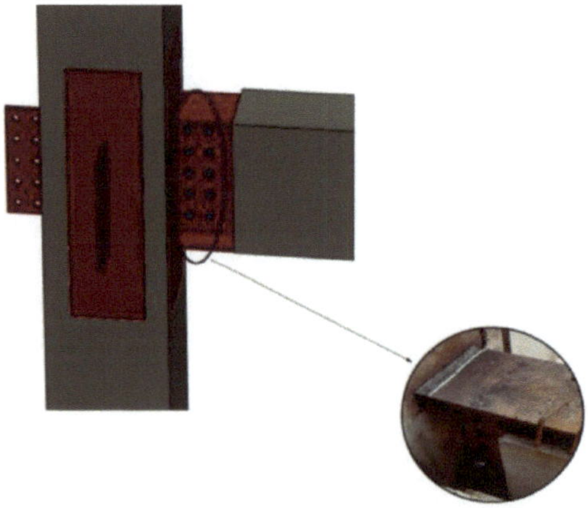

Figure 10. Assembly of the column-beam unit.

Figure 11. Building construction drawing.

The connection between reinforced concrete members is changed into the connection mode of steel structures, so it is easy to figure out the inconvenient connection between prefabricated reinforced concrete members using wet connections. For this fabricated structure system, the welding and high-strength bolt connection construction technology of steel structures is mature and reliable, which is convenient for the inspection and detection of the connection joint. Meanwhile, the original professional team of steel structure construction can be used for construction, which does not exceed the existing construction installation and construction management level. Hence, it is suitable for the quality of domestic existing construction and management and can effectively reduce the project cost.

2.4. Key Technologies in Construction

The construction process and key technologies of this steel-joint precast frame system were shown in Figure 12.

(1) The reinforced concrete precast column members used in the project have large slenderness ratios and deadweight, which easily produce brittle fractures in the hoisting process. Hence, the precast column members are hoisted smoothly into the corresponding foundation of the cup mouth by two truck cranes.
(2) The accuracy control of precast column installation in the project is ensured by the installation control lines placed on the cup base and column before the construction stage. During the construction process, the truck crane was used to pull the precast column and cooperate with the on-site workers to finely adjust the installation position of the precast column in the cup foundation. After the adjustment, steel wedges were built into the cup foundation to constrain the horizontal displacement of the precast column. In addition, verticality control of the precast column by installing inclined support adjustment in the scape, oblique support that is built in on both ends can be adjusted by adjusting the screw support length on the transverse and longitudinal axis of the installation of the prefabricated column, respectively; the verticality control line on the scape erects theodolite observation, and on the basis of the scape verticality control line in the theodolite eyepiece deviation in adjusting inclined support, verticality controls scape.
(3) In order to improve the fault tolerance rate of precast components, the connecting plate of precast column steel joints is prepared during construction, the connecting plate is welded during the construction of beam and column components, and the connection deviation of the beam and column components can be partially adjusted through the weld.

(**a**)

(**b**)

Figure 12. *Cont.*

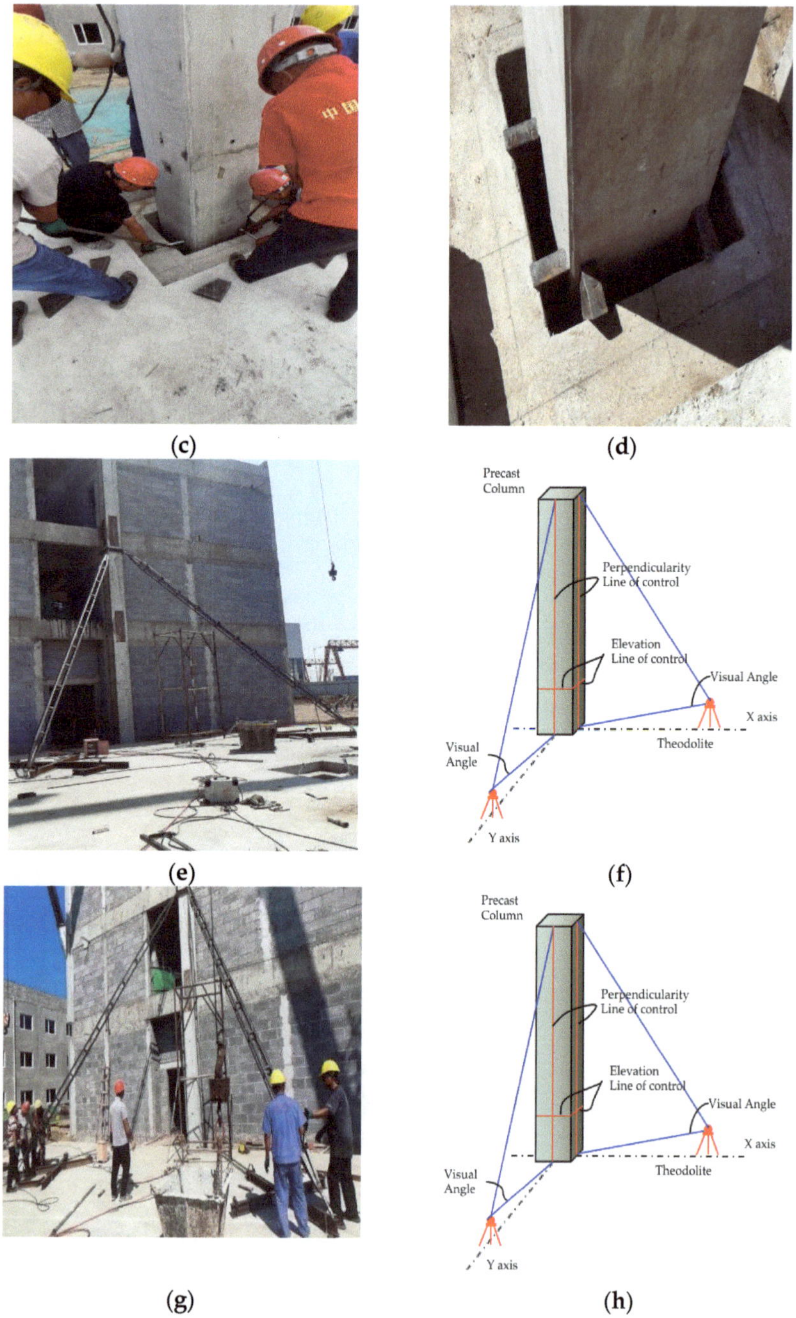

Figure 12. (**a**) Hoisting; (**b**) leveling control; (**c**) horizontal plane positioning; (**d**) steel wedge fixation; (**e**) diagonal support fixation; (**f**) verticality observation; (**g**) diagonal support adjust the verticality; (**h**) pouring.

3. FEM Simulation for Staged Construction and Monitoring Scheme

In the construction stage, the stress and deformation status monitoring of key parts of the structure can provide a timely grasp on the deformation trend of the structure, visualize the structural damage and the parts with hidden risks, and effectively reduce the construction risk. This section conducts finite element simulation analysis on the construction process of the test project. According to the simulated data, it analyses the stress variation law of precast concrete frame structure with steel joints during construction, detects the structural parts with construction risks and hidden dangers, and warns the actual construction. On the other hand, the secondary structure construction of steel-joint prefabricated concrete frame structure is monitored, and the stress and strain changes in the beam-column joint area are analyzed during the construction. By comparing the monitoring data with the simulated data, it can provide a safety guarantee for the construction and provide scientific experience for the construction monitoring of similar building structures in the future.

3.1. FEM Simulation for Staged Construction

In this study, the "life-and-death unit" technology in ABAQUS finite element software was used to simulate the construction process of the prefabricated concrete frame structure building with steel joints. Due to the rapid construction speed of steel-joint prefabricated concrete structures, the simulation process does not consider the influence of time-varying factors, such as concrete creep and temperature on the construction. During the simulation, the working load is divided into constant load and construction live load, and the engineering entity (filled wall) that does not participate in the calculation is equivalent to the constant load applied to the main frame of the building. The construction live load is 2 kN/m^2, which is applied to the building floor in the way of uniform load distribution. An 8-node linear 6-dihedral reduction and integration element (C3D8R) was used for concrete and steel, and a 2-node linear 3-dimensional truss element (T3D2) was used for reinforcement.

The plastic damage constitutive is considered for concrete, and the reinforced double oblique constitutive is considered for steel members and rebar materials. The connection between the upper and lower flange of H-shaped steel at the beam end and the square steel pipe of the column are defined by the Tie constraint instruction. The tangential action of face-to-surface contact is defined as a penalty, and the normal action of face-to-surface contact is defined as hard contact.

In the simulation of this project, the main frame model of the building is first established, and then the components that have not been built in the current analysis step are "killed" from the current analysis step of the model, according to the real working conditions, and then the corresponding units are gradually "activated", according to the actual construction sequence. The construction simulation process of the main frame of the building is shown in Figures 13 and 14.

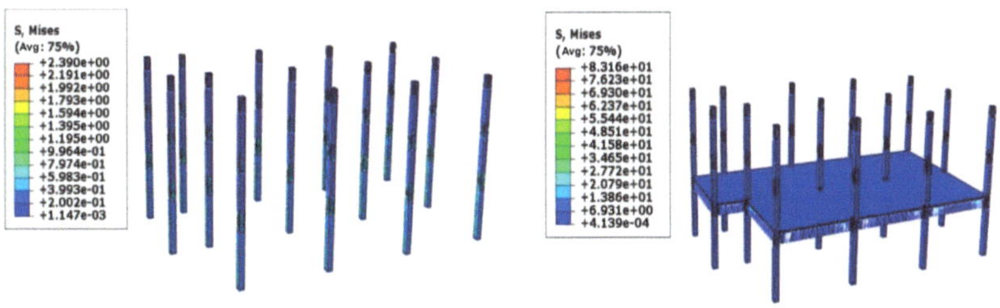

(a) (b)

Figure 13. *Cont.*

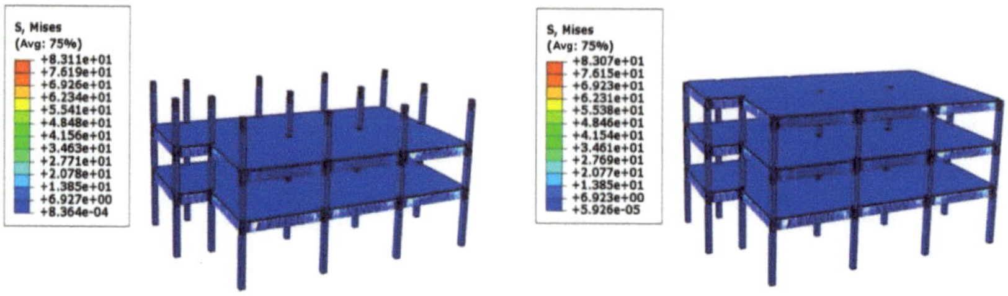

(c) **(d)**

Figure 13. Construction simulation process diagram of the main frame of the building: (**a**) Activate the first analysis step—construction of precast columns; (**b**) Activate the second analysis step—the construction of a layer of precast beams and slabs; (**c**) Activate the third analysis step—construction of two-story precast beams and slabs; (**d**) Activate the fourth analysis step—construction of three-story precast beams and slabs.

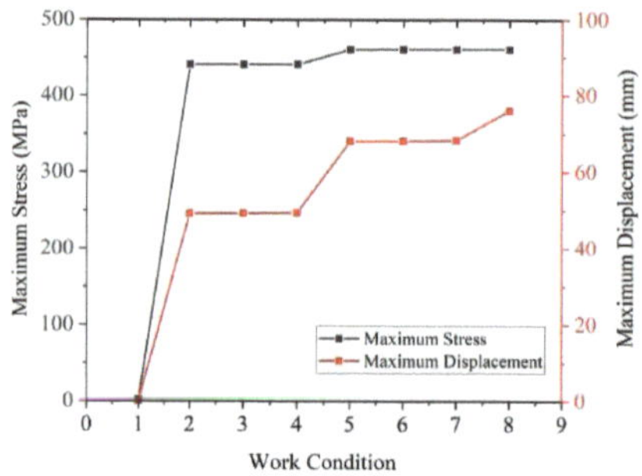

Figure 14. Changes of maximum stress and displacement values for each working condition.

The stress cloud diagram of the final stage of construction is shown in Figure 15. In the final stage, the stress value of each component reaches the maximum, so as to measure the safety of construction.

The stress of reinforcement is generally smaller than the tensile strength of 360MPa to HRB400 steel bars, and only the yield strength of the first-floor slab is slightly higher than that of HRB400. According to the maximum principal stress cloud diagram of concrete, the compressive stress of concrete components is much lower than the compressive strength of C30 concrete material, and most components are in the linear elastic stage during the construction process, which indicates that the construction process is safe and reliable.

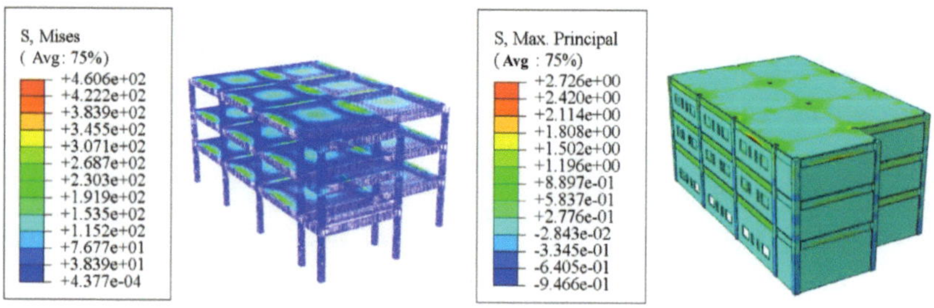

(a) **(b)**

Figure 15. *Cont.*

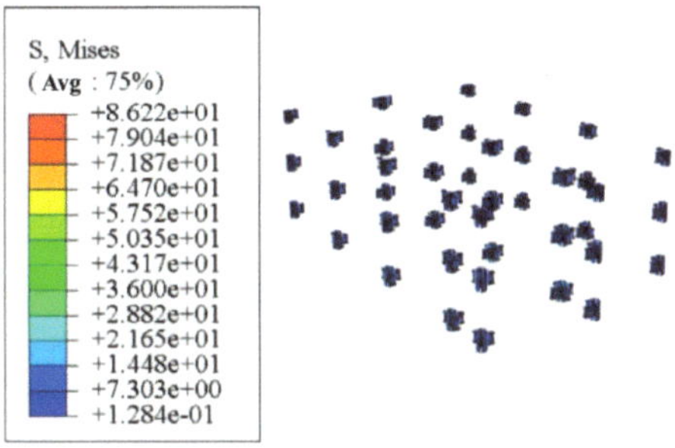

(c)

Figure 15. (**a**) The stress cloud diagram of steel bars in the final stage; (**b**) the stress cloud diagram of concrete in working condition 8; (**c**) the stress cloud diagram of steel components in working condition 8.

3.2. Monitoring Scheme

In this project, according to the different monitoring positions of strain gauges, two kinds of resistance strain gauges with different applicability are selected, which are, respectively, YBM tool-type surface strain sensors welded on the surface of the steel joints of beam-column members and YNM-type embedded strain sensors tied together with the steel cage of beam-column members, as shown in Figure 16.

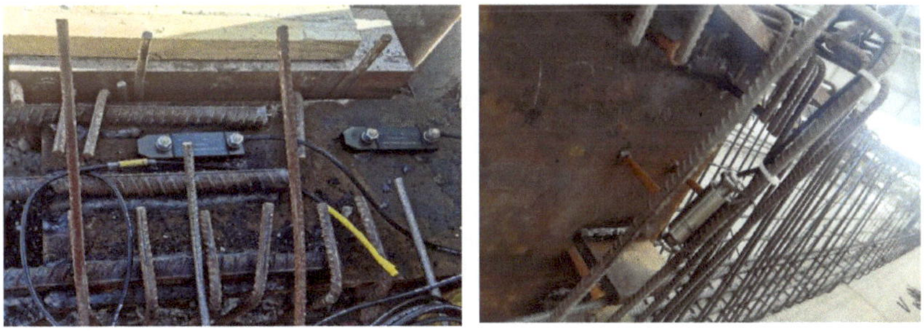

(**a**) (**b**)

Figure 16. (**a**) YBM-type tool-type surface strain sensor (**b**) YNM-type concrete-embedded strain sensor.

The measurement point arrangement in the scheme is to select the connected corner column, side column, and middle column in the building to form a small frame structure instead of the overall arrangement of measurement points in the building. The beam-column connection points with the largest stress in the small frame structure are selected as the representative measurement point, which is used to study the stress state of the whole construction process of the building structure.

The plan of the first floor is shown in Figure 17a. In this project monitoring scheme, the frame system surrounded by four axes of A, B, 1, and 2 in the plan is selected. The number of beam-column connection points monitored in the frame is three, all on the first floor of the building, and the location of monitored nodes are marked by a circle in Figure 17a. In the monitored beam-column connection joints, a total of eight YBM tool-type surface strain sensors are welded on the square steel tube embedment of each joint and the beam-end H-shaped steel embedment is connected with it. In addition, three YNM-embedded strain sensors are embedded in each point area on the reinforcement skeleton. There are 33 visual inspection points in this project. The number of measuring points is shown in the following table. The letters represent the position of measuring points; the details are as follows.

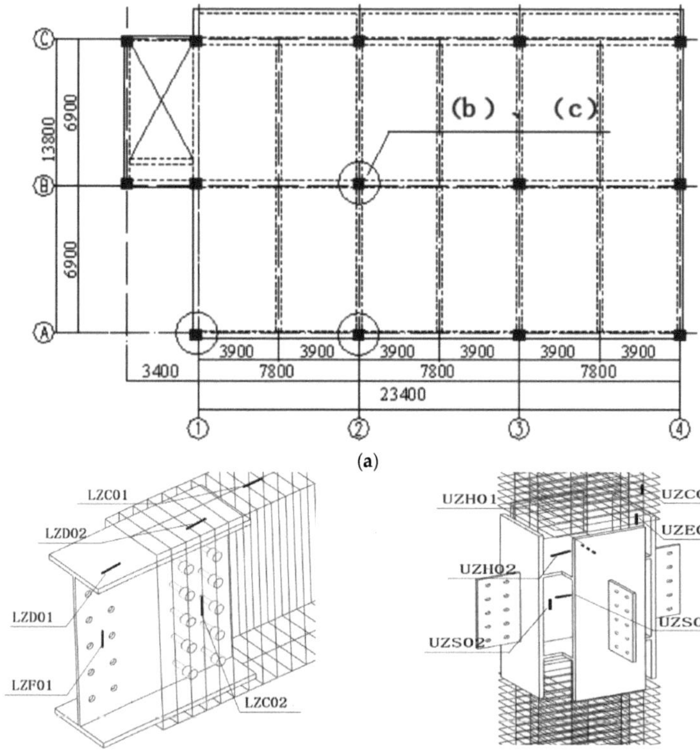

(b) (c)

Figure 17. Schematic diagram of the location of measuring points: (**a**) Floor plan of the building; (**b**) Layout of the sensor position of the connecting beam end of the center column; (**c**) Layout of the sensor position of the connecting node in the center column.

U stands for the column, L stands for beam, Z stands for middle column point, B stands for side column point, J stands for corner column point, H stands for horizontal plate, S stands for cross plate, E stands for vertical plate, C stands for measuring point in concrete, D stands for top of the upper flange, F stands for web, and the number of measuring points is shown in Table 2.

The sensor arrangement position in each monitoring point area is the same. Taking the beam-column connection point of the middle column as an example [28], the detailed arrangement of measuring points in the joint is shown in Figure 17b,c.

Table 2. Number of measuring points.

Installation Site	Measuring Points
Steel embedment of middle column joint	UZH01, UZH02, UZS01, UZS02, UZE01, UZC01
Steel embedment of side column joint	UBH01, UBH02, UBS01, UBS02, UBE01, UBC01
Steel embedment of corner column joint	UJH01, UJH02, UJS01, UJS02, UJE01, UJC01
Steel embedment of middle column-beam connection	LZD01, LZD02, LZF01, LZC01, LZC02
Steel embedment of side column-beam connection	LBD01, LBD02, LBF01, LBC01, LBC02
Steel embedment of corner column-beam connection	LJD01, LJD02, LJF01, LJC01, LJC02

From the completion of the main structure to the completion of the project, daily periodic monitoring is carried out to analyze the structural state of the strain and stress changes subject to the gradually increasing construction load. The construction schedule node is shown in Table 3.

Table 3. Construction schedule node table.

Construction Step	Time	Progress Schedule
0	1	The second and third floors have the building materials placed
1	5	The second and third floors of the infill wall masonry was completed and the second floor of the building was tiled
2	9	Beams and parapets are poured on the top floor
3	16	The third floor was tiled and the top floor infill wall masonry was completed
4	24	The construction of the thermal insulation layer of the top floor is completed

By comparing the stress changes at the same measuring point of each precast beam monitored on the first floor of the building, as shown in Figure 18, the analysis shows that:

(1) The stress values of the measured points at each beam end are relatively discrete and the stress curves deviate significantly, which is mainly caused by the uneven distribution of the construction load on the building floor and the irregular loading and unloading during the construction. The stress fluctuation intervals of the measuring points of the steel section on the upper flange, reinforced concrete section on the upper flange, and web of H-shaped steel at each beam end are −2–12 MPa, −0.5~3.5 Mpa, and −3.5~1.5 MPa, respectively. The stress fluctuation intervals of the measuring points on the top of the concrete section at each beam end are −4.5~0.5 MPa. It can be seen that the stress change of the beam end in the construction of the secondary structure of the building is far less than the yield strength of the steel and concrete materials used in the building.

(2) The stress variation at the same measurement point of each beam end during construction is basically the same, and the absolute value of stress shows a deviating upward trend, indicating that the precast beams of the building can work well together.

(3) The stress value of steel section on the upper flange of H-shaped steel at the end of each beam is generally higher than that of the reinforced concrete section, and both of them show tensile stress. The compressive stress was observed at the top of the concrete section at each beam's end. It can be seen that the stress state of the precast beam of the steel-joint prefabricated concrete frame structure during construction is that the beam end presents negative bending moment, and the bending moment gradually becomes positive when extending from the beam end to the middle of the beam span. At the same time, it is found that the stress value of each measuring point at the beam end has no obvious mutation, indicating that the connection state of the monitored frame nodes under the action of construction load is good, and no risks, such as unwelding or bolt slip, occur.

By comparing and analyzing the stress data of the measuring points of the steel joints of the middle column, side column, and corner column in the construction process, as shown in Figure 19, it can be seen that:

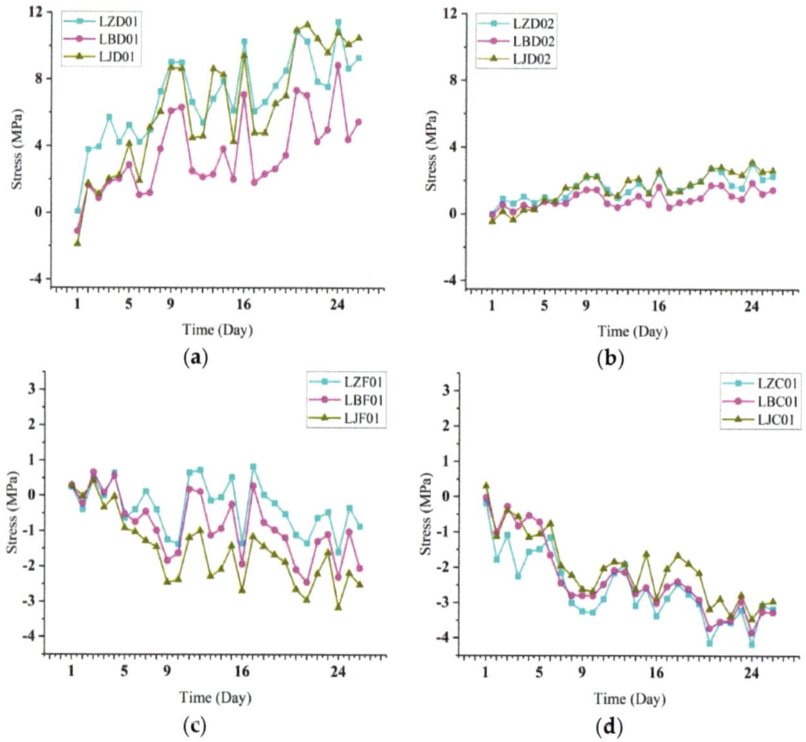

Figure 18. Comparison of stress of measured points at each beam end: (**a**) The stress at the measuring point of the pure steel section of the upper flange of the H-beam at the beam end; (**b**) the stress at the measuring point of the mixed section of the upper flange of the H-beam at the beam end; (**c**) the stress at the measuring point of the H-beam web at the beam end; (**d**) stress at the measuring point at the top of the concrete section at the beam end.

(1) The stress curve of each measurement point is rather meandering, but the range of stress fluctuation is small, which is caused by construction material handling and stacking on the building floor. The stress fluctuation range of horizontal plate 01, horizontal plate 02, cross partition 0°, cross partition 90°, and a vertical plate of each column steel joint in the construction process of the secondary structure is −2~8 MPa, −2~7 MPa, −1~4 MPa, −8~2 MPa, and −8~1 MPa, respectively. The maximum stress variation is about 10 MPa, which is much lower than the yield strength of Q345 steel used for steel joints. Combined with the analysis of the stress fluctuation range of each beam end measurement point, it is concluded that in the construction of the secondary structure of the building, the stress of the frame node position does not change much, and the safety of the building is not greatly affected.

(2) Each column steel measuring point with the same location has almost the same stress changes; the absolute value of the stress is winding up trends, which, combined with the analysis of the stress curve of the beam end of the same measuring point, indicates that the prefabricated concrete frame structure in steel point and edge point have the same angle in the same state of stress changes in the process of construction, which further indicates that the construction of each point can work better together. At the same time, according to the stress data of each column's steel joint in the building, the stress value of the middle point is generally the largest, while the stress value of the corner point is generally the smallest. This is because the force transmission path of the frame structure is from the floor to the column and then from the beam to the

column. Therefore, the number of columns connecting the beams is large, and the stress is also generally large. The data show that the stress value of some measuring points of the corner column steel joints is greater than that of middle column steel joints, which is due to the occurrence of construction loads, such as building filling walls, concentrated in the measuring point area of the corner column steel joints, resulting in a larger stress value of some measuring points of the corner column steel joints.

(3) Through the monitoring data, it is found that the stress change trend of measurement points 01 and 02 of the steel-joint horizontal plate of each column is the same, which is the tensile stress of tortuous growth. At the same time, through the monitoring data of each column steel-joint cross partition and vertical plate, it can be seen that the cross partition is subjected to horizontal tensile stress and vertical compressive stress, while the vertical plate is mainly subjected to compressive stress. The results show that steel joints can play a positive part in the stress work in the construction of building structures and play a positive role in improving the bearing capacity of joints.

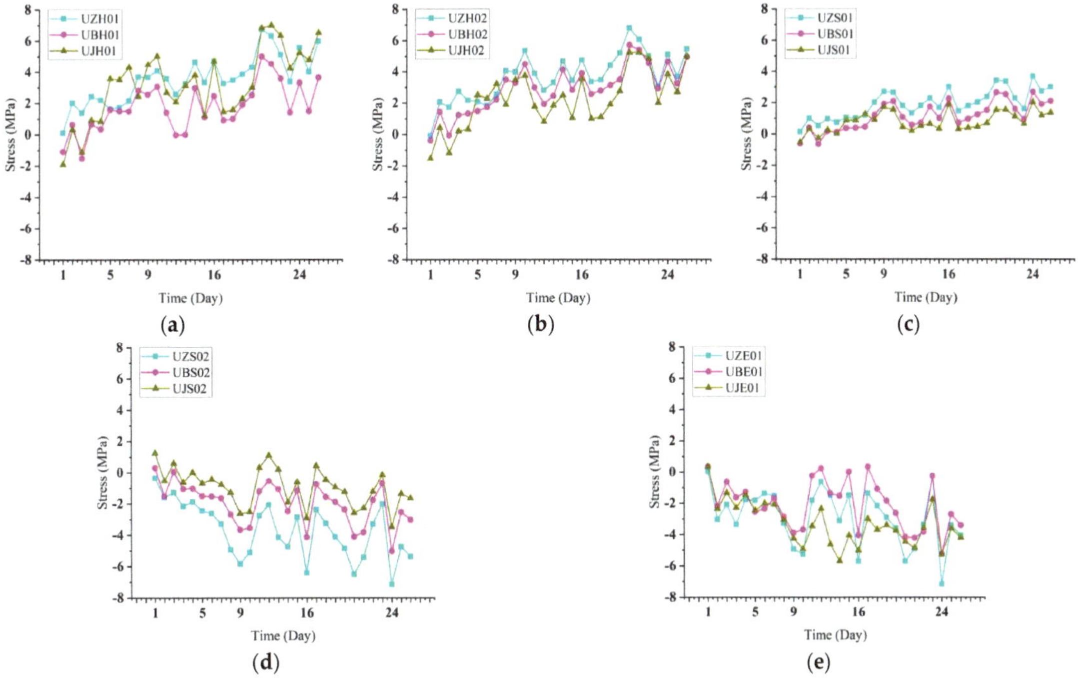

Figure 19. The stress comparison diagram of the measured points on the column members: (**a**) Stress at measuring point 01 of steel-joint transverse plate; (**b**) Stress at measuring point 02 of steel-joint transverse plate; (**c**) 0° axis measuring point stress of steel-joint cross diaphragm; (**d**) 90° axis measuring point stress of steel-joint cross diaphragm; (**e**) Stress at the measuring point of the vertical plate of the steel joint.

3.3. Discussion

The stress simulation data of the measuring point unit under each construction step is obtained through the field variable output function of the software.

By comparing the monitored stress data and simulated stress data of measuring points under different construction steps, it is found that the simulated calculated value of each measuring point under different working conditions is generally less than the actual monitored value. Figure 20 shows the comparison diagram of stress data of some measuring points. The main reason for the deviation between the monitored value and the calculated value is that several construction steps are carried out simultaneously in the actual construction, and the actual construction step is larger than the construction load in the simulated construction step. The site of the construction process with measured value and the simulation calculation of stress have the same change trend, numerical deviation within ±2 MPa, monitoring data, and simulated data in which the coincidence degree is higher; it shows that the finite element simulation analysis simulation steel-joint prefabricated reinforced concrete frame structure building in the construction phase of the stress change of the method is feasible, which provides theoretical support for the construction of the building.

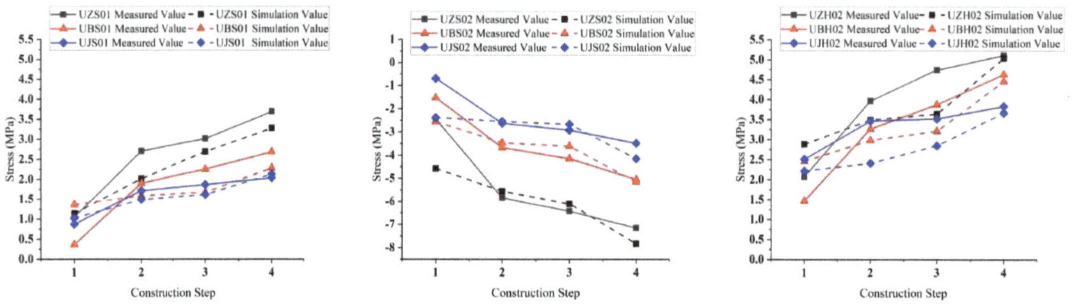

(a) (b) (c)

Figure 20. Comparison of stress data of some measuring points: (**a**) 0° axis measuring point stress of the cross diaphragm; (**b**) 90° axis measuring point stress of the cross diaphragm; (**c**) Measurement point of the transverse plate.

The stress variation trend reflects that the maximum tensile stress and the maximum compressive stress of the building model occur when the insulation layer of the top floor is laid, and the maximum stress value of reinforcement, steel, and concrete materials in the simulation process does not reach the yield strength of each material. It shows that the building structure has high safety in the construction process.

4. Construction Efficiency Analysis

The construction technology of conventional reinforced concrete frame structure (hereinafter, RC), precast concrete frame structure (hereinafter, PC), and steel-joint precast concrete frame structure (hereinafter, SPC) is analyzed based on the actual construction of the steel-joint precast concrete frame structure. The construction periods of the three structural systems are compared. Consequently, the construction period of the steel-joint precast concrete frame structure system is found to be the shortest. Compared with the conventional reinforced concrete frame structure, the construction period is reduced by 50–60%; compared with the precast concrete frame structure, the construction period is reduced by nearly 40%.

4.1. Comparative Analysis of Construction Process and Cycle

Based on the single-layer construction of the same frame structure building. The construction process of each type of structure is shown in Figure 21.

The conventional construction procedure of a reinforced concrete frame structure building, in turn, includes surveying, steel works of columns and walls, formwork installation of columns and walls, formwork installation of beams and floor slabs, steel works of beams and floor slabs, and concrete works. The construction period of a single layer of a frame structure building is 9 days.

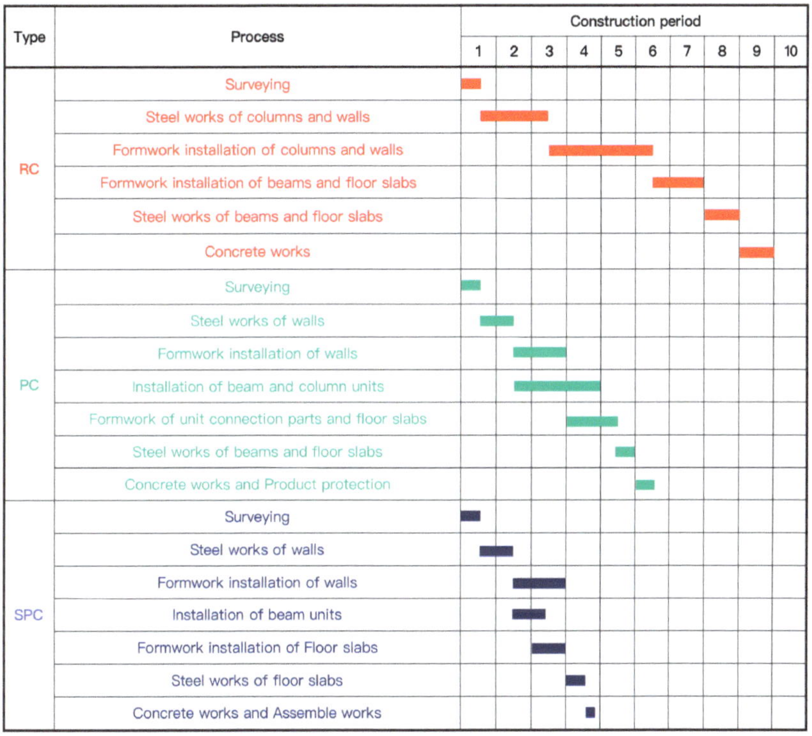

Figure 21. Construction process of each type of structure.

The construction process of a precast concrete frame structure building, in turn, includes surveying, steel works of walls, formwork installation of walls, installation of beam and column units, formwork of unit connection parts and floor slabs, steel works of beams and floor slabs, concrete works, and product protection. The construction process of this structure system frame building is tedious, but the mutual interference of each construction step is not strong, so multiple steps can be carried out simultaneously. Therefore, the construction period for completing a single layer of a frame structure building is 6.5 days.

The construction process of the steel-joint precast concrete frame structure is as follows in sequence: surveying, steel works of walls, formwork installation of walls, installation of beam unit, formwork installation of floor slabs, steel works of floor slabs, assembly works, and concrete works. The construction period for completing a single layer of a frame structure building is 3.8 days. Because precast column members need to be hoisted and installed during the first layer of construction of this structure system, the construction period for completing the first layer of the building is 4.3 days.

To compare the construction periods of the three structural systems more intuitively, this study introduces a four-story standard frame building as a calculation model. The construction of a model building with a steel-joint precast concrete frame structure system can be completed in 15.7 days, compared with the construction period of 26 days for the precast concrete frame structure system. Therefore, the construction period is reduced by 39.6%. However, the conventional reinforced concrete frame structure system needs 36 days to complete the project, and the construction period is reduced by 56.4%. Furthermore, compared with the conventional reinforced concrete structure system, the precast structure system mainly saves time in the construction steps of steel binding and formwork support, in addition to saving labor and building materials. As an update of the precast concrete structure system, the steel-joint precast concrete structure system mainly reduces the construction period from the assembly of precast components. Moreover, it reduces the use of building materials, such as formwork and concrete, and gradually evolves into a construction mode based mainly on machines and tools and assisted by manpower.

4.2. Anticipation of Perfecting Steel-Joint Precast Concrete Structure System

According to the construction period of the steel-joint precast concrete structure system given in Table 1, the construction of walls and floors of the building accounts for approximately 70% of the total construction period. Therefore, when the steel-joint precast concrete structural system is applied to the framework building, the precast wall panels and floors are formed with the precast building, fully utilizing the advantages of the structural system. This further improves construction efficiency and reduces labor consumption. The fully precast components are more consistent with the steel-joint precast concrete structural system because the main frame of the structure building is formed quickly during construction. This is sufficient for providing the bearing capacity for the construction load introduced by the rapid installation of the precast components.

5. Building Energy Efficiency

Large amounts of energy are consumed in building construction as cities grow. According to the number of building materials consumed during the construction of different structural types of buildings, the basic energy consumption of each structural type of building is estimated with the energy amount consumed by the production of building materials as the basic unit [36–40]. Taking a three-story framework structure as an example, the building energy consumption of the steel-joint precast concrete, conventional concrete, and precast concrete is calculated and compared. The calculation results are presented in Table 4, indicating that the building energy consumption of the steel-joint precast concrete structural system is reduced by 22.2% and 19.1%, compared with the conventional concrete structure system and the conventional precast concrete structure system, respectively. The energy consumption of formwork is reduced by 54.9% and 22.5%, respectively, because the steel-joint precast concrete structure buildings only lay formwork in the floor construction, which greatly shortens the usage of formwork. When the structural system is perfect, the building adopts fully precast components, with a formwork energy consumption of zero, which further reduces the energy consumption of the building. Compared with the energy consumption of the conventional concrete structure system and conventional precast concrete structure system, that of reinforcement is reduced by 34.9% and 37.6%, respectively. In the beam-column connection joints, the dense steel bars are replaced with steel members due to the resistance to the shear forces, resulting in a reduction in the energy consumption of steel bars. Compared with the conventional concrete structure and precast concrete structure systems, the concrete energy consumption is reduced by 25.5% and 31.6%, respectively. This is because steel members embedded in the precast concrete structure of steel joints enhance the stiffness of each precast member, reducing the cross-section area of the members and reducing the usage of concrete in the building. In summary, the steel-joint precast concrete structure has certain advantages in building energy consumption.

Table 4. Comparison of building energy consumption of various structures.

Material	Energy Consumption per Original Unit	Amount of Materials			Energy Consumptions		
		RC	PC	SPC	RC	PC	SPC
Formwork	241 MJ/m^2	6251 m^2	3641 m^2	2821 m^2	1505 GJ	876 GJ	679 GJ
Steel section	4066 MJ/kN	0	0	257 kN	0	0	1045 GJ
Reinforcement	3522 MJ/kN	1107 kN	1156 kN	721 kN	3899 GJ	4073 GJ	2540 GJ
Concrete	2046 MJ/m^3	931 m^3	1013 m^3	693 m^3	1905 GJ	2073 GJ	1419 GJ
Sum					7308 GJ	7023 GJ	5683 GJ

6. Conclusions

This research elucidates the steel-joint precast concrete structure system, considering engineering practices. This structure system solves the assembly difficulties of the conventional connection mode for precast concrete components using steel connection. The key technologies for construction have been investigated for steel-joint fabricated concrete frame structures by using the methods of on-site supervision, stress monitoring, and numerical simulation. The main research contents and conclusions are as follows:

(1) The steel-joint precast concrete structure combines the steel and concrete structures effectively. With the quick and simple connection of beam-column members, the main framework may be completed at one time without wet construction. This building structure system can be widely used with high practical utility.
(2) Construction techniques, such as "piecewise casting", "theodolite & oblique support" adjustment installation, and "lag-connect & lag-weld" installation of connecting plates, are proposed, which can solve engineering difficulties and improve construction efficiency.
(3) The main structure can work coordinately and can be better at forcing transmission performance works at an elastic state under the normal service condition. The comparison and analysis of simulation results and monitoring values show that the two development trends are basically the same, indicating that the construction monitoring and simulation method adopted is feasible, and the building has high safety during construction.
(4) Construction pile load has a significant influence on stress distribution. It is recommended to reasonably plan the placement of construction materials and tools in the construction process to avoid excessive concentrated load on the floor slab and safety accidents.
(5) For a single-story framework building, the construction cycle of the steel-joint precast concrete structure is only 3.8 d, which is a 39.6% and 56.4% reduction, compared with that of the precast concrete and reinforced concrete structures, respectively. The construction progress analysis shows that the precast system composed of precast floors and walls may further improve construction efficiency.
(6) Compared with the precast concrete and conventionally reinforced concrete structures, the steel-joint precast concrete structure generally reduces the amount of formwork, reinforcement, and concrete, lowering the building energy consumption by 19.1% and 22.2%, respectively.

The steel-joint precast concrete structure features low consumption of building materials, high assembly efficiency of precast components, and reduced energy consumption of construction equipment. This research on steel-joint precast concrete structure systems provides a reference for the development of precast concrete structures.

Author Contributions: Conceptualization, X.R. and J.Z.; methodology, X.S., X.R. and J.Z.; software, L.N. and L.W.; validation, X.S. and J.Z.; formal analysis L.N.; investigation, X.S.; resources, L.W.; data curation, L.N.; writing—original draft preparation, X.S. and L.N.; writing—review and editing, J.Z.; supervision, X.R.; funding acquisition, X.R. All authors have read and agreed to the published version of the manuscript.

Funding: This research was funded by the Natural Science Foundation of Hebei Province, China, grant number E2020202134; the Natural Science Foundation of Tianjin, China, grant number: 20JCZDJC00370.

Data Availability Statement: The data presented in this study are available on request from the corresponding author.

Conflicts of Interest: The authors declare no conflict of interest.

References

1. Navaratnam, S.; Tuan, N.; Gunawardena, T.; Henderson, D. Performance Review of Prefabricated Building Systems and Future Research in Australia. *Buildings* **2019**, *9*, 38. [CrossRef]
2. Wang, B. Structural Concrete: Theory and Design, 7th ed. *Proc. Inst. Civ. Eng.-Civ. Eng.* **2021**, *174*, 12.
3. Aninthaneni, P.K.; Dhakal, R.P.; Marshall, J.; Bothara, J. Nonlinear Cyclic Behaviour of Precast Concrete Frame Sub-Assemblies With "Dry" End Plate Connection. *Struct* **2018**, *14*, 124–136. [CrossRef]
4. Savoia, M.; Buratti, N.; Vincenzi, L. Damage and collapses in industrial precast buildings after the 2012 Emilia earthquake. *Eng. Struct.* **2017**, *137*, 162–180. [CrossRef]
5. Magliulo, G.; Ercolino, M.; Manfredi, G. Influence of cladding panels on the first period of one-story precast buildings. *Build. Earthq. Eng.* **2015**, *13*, 1531–1555. [CrossRef]
6. Magliulo, G.; Ercolino, M.; Cimmino, M.; Capozzi, V.; Manfredi, G. FEM analysis of the strength of RC beam-to-column dowel connections under monotonic actions. *Constr. Build. Mater.* **2014**, *69*, 271–284. [CrossRef]
7. Kramar, M.; Isakovic, T.; Fischinger, M. Seismic collapse risk of precast industrial buildings with strong connections. *Earthq. Eng. Struct. D* **2010**, *39*, 847–868. [CrossRef]
8. Magliulo, G.; Fabbrocino, G.; Manfredi, G. Seismic assessment of existing precast industrial buildings using static and dynamic nonlinear analyses. *Eng. Struct.* **2008**, *30*, 2580–2588. [CrossRef]
9. Biondini, F.; Toniolo, G. Probabilistic calibration and experimental validation of the seismic design criteria for one-story concrete frames. *J. Earth Eng.* **2009**, *13*, 426–462. [CrossRef]
10. Huang, Y.; Mazzarolo, E.; Briseghella, B.; Zordan, T.; Chen, A. Experimental and numerical investigation of the cyclic behavior of an innovative prefabricated beam-to-column joint. *Eng. Struct.* **2017**, *150*, 373–389. [CrossRef]
11. Guan, D.; Guo, Z.; Xiao, Q.; Zheng, Y. Experimental study of a new beam-to-column connection for precast concrete frames under reversal cyclic loading. *Adv. Struct. Eng.* **2016**, *19*, 529–545. [CrossRef]
12. Restrepo, J.I.; Park, R.; Buchanan, A.H. Tests on connections of earthquake resisting precast reinforced concrete perimeter frames of buildings. *PCI J.* **1995**, *40*, 44–61. [CrossRef]
13. Xue, W.; Zhang, B. Seismic behavior of hybrid concrete beam-column connections with composite beams and cast-in-place columns. *ACI Struct. J.* **2014**, *111*, 617–627. [CrossRef]
14. Ghayeb, H.H.; Razak, H.A.; Sulong, N.H.R. Development and testing of hybrid precast concrete beam-to-column connections under cyclic loading. *Constr. Build. Mater.* **2017**, *151*, 258–278. [CrossRef]
15. Parra-Moritesinos, G.J.; Dasgupta, P.; Goel, S.C. Development of connections between hybrid steel truss-FRC beams and RC columns for precast earthquake-resistant framed construction. *Eng. Struct.* **2005**, *27*, 1931–1941. [CrossRef]
16. Dogangun, A. Performance of reinforced concrete buildings during the 1 May 2003, Bingol Earthquake in Turkey. *Eng. Struct.* **2004**, *26*, 841–856. [CrossRef]
17. Ghayeb, H.H.; Razak, H.A.; Sulong, N.H.R. Performance of dowel beam-to-column connections for precast concrete systems under seismic loads: A. review, Constr. *Build. Mater.* **2020**, *237*, 117582. [CrossRef]
18. Ha, S.; Kim, S.; Lee, M.S.; Moon, J. Performance Evaluation of Semi-Precast Concrete Beam-Column Connections with U-Shaped Strands. *Adv. Struct. Eng.* **2014**, *17*, 1585–1600. [CrossRef]
19. Parastesh, H.; Hajirasouliha, I.; Ramezani, R. A new ductile moment-resisting connection for precast concrete frames in seismic regions: An experimental investigation. *Eng. Struct.* **2014**, *70*, 144–157. [CrossRef]
20. Eom, T.; Park, H.; Hwang, H.; Kang, S. Plastic Hinge Relocation Methods for Emulative PC Beam-Column Connections. *J. Struct. Eng.* **2016**, *142*, 04015111. [CrossRef]
21. Im, H.; Park, H.; Eom, T. Cyclic Loading Test for Reinforced-Concrete-Emulated Beam-Column Connection of Precast Concrete Moment Frame. *ACI Struct. J.* **2013**, *110*, 115–125.
22. Morgen, B.G.; Kurama, Y.C. Seismic design of friction-damped precast concrete frame structures. *J. Struct. Eng.* **2007**, *133*, 1501–1511. [CrossRef]
23. Erkmen, B. Effects of unbonded steel layout on seismic behavior of post-tensioned precast concrete shear walls. *B. Earthq. Eng.* **2021**, *19*, 179–201. [CrossRef]
24. Song, M.; He, J.; Liu, Y.; Zhang, H.; Ge, C.; Jin, Y.; Liu, B.; Huang, S.; Liu, Y. Seismic Behavior of Three-Story Prestressed Fabricated Concrete Frame under Dynamic and Low Reversed Cyclic Loading. *Adv. Civ. Eng.* **2018**, *2018*, 7876908. [CrossRef]
25. Kim, J.H.; Cho, Y.S.; Lee, K.H. Structural performance evaluation of circular steel bands for PC column-beam connection. *Mag. Concr. Res.* **2013**, *65*, 1377–1384. [CrossRef]
26. Ertas, O.; Ozden, S.; Ozturan, T. Ductile connections in precast concrete moment resisting frames. *PCI J.* **2006**, *51*, 66. [CrossRef]
27. Vidjeapriya, R.; Jaya, K.P. Experimental Study on Two Simple Mechanical Precast Beam-Column Connections under Reverse Cyclic Loading. *J. Perform. Constr. Fac.* **2013**, *27*, 402–414. [CrossRef]
28. Zhang, J.; Ding, C.; Rong, X.; Yang, H.; Li, Y. Development and experimental investigation of hybrid precast concrete beam-column joints. *Eng. Struct.* **2020**, *219*, 110922. [CrossRef]
29. Brownjohn, J.M.W. Structural health monitoring of civil infrastructure. *J. Philosofical Trans. A Math. Phys. Eng. Sci.* **2006**, *365*, 589–622. [CrossRef]
30. Brault, A.; Hoult, N.A. Monitoring reinforced concrete serviceability performance using fiber-optic sensors. *ACI Struct. J.* **2019**, *116*, 57–70. [CrossRef]

31. Ni, Y.Q.; Xia, Y.; Liao, W.Y.; Ko, J.M. Technology innovation in developing the structural health monitoring system for Guangzhou New TV Tower. *Struct. Control Health Monit.* **2009**, *16*, 73–98. [CrossRef]
32. Jiang, X.; Titi, H.; Ma, Y.; Polaczyk, P.; Zhang, M.; Gabrielson, J.; Huang, B. Evaluating the performance of inverted pavement structure using the accelerated pavement test (APT). *Constr. Build. Mater.* **2022**, *346*, 128489. [CrossRef]
33. Barrias, A.; Rodriguez, G.; Casas, J.R.; Villalba, S. Application of distributed optical fiber sensors for the health monitoring of two real structures in Barcelona. *Struct. Infrastruct. Eng.* **2018**, *14*, 967–985. [CrossRef]
34. Gao, F.; Zhou, H.; Liang, H.; Wen, S.; Zhou, H. Structural deformation monitoring and numerical simulation of a supertall building during construction stage. *Eng. Struct.* **2020**, *209*, 110033. [CrossRef]
35. Bado, M.F.; Casas, J.R. A Review of Recent Distributed Optical Fiber Sensors Applications for Civil Engineering Structural Health Monitoring. *Sensors* **2021**, *21*, 1818. [CrossRef] [PubMed]
36. Hong, W.K.; Lim, G.T.; Park, S.C.; Kim, J.T. Energy efficiencies of linear-shaped multi-residential apartment buildings built with hybrid structural systems. *Energy Build.* **2012**, *46*, 30–36. [CrossRef]
37. Song, Q.; Duan, H.; Yu, D.; Li, J.; Wang, C.; Zuo, J. Characterizing the essential materials and energy performance of city buildings: A case study of Macau. *J. Clean. Prod.* **2018**, *194*, 263–276. [CrossRef]
38. Kim, S.; Hong, W.; Kim, J.; Kim, J.T. The development of modularized construction of enhanced precast composite structural systems (Smart Green frame) and its embedded energy efficiency. *Energy Build.* **2013**, *66*, 16–21. [CrossRef]
39. Won, I.; Na, Y.; Kim, J.T.; Kim, S. Energy-efficient algorithms of the steam curing for the in situ production of precast concrete members. *Energy Build.* **2013**, *64*, 275–284. [CrossRef]
40. Lee, S.; Kim, S.; Na, Y. Comparative analysis of energy-related performance and construction cost of the external walls in high-rise residential buildings. *Energy Build.* **2015**, *99*, 67–74. [CrossRef]

buildings

MDPI

Article

Characteristics and Mechanism of Fire Spread between Full-Scale Wooden Houses from Internal Fires

Shasha Yuan *, Kun Xiang, Feng Yan, Qing Liu, Xuan Sun, Yinqing Li and Peng Du

Institute of Building Fire Research, China Academy of Building Research, Beijing 100013, China; xk0807@163.com (K.X.); yf13910598628@126.com (F.Y.); alecaza@163.com (Q.L.); 13911365611@126.com (X.S.); cabr2582@126.com (Y.L.); 18810503116@163.com (P.D.)
* Correspondence: yuanshasha1988@126.com

Abstract: In ancient villages, the spread of uninterrupted fires caused great damage to clustered wooden houses. Thus, the spread of fire among wooden houses should be systematically studied to explore its characteristics. Statistical analysis is a feasible way to study the characteristics and underlying mechanisms of fire in full-scale wooden houses. In this study, 4 full-scale wooden buildings were built in an ethnic village in Guizhou Province, and the fire spread test was conducted by igniting a 0.63-MW power wood crib. To investigate the fire spread, the visual characteristics were observed, and the temperatures and heat radiation at special locations were monitored with thermocouples and radiation flowmeters, respectively. The effect of relative slope, heat radiation, and wind direction on fire spread characteristics was established by mathematical statistics, and the measured temperatures were used to verify the statistics' regularity. The results showed that in wooden houses, fire spread was mainly influenced by the slope, the distance between houses, and wind direction. When the inner wall of a wooden house is protected by a fireproof coating, the thermal radiation spread and fire spread are both slower. The slope and distance had the same influence weight (0.41) on fire spread; however, since they affect the process in different ways, they should be analyzed separately for fire risk evaluation. The findings of this study provide a theoretical foundation for understanding the fire spread process in wooden buildings.

Keywords: fire spread; full-scale experiment; wooden houses; fire behavior; flame temperature

Citation: Yuan, S.; Xiang, K.; Yan, F.; Liu, Q.; Sun, X.; Li, Y.; Du, P. Characteristics and Mechanism of Fire Spread between Full-Scale Wooden Houses from Internal Fires. *Buildings* **2022**, *12*, 575. https://doi.org/10.3390/buildings12050575

Academic Editors: Karim Ghazi Wakili and Jorge Manuel Branco

Received: 8 March 2022
Accepted: 28 April 2022
Published: 29 April 2022

1. Introduction

1.1. Introduction to Fire Spread of Wooden Houses

Due to China's long history and ethnic diversity, ancient wooden buildings with regional characteristics are widespread throughout the country [1–3]. These buildings have not only residential value but also are integrated into the local folklore, culture, history, economy, and natural scenery and have both historical and touristic value [4–6].

Figure 1 depicts one type of stilt-style architecture, which is a common style of ancient rural wooden buildings in southwest China. This type of house is often built near the mountains and is supported by wooden columns. Since this type of building is composed of stacked wood, they are at high risk of fire [7]. When uninterrupted fires occurred in ancient villages, the fire spread caused great damage to clustered wooden houses [8]. Deteriorated wooden buildings might have a greater possibility of catching fire when they contact open flame [9–11]. However, the current fire protection force in rural areas is still relatively weak, mainly due to a lack of public firefighting facilities, a lack of firefighting management, low firefighting awareness among villagers, and difficulties in external firefighting rescue [12]. Unlike wildfire spread in grass and forests, wooden houses are prone to igniting from the building's interior [13]. The internal fire process consists of ignition, flashover, full development, collapse, and extinguishment. This happens quickly, causes great damage, and harms people's safety and lives [14,15]. Traditional small-scale experiments cannot objectively and accurately reproduce the fire spread characteristics of wooden building clusters [5,16]. Therefore, it is important to adopt a full-scale experimental approach for fire spread research on wooden houses.

Figure 1. Model of an ancient wooden building in southwest China.

1.2. Literature for Wooden Houses Research

Several studies have investigated the fire spread characteristics of wooden building clusters in China and other countries. Hasemi [17] discussed the fire spread characteristics of three-story wood houses using a full-scale experimental approach, while Matthew et al. [18] found that fires of intermediate size were negatively associated with relative humidity. Zhang et al. [19] studied the thermo-mechanism behavior of wooden joints under fire exposure. Kristoffersen et al. [11] summarized improving the fire protection of wooden buildings with limited additional costs. Bartlett et al. [11] presented the pyrolysis, ignition, and combustion processes associated with wood products. Li et al. [20] conducted an in-depth study on fire safety planning and strategies for rural ancient building groups in Japan; moreover, Maraveas et al. [21] studied the factors influencing the fire resistance of wood-based houses. Stubbs et al. [22] found that flame height and surface area increased significantly with wood materials but also approached asymptotic values.

Until now, only a few full-scale studies have focused on the fire hazard of multiple wooden houses in China. Notably, the smoke concentration and the temperature distribution during fire spread are closely related to a series of conditions [23,24], such as the fire separation between adjacent buildings [25], the combustion structure [26], the relative slope [27], the roof temperature [28], the external wind speed at the time of fire [29], the moisture content of wood [30], and the atmospheric temperature [31,32]. Therefore, the main factors influencing fire spread need to be analyzed by actual fire tests [33,34]. Traditionally, fire spread tests have been based on small-sized models of wooden structures [35,36]. Although these studies have provided a basic theoretical understanding of the fire spread phenomenon, they focused less on wind speed and space temperature; therefore, their results were not completely realistic [37,38].

Fire spread in full-size buildings have been studied mainly considering the material [39], geometry [39], ventilation [40], boundary properties [26,41,42], and wood structures [43,44]. Their studies have systematically analyzed how these factors affected the spread of fire in full-scale buildings. However, these factors were analyzed by a simple variable method, and only a few studies have investigated the main factors influencing the fire spread process by statistical theory, which could further study the inter-correlations among these factors. Additionally, the tendencies of fire spread when these factors are simultaneously varied are not yet known. Thus, the aim of this study was to (i) investigate the fire spread in wooden houses when multiple factors simultaneously varied; (ii) analyze the main factors influencing fire spread by statistical theory; (iii) provide fire protection suggestions for wooden houses in the ethnic villages of China.

1.3. Brief Description of This Experiment

In this study, we studied the actual fire spread process in 4 full-scale wooden houses by igniting a 0.63-MW power wood crib. Four full-size wooden houses were built on a vacant lot in an ethnic village in Guizhou Province. The fire spread, temperature, and heat radiation were monitored, and the factors, i.e., relative distance, height, relative slope, and wind direction, influencing fire spread were investigated. Subsequently, the statistical software Statistical Product Service Solutions (SPSS) was used on the measured temperature and radiation; analysis of cluster and dimension reduction were used as the main methods. Finally, the analysis result by SPSS was verified by the detection results. The findings of this study will provide a theoretical foundation for understanding fire spread in wooden houses.

2. Materials and Methods

2.1. Construction and Test Environment of the Wooden Houses

The test site was located on a vacant lot characterized by yellow mud in a village in Southeastern Miao and Dong Autonomous Prefecture (Guizhou Province, China). A 6.0 m high platform, backed by a wooded area, was positioned north of the site (Figure 2a).

Figure 2. *Cont.*

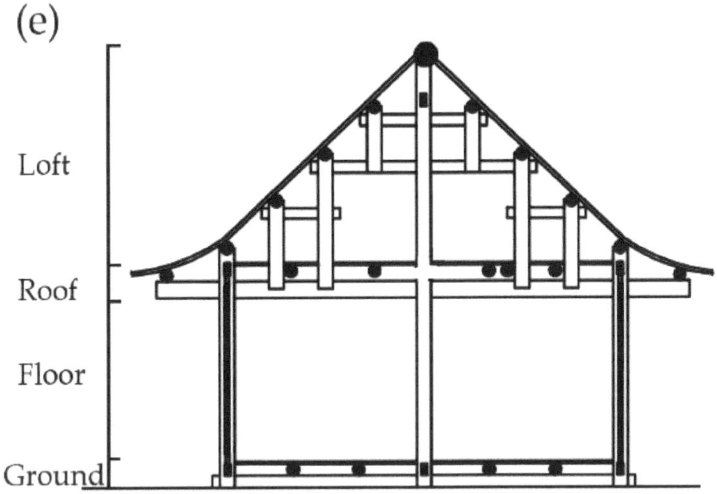

Figure 2. Construction process of the wooden houses and construction profile of the buildings. (**a**) Situation of test platform, (**b**) Prepared test wood, (**c**) Construction of buildings, (**d**) Completion of buildings, (**e**) Cross-section of the buildings' structure.

In order to improve the accuracy of the experiment, after field investigation, four representative adjacent wooden houses were selected as models for the experimental wooden houses. The building size, separation distance between houses, and architectural structure were based on the models. The wood used for the test construction was taken from a wooden building that had been inhabited for more than 10 years (Figure 2b). A total of 4 wooden buildings (1–4) were erected for the test. Their structural frames, roof covers, floor slabs, beams, and other structural elements were composed mainly of wood (Figure 2c,d). Figure 2d shows the side structure of one of the buildings. The space between the floor and the ceiling of the living space was 2.3 m, and the space between the ceiling and the roof (i.e., the loft) was 2.7 m. Notably, the loft was partially open (i.e., the building created a non-completely confined space); moreover, the rooms and the loft of all the buildings had the same height (Figure 2d). The four buildings were exactly made of the same materials. The walls were composed of two parts, outer and inner wood panel walls. In addition, Houses 1 and 4 had 2 floors (floor areas of 75.52 m^2 and 42.24 m^2, respectively), while Houses 2 and 3 had only 1 floor (floor areas of 20.67 m^2 and 42.24 m^2, respectively).

2.2. Arrangement of the Test Points

A layout of the experimental site and the arrangement of test points are shown in Figure 3. House 1 (i.e., the ignition building) had 2 floors, with 4 rooms on the first floor; notably, the rooms on the second floor were intercommunicating. For free burning conditions, these cribs would have had a maximum surface-controlled mass loss rate (MLR) of 0.39 kg/s combined, according to [19].

A total of 80 wood cribs (about 25.0 Kg) with the dimension of 250 mm × 25 mm × 25 mm (length × weight × height) were set as the fire source according to [15]. The heat of combustion of wood cribs used was 17 MJ/kg as measured by a bomb calorimeter, thus giving a maximum surface-controlled heat release rate of 0.63 MW.

The location of rooms 1, 2, 3, and 4 is shown in Figure 3. The center of room 1 was set as the ignition point. Temperature measuring thermocouples (type K, Omega, New York, NY, USA) were installed on the east, south, west, and north walls of room 1, with correspondent measuring points named D, N, X, and B, respectively. Additional temperature measuring thermocouples were installed at heights of 0.5, 1.0, 1.5, and 2.0 m from the floor and on the floor of room 1: the corresponding points were 1-1-1, 1-1-2, 1-1-3, 1-1-4, and

L2, respectively. In room 1, at the height of 1.0 m, we positioned another temperature measuring thermocouple (E-1). The lobby, rooms 2, 3, and 4 were equipped with temperature measuring thermocouples 1.0 m above the floor, named points T, 1-2, 1-3, and 1-4, respectively. Additional temperature measuring thermocouples, named points E-2 and E-3, were located 1.0 m above the ground on the second floor directly above the hall and 1-3. Another temperature measuring thermocouple, located in the middle of the staircase against the wall, was identified as point L. At the height of 1.0 m, water-cooled radiant heat flowmeters (radiant heat flow dual-use type, Shanghai Tuxin Company, Shanghai, China) were set at the front door, at a horizontal distance of 3.0 m and 6.0 m, respectively, from the front door, named points G1, G2, and G3. The total number of thermocouples and heat flowmeter were 31 and 5, respectively, as listed in Table A1.

Figure 3. Plan of wooden houses, showing the test point locations.

The distance between Houses 1 and 2 was 0.5 m (based on the position of the respective inner walls). In particular, House 2 was located to the east of House 1. In House 2, temperature measuring thermocouples were set at the height of 1.0 m from the base of the first floor, in correspondence to the inner and outer walls (points Q1 and Q2, respectively), as well as 1.0 m below the eaves (point R1). There was also a water-cooled radiant heat flow meter at the same location on the exterior wall Q1, named point G4. Notably, the interior wall of House 2 was sprayed with a halogen-free, efficient, and environmental fireproof paint (commercial products from Cuizhixin New Technology Development Co., Ltd., Suzhou, China), which has been independently fire-tested to a standard and applied to materials made of wood, paper, and plastic. Detailed information about the fireproof paint is listed in Appendix A. The dose of brushing the paint on the wall was 350 g/m^2, while the exterior wall was not sprayed.

House 3 was constructed to the west of House 1, and the distance between them (measured from the respective inner walls) was 1.0 m. The interior and exterior walls of House 3 were not treated with fireproof paint. Here, 1 temperature measuring thermocouple was set at the height of 1.0 m from the floor (point 3-1), while the other was 1.0 m below the eaves (point R2), respectively.

House 1 was attached to a 6.0 m platform on the north side. This platform had a slope inclination of about 80° and a slope length of 6.1 m. Six temperature measuring thermocouples were uniformly distributed along the straight slope to measure the fire spread temperature along its length. From bottom to top, the locations of the instruments were marked as S1, S2, S3, S4, S5, and S6. House 4, whose interior and exterior walls were not fireproof, was built on the platform. This house had two floors, each containing intercommunicating rooms. Temperature measuring thermocouples were set at the height of 1.0 m from the floor and 1.0 m below the eaves on the first and second floors, named 4-1, E-4, and R3 points, respectively. The exterior wall was equipped with a water-cooled radiant heat flow meter located at the height of 1.0 m from the floor (point G5).

2.3. Test Procedure

The test took place on 23 August 2020, between 11:00 and 13:00, the temperature ranged from 21 to 28 °C, and the humidity was 91%. The weather on that day was sunny, with an east-southeast wind at the speed of 2.4–5.4 m/s. At 11:30 a.m., all the windows and doors of House 1 were opened, and a fire was lit in the center of room 1 (Figure 3). A video camera was used to film the spread of the fire in all four houses; meanwhile, the temperature measuring thermocouples monitored the temperature at different points. To ensure the accuracy of the measurements, two measuring thermocouples were installed horizontally at each measurement point, and their results were averaged. Additionally, radiation flowmeters were installed at special locations, shown in Figure 4, to monitor the heat radiation power. In the middle and late stages of the test (20 min after ignition), a professional team was asked to extinguish the fire in Houses 2 and 4 to prevent the test-fire from getting out of control.

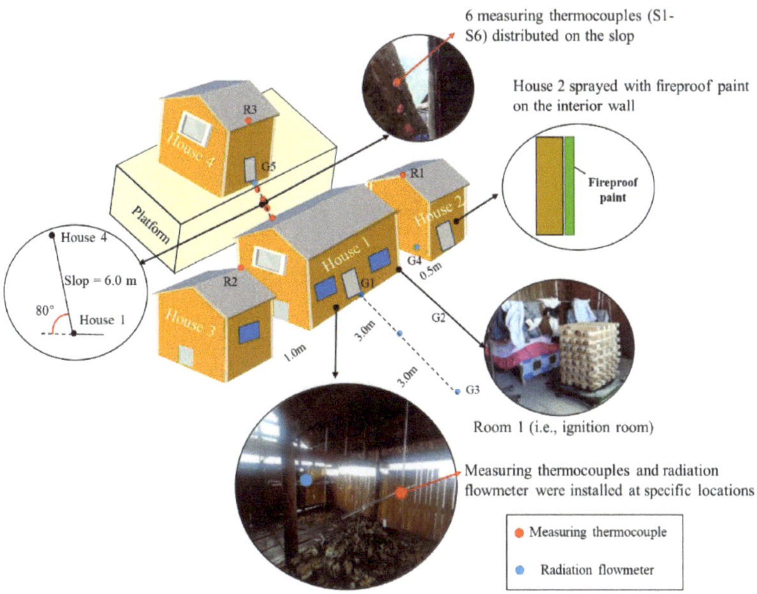

Figure 4. Diagram of the experimental model.

2.4. Statistical Analysis

As shown in Figure 4, the temperature and thermal radiation results obtained from the test were verified using the software IBM SPSS 25.0 (with the Python 2.2.17 extension package); statistical analysis was conducted on the detection results using the actual measured temperature and radiation during the whole experimental process. Following the previous method [45], cluster analysis and dimension reduction analysis was applied to analyze the correlation among measured temperatures, while the algorithm of the bootstrapping method was used to obtain the confidence intervals. The cluster analysis was based on the algorithm of K-means, while the dimension reduction analysis was based on the algorithm of principal component analysis.

3. Test Results and Discussion

3.1. Analysis of the Fire Spread Patterns in the Igniting House

3.1.1. Analysis of the Visual Characteristics

Figure 5 shows a series of photos taken from outside House 1 during the test. Fire ignition occurred at 0 min in room 1-1. At 3.1 min, smoke began to overflow from room 1-1. At 6.6 min, the floor of the room floor was lit, and the fire burned on the room's inner walls. At 8.9 min, a small amount of open fire spilled out of the window. At 10.2 min, a large number of open flames came out of the window and gradually went through the window to the second floor. At 10.8 min, the ceiling of room 1-1 collapsed, the fire quickly reached the second floor, and the beams were burned. At 12.3 min, the roof of House 1 was completely burned, and a large amount of smoke was produced. At 12.7 min, the foundation of the house (near room 1-1) was burned. At 16 min, only the main frame of House 1 was left. At 23 min, the house completely collapsed, and the walls and floors were completely consumed by the fire.

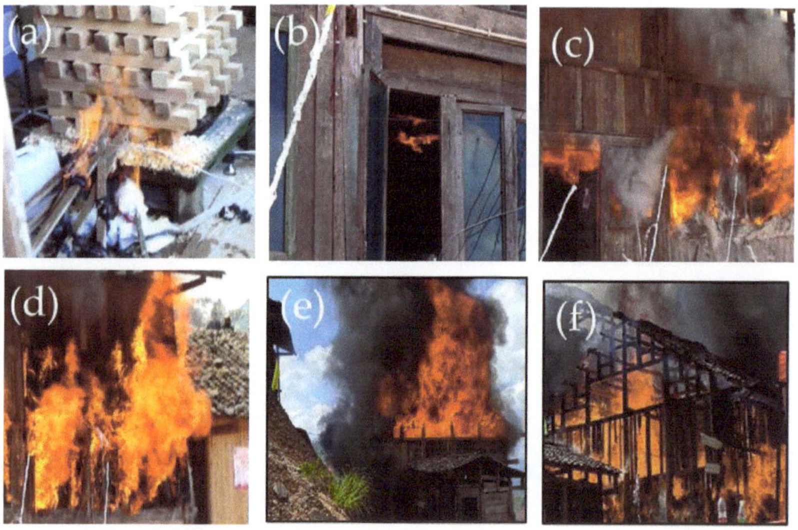

Figure 5. Visual observation of the fire spread phenomenon (House 1), at times of (**a**) 0 min, (**b**) 6.6 min, (**c**) 10.2 min, (**d**) 10.8 min, (**e**) 12.3 min, and (**f**) 16 min.

3.1.2. Temperature Changing Rule

Figure 6 shows the temperature of each test point in room 1. Figure 6a shows how, before 10.00 min, the detection height and the room temperature were significantly correlated, and the highest temperature was registered at point 1-1-4. Between 10–12.5 min, the temperature increased rapidly at all 4 detection points in room 1. At 12.5 min, the temperature at points 1-1-2, 1-1-3, and 1-1-4 was similar and slightly higher than that at point 1-1-1; the correlation between height and room temperature decreased (in accordance with the observations reported by Huang et al. [46]). Figure 6b shows how, before 10 min, the temperature at the measurement points on all 4 walls of room 1 showed gentle change. At 10 to 11 min, the temperature of all points increased significantly, showing the room had reached flashover. This phenomenon may be due to the amount of oxygen provided by the vacancy structure of lofts in ancient wooden buildings; a similar phenomenon could be found in a previous study [15]. The temperature began to rise at 12.5 min at point B and at 14 min at points N and D. After 15 min, all 4 points reached a stable temperature; the highest values were measured at point X (west wall), followed by point B (north wall) and points N and D (south and east walls, respectively). Compared with the other three walls, the west wall showed the earliest temperature rise and the highest final temperature; in comparison, the north wall showed a slightly later temperature rise and a slightly lower final temperature. This difference was related to the non-completely confined nature of the experimental wood-frame building space. The outdoor east-southeast wind will, in fact, enter the interior of a non-completely confined building, affecting the fire spread direction. In our test, after 10 min from fire ignition, the ventilation direction significantly affected the fire spread direction.

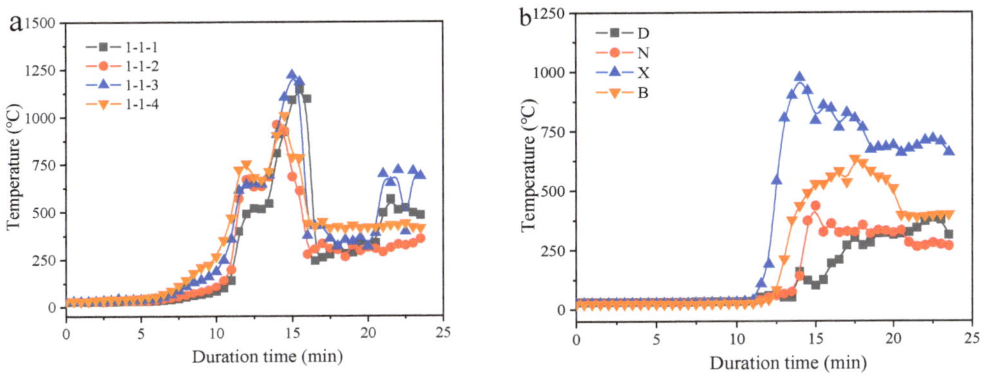

Figure 6. Correlation among (**a**) relative height, (**b**) direction, and temperature in room 1 of House 1. (**a**) Temperatures measured at the height of 0.5–2.0 m from the room floor, (**b**) Temperatures measured on the four walls of the room.

Figure 7 shows the temperatures registered at the height of 1.0 m from the floor in each room of House 1 during the fire spread. The increase in temperatures followed the order of 1-1-2 (room 1) > T (hall) = L (stairs) = L2 (floor) = 1-2 (room 2) > 1-4 > 1-3. Specifically, the point 1-1-2 increased rapidly at 9.5 min, and T, L, L2, and 1-2 increased at 11.5 min, while 1-4 and 1-3 increased at 12 min and 12.5 min. The staircase was at a certain distance from room 1, and it heated up at the same time as the hall, possibly because the staircase was connected to the second floor and had sufficient air (not a completely airtight loft). After 11 min, the fire spread phenomenon was influenced by the direction of the wind coming from outside. In addition, Room 3 was located northwest of the burning room and intercepted the wind; here, the fire spread and heated up rapidly. Finally, in Room 4 (to the west of the burning room), the fire spread and heated up slowly. It showed that the order of room heating was similar to the ventilation direction. However, the order of increased temperature could not verify this assumption; thus, the measured temperatures were analyzed using SPSS.

3.1.3. Principal Component Analysis and Validation

Figure 8 and Table 1 show the correlation of the temperatures during the whole experimental process at each detection point in the building where the fire started. The confidence intervals of measured temperatures are listed in Table A2. In particular, Figure 8a shows that the detection points on the first floor of the ignition building can be classified into two clusters; inside the ignition building, fire spread was mainly influenced by two factors. These two factors were further analyzed (Figure 8b and Table 1). Component 1 (Factor 1) and Component 2 (Factor 2) likely represented the wind direction (influence weight = 0.47) and the relative height (influence weight = 0.43), respectively. Combining this information with that in Figures 6–8, it was speculated that the relative height and the wind direction significantly influenced both the fire spread phenomenon and the temperature change in the burning building. Similar studies have investigated fire spread in forests and found that 76% of fire behavior is linked to wind direction, wind speed, altitude (height), and other factors [46]. Specifically, before the fire started (at 12 min), the correlation between fire spread and the relative height was greater than that between the wind direction and the spread direction, and relative height and spread rate increased together; however, after the fire started (11 min), the correlation between the wind direction and the spread direction was greater than that between fire spread and the relative height.

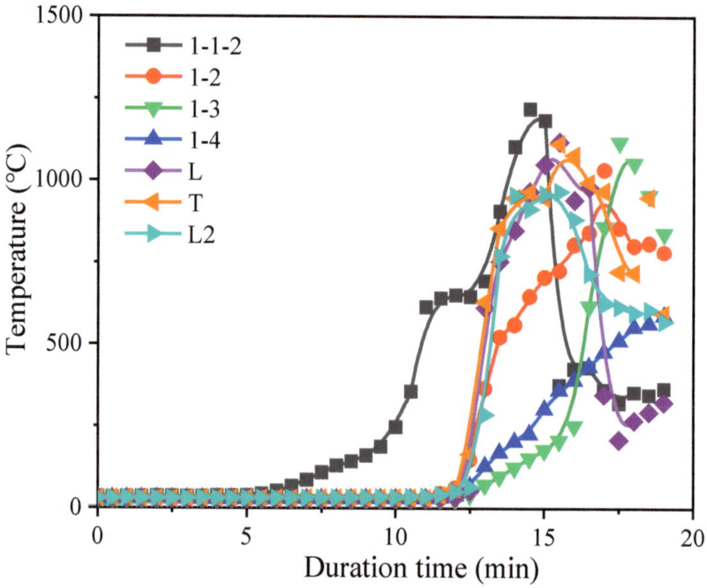

Figure 7. Variation of temperature at the test points on the first floor of House 1.

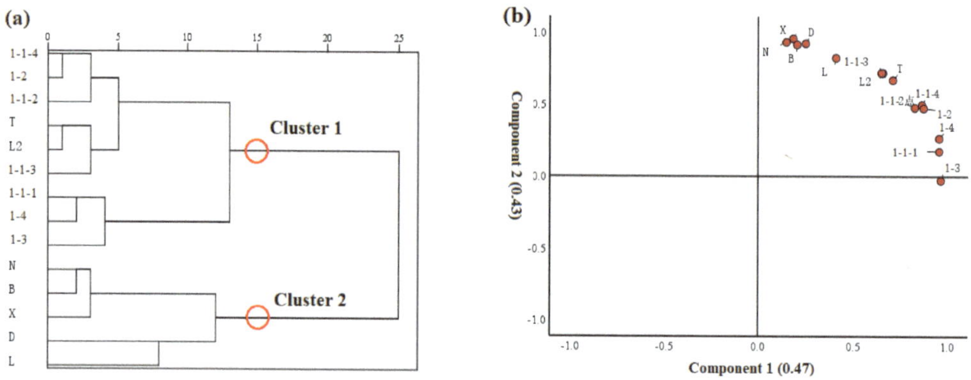

Figure 8. Statistical correlation between the temperatures at different test points in House 1. (**a**) Cluster analysis, (**b**) Dimension reduction analysis.

Table 1. Dimension reduction analysis of temperature at the test points in House 1 during the whole experiment process.

Point Number	Principal Component 1	Principal Component 2
1-3	0.966	
1-1-1	0.957	
1-4	0.956	
1-2	0.873	0.475
1-1-4	0.864	0.497
1-1-2	0.825	0.478
T	0.709	0.669
X		0.955
N		0.930
D		0.921
B		0.914
L	0.408	0.821
L2	0.648	0.719
1-1-3	0.657	0.718

Note: The principal component matrix converged after three iterations by rotation.

Figure 9 shows the temperature results obtained at the measurement points on the second floor of House 1. The point points E-1 (directly above room 1) and E-2 (directly above the hall) almost simultaneously rose. Compared with E1, E2 was far away from the fire source, which was supposed to warm up early. The simultaneous warming indicated that both the wind direction and height strongly influenced the fire spread.

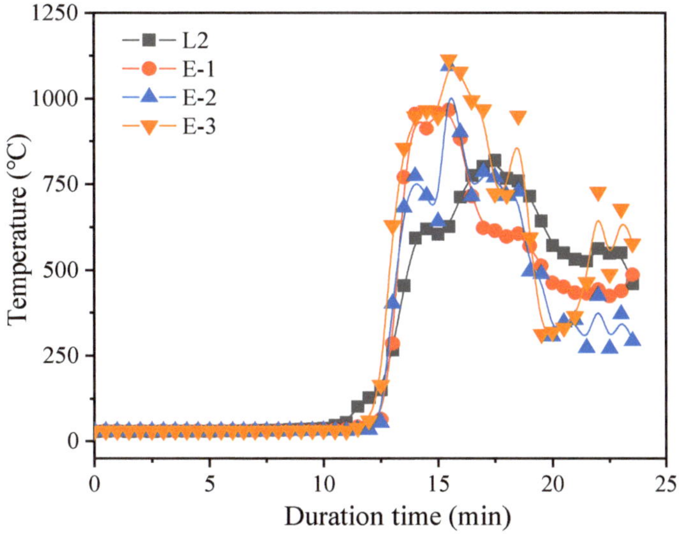

Figure 9. Variations of temperature at the test points on the second floor of House 1.

The above results show that fireproofing measures need to be taken on the ceilings of wooden buildings to reduce the influence of relative height. In addition, placing windows in the area of the perennial prevailing wind direction should be limited.

3.2. Analysis of Fire Spread Rule in Wooden Houses

3.2.1. Analysis of the Visual Characteristics

Figure 10 contains photos taken from outside the buildings during the fire spread. Between 10 and 12 min, House 1 was burning in its entirety (large amounts of fire and smoke were visible from the outside), while Houses 2–4 showed no signs of ignition.

Figure 10. Fire spread among the Houses 1-4 (**a**,**b**), of which (**a1**,**b1**,**c1**,**d1**) were during the time of 10–12 min, (**a2**,**b2**,**c2**,**d2**) were at the time of 16 min, (**a3**,**b3**,**c3**,**d3**) were at the time of 19 min, and (**a4**,**b4**,**c4**,**d4**) were during the time of 23–33 min.

At 16 min, only the foundation structure of House 1 was left; moreover, the west eaves of House 2 and the south eaves of House 4 had ignited. At 19 min, the foundation of House 1 was burning, the west eaves of House 2 were burning, the east eaves of House 3 had ignited, and the south exterior wall and eaves of House 4 were burning. At 20 min, firefighting operations were carried out on Houses 2 and 4. In the final part of the test (between 23 and 33 min), Houses 1 and 3 completely collapsed, the exterior walls and eaves of House 2 were completely burned (with no obvious changes in the interior), and the south exterior wall and eaves of House 4 were completely burned.

3.2.2. Effect of Fireproof Materials

The correlations between temperatures at points 1-1 (room 1), 1-2 (room 2), T (hall) (House 1), Q1 (exterior wall), and Q2 (interior wall) (1 m above the floor in House 2) are shown in Figure 11. In the case of House 1, the p-value between the temperatures of room 1 and room 2 (or the hall) was >0.05. Notably, there was a low significant relationship between the temperatures of the 2 tested points and the external wall of House 2 ($0.01 < p < 0.05$). The p-value between the temperature of room 1 (or the internal wall of House 2) and the internal wall of House 2 was instead <0.01. The above results proved that the fireproof paint sprayed on the inner wall of House 2 had a good fireproof effect, effectively slowing down the heating rate and the spread of fire.

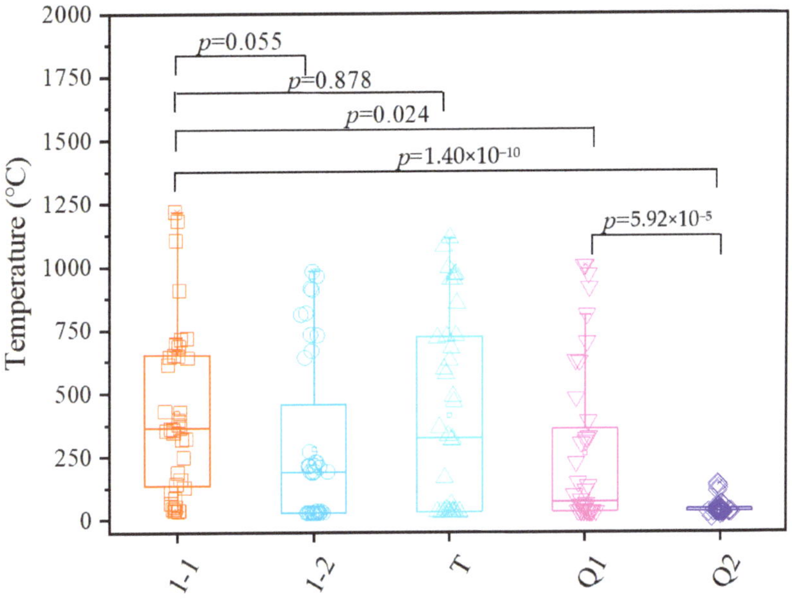

Figure 11. Correlations between temperatures registered in Houses 1 and 2.

3.2.3. Effect of Slope

The temperatures measured at points S1–S6 (along the slope between House 1 and House 4) are shown in Figure 12. At points 1–4, the temperature did not significantly change between 0 and 14 min; however, it started to rise from 14 min onwards. At points 5–6, the temperature gradually increased between 0 and 14 min, but it was higher at point 6; then, from 14 min onwards, it rose faster. The calculation followed Equations (1) and (2) and showed that F and R were 1.79 and 0.23 km/h, respectively. Thus, the rate of forward spread of fire on level to undulating ground was 0.233 km/h. The above results indicate that the slope had the greatest effect on fire spread. Before the fire spreads, a large amount of fire and smoke spread upward along the slope due to thermal pressure and the wall-hugging effect. This explains why the highest points (5 and 6) warmed up first.

$$F = 2.0 \times e^{(-23.6+5.01\times \ln(C)+0.0281\times T-0.226\sqrt{H}+0.633\sqrt{V})} \quad (1)$$

$$R = 0.13 \times F \quad (2)$$

where F was the fire danger index; C was the degree of curing (percent), 89%; T was the air temperature (°C), 24.5 °C; H was the relative humidity (percent), 91%; V was the average wind velocity in the open at the height of 10 m (km/h), 14.04 km/h; R was the rate of forward spread of fire on level to undulating ground.

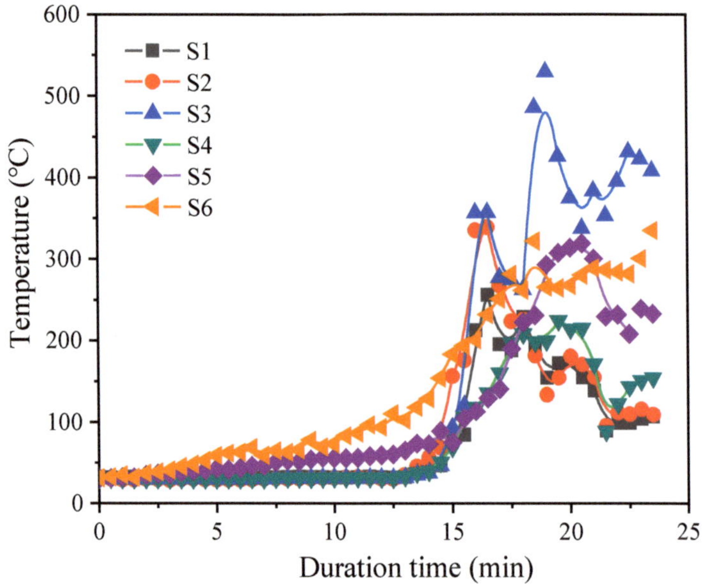

Figure 12. Temperatures registered at different test points along the slope.

3.2.4. Effect of Distance

The thermal radiation results obtained from points G1–G5 are shown in Figure 13. At point G1, thermal radiation started to increase at 8 min; then, from 10 min, it increased exponentially. Boom combustion occurred due to the non-completely confined configuration of the space fueling the combustion. Meanwhile, thermal radiation began to rise also at G2 and G3; however, the amount of thermal radiation at G2 was twice as high as that at G3. These results are consistent with those of Nishino et al. [47]; thermal radiation decreased as the distance between the monitoring point and the heat center increased. At G4 and G5, thermal radiation started to increase at 12 min and reached its maximum (9.53 kW m^{-2} and 15.29 kW m^{-2}, respectively) at 14 min. The low thermal radiation of House 2 was related to the relative height and slope compared to House 4.

3.2.5. Principal Component Analysis Results and Validation

Figure 14 and Table 2 show the temperatures registered in Houses 1, 2, 3, and 4 and elaborated using the SPSS software, and the confidence intervals of measured temperatures are listed in Table A3. Figure 15a shows that the test points could be classified into three clusters, indicating that the fire spread phenomenon in the wooden houses was mainly influenced by three factors.

These three factors were further analyzed (Figure 15b and Table 1). Component 1 (Factor 1, identified as the slope) had an influence weight of 0.41, Component 2 (Factor 2, identified as the distance) had an influence weight of 0.41, and Component 3 (Factor 3, identified as the wind direction) had an influence weight of 0.08. Thus, distance and slope had the same influence weights on fire spread in the wooden houses, but thermal radiation was lower at G5 (where the slope was high) than at G3 (where the slope was low) (Figure 13). This indicates that, despite the high correlation between slope and distance, these two parameters influenced fire spread in different ways. In the fire risk assessment of wooden houses, safety analyses and calibrations of both distance and slope are required. Fireproofing materials achieve fire protection by slowing down heat radiation (Figure 11). Compared to slope and distance, wind direction had a lower influence weight in the test.

This may have been due to the absence of adjacent buildings to the northwest (downwind side) of the burning building.

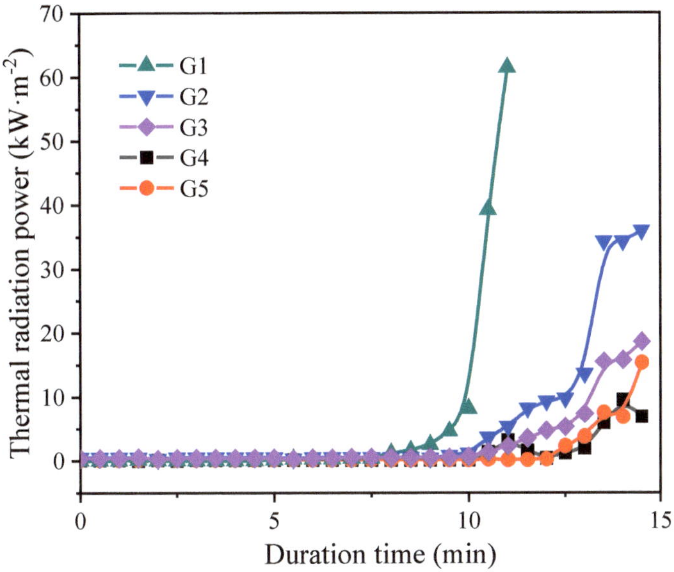

Figure 13. Thermal radiation at the test points.

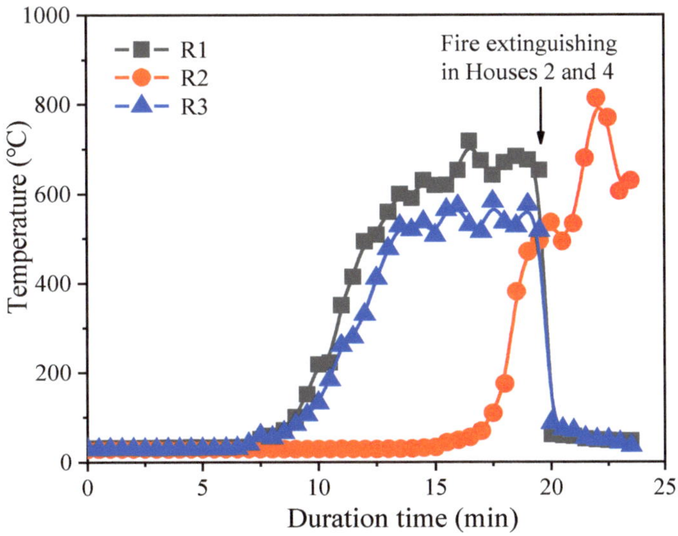

Figure 14. Building eaves temperature of House 2–4.

Table 2. Dimension reduction analysis results for temperature and thermal radiation at the test points.

Point Number	Principal Component 1	Principal Component 2	Principal Component 3
G2	0.952		
G4	0.947		
G1	0.945	0.315	
G3	0.920	0.320	
Q2	0.806	0.567	
Q1	0.739	0.653	
1-2	0.723	0.606	
G5	0.456		
S5		0.932	
S3		0.893	0.324
S4	0.333	0.871	
1-1	0.496	0.852	
S6	0.321	0.834	
S2	0.506	0.769	
T	0.540	0.741	
S1			0.944

Note: The principal component matrix converged after five iterations by rotation.

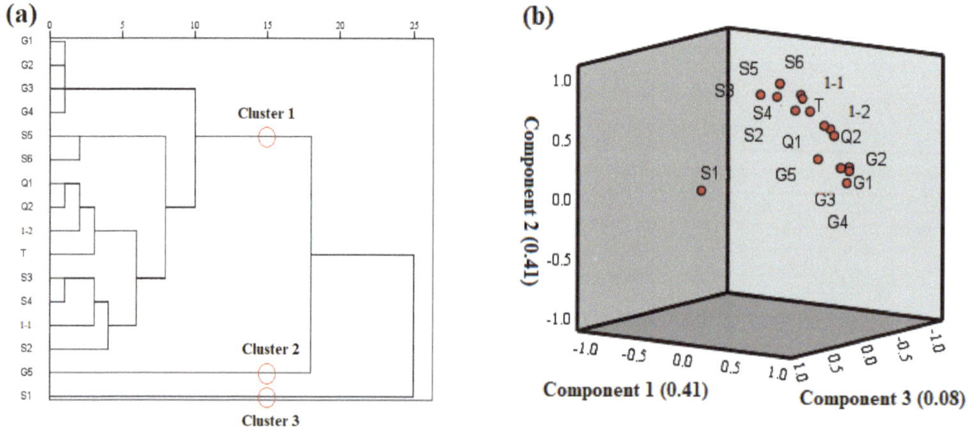

Figure 15. Results of the statistical correlations between temperature and thermal radiation at the whole test points. (**a**) Cluster analysis, (**b**) Dimension reduction analysis.

Since floors 2, 3, and 4 were first lit in correspondence with the eaves, we monitored the temperature at these locations (Figure 15). The temperatures of both R2 (eaves of House 2) and R4 (eaves of House 4) started to increase rapidly from 7.5 min; then, between 13 and 20 min, they remained relatively stable (averages of 650 and 550 °C, respectively). Finally, they started to decrease after 20 min due to fire extinguishing.

The temperature at R3 (eaves of House 3) started to rise at 15 min, reaching an average value of 800 °C at 21.5 min. Notably, R2 was ignited earlier due to its closer proximity to the fire source and the higher heat radiation compared to R4. Additionally, R4 was ignited earlier than R3 due to the higher slope of Houses 2 and 4. R2 and R4 were ignited almost simultaneously, indicating that distance and slope had a similar influence weight on fire spread. The temperatures shown in Figure 14 confirmed that distance and slope were the main factors influencing fire spread in the wooden houses, which was consistent with previous studies [48].

Therefore, while fire safety and cost should be simultaneously considered in a project, the fire damper and fireproof coating were preferentially chosen since the main factors are distance and slope. Additionally, further expansion of fire prevention space needs to be considered in rural planning due to the coupling effect of multi-factors.

4. Conclusions

The following conclusions were inferred from our experimental study.

(1) Fire spread was significantly correlated with relative height and wind direction. Before the fire started at 12 min, the correlation between height and fire spread was great (influence weight of 0.42); the higher the relative height, the greater the spread rate. At 11 min after the fire started, the correlation between wind direction and fire spread was great (influence weight of 0.47); the wind direction influenced the spread direction.

(2) The fire spread between adjacent buildings was mainly influenced by the distance, wind direction, and slope; specifically, the rate of forward spread of fire increased to 0.06 m/s when at the slope of 80°.

(3) When fire risk assessments are conducted, the relative slope and the distance between buildings need to be calibrated and analyzed separately.

(4) In this study, the findings provide a theoretical foundation for understanding the fire spread characteristics of wooden buildings. Further experiments should set varied wind velocities and directions to compare the influence weights of slope, distance, and wind properties.

Author Contributions: S.Y. and Y.L. were responsible for the experimental work, K.X., F.Y. and X.S. supervised the laboratory work, Q.L. and P.D. led the research. All authors have read and agreed to the published version of the manuscript.

Funding: This research and APC was funded by the National Key R&D Program Project [2018YFD1100403], and Key Technology Research on Fire Inspection for Construction Project Completion and Acceptance [2021-K-029].

Conflicts of Interest: The authors declare no conflict of interest.

Appendix A

Appendix A.1 Previous Fire Spread Tests for Fireproof Paint

Appendix A.1.1 In the Treatment of Chinese Fir Board (Nominal Density of 500 kg/m^3, Nominal Density of 15 mm)

The dose of brushing the paint was 70 kg/m^3 or 350 g/m^2. After brushing, the pressure of 1 bar on board lasted for 10 min, followed by 12–14 bar for 150 min. After fire spread test, the woods with the paint brushing had a better property of fireproof than that without brushing. In addition, the board after painting had the corrosion protein.

Appendix A.1.2 In the Treatment of Other Materials

The dose of brushing the paint when the materials:

Bafta: 0.15 g paint per gram
Wool fabric: 0.12 g paint per gram
Polyamide: 0.12 g paint per gram
Polyester: 0.27 g paint per gram
Paper: 0.06 g paint per gram

Table A1. Description of site of measured thermocouple.

Location		Number of Thermocouples	Sum
Inside the House 1	Room 1	D, N, X, B, 1-1-1, 1-1-2, 1-1-3, 1-1-4	8
	Room 2	1-2	1
	Room 3	1-3	1
	Room 4	1-4	1
	Hall	T	1
	Staircase	L	1
	Second floor	L2, E1, E2, E3	4
Outside the House 1 and House 4		S-1, S-2, S-3, S-4, S-5, S-6	6
Inside the House 2		Q2, 2-1, R1	3
On the wall of House 2		Q1	1
Inside the House 3		3-1, R2	2
Inside the House 4		4-1, R3	2
Total numbers			31

Appendix A.2 Principal Component Analysis and Validation

Table A2. Sample description [a].

Points		Statistics	Deviation	Standard Error	Inferior Limit	Upper Limit
R112	Average	356.63	0.56	75.71	212.80	507.03
	Standard deviation	371.01	−11.06	53.512	240.99	457.53
	Case number	23	0	0	23	23
R12	Average value	132.05	0.98	46.17	47.68	232.26
	Standard deviation	225.76	−10.89	54.87	61.47	299.22
	Case number	23	0	0	23	23
T	Average value	219.85	1.40	71.73	86.43	375.76
	Standard deviation	355.68	−11.51	61.94	195.57	434.16
	Case number	23	0	0	23	23
1	Average value	44.79	0.07	4.69	36.65	54.88
	Standard deviation	22.84	−1.14	5.49	9.75	31.69
	Case number	23	0	0	23	23
Q2	Average value	30.06	0.01	0.29	29.55	30.72
	Standard deviation	1.43	−0.08	0.39	0.44	2.01
	Case number	23	0	0	23	23
G4	Average value	0.29	0.01	0.14	0.10	0.61
	Standard deviation	0.67	<-0.10	0.31	0.03	1.06
	Case number	23	0	0	23	23
G5	Average value	0.12	0.01	0.01	0.11	0.13
	Standard deviation	0.02	<-0.01	0.01	0	0.04
	Case number	23	0	0	23	23
G1	Average value	5.24	0.05	3.03	0.56	12.21
	Standard deviation	14.74	−1.52	5.75	1.08	22.58
	Case number	23	0	0	23	23

Table A2. *Cont.*

	Points	Statistics	Deviation	Standard Error	Inferior Limit	Upper Limit
G2	Average value	0.84	<0.01	0.24	0.47	1.40
	Standard deviation	1.18	−0.12	0.45	0.10	1.80
	Case number	23	0	0	23	23
G3	Average value	0.52	<0.01	0.09	0.39	0.73
	Standard deviation	0.44	−0.05	0.17	0.07	0.68
	Case number	23	0	0	23	23
S1	Average value	31.72	−0.01	0.15	31.41	32.00
	Standard deviation	0.72	−0.02	0.10	0.50	0.88
	Case number	23	0	0	23	23
S2	Average value	29.79	<0.01	0.20	29.43	30.21
	Standard deviation	0.99	−0.03	0.14	0.64	1.22
	Case number	23	0	0	23	23
S3	Average value	30.08	<−0.01	0.14	29.80	30.37
	Standard deviation	0.70	−0.02	0.07	0.52	0.80
	Case number	23	0	0	23	23
S4	Average value	30.25	<0.01	0.12	30.02	30.50
	Standard deviation	0.57	−0.01	0.06	0.42	0.66
	Case number	23	0	0	23	23
S5	Average value	43.40	<0.01	1.81	39.82	47.07
	Standard deviation	8.93	−0.22	0.82	7.06	10.28
	Case number	23	0	0	23	23
S6	Average value	55.53	<0.01	3.42	48.83	62.24
	Standard deviation	16.97	−0.45	1.80	12.80	19.92
	Case number	23	0	0	23	23
R1	Average value	74.48	0.15	17.02	44.37	111.60
	Standard deviation	83.25	−4.99	22.00	28.67	115.98
	Case number	23	0	0	23	23
R2	Average value	28.62	<0.01	0.04	28.53	28.70
	Standard deviation	0.21	−0.01	0.05	0.12	0.28
	Case number	23	0	0	23	23
R3	Average value	60.51	0.13	12.11	39.22	87.05
	Standard deviation	59.00	−3.59	16.16	20.77	84.62
	Case number	23	0	0	23	23

[a.] Unless otherwise noted, Self-help sampling results were based on 1000 samples, and the confident interval was 95%.

Appendix A.3 Principal Component Analysis and Validation

Table A3. Sample description [a].

	Points	Statistics	Deviation	Standard Error	Inferior Limit	Upper Limit
D	Average	262.31	−2.65	53.74	165.11	368.37
	Standard deviation	335.60	−10.61	50.20	220.97	408.14
	Case number	39	0	0	39	39
N	Average value	250.54	−2.28	44.26	166.79	339.38
	Standard deviation	279.74	−7.99	33.04	201.12	330.23
	Case number	39	0	0	39	39
X	Average value	314.05	−2.77	54.76	210.00	427.83
	Standard deviation	344.22	−11.10	47.37	235.27	415.66
	Case number	39	0	0	39	39
B	Average value	324.40	−2.03	47.03	230.22	418.31
	Standard deviation	297.74	−7.38	27.38	230.42	339.12
	Case number	39	0	0	39	39
R111	Average value	86.48	−0.34	14.33	59.12	114.66
	Standard deviation	90.44	−2.20	12.46	60.02	109.63
	Case number	39	0	0	39	39
R112	Average value	117.78	−1.00	22.48	75.21	162.56
	Standard deviation	139.10	−3.23	14.52	101.41	158.28
	Case number	39	0	0	39	39
R113	Average value	310.65	−2.75	59.01	196.78	429.91
	Standard deviation	375.38	−6.98	23.74	311.73	404.59
	Case number	39	0	0	39	39
R114	Average value	187.86	−1.61	38.05	114.45	259.38
	Standard deviation	239.63	−4.34	19.48	189.59	264.50
	Case number	39	0	0	39	39
R12	Average value	266.33	−2.04	54.68	163.47	370.59
	Standard deviation	344.25	−6.09	30.72	267.89	387.57
	Case number	39	0	0	39	39
R14	Average value	145.88	−0.97	29.93	89.22	203.47
	Standard deviation	188.68	−4.27	22.72	132.36	221.88
	Case number	39	0	0	39	39
R13	Average value	184.71	−0.78	51.86	90.33	294.16
	Standard deviation	323.33	−9.97	57.55	184.45	413.93
	Case number	39	0	0	39	39
L	Average value	243.00	−2.88	55.53	134.22	358.23
	Standard deviation	360.46	−9.12	46.43	249.06	426.73
	Case number	39	0	0	39	39
T	Average value	318.63	−3.34	64.80	194.88	446.98
	Standard deviation	415.31	−7.93	32.63	335.43	461.11
	Case number	39	0	0	39	39

Table A3. *Cont.*

	Points	Statistics	Deviation	Standard Error	Inferior Limit	Upper Limit
L2	Average value	264.16	−3.05	55.37	157.41	377.83
	Standard deviation	350.73	−7.89	33.36	268.41	396.39
	Case number	39	0	0	39	39

[a] Unless otherwise noted, Self-help sampling results were based on 1000 self-help samples, and the confident interval was 95%.

References

1. Chen, L.; Tang, F.; Pang, H. Ceiling heat flux and downward received radiation heat flux induced by weak and relative strong fire plume in ventilation tunnels. *Appl. Therm. Eng.* **2020**, *169*, 114924. [CrossRef]
2. Bedon, C.; Fragiacomo, M. Fire Resistance of In-Plane Compressed Log-House Timber Walls with Partial Thermal Insulation. *Buildings* **2018**, *8*, 131. [CrossRef]
3. Khidmat, R.P.; Fukuda, H. Kustiani Design Optimization of Hyperboloid Wooden House Concerning Structural, Cost, and Daylight Performance. *Buildings* **2022**, *12*, 110. [CrossRef]
4. Chorlton, B.; Gales, J. Fire performance of cultural heritage and contemporary timbers. *Eng. Struct.* **2019**, *201*, 109739. [CrossRef]
5. Lange, D.; Boström, L.; Schmid, J.; Albrektsson, J. The Reduced Cross Section Method Applied to Glulam Timber Exposed to Non-standard Fire Curves. *Fire Technol.* **2015**, *51*, 1311–1340. [CrossRef]
6. Allaire, F.; Mallet, V.; Filippi, J.-B. Novel method for a posteriori uncertainty quantification in wildland fire spread simulation. *Appl. Math. Model.* **2021**, *90*, 527–546. [CrossRef]
7. Zhao, S. GisFFE—An integrated software system for the dynamic simulation of fires following an earthquake based on GIS. *Fire Saf. J.* **2010**, *45*, 83–97. [CrossRef]
8. Huang, X.; Sun, J.; Ji, J.; Zhang, Y.; Wang, Q.; Zhang, Y. Flame spread over the surface of thermal insulation materials in different environments. *Chin. Sci. Bull.* **2011**, *56*, 1617–1622. [CrossRef]
9. Johnson, M.C.; Kennedy, M.C.; Harrison, S.C.; Churchill, D.; Pass, J.; Fischer, P.W. Effects of post-fire management on dead woody fuel dynamics and stand structure in a severely burned mixed-conifer forest, in northeastern Washington State, USA. *For. Ecol. Manag.* **2020**, *470*, 118190. [CrossRef]
10. Jones, N.; Peck, G.; McKenna, S.T.; Glockling, J.L.D.; Harbottle, J.; Stec, A.A.; Hull, T.R. Burning behaviour of rainscreen façades. *J. Hazard. Mater.* **2021**, *403*, 123894. [CrossRef] [PubMed]
11. Kristoffersen, M.; Log, T. Experience gained from 15 years of fire protection plans for Nordic wooden towns in Norway. *Saf. Sci.* **2021**, *146*, 105535. [CrossRef]
12. Martín-Garín, A.; Millán-García, J.A.; Terés-Zubiaga, J.; Oregi, X.; Rodríguez-Vidal, I.; Baïri, A. Improving Energy Performance of Historic Buildings through Hygrothermal Assessment of the Envelope. *Buildings* **2021**, *11*, 410. [CrossRef]
13. Himoto, K.; Suzuki, K. Computational framework for assessing the fire resilience of buildings using the multi-layer zone model. *Reliab. Eng. Syst. Saf.* **2021**, *216*, 108023. [CrossRef]
14. Cicione, A.; Walls, R.; Sander, Z.; Flores, N.; Narayanan, V.; Stevens, S.; Rush, D. The Effect of Separation Distance Between Informal Dwellings on Fire Spread Rates Based on Experimental Data and Analytical Equations. *Fire Technol.* **2021**, *57*, 873–909. [CrossRef]
15. Cicione, A.; Walls, R.; Kahanji, C. Experimental study of fire spread between multiple full scale informal settlement dwellings. *Fire Saf. J.* **2019**, *105*, 19–27. [CrossRef]
16. Friedlander, S.K. *Smoke, Dust and Haze: Fundamentals of Aerosol Behavior*; Wiley-Interscience: New York, NY, USA, 1977; p. 317.
17. Hasemi, Y. Full-Scale Burn Test of Wooden Three-Story Apartment Building. *Fire Sci. Technol.* **1997**, *17*, 78–92. [CrossRef]
18. Slocum, M.G.; Beckage, B.; Platt, W.J.; Orzell, S.L.; Taylor, W. Effect of Climate on Wildfire Size: A Cross-Scale Analysis. *Ecosystems* **2010**, *13*, 828–840. [CrossRef]
19. Zhang, J.; Wang, Y.; Li, L.; Xu, Q. Thermo-mechanical behaviour of dovetail timber joints under fire exposure. *Fire Saf. J.* **2019**, *107*, 75–88. [CrossRef]
20. Li, M.; Hasemi, Y.; Nozoe, Y.; Nagasawa, M. Study on strategy for fire safety planning based on local resident cooperation in a preserved historical mountain village in Japan. *Int. J. Disaster Risk Reduct.* **2021**, *56*, 102081. [CrossRef]
21. Maraveas, C.; Miamis, K.; Matthaiou, C.E. Performance of Timber Connections Exposed to Fire: A Review. *Fire Technol.* **2013**, *51*, 1401–1432. [CrossRef]
22. Stubbs, D.C.; Humphreys, L.H.; Goldman, A.; Childtree, A.M.; Kush, J.S.; Scarborough, D.E. An experimental investigation into the wildland fire burning characteristics of loblolly pine needles. *Fire Saf. J.* **2021**, *126*, 103471. [CrossRef]
23. Zekri, N.; Zekri, L.; Lallemand, C.; Pizzo, Y.; Kaiss, A.; Clerc, J.; Porterie, B. Fire spread and percolation in polydisperse com-partment structures. *J. Phys. Conf. Ser.* **2012**, *395*, 12010. [CrossRef]
24. Bilyaz, S.; Buffington, T.; Ezekoye, O.A. The effect of fire location and the reverse stack on fire smoke transport in high-rise buildings. *Fire Saf. J.* **2021**, *126*, 103446. [CrossRef]

25. Gerzhova, N.; Blanchet, P.; Dagenais, C.; Côté, J.; Ménard, S. Heat Transfer Behavior of Green Roof Systems Under Fire Condition: A Nu-merical Study. *Buildings* **2019**, *9*, 206. [CrossRef]
26. Suzuki, S.; Manzello, S.L. Understanding structure ignition vulnerabilities using mock-up sections of attached wood fencing assemblies. *Fire Mater.* **2019**, *43*, 675–684. [CrossRef]
27. Zhang, X.; Hu, L.; Sun, X. Temperature profile of thermal flow underneath an inclined ceiling induced by a wall-attached fire. *Int. J. Therm. Sci.* **2019**, *141*, 133–140. [CrossRef]
28. Zhang, X.; Tao, H.; Zhang, Z.; Liu, J.; Liu, A.; Xu, W.; Liu, X. Flame extension area of unconfined thermal ceiling jets induced by rectangular-source jet fire impingement. *Appl. Therm. Eng.* **2018**, *132*, 801–807. [CrossRef]
29. Anderson, K.; Reuter, G.; Flannigan, M.D. Fire-growth modelling using meteorological data with random and systematic per-turbations. *Int. J. Wildland Fire* **2007**, *16*, 174. [CrossRef]
30. Anderson, K. A climatologically based long-range fire growth model. *Int. J. Wildland Fire* **2010**, *19*, 879–894. [CrossRef]
31. Ding, L.; Ji, J.; Khan, F.; Li, X.; Wan, S. Quantitative fire risk assessment of cotton storage and a criticality analysis of risk control strategies. *Fire Mater.* **2020**, *44*, 165–179. [CrossRef]
32. Lönnermark, A.; Ingason, H. Fire Spread and Flame Length in Large-Scale Tunnel Fires. *Fire Technol.* **2006**, *42*, 283–302. [CrossRef]
33. Ciri, U.; Garimella, M.M.; Bernardoni, F.; Bennett, R.L.; Leonardi, S. Uncertainty quantification of forecast error in coupled fire–atmosphere wildfire spread simulations: Sensitivity to the spatial resolution. *Int. J. Wildland Fire* **2021**, *30*, 790. [CrossRef]
34. Raposo, J.R.; Viegas, D.X.; Xie, X.; Almeida, M.; Figueiredo, A.R.; Porto, L.; Sharples, J. Analysis of the physical processes associated with junction fires at laboratory and field scales. *Int. J. Wildland Fire* **2018**, *27*, 52–68. [CrossRef]
35. Schulz, J.; Kent, D.; Crimi, T.; Glockling, J.L.D.; Hull, T.R. A Critical Appraisal of the UK's Regulatory Regime for Combustible Façades. *Fire Technol.* **2021**, *57*, 261–290. [CrossRef]
36. Borodinecs, A.; Geikins, A.; Barone, E.; Jacnevs, V.; Prozuments, A. Solution of Bullet Proof Wooden Frame Construction Panel with a Built-In Air Duct. *Buildings* **2021**, *12*, 30. [CrossRef]
37. De Koker, N.; Walls, R.S.; Cicione, A.; Sander, Z.R.; Löffel, S.; Claasen, J.J.; Fourie, S.J.; Croukamp, L.; Rush, D. 20 Dwelling Large-Scale Experiment of Fire Spread in Informal Settlements. *Fire Technol.* **2020**, *56*, 1599–1620. [CrossRef]
38. Rossa, C.G.; Fernandes, P.M. Empirical Modeling of Fire Spread Rate in No-Wind and No-Slope Conditions. *For. Sci.* **2018**, *64*, 358–370. [CrossRef]
39. Suh, H.-W.; Im, S.-M.; Park, T.-H.; Kim, H.-J.; Kim, H.-S.; Choi, H.-K.; Chung, J.-H.; Bae, S.-C. Fire Spread of Thermal Insulation Materials in the Ceiling of Piloti-Type Structure: Comparison of Numerical Simulation and Experimental Fire Tests Using Small- and Real-Scale Models. *Sustainability* **2019**, *11*, 3389. [CrossRef]
40. Gamba, A.; Charlier, M.; Franssen, J.-M. Propagation tests with uniformly distributed cellulosic fire load. *Fire Saf. J.* **2020**, *117*, 103213. [CrossRef]
41. Spearpoint, M.; Quintiere, J. Predicting the piloted ignition of wood in the cone calorimeter using an integral model—Effect of species, grain orientation and heat flux. *Fire Saf. J.* **2001**, *36*, 391–415. [CrossRef]
42. Rackauskaite, E.; Bonner, M.; Restuccia, F.; Anez, N.F.; Christensen, E.G.; Roenner, N.; Wegrzynski, W.; Turkowski, P.; Tofilo, P.; Heidari, M.; et al. Fire Experiment Inside a Very Large and Open-Plan Compartment: X-ONE. *Fire Technol.* **2021**, *58*, 905–939. [CrossRef]
43. Nishino, T.; Tanaka, T.; Hokugo, A. An evaluation method for the urban post-earthquake fire risk considering multiple scenarios of fire spread and evacuation. *Fire Saf. J.* **2012**, *54*, 167–180. [CrossRef]
44. Poulsen, A.; Bwalya, A.; Jomaas, G. Evaluation of the Onset of Flashover in Room Fire Experiments. *Fire Technol.* **2013**, *49*, 891–905. [CrossRef]
45. Daşdemir, O.; Aydın, F.; Ertuğrul, M. Factors Affecting the Behavior of Large Forest Fires in Turkey. *Environ. Manag.* **2021**, *67*, 162–175. [CrossRef]
46. Huang, X.; Zhu, H.; Peng, L.; Zheng, Z.; Zeng, W.; Bi, K.; Cheng, C.; Chow, W. Thermal Characteristics of Vertically Spreading Cable Fires in Confined Compartments. *Fire Technol.* **2019**, *55*, 1849–1875. [CrossRef]
47. Nishino, T.; Kagiya, K. A multi-layer zone model including flame spread over linings for simulation of room-corner fire behavior in timber-lined rooms. *Fire Saf. J.* **2019**, *110*, 102906. [CrossRef]
48. Viegas, D.X. Slope and wind effects on fire propagation. *Int. J. Wildland Fire* **2004**, *13*, 143–156. [CrossRef]

buildings

MDPI

Article

Study of the Axial Compressive Behaviour of Cross-Shaped CFST and ST Columns with Inner Changes

Zhong Tao [1,2,*], Md Mehedi Hasan [1,2,*], Dongji Han [1,2], Qiudong Qin [1,2] and Wahab Abdul Ghafar [1,2]

[1] Civil Engineering and Architecture Faculty, Kunming University of Science and Technology, Kunming 650500, China
[2] Yunnan Earthquake Engineering Research Institute, Kunming 650000, China
* Correspondence: taozhong@kust.edu.cn (Z.T.); mehedihasan.cen@gmail.com (M.M.H.)

Abstract: In this study, novel cross-shaped concrete-filled steel tube (CFST) and steel tube (ST) columns were developed. CFST columns have a high load-carrying capacity and excellent performance under seismic conditions, and the construction process is fast. In order to investigate the axial load bearings and failure mechanisms, six specimens of CFST and ST columns were tested under the axial load. Three different forms of CFST were employed in this study; one was an ordinary cross-shaped CFST (OC-CFST), while the other two were executed with significant inner changes; namely, stiffeners cross-shaped CFST (SC-CFST), and multi-cell cross-shaped CFST (MC-CFST) filled with concrete. The other group has the same OC-ST, SC-ST, and MC-ST, but these test subjects were without filled concrete. Through discussion of the failure mechanism, load displacement and load strain correlations are determined. The effects of parameters on ultimate resistance, failure pattern, and ductility index were studied. The axial load-carrying performance of the cross-shaped CFST columns was 75–80% better than that of ST columns; and each ST column displayed cooperative behavior. The finite element model (FEM) was simulated, and the outcomes of the experiments were used to validate it. The load–displacement relationships were established using parametric analysis. Existing design standards were used to calculate CFST column loading capacity. Finally, mathematical formulas were improvised to determine the ultimate load of the cross-shaped CFST column.

Keywords: cross-shaped CFST column; multi-cell CFST column; stiffened column; steel tube column; FEM analysis; structural application

Citation: Tao, Z.; Hasan, M.M.; Han, D.; Qin, Q.; Abdul Ghafar, W. Study of the Axial Compressive Behaviour of Cross-Shaped CFST and ST Columns with Inner Changes. *Buildings* **2023**, *13*, 423. https://doi.org/10.3390/buildings13020423

Academic Editors: Jingxuan Wang, Dewen Kong, Yong Liu and Shan Gao

Received: 24 November 2022
Revised: 20 January 2023
Accepted: 22 January 2023
Published: 3 February 2023

1. Introduction

Applying specially shaped CFSTs, as the columns that protrude from walls, can increase the amount of floor space accessible in structures. Special-shaped CFST columns are widely used in industrial sites, tall buildings, bridges, substantial transmission towers, and other structures due to their high load-bearing capacity, outstanding ductility, compatible construction, better construction technique and economic advantages [1]. They are currently being used in several buildings in China, including the Fukang Home, the Guangzhou new China Mansion, the Guangzhou Mingsheng Plaza, and the residences in Yuzixi village [2,3]. CFST columns are currently primarily classified into groups possessing square, rectangular, circular, and special-shaped cross-sections. Implementing specially shaped CFST columns can improve the internal usage area of a high-rise structure significantly [4–6]. A proper theoretical and specification system for CFSTs was created as a result of significant laboratory experiments and theoretical study that revealed the mechanical characteristics of square and circular CFST columns [7–13]. Commonly, a square, rectangular, or circular cross-section is used for the steel column in a typical residential building. In these situations, the column can protrude out from the wall and affect the building structure [14,15]. There are several conventional section forms. Corner columns for architectural construction frequently use cross-sections which are T-shaped, L-shaped, and cross-shaped.

Researchers have previously examined the achievement of multi-cell, L-shaped CFST columns that have been stiffened while loaded in either pure bending or axial compression conditions [3,14,16–19], and the currently available design methodologies have been suggested. Qiguang et al. [20] presented a calculation method for CFST columns with cross-shaped sections; it investigated how factors such as cross-section dimensions, width-to-thickness ratio, and yield stress affected loading capacity, displacement form, and stress formation for the steel tube. According to experimental findings, lowering the tube's width-to-thickness ratio decreased the overall out-of-plane displacement at the corner. As the steel grade was raised, the ductility increased while the bearing load and deformation decreased inversely. Increased premature load buckling and overall outward extrusion were seen when the tube's length-to-width ratio was increased. A CFST column that has been exposed to axial compression can have its ultimate load-bearing capacity and lateral displacement impact of a wide range of factors, however, it has been discovered that one of the more effective factors, in this case, is the concrete confinement [21–26]. Multi-cell CFST columns with axial loading were introduced by Song et al. [27]; these researchers carried out both experimental and FEM research. The strength of the concrete and specimen width–thickness ratio were the experiment criteria. A parametric investigation was carried out using ABAQUS once the experimental and numerical data had been validated. China and European standards were the most accurate in calculating the specimen's loading capacity according to a comparison of experimental data with those produced using various standard methodologies.

In order to create a multi-cell T-shaped CFST column, Cao et al. [28] and Xu et al. [29] invented welding techniques for rectangular tubes. The proposed method improved confinement in infilled concrete by increasing load carrying capacity and recommending design strategies based on parametric analysis. Liu et al. [2] established a numerical model for ordinary, stiffened, multi-cell T and L-shaped CFST columns with concentric and eccentric loads. According to the investigation, multi-cell special-shaped CFST excelled over other types of special-shaped CFST columns in mechanical behaviour. Earlier studies have shown that adding longitudinal stiffeners or implementing a multi-cell cross-section can improve the mechanical behaviour of the special-shaped CFST columns.

In comparison to previous studies' findings on the behavior of peculiar cross-shaped CFST and ST columns, the number of research investigations is limited and insufficient. This research is an experimental, numerical, and analytical study of six cross-shaped CFST and ST columns subjected to concentric axial compression. The main goal of the test system is to concentrate on the behavior of the columns with internal changes, like MC-CFST, MC-ST, SC-CFST, SC-ST, OC-CFST, and OC-ST. Additionally, parametric analysis is carried out by the FEM approach using the ABAQUS software while considering different geometric cross-sections and material properties. A mathematical formula for structural engineering practice in the future has also been proposed, along with a discussion of the best way to estimate how much resistance particular cross-shaped CFST columns will exprerience.

2. Experimental Research

2.1. Specimens Details

Three types of special-shaped CFST and ST columns cross-sections are made up of steel tubes with 300 mm × 300 mm dimensions. All the study specimens had identical overall cross sections with internal changes. In Figure 1 and Table 1, the cross-sectional information of all specimens is presented. The six specimens were identified as MC-CFST, SC-CFST, OC-CFST, MC-ST, SC-ST, and OC-ST by symbol codes. Concentric axial loading was performed on all specimens. Respectively, l and b represent the length and width of the column, t represents the thickness of the steel tube, and ξ is the confinement factor. Montuori et al. [30] analyze the finding of a study, looking at several fundamental laws for the confinement factor. Among them, the square section constitutive laws are shown in Table 2.

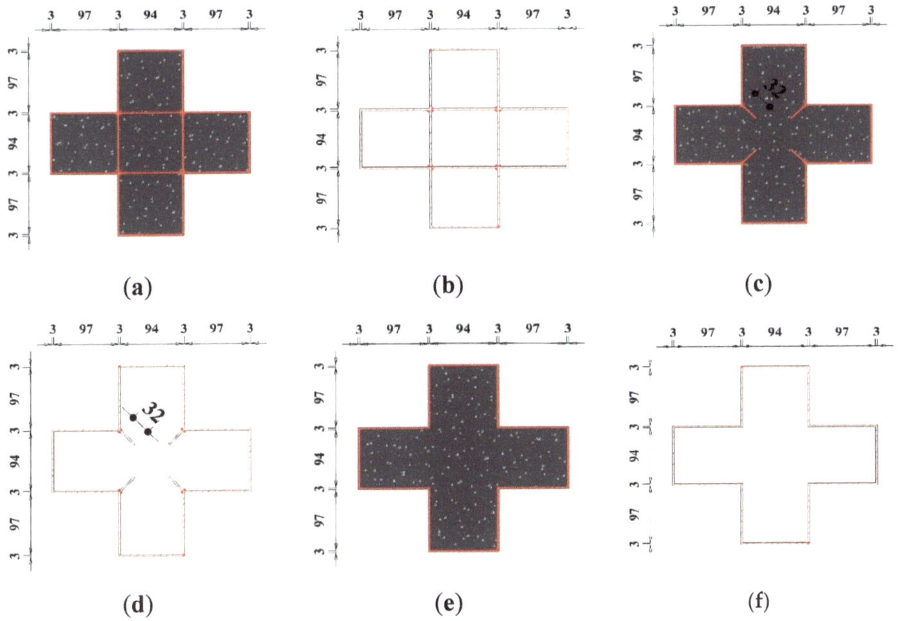

Figure 1. Typical specimen cross-section: (**a**) MC-CFST; (**b**) MC-ST; (**c**) SC-CFST; (**d**) SC-ST; (**e**) OC-CFST; (**f**) OC-ST.

Table 1. The specimen's parameters.

No.	Specimens	*l* (mm)	b (mm)	t (mm)	Concrete Strength (MPa)	Steel Strength (MPa)	*l*/b	ξ
1	MC-CFST	900	300	3	45.28	330	3	1.136
2	SC-CFST	900	300	3	45.28	330	3	0.936
3	OC-CFST	900	300	3	45.28	330	3	0.842
4	MC-ST	900	300	3	-	330	3	-
5	SC-ST	900	300	3	-	330	3	-
6	OC-ST	900	300	3	-	330	3	-

Table 2. Constitutive laws of Confinement.

Method	Formula	Reference
01	$f_{cc} = f_{c0} \times \left[2 \times \left(\frac{\rho_f \cdot E_f}{f_{c0}}\right) \times \left(-0.4142 \cdot E_f \cdot 10^{-7} + 0.0248\right) \times \left(\frac{2r}{d}\right) + 1\right]$	[30,31]
02	$f_{cc} = f_{cc,s} + f_{cc,c}$	[30,32]
03 (Used method)	$\xi = \frac{A_s}{A_c} \times \frac{f_y}{f_{ck}}$	[15]

Here, f_y and f_{ck} are the steel and concrete strengths. A_s and A_c represent steel and concrete cross-sectional area. The length–width ratio (l/b) is maintained at 3 mm for all cross-shaped columns to avoid the general buckling reaction and boundary condition, culminating in a physical column length of 900 mm.

2.2. Material Properties

During the experiment, a Q235-grade steel sheet was used. Five tensile test samples were taken and inspected to look at the mechanical properties of a steel sheet with a thickness of 7.43 mm inner lengths of 111 mm and 80 mm. An electronic universal testing machine based on (GBT228.1-2010) [33] was used to conduct the tensile test.

Figure 2 displays the typical stress–strain curve and tensile coupon sample. Table 3 shows the average results for the inspected yield strength, ultimate strength, elastic modulus, and Poisson's ratio, denoted as f_y, f_u, E_s, and μ_s, respectively.

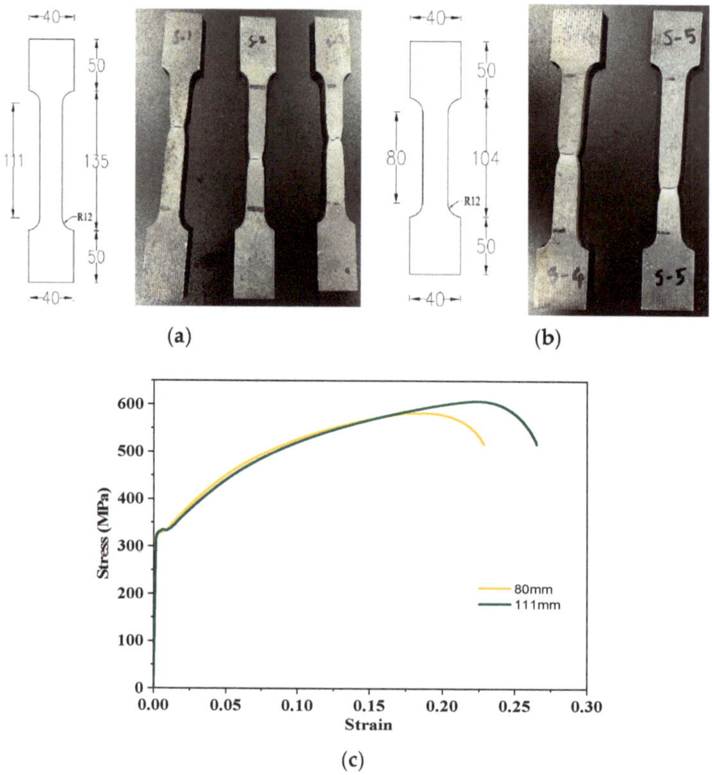

Figure 2. Tensile test results of the Steel: (**a**) 111 mm length sample; (**b**) 80 mm length sample; (**c**) Steel tensile in tested properties graph.

Table 3. Steel inspected properties.

Inner Length	f_u (MPa)	f_y (MPa)	E_s (MPa)	ν_s
111 mm	587	328	201,059	0.30
80 mm	607	333	201,426	0.30

There, concrete cube samples of 150 × 150 × 150 mm were manufactured, self-made core concrete was utilized for the concrete material property test, and the concrete was poured into the steel tube columns. The proportions of the concrete mixtures were Cement:Sand:Stone:Water (1:1.11:2.25:0.4). To optimize the curing state of the concrete core in the steel tube, the concrete cubes were first covered in aluminum foil and subsequently in translucent protective plastic. According to the standard concrete cube tests, the average cubic compressive strength (f_{cu}) was 45.28 MPa, and the elastic modulus of concrete (Ec) was 28023 MPa.

2.3. Specimens' Construction

The experiment specimen may sustain damage or develop flaws throughout the manufacturing and processing of steel tubes due to welding and other procedures. As a result, logically planning the manufacture and processing method, shortening the welding method, and optimizing welding accuracy can raise the experiment specimen performance. Cold-formed empty tubes made from the same steel sheet were created to preserve the comparable qualities of all specimens. Two bent-over steel sheets were used to construct each OC-CFST, OC-ST specimen, for MC-CFST, SC-CFST, MS-ST, and SC-CFST specimens welded with the extra steel sheet in mid-portion, and they were linked by vertical welding, as illustrated in Figure 3.

Figure 3. Steel tube preparation of the specimen.

At the bottom of the steel tube, a loading cover plate with a 20 mm thickness and measurements of 300 × 300 mm was welded, providing the framework for pouring concrete. Before welding the loading plate, we checked the leveling to ensure that the load was distributed accurately on the specimens, as illustrated in Figure 4. Following vertical filling, concrete was compacted, utilizing a poker vibrator on the outside of the steel tubes and a vibrating rod on the inside. After pouring, the top portion of the concrete in steel tubes was covered with aluminum foil and a thin plastic sheet to prevent drying shrinkage. After that, all specimens were left in the room for 28 days. To ensure that the concrete core and steel tube could bear loads combinedly, another steel loading plate with a thickness (of 20 mm) was welded at the top of the specimen. Figure 3 demonstrates several crucial steps that must be taken before and during the concrete pouring.

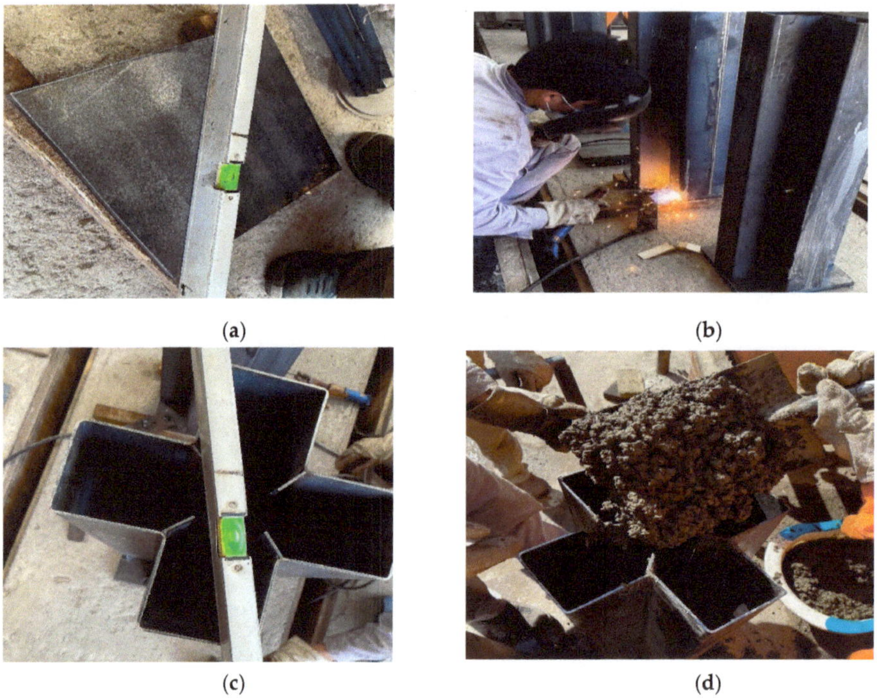

Figure 4. Specimen before and during concrete pouring: (**a**) Loading plate leveling; (**b**) Loading plate welding with the specimen; (**c**) Levelling the specimen; (**d**) Concrete pouring.

2.4. Measuring Instrument and Timeline for Loading

Using 72 resistance strain gauges, 36 strain gauges were installed longitudinally, and the other 36 were installed horizontally. All the strain gauges were glued to the column's middle, upper-middle, and lower-middle 225 mm surface to acquire the axial load. Figure 5 illustrates the experimental configuration and the arrangement of the measuring instruments. The CFST and ST columns were placed up in a similar manner.

Nine displacement transducers (LVSTs) were placed at different places to obtain the vertical and lateral deformation of the specimen. The compression-controlling device and the computer automatically measure and record axial load and vertical deformation. Besides this, we installed LVDTs vertically under the loading plate to get the exact vertical deformation. Each specimen had eight lateral deformation LVSTs installed on every outer face and corner face. For OC-CFST, all lateral LVSTs were in the center, but after seeing the actual effect on the OC-CFST column, the rest of all columns, D-2, 4, 6, and 8, were installed from the upper-inner part to the 225 mm center.

A hydraulic machine with a 10,000 kN capacity was used to evaluate the cross-shaped CFST and ST columns under axial compression. The compressive axial load was imposed at the cross-shaped CFST and ST column centers. The information from the strain gauges and LVDTs was recorded by an automatic data recorder system that was connected to a computer.

The load was first discharged to zero and then reloaded for formal loading after the specimen had completed its preliminary work. The formal load application stage implements the rule of applying the load in stages regulated by force and establishes the size of the load applied in stages.

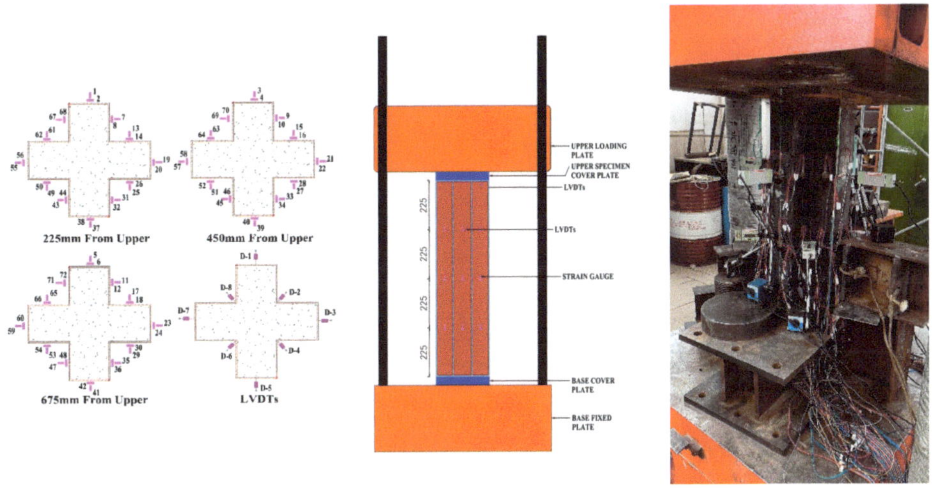

(a) (b) (c)

Figure 5. Test setup of cross-shaped CFST and ST columns: (**a**) Measuring instrument setting; (**b**) Loading simple diagram; (**c**) Loading machine setup.

The force-controlled load was applied at this phase loading at a speed of 120 kN/min. This loading motion continued until the maximum load was reached. Once the maximum load has been attained, the force load is maintained to assure the experiment's safety.

Now the experiment loading stage has entered the failure stage. The experiment is considered to have failed when the load falls below 75% of the observed limit load, or if the experiment specimen deforms excessively. The local buckling's location was identified, and the associated load data were recorded. The specimen failed due to the concrete crashing and steel tubes' outward expansion, which ended the loading procedure.

3. Discussion and Analysis

3.1. Failure Occurrence

Effective ductile behavior was displayed in every cross-shaped CFST column. The failure system of the specimens is illustrated in Figures 6 and 7. There was no welding failure at the initial loading stage, and the welded seams held up well while loading. During the failure phase, most specimens made the sound. Each type was carefully inspected due to the various failure occurrences and specimen-related modes.

3.1.1. Cross-Shaped CFST Specimen's Failure Modes

Similar failure processes were displayed by the specimens MC-CFST, SC-CFST, and OC-CFST when loaded concentrically. No obvious occurrence was seen during the initial loading process on the specimen's face. For specimen MC-CFST, the load–displacement graph increased linearly until 3082 kN and then became nonlinear as the load increased. Local buckling started on two faces when specimen MC-CFST's average load declined to 3652 kN from an ultimate load of 3805 kN. When the MC-CFST specimen reached its maximum load, there was a crack in the joint, as seen in Figure 6a. After that, during the load-dropping phase, significant local buckling happened randomly at adjusted faces.

However, significantly less local buckling can be detected in specimen SC-CFST, demonstrating the cooperation and development between the core concrete and steel tube and also the different cross sections. A reduced steel ratio is to blame for this, which enhances the restraining effect of the infilled core concrete and the control of local buckling. There was no bulging on the specimen's MC-CFST and SC-CFST concave portions.

(**a**) (**b**) (**c**)

Figure 6. Cross-shaped CFST specimen's failure modes: (**a**) MC-CFST specimen; (**b**) SC-CFST specimen; (**c**) OC-CFST specimen.

(**a**) (**b**) (**c**)

Figure 7. Cross-shaped ST specimen's failure modes: (**a**) MC-ST specimen; (**b**) SC-ST specimen; (**c**) OC-ST specimen.

The specimen OC-CFST was in the initial stage during the first loading, and no deformation was visible before the specimen reached its ultimate load. In specimen OC-CFST, local buckling started as the load decreased from its ultimate load of 2824 kN to roughly 2767 kN, as shown in Figure 6b. There was no sign of local buckling waves on the faces in the zone, in contrast to the MC-CFST and SC-CFST specimens. Figure 6c shows that subsequent local buckling in specimens OC-CFST was less pronounced than in specimens

MC-CFST and SC-CFST due to changes in the steel confinement value and cross section following the failure load.

3.1.2. Cross-Shaped ST Specimen's Failure Modes

During the initial stage of the experiments, the performance of the specimen's MC-ST, SC-ST, and OC-ST, loaded concentrically, was consistent. Before the ultimate load, lateral deformation did not immediately enter the failure stage. In specimen MC-ST, the first wave of local buckling was seen when the load decreased from an ultimate load of 1195 kN to roughly 1183 kN. In the compression area, there was noticeable buckling. After that, the displacement quickly dropped. While there was no noticeable buckling on the bottom of the specimen, the steel surface displayed significant buckling close to the top, as illustrated in Figure 7. It is possible that this occurred because the inner part of the steel tube lacked a sufficient solid area, preventing the load from properly transferring to the lower portion of the column. Although the local buckling shape was perfect, it undoubtedly became deformed, possibly due to an empty steel tube.

3.2. Axial Load–Displacement Relationship

To examine the specimens MC-CFST, SC-CFST, and OC-CFTS, the ultimate load-bearing capacity and ductility were improved in specimen MC-CFST compared to SC-CFST and OC-CFST. On the other hand, SC-CFST's ultimate loading capacity is better than ha of OC-CFST, as illustrated in Figure 8a,c.

Table 3 shows the ultimate loading capacities (N_{Emax}) of cross-shaped CFST and ST group columns that are compressed axially. As shown in Figure 8b,d, when comparing the specimens MC-ST, SC-ST, and OC-ST, the specimen MC-ST had enhanced load-carrying capacity. Due to the inner cross-sectional difference, the SC-ST and OC-ST specimens' loading capacities were lower than those of the MC-ST specimen. Compared to cross-sectional changes, MC-CFST and MC-ST specimen capacity improved by 22.48% and 18.26% more than SC-CFST and SC-ST specimens, respectively. On the other side, SC-CFST and SC-ST specimens improved by 7.23%, and 19.04% more than OC-CFST and OC-ST specimens. The confinement factor benefit for infilled concrete, provided by the steel tubes, prevented potential local buckling of the steel tube and raised ductility. Due to the cross-sectional changes and increased steel ratio in the MC-CFST and MC-ST specimens, the steel tubes played a significant role in the ultimate loading capacity and the outside swelling of the steel surface. The confinement effect for infilled concrete provided by the steel tubes prevents potential local buckling of the tube and improves ductility. All CFST columns perform better than ST columns; CFST columns have shown an almost three times batter performance.

3.3. Ductility Index

According to the geometric graphic approach, as illustrated in Figure 9, some loading and Vertical displacement highlighted position, such as the yielding experimental location (N_{Ey}, δ_{Ey}), ultimate experimental place (N_{Emax}, δ_{Emax}), 0.85% dropped location ($N_{0.85Emax}$, $\delta_{0.85Emax}$) [34–36]. A member's ability to resist significant plastic deformation without a noticeable loss of strength is referred to as mechanical ductility. In order to create the ductility index (η), the ratio of the compressive strain must be equal to the residual load-bearing capacity (average $0.85N_{Emax}$) of the strain, equivalent to N_{Emax}, used,

$$\eta = \frac{\varepsilon_{0.85N_{Emax}}}{\varepsilon_{N_{Ey}}} \quad (1)$$

There is a list of the determined ductility index (η) values in Table 4 as well. Although the increase was relatively small, η improved when the confinement increased; due to the decrease in confinement, OC-CFST had a lower value.

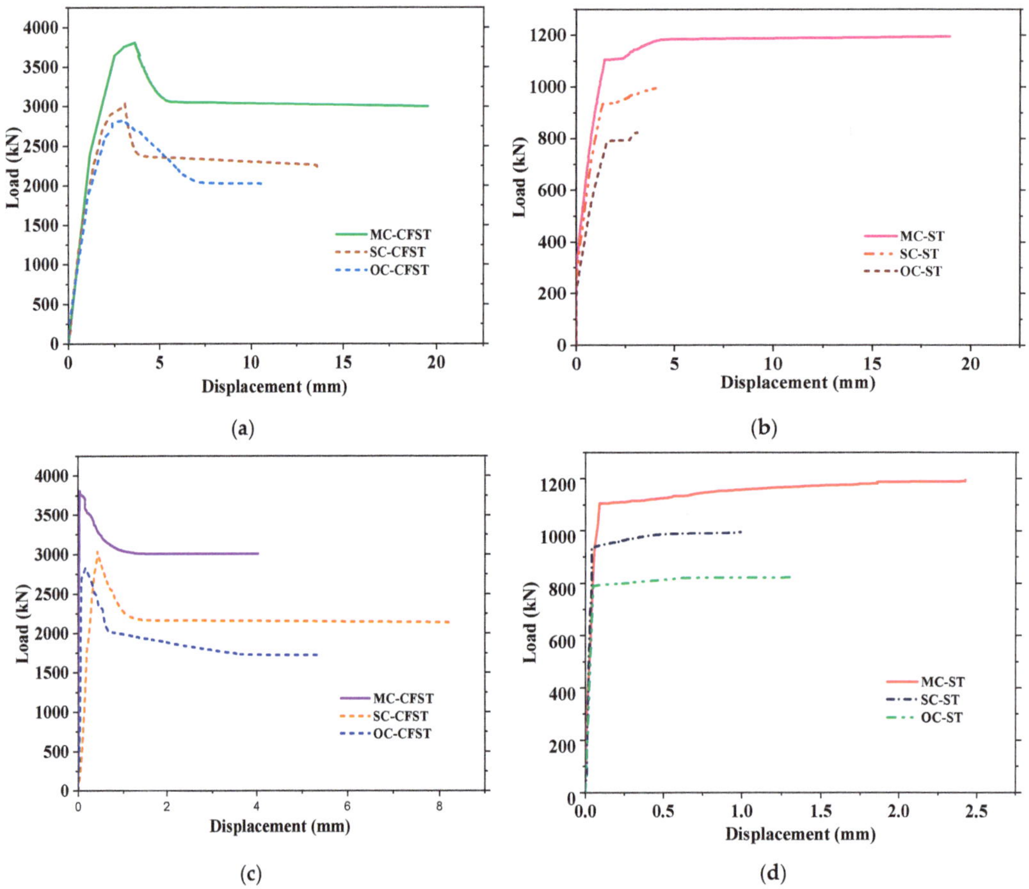

Figure 8. Load–displacement graph, cross-shaped CFST, and ST specimens: (**a**) load–vertical displacement; (**b**) load–vertical displacement; (**c**) load–lateral displacement; (**d**) load–lateral displacement.

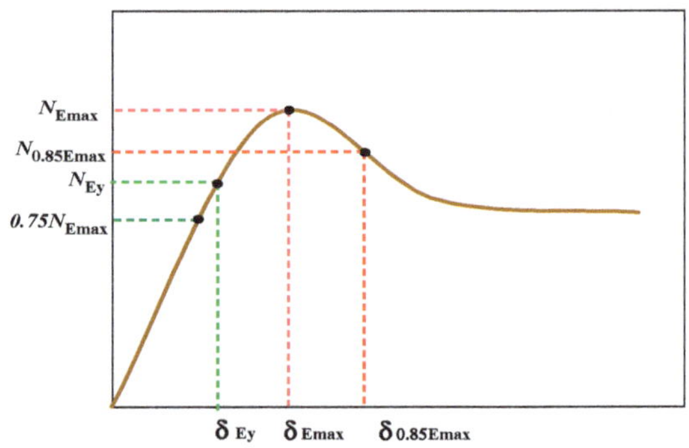

Figure 9. Graphical geometric method.

Table 4. Summary of the experimental and FEM results.

Specimen	$N_{0.85E}$ (kN)	N_{Ey} (kN)	δ_{Ey} (mm)	N_{Emax} (kN)	δ_{Emax} (mm)	N_{FEM}	N_{FEM}/N_{Emax}	η
MC-CFST	3234.24	3306	2.138	3805	3.581	3688	0.97	0.93
SC-CFST	2580.6	2735	2.067	3036	3.063	2948	0.97	0.862
OC-CFST	2400.4	2627	2.045	2824	2.936	2845	1.007	0.810
MC-ST	939.25	1105	1.445	1195	1.445	1210	1.012	-
SC-ST	794.75	935	1.224	995	1.224	974	0.979	-
OC-ST	698.7	791	1.1853	822	1.185	843	1.026	-
AVP							0.99	
AVE							0.01	

4. Numerical Model Analysis

4.1. Type of Elements

A three-dimensional (3D) FEM model was created in the ABAQUS application to more closely research the impact of different parameters on the axial bearing capacity of the cross-shaped CFST and ST column specimens. Both steel tube and core concrete were modeled by solid 8-node model variant (C3DR), and the loading cover plate had a stiff body.

4.2. Materials, Interaction, Boundary Condition, and Meshing

In the steel stress–strain relationship graph, that graph separated into five stages: the elastic stage; the elastic–plastic step; the plastic stage; the strengthening stage; and the second plastic flow stage, as illustrated in Figure 10c, and the equation that describe the related theoretical expression is presented below [37].

$$\sigma_s = \begin{cases} E_s\varepsilon_s \ \varepsilon_s \leq \varepsilon_e \\ -A\varepsilon_s^2 + B_s\varepsilon_s + C \ \varepsilon_e < \varepsilon_s \leq \varepsilon_{y1} \\ f_y\left[1 + \dfrac{0.6(\varepsilon_s - \varepsilon_{y2})}{(\varepsilon_u - \varepsilon_{y2})}\right] \ \varepsilon_{y2} < \varepsilon_s \leq \varepsilon_u \\ 1.6f_y \ \varepsilon_s > \varepsilon_u \end{cases} \tag{2}$$

Here,

$$\varepsilon_c = \frac{0.8f_y}{E_s},\ \varepsilon_{y1} = 1.5\varepsilon_e,\ \varepsilon_{y2} = 10\varepsilon_{y1},\ \varepsilon_u = 10\varepsilon_{y2},\ A = \frac{0.2f_y}{(\varepsilon_{y1} - \varepsilon_e)^2},\ B = 2A\varepsilon_{y1},\ C = 0.8f_y + A\varepsilon_e^2 - B\varepsilon_e$$

where σ_s, ε_s represent the Steel stress and strain, respectively, Figure 10a,b demonstrate the model suggested in the standard for the design of concrete structures [37,38]. The theoretical statement is as follows:

$$y = \begin{cases} 2x - x^2\ ,\ x \leq 1 \\ \dfrac{x}{\beta_0(x-1)^\eta + x}\ ,\ x > 1 \end{cases} \tag{3}$$

$$x = \frac{\varepsilon}{\varepsilon_0}\ ,\ y = \frac{\sigma}{\sigma_{c0}},\ \sigma_{co} = f_c',\ f_c' = 0.76f_{cu},\ \varepsilon_{co} = \varepsilon_c + 800\xi^{0.2} \times 10^{-6}$$

$$\varepsilon_c = \left(1300 + 12.5f_c'\right) \times 10^{-6},\ \eta = 1.6 + \frac{1.5}{x},\ \beta_0 = \frac{\left(f_c'\right)^{0.1}}{0.2\sqrt{1+\xi}},\ \xi = \frac{f_y A_s}{f_{ck} A_c}$$

Here,

σ_{c0} —Compressive stress and strain relationship of the ultimate stress;
ε_{c0} —The ultimate strain of the compressive stress and strain relationship;

f_c', f_{ck}, f_{cu} —Compressive strength of concrete cylinders, and cube;
A_s, A_c —The cross-sectional area of the steel tube and the concrete in the core area;

$$y = \begin{cases} 1.2x - 0.2x^6, \ x \le 1 \\ \dfrac{x}{\alpha_t(x-1)^{1.7} + x}, \ x > 1 \end{cases} \tag{4}$$

$$x = \frac{\varepsilon}{\varepsilon_{t0}},\ y = \frac{\sigma}{\sigma_{t0}},\ \sigma_{to} = 0.26 f_{cu}^{2/3},\ \varepsilon_{to} = 65 \times 10^{-6} \times \sigma_{t0}^{0.54},\ \varepsilon_{to} = 65 \times 10^{-6} \times \sigma_{t0}^{0.54}$$

Here

σ_{t0} —The ultimate stress of the tensile stress and strain relationship;
ε_{t0} —The ultimate strain of the tensile stress and strain relationship;

The material plasticity parameter was chosen, as illustrated in Table 5. Surface-to-surface contact establishes the interface between the concrete and the steel tube. The slave and enslaver surface, respectively, correspond to the interior side of the steel plant and the outer faces of the core concrete plan. The definitions of hard contact and penalty friction were used to characterize normal and tangential contact, respectively. Based on earlier studies, the friction coefficient was expected to be 0.3 [39].

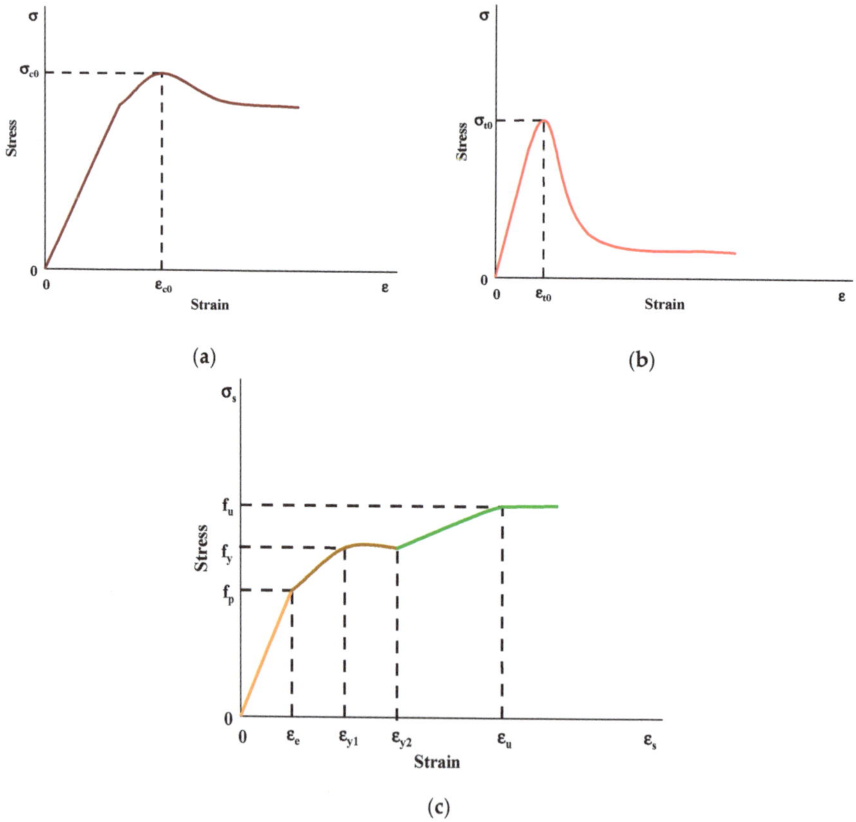

Figure 10. Concrete and steel, simplified structural relationship: (**a**) Compressive stress–strain of concrete; (**b**) Tension stress–strain of concrete; (**c**) Steel stress–strain.

Table 5. Concrete plasticity parameter.

Dilation Angle	Eccentricity	f_{b0}/f_{c0}	Shape Factor, K	Viscosity Parameter
30°	0.1	1.16	0.667	0.0001

The two end sides of the cross-shaped CFST and ST columns were given two reference points: the top and bottom. The boundary conditions were implemented using the experiment circumstances as a guide. The top was subjected to loading by using displacement control models. The tie function applied rigid constraints between the top and bottom surfaces and corresponding reference points. End plates were not used in the studies since the end faces were uniform due to strict constraints. The element sizes used for meshing were all 20 mm through the cross-section.

4.3. FEM Model Verification

In Table 4, the ultimate load N_{Emax} from the experiment is compared with the maximum load N_{FEM} from the FEM. The two groups of results are harmonious, and N_{FEM} has an average percentage (AVP) value of 0.99; this average error percentage (AVE) is 0.01. Figures 11 and 12 display a comparison between the experimental and FEM simulation results of the failure characteristics of the MC-CFST, SC-CFST, OC-CFST, MC-ST, SC-ST, and OC-ST specimens. Figure 11 shows that the two kinds of buckling for MC-CFST and SC-CFST specimen include outward and local buckling, but OC-CFST just has outward buckling. In the ST specimen all local buckling in the upper area of the specimen as illustrated Figure 12. As can be observed, the FEM simulation failure pattern was essentially consistent with the experimental pattern. In the experiments on cross-shaped CFST and cross-shaped ST specimens, the axial load–vertical and –lateral strain graphs generated from the FEM were compared in Figures 13 and 14.

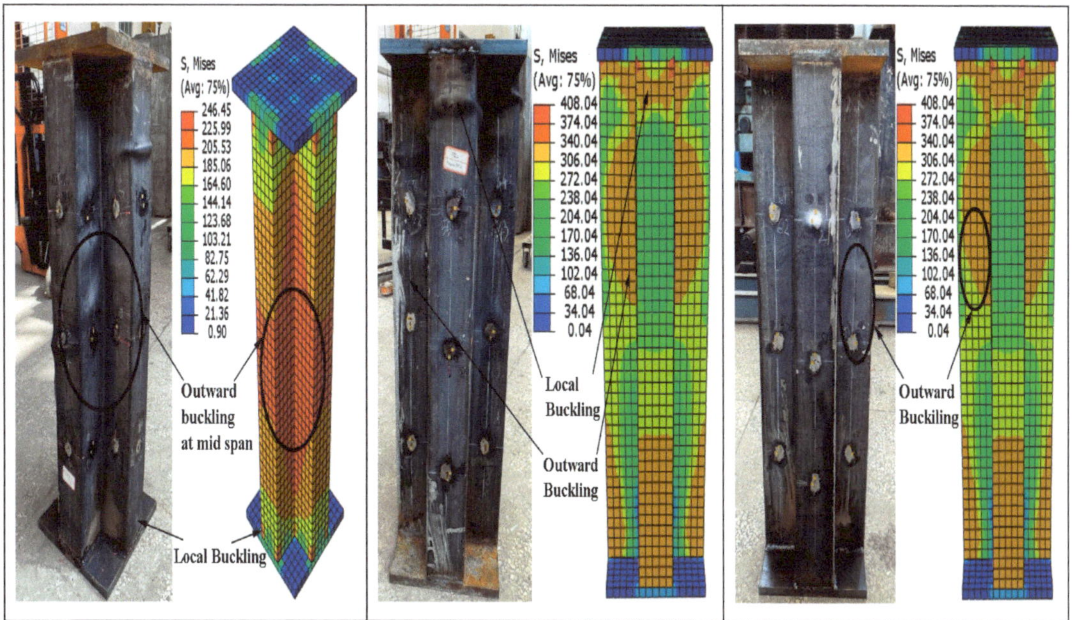

(a)	(b)	(c)

Figure 11. FEM and experiment comparison of the cross-shaped CFST columns: (**a**) MC-CFST; (**b**) SC-CFST; (**c**) OC-CFST.

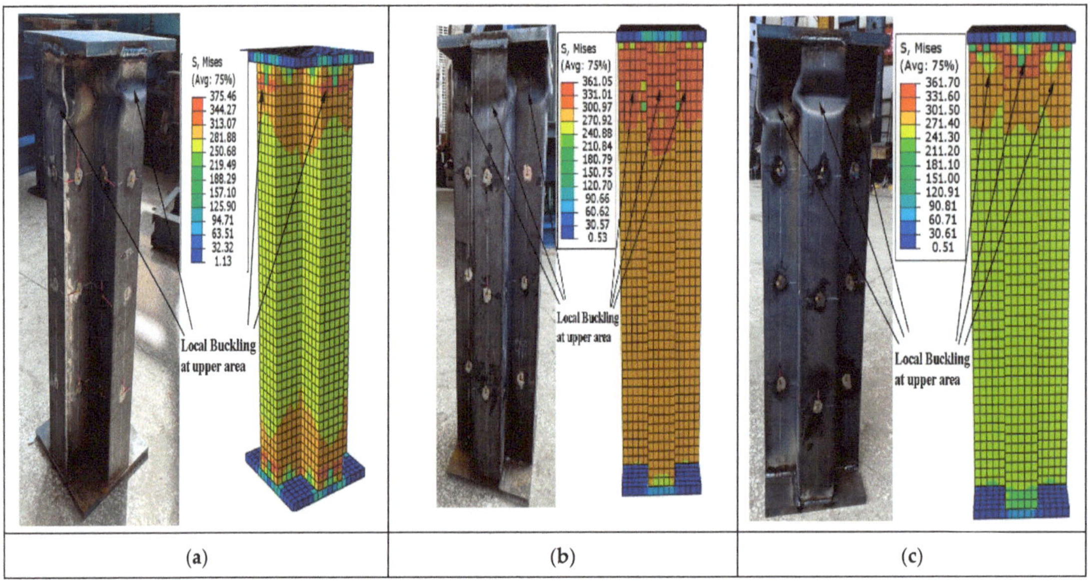

Figure 12. FEM and experiment comparison of the cross-shaped ST columns: (**a**) MC-ST; (**b**) SC-ST; (**c**) OC-ST.

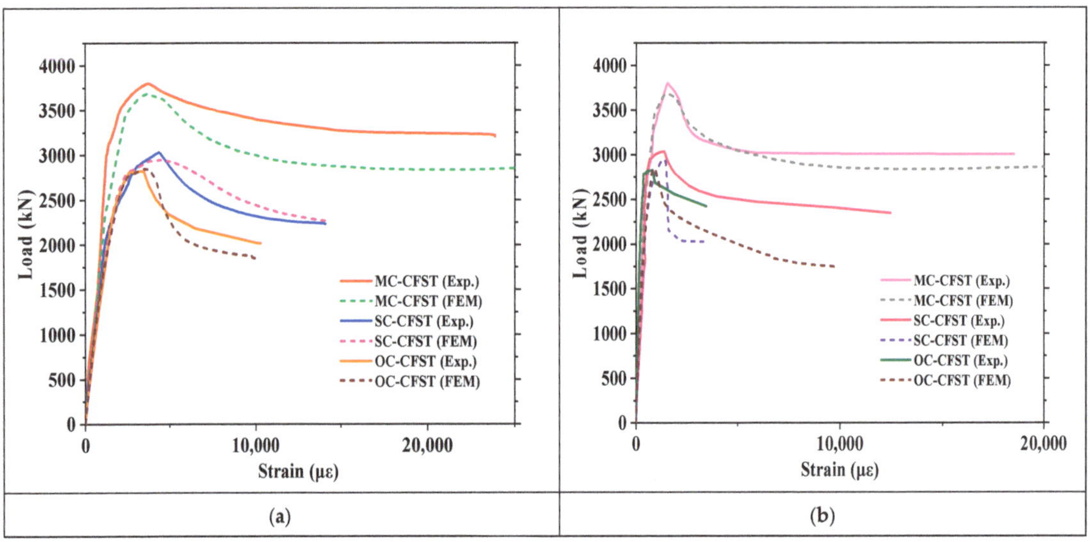

Figure 13. Axial load–strain graph FEM and experimental CFST Specimen: (**a**) Axial load–vertical strain; (**b**) Axial load–lateral strain.

It can be seen that the FEM strains match the experiment results in the elastic stage; the load–strain graphs are almost identical. The FEM and experiment load–strain graphs are not entirely as accurate for all specimens but are still close in the elastic–plastic and failure stages. This comparison further confirms the accuracy of the FEM. It is possible to say that the axial compression properties of the cross-shaped CFST and cross-shaped ST column specimens can be accurately reflected by the FEM simulation model developed in this research.

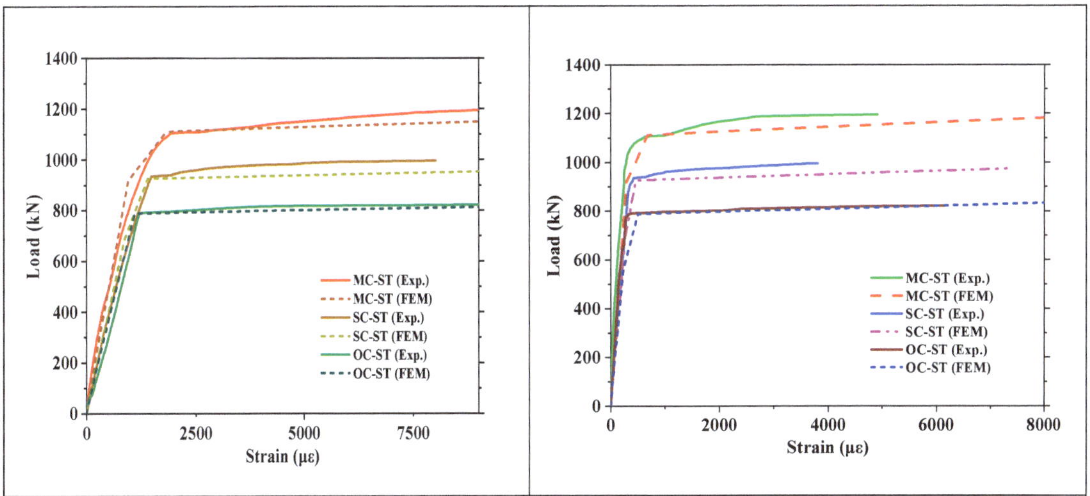

(**a**) (**b**)

Figure 14. Axial load–strain graph FEM and experimental ST Specimen: (**a**) Axial load–vertical strain; (**b**) Axial load–lateral strain.

4.4. Parametric Analysis of CFST Columns

The parametric analysis investigates the impact of the steel tube thickness (t), steel yield strength (f_y), and concrete strength (f_{cu}), on the mechanical performance of cross-shaped CFST column specimens. Table 6 shows the chosen mechanical performance for parametric analysis. Figure 15 illustrates the specimen numbering guidelines for the parameter analysis.

Table 6. The values of analysis parameters.

Parameter	t (mm)	f_y (MPa)	f_{cu} (MPa)
Concrete strength	3	330	40
	3	330	60
Steel strength	3	300	60
	3	390	60
	3	420	60
Steel thickness	2	330	45
	5	330	45

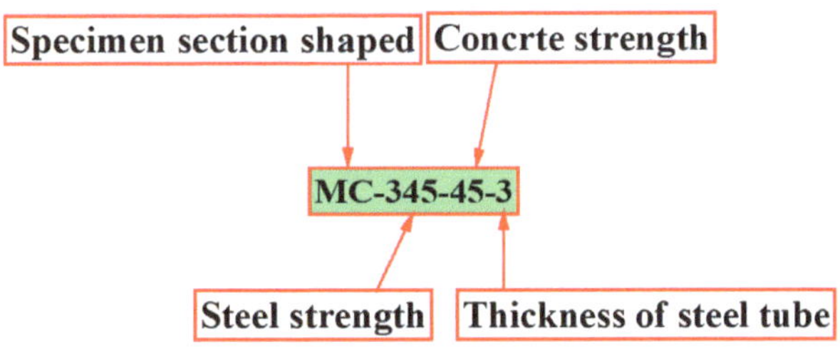

Figure 15. Parameter specimen numbering procedure.

4.4.1. Impact of Concrete Strength

As illustrated in Figure 16a, the impacts of concrete strength on the axial load–vertical displacement graphs for MC-CFST, SC-CFST, and OC-CFST columns were evaluated. The chart matches up during the initial phase, showing that altering the concrete's strength minimizes the initial compression flexibility. The load-bearing capacity seems to rise as concrete strength rises until it reaches the ultimate stage. High-strength concrete displays a faster reduction in load-bearing capacity during the failure stage, and low-strength concrete, compared to high-strength concrete, displays a slight decrease in load-bearing capacity. This is due to the steel tube's fundamental role in supporting the load placed on the core of the crushed concrete.

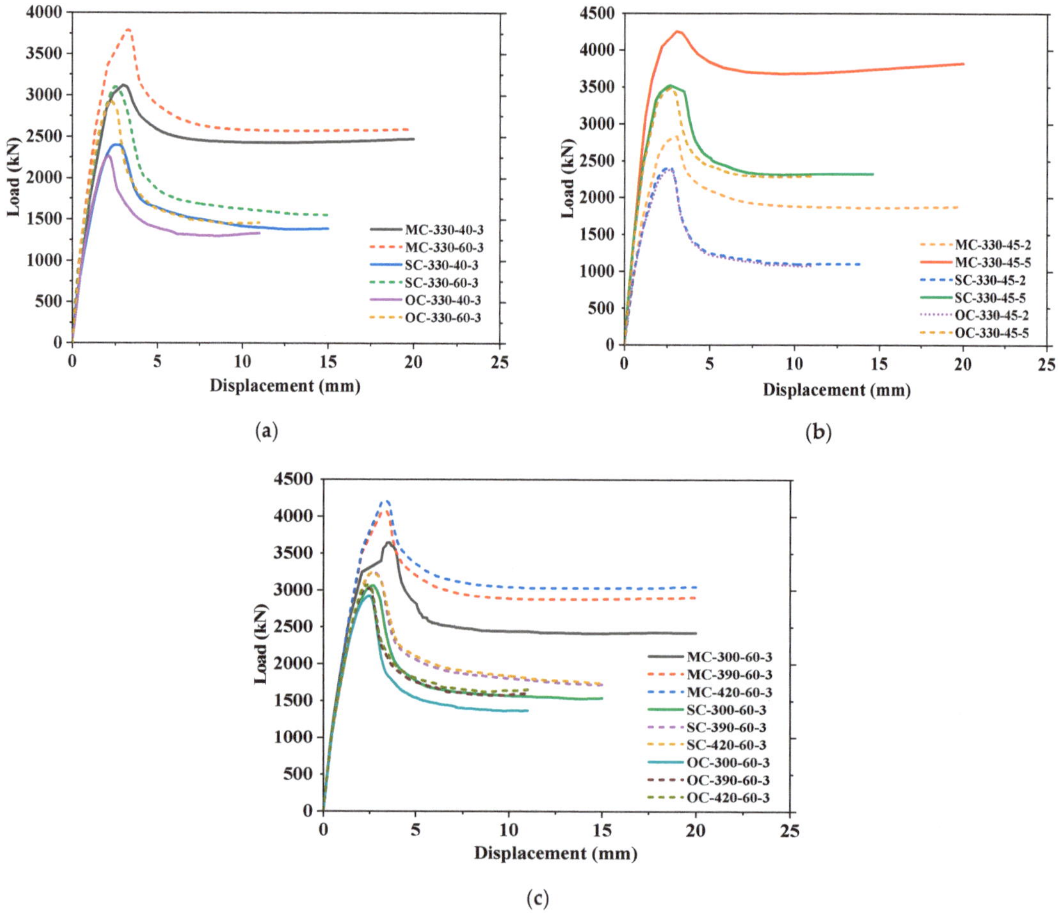

Figure 16. Axial load–displacement graph of CFST specimen: (**a**) different concrete strength; (**b**) different steel tube thickness; (**c**) different steel strength.

4.4.2. Impact of Steel Thickness

The axial load–displacement graph of the MC-CFST, SC-CFST, and OC-CFST columns is shown as a function of steel tube thickness in Figure 16b. Based on thickness, the SC-CFST and OC-CFST column's ultimate load is almost the same, but the MC-CFST column shows a very high maximum loading capacity. However, in initial loading, all columns show the same growth. Based on steel tube thickness in the failure phase, MC-CFST loading capacity decreases slowly compared to SC-CFST and OC-CFST. This happened because of the steel confinement factor. Better confinement factors showed better performance, but to reach the ultimate loading capacity, all CFST columns follow almost the same slope with the same thickness.

4.4.3. Impact of Steel Strength

Figure 16c illustrates the axial load—displacement graph for MC-CFST, SC-CFST, and OC-CFST columns of various steel strengths. The axial compression flexibility of all CFST specimens appears to be unaffected by changes in steel strength with the same cross-section. A more substantial restriction of the steel tube to the core concrete is suggested by higher steel strength. As all CFST specimens' ultimate load-bearing capacity rises, the axial load–vertical displacement graph grows softer and softly reaches a point of failure.

5. Cross-Shaped CFST Column Calculation Method

A particular design standard cannot be used to directly calculate the axial ultimate load-bearing capacity of cross-shaped CFST columns. The CFST column's load-bearing capacity is determined in Table 7, using the different standards, AIJ [40], GB50936-2014 [41], ACI and AS [42], EC-4 [43], and CECS-2004 [44], made for square and rectangular CFST columns. Table 7 shows N_{us} standards ultimate loading capacity of cross-shaped CFST columns and $\omega = N_{Emax}/N_{us}$.

Table 7. Cross-shaped CFST column load-bearing capacity with different standards.

Specimen	AIJ		GB50936-2014		ACI & AS		EC-4		CECS-159:2004	
	N_{us} (kN)	ω	N_{us} (kN)	ω	N_{us} (kN)	ω	N_{us} (kN)	ω	N_{us} (kN)	ω
MC-CFST	3699.45	1.03	3321.96	1.15	3281.40	1.16	2766.84	1.38	3340.45	1.14
SC-CFST	3410.40	0.89	3126.67	0.97	3059.95	0.992	2561.91	1.18	3143.61	0.67
OC-CFST	3269.85	0.86	3033.11	0.93	2952.30	0.957	2462.28	1.15	3051.20	0.93

Different standards have few errors to get accurate results, as shown in Table 7. In this research developed GB50936-2014 [41] formula to obtain the more precise ultimate loading capacity of cross-shaped CFST columns,

$$N_u = \varphi N_0 \tag{5}$$

N_0 is the CFST column design value under axial strength load that is calculated by

$$N_0 = f_{sc}A_{sc} \tag{6}$$

$$f_{sc} = (1.212 + \theta B_s + \theta^2 C_c)f_c \tag{7}$$

$$\theta = \alpha_{sc}\frac{f_y}{f_c} \tag{8}$$

$$\alpha_{sc} = \frac{A_s}{A_c} \tag{9}$$

$$B_s = \left(\frac{0.131 \times f_y}{213}\right) + 0.723 \tag{10}$$

$$C_c = -\left(\frac{0.070 \times f_c'}{14.4}\right) + 0.026 \tag{11}$$

φ is dependent on the confinement factor value, that is,

$$\xi \geq 1.5,\ \varphi = 0.95$$

$$\xi = 1.2,; \ \varphi = 0.90;$$

$$\xi = 1, \ \varphi = 0.85;$$

$$\xi \leq 0.75; \ \varphi = 0.80;$$

The section capacities were calculated using the developed mathematical Equation (5), applying representative models from the different parametric analyses. As demonstrated in Figure 17, there is a logical correlation between parametric and mathematical results and the mean (τ) and standard deviation (μ), which shows a better relationship between them.

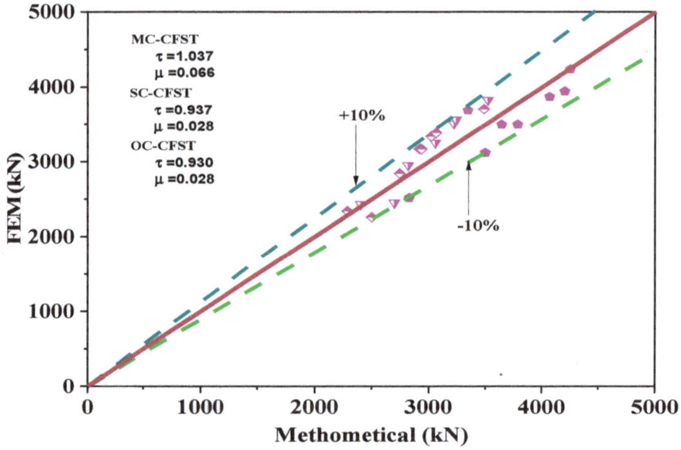

Figure 17. FEM and mathematical results for axial load results comparison.

6. Conclusions

An in-depth analysis of the structural behavior of MC-CFST, SC-CFST, OC-CFST, MC-ST, SC-ST, and OC-ST columns under axial compression is discussed in this study. The following is a list of the main findings from the research that is being presented:

1. The cross-shaped CFST showed more significant load-carrying capacity than cross-shaped ST columns. The cross-shaped columns' load-bearing capacity can be increased by increased CFST column confinement factor. A decrease the in the confinement factor ratio will decrease the local buckling. Additionally, there was a crack indication in the corner area on the MC-CFST and MC-ST specimens.
2. The results of the experiments were used to establish and validate the FEM model. The load-bearing capacity and stiffness of the specimen might all be accurately simulated by the FEM model. The FEM model has been used to analyze the steel tube's local buckling and the concrete's stress concentration.
3. The MC-CFST column loading capacity under compression is underestimated by the design code GB50936-2014, ACI & AS, EC-4, and CECS-159:2004. The SC-CFST and OC-CFST columns' capacity was underestimated by AIJ and EC-4 design codes. However, the cross-shaped CFST column is unsafe in those cases.
4. Improved calculation procedures were proposed for estimating the ultimate load-bearing capacity of cross-shaped CFST columns under axial compression. The calculation technique for the factor coefficient, φ, has been provided by introducing the influence of the confinement factor, ξ, for the axial compressive load. With the conclusion of experimental and FEM data, the mathematical calculation values are in better agreement.

Author Contributions: M.M.H., Methodology, Supervision, Writing—review and editing, Software, Validation. Z.T., Conceptualization, Investigation, Funding acquisition. D.H. and Q.Q. Project administration & supervision, W.A.G., review and editing. All authors have read and agreed to the published version of the manuscript.

Funding: This research was funded by Key Research and Development programs (Key R&D programs) department of science and technology of Yunnan province, the Faculty of Civil Engineering and Architecture, Kunming University of Science and Technology, and the Yunnan Earthquake Engineering Research Institute (YEERI) grant number 202003AC100001.

Institutional Review Board Statement: Not applicable.

Informed Consent Statement: Not applicable.

Data Availability Statement: The data presented in this research are available on request from the corresponding authors.

Acknowledgments: Authors delightfully acknowledge the Key Research and Development programs (Key R&D programs) department of science and technology of Yunnan province, the Faculty of Civil Engineering and Architecture, Kunming University of Science and Technology, and the Yunnan Earthquake Engineering Research Institute (YEERI).

Conflicts of Interest: This manuscript, based on independent research, has been conceptually prepared and written by the authors. The authors report no declarations of interest and the views expressed in this manuscript are solely those of the authors.

References

1. Han, L.-H. *Concrete Filled Steel Tubular Structures-Theory and Practice*; Science Press: Beijing, China, 2007.
2. Liu, X.J.L.; Yang, Y.; Cheng, G.; Lanning, J. Resistance of special-shaped concrete-filled steel tube columns under compression and bending. *J. Constr. Steel Res.* **2020**, *169*, 106038. [CrossRef]
3. Yongqian, Z.S.Z. Design of L-shaped and T-shaped concrete-filled steel tubular stub columns under axial compression. *Eng. Struct.* **2020**, *207*, 110262. [CrossRef]
4. Ahmeda, M.; Liang, Q.Q.; Patelb, V.I.; Hadi, M.N.S. Experimental and numerical studies of square concrete-filled double steel tubular short columns under eccentric loading. *Eng. Struct.* **2019**, *197*, 109419. [CrossRef]
5. Anatoly Leonidovich Krishan, M.A.A.; Chernyshova, E.P. Strength Calculation of Short Concrete-filled Steel Tube Columns. *Int. J. Concr. Struct. Mater.* **2018**, *12*, 84. [CrossRef]
6. Liang, Q.Q. Numerical simulation of high strength circular double-skin concrete-filled steel tubular slender columns. *Eng. Struct.* **2018**, *168*, 205–217. [CrossRef]
7. Farid Abed, M.A.; Abdalla, S. Experimental and numerical investigations of the compressive behavior of concrete filled steel tubes (CFSTs). *J. Constr. Steel Res.* **2012**, *80*, 429–439. [CrossRef]
8. Ren, Q.X.; Lam, D.; Hou, C. Experiments on special-shaped CFST stub columns under axial compression. *J. Constr. Steel Res.* **2014**, *98*, 123–133. [CrossRef]
9. Bin, Z.; Wang, Z.T.; Han, L.-H.; Uy, B.; Lam, D.; Kang, W.-H. Strength, stiffness and ductility of concrete-filled steel columns under axial compression. *Eng. Struct.* **2017**, *135*, 209–221. [CrossRef]
10. Deren, L.Y.G.; Ding, F.; Wang, L.; Deng, C.; Yuan, T.; Clae, R. Experimental Study of Square CFST Stub Columns With a Low Steel Ratio Under Axial Loading. *Struct. Mater. A Sect. J. Front. Mater.* **2021**, *8*, 629819. [CrossRef]
11. Wen, X.J.C.; Xie, F.; Ye, C.; Liu, C. Behavior of Corroded Thin-Walled Concrete-Filled Steel Tubular Stub Columns. *Buildings* **2022**, *12*, 481. [CrossRef]
12. Jiang, L.; Ji, J.; Ren, H.; Wang, Q.; Sun, R.; Yu, C.; Zhang, H.; Luo, G. Bearing behavior of high-performance concrete-filled high-strength steel tube composite columns subjected to eccentrical load. *Struct. Mater. A Sect. J. Front. Mater.* **2022**, *9*, 972811. [CrossRef]
13. Payam, S.; Asteris, P.G.; Formisano, A.; Armaghani, D.J. Iterative Finite Element Analysis of Concrete-Filled Steel Tube Columns Subjected to Axial Compression. *Buildings* **2022**, *12*, 2071. [CrossRef]
14. Liu, X.; Liu, J.; Yang, Y. Research on special-shaped concrete-filled steel tubular columns under axial compression. *J. Constr. Steel Res.* **2018**, *147*, 203–223. [CrossRef]
15. Wang, Z.; Wei, F.; Li, M. Performance of Special-Shaped Concrete-Filled Square Steel Tube Column under Axial Compression. *Adv. Civ. Eng.* **2020**, *2020*, 1763142. [CrossRef]
16. Zheng, Y. Flexural behaviour of stiffened and multi-cell L-shaped CFSTs considering different loading angles. *J. Constr. Steel Res.* **2021**, *178*, 106520. [CrossRef]
17. Zhu, Y. Behavior of Concrete Filled L-Shaped Stiffened Steel Tubes under Axial Compression. Master's Thesis, East China Jiaotong University, Jiangxi, China, 2017.

18. Song, H. Research on Mechanical Behavior of Multi-Cell Special-Shaped Concrete-Filled Steel Tubular Stub Columns Under Axial Compression. Master's Thesis, Chongqing University, Chongqing, China, 2017.
19. Zhang, W.; Xiong, Q. Performance of L-shaped columns comprising concrete-filled steel tubes under axial compression. *J. Constr. Steel Res.* **2018**, *145*, 573–590. [CrossRef]
20. Qiguang, C.; Liang, Z.Z.; Jinxuan, H.; Jieying, W. Study on Cross-shaped Concrete Filled Steel Tubular Stub Columns Subjected to Axial Compression: Experiments and Design Method. *Open Civil Eng. J.* **2017**, *11*, 1–13. [CrossRef]
21. Hossain, K.M.A. Axial load behaviour of thin walled composite columns. *Compos. Part B* **2003**, *34*, 715–725. [CrossRef]
22. Vrcelj, Z.B.U. Strength of slender concrete-filled steel box columns incorporating local buckling. *J. Constr. Steel Res.* **2002**, *58*, 275–300. [CrossRef]
23. Roeder, C.W.; Lehman, D.E.; Bishop, E. Strength and Stiffness of Circular Concrete-Filled Tubes. *J. Struct. Eng.* **2010**, *136*, 1545–1553. [CrossRef]
24. de Oliveira, W.L.A.; de Cresce El Debs, A.L.H.; Debs, M.K.E. Influence of concrete strength and length/diameter on the axial capacity of CFT columns. *J. Constr. Steel Res.* **2009**, *65*, 2103–2110. [CrossRef]
25. Zeghichea, K.C. An experimental behaviour of concrete-filled steel tubular columns. *J. Constr. Steel Res.* **2005**, *61*, 53–66. [CrossRef]
26. Lyu, H.M.; Zhou, A.; Yang, J. Perspectives for flood risk assessment and management for mega-city metro system. *Tunn. Undergr. Space Technol.* **2019**, *84*, 31–44. [CrossRef]
27. Song, H.; Yang, Y.; Chen, Y.F. Study on Mechanical Behavior of Integrated Multi-cell Concrete-filled Steel Tubular Stub Columns Under Concentric Compression. *Int. J. Civ. Eng.* **2019**, *17*, 361–376. [CrossRef]
28. Bing, C.X.Z.; Liang, N.; Yang, Y.; Shen, D.; Huang, B.; Du, Y.-H. Bearing capacity of welded composite T-shaped concrete-filled steel tubular columns under axial compression. *Adv. Mech. Eng.* **2020**, *12*, 3102. [CrossRef]
29. Xu, L.; Du, G.; Wen, F.; Xu, H. Experimental study on normal section compression bearing capacity of composite T-shaped concrete-filled steel tubular columns. *China Civ. Eng. J.* **2009**, *42*, 14–21.
30. Montuori, R.; Tisi, A. Comparative analysis and critical issues of the main constitutive laws for concrete elements confined with FRP. *Compos. Part B* **2012**, *43*, 3219–3230. [CrossRef]
31. Richart, F.E.; Brandtzaeg, A.; Brown, A. The Failure of Concrete of Plain and Spirally Reinforced Concrete in Compression. *Univ. Ill. Eng. Exp. Stn. Bull.* **1929**, *190*.
32. Karabinis, A.I.; Rousakis, T.C. FRP Confining Effects on Steel Reinforced Concrete Square Sections Subjected to Axial Load. In Proceedings of the 2nd International fib Congress, Naples, Italy, 5–8 June 2006.
33. General Administration of Quality Supervision. *Tensile Test of Metal Materials-Part 1: Test Method at Room Temperature*; China Construction Industry Press: Beijing, China, 2010.
34. Park, R. Ductility Evaluation from Laboratory and Analytical Testing. In Proceedings of the Ninth World Conference on Earthquake Engineering, Tokyo-Kyoto, Japan, 2–9 August 1988.
35. Huang, Z.; Li, D.; Wang, J. Behaviour and design of ultra-high-strength CFST members subjected to compression and bending. *J. Constr. Steel Res.* **2020**, *175*, 6351. [CrossRef]
36. Li, W.; Han, L.-H. Seismic performance of concrete-filled double-skin steel tubes after exposure to fire: Experiments. *J. Constr. Steel Res.* **2018**, *154*, 209–223. [CrossRef]
37. Hassam, M.; Jia, C. Concentric and eccentric compression performance of multiple-cell cruciform CFSTs. *J. Constr. Steel Res.* **2022**, *192*, 7205. [CrossRef]
38. Hana, L.-H.; Zhao, X.-L. Tests and calculations for hollow structural steel (HSS) stub columns filled with self-consolidating concrete (SCC). *J. Constr. Steel Res.* **2005**, *61*, 1241–1269. [CrossRef]
39. Chen, S.; Zhang, H. Numerical analysis of the axially loaded concrete filled steel tube columns with debonding separation at the steel-concrete interface. *Steel Compos. Struct.* **2012**, *13*, 277–293. [CrossRef]
40. Japan, A.I. *Recommendations for Design and Construction of Concrete Filled Steel Tubular Structures*; Architectural Institute of Tokyo: Tokyo, Japan, 1997. (In Japanese)
41. *GB 50936-2014*; Technical Code for Concrete Filled Steel Tubular Structures. MOHURD: Beijing, China, 2014.
42. Güneyisi, E.M.; Mermerdaş, K. Ultimate Capacity Prediction of Axially Loaded CFST Short Columns. *Int. J. Steel Struct.* **2016**, *16*, 99–114. [CrossRef]
43. *EN 1994-1-1 E.C.f. Eurocode 4 (EC4)*; Design of Composite Steel and Concrete Structures, in Part 1-1: General Rules and Rules for Buildings. European Council: Brussels, Belgium, 2005.
44. *CECS 159: 2004*; Technical Specifications for Structures with Concrete-Filled Rectangular Steel Tube Members. ICECS Association: Glasgow, UK, 2004.

buildings

MDPI

Article

The Investigation on Static Stability Analysis for Reticulated Shell with Initial Defect Value Using Stochastic Defect Mode Method

Sheng He [1,2], Xinheng Huang [1], Peng Yu [1,*] and Weijing Yun [1]

[1] Key Laboratory of Disaster Prevention and Structural Safety of Ministry of Education, Guangxi Key Laboratory of Disaster Prevention and Structural Safety, College of Civil Engineering and Architecture, Guangxi University, Nanning 530004, China; hesheng@gxu.edu.cn (S.H.); 1910391025@st.gxu.edu.cn (X.H.); 1910302079@st.gxu.edu.cn (W.Y.)

[2] Guangxi Bossco Environmental Protection Technology Co., Ltd., Nanning 530007, China

* Correspondence: py@gxu.edu.cn

Abstract: Regarding the effect of the initial geometric defect (IGD) on the static stability of single-layer reticulated shells, its distribution pattern and magnitude are the main concerns to researchers. However, the suitable selection of the initial geometric defect magnitude (IGDM) is still a controversial topic. Therefore, it is intended to study the determination of the proper IGDM based on the structure force state (SFS) and the defect coefficient. In order to find out a qualified IGDM, more than 5200 numerical cases are carried out for four types of commonly used single-layer reticulated shells with the span ranging from 40 to 70 m and the rise–span ratio from 1/4 to 1/7, within the random defect mode method, by taking both geometric and material nonlinearity into account. The results show that it is more feasible to set the $L/500$ as IGDM when evaluating the stability of the single-layer reticulated shell. In addition, an updated criterion to identify the SFS at the stability critical state (SCS) is developed.

Keywords: single-layer reticulated shells; stability analysis; random defect mode method; initial geometric defect magnitude; structural force state; defect coefficient

Citation: He, S.; Huang, X.; Yu, P.; Yun, W. The Investigation on Static Stability Analysis for Reticulated Shell with Initial Defect Value Using Stochastic Defect Mode Method. *Buildings* **2022**, *12*, 615. https://doi.org/10.3390/buildings12050615

Academic Editor: Emilio Bastidas-Arteaga

Received: 1 March 2022
Accepted: 26 April 2022
Published: 7 May 2022

1. Introduction

Highlights:

- An accurate calculation method for the stable critical load of the single-layer reticulated shell is proposed.
- Put forward the calculation formula to evaluate the structural force state at stable critical state.
- An updated criterion to identify the structural force state at the stability critical state is developed.
- The initial geometric defect magnitude is determined by the structural force state and the defect coefficient.

Due to the light weight, excellent mechanical properties and economic benefits, single-layer reticulated shells are widely used in long-span space structures [1,2]. In our previous study, the static stability of single-layer reticulated spherical shells with Kiewitt–Sunflower type was discussed [3], and the damage constitutive model for circular steel tube of reticulated shells was proposed [4,5]. However, because of the initial eccentricity of member, the initial bending of member [6,7], the installation deviation of the reticulated shell node [8], etc., the phenomenon that IGD occurs in single-layer reticulated shells is nearly inevitable. In the early documented work, it is demonstrated that the single-layer reticulated shell is quite sensitive to IGD [9–12]. For instance, even a small IGD may lead to a substantial reduction in bearing capacity of the structure [13–16]. SFS regard as one of the reflections of structure stability; the relationship between SFS and IGDM is required for further investigation. Zhu [17] discussed the non-linear buckling load of aluminum alloy reticulated shells with gusset joints, and the result showed that the non-linear buckling load is not highly associated with the bending strain energy ratio and the total strain energy. Shan [18] examined the effect of joint stiffness on the dynamic response of single-layer reticulated shells jointed with bolt–column join, and the joint with large stiffness displayed deep plastic development. In order to exhibit the SFS in essence and ensure the structure safety, it is essential to consider the IGD in the stability analysis of the single-layer reticulated shells.

The influence of IGD on the stability of single-layer latticed shells are primarily charactered from two aspects: the distribution and the magnitude of IGD. With respect to various numerical methods to calculate IGD distribution, the consistent imperfection mode method [19] and random defect modal method [20] are the most popular. In consistent imperfection mode method, the lowest buckling mode is used as the IGD distribution mode, and the stability critical load (SCL) of the structure can be obtained through one direct calculation. However, the SCL obtained by this approach may not be reliable. Within the random defect mode method, the IGD can be allocated stochastically on the reticulated shell structure. Using this strategy, the obtained SCL is more accurate [21,22], though this approach is more expensive. With respect to IGDM, it was adopted as a certain value in early work [23–25]. Until recent decades, it was suggested by the current standard [26] that IGDM should be related to the span of the structure and could be assigned as $L/300$ (L is the structural span). Moreover, He et al. [16] investigated the elastic and elastic–plastic stability of the single-layer inverted catenary cylindrical reticulated shell, then proposed that it is appropriate to assign IGDM as an amplitude of $1/300$ of the structural span in terms of the stability research on latticed shells. A similar conclusion was presented by Liu et al. [27] Meanwhile, Cui [28] evaluated the critical load capacity for the global instability of a spherical latticed shell, the IGDM of $L/300$ exhibited the ability to prevent the rapid decrease in critical load factor for a range of values of maximum IGD. Xiong [29] carried out an elasto-plastic stability analysis of the K6 shell with six different IGDMs, and the result indicated that when the IGDM was larger than $L/300$, the ultimate buckling load of K6 shells tended to be stable. However, according to the standard [26] and refs. [16,27], the lowest buckling mode is employed as the IGD distribution mode. Therefore, the obtained SCL of the structure is usually not the most critical, which leads to inaccurate analysis results. Guo [30] carried out a stability analysis of a single-layer latticed shell and three type of suspen-dome, then assumed that the installation deviation arranging from $L/500$ to $L/300$ can be regarded as the maximum allowable defect value of the reticulated shell. Chen et al. [12] conducted an experimental measurement of IGD of a real reticulated shell and insisted that the designed value of $L/300$ for IGD appears to be somewhat conservative. Su [31] proposed a new type of joint with the angled slotted-in steel plates, and the numerical result illustrated that the amplitude of IGD larger than $L/1000$ led to a great loss of its ultimate bucking capacity. Shen and Chen [32] pointed out that if the IGDM is too large, the structure may become a distorted structure. As mentioned above, a further discussion on the identification of suitable IGDM is required.

This paper intends to select a more reasonable IGDM which is suitable for reticulated shells with a different type, span, and rise–span ratio. Based on the comprehensive consideration of the SFS, the influence of IGD on the structural stability and structural defect coefficient, we intend to suggest $L/500$ as a more feasible value for IGDM in the stability investigation of single-layer reticulated shells in this work. This paper is organized as follows: in Section 2, the governing equation and constitutive model are described. Meanwhile, the calculation procedure of the SCL and analysis method of a single-layer latticed shell is introduced. In Section 3, the numerical analysis is carried out. In Section 4, the stress of the single-layer reticulated shell is analyzed, and the force state of the members and the structure is defined. Then, an updated criterion to identify the SFS at the SCS is proposed, and it is applied to determine the suitable IGDM. Furthermore, the IGDM is characterized using the defect coefficient. The conclusions are drawn in Section 5, and the recommendations for future research and engineering practice are put forward in Section 6.

2. Material and Methods

2.1. The Governing Equation and Constitutive Model

2.1.1. The Governing Equation

Motivated by the investigation on elastic–plastic bending of beams conducted by Štok and Halilovič [33], we assume that a straight circular steel tube with an area of cross section $A(x)$ is mainly subjected to the bending moment $M_y(x)$, $M_z(x)$ and uniaxial force $N(x)$, as shown in Figure 1a. As the tube studied in this work is slender (ratio $l/D > 5$, where l is the length of the tube and D is outer diameter for cross section), the shear stress is too small compared with the normal stress such that it can be ignored (i.e., $\tau_{xy}(x, y, z) = \tau_{xz}(x, y, z) = 0$). Then the governing equations of stress state distributed on the cross section can be identified according to the Bernoulli–Navier hypothesis on planarity and respective perpendicularity for cross section [34], which reads:

$$\int_{A(x)} \sigma_{xx} \mathrm{d}A = N(x), \tag{1}$$

$$\int_{A(x)} z\sigma_{xx} \mathrm{d}A = M_y(x), \tag{2}$$

$$\int_{A(x)} y\sigma_{xx} \mathrm{d}A = M_z(x), \tag{3}$$

$$\int_{A(x)} \tau_{xy} \mathrm{d}A = 0, \tag{4}$$

$$\int_{A(x)} \tau_{xz} \mathrm{d}A = 0, \tag{5}$$

$$\int_{A(x)} (\tau_{xz}y - \tau_{xy}z) \mathrm{d}A = 0. \tag{6}$$

Here, normal stress $\sigma = \sigma_{xx}(x, y, z)$ is the sum of the stress caused by uniaxial force $N(x)$, bending moments $M_y(x)$ and $M_z(x)$, respectively. The stress distribution along the y and z directions, for instance, is presented in Figure 1b, and it can be observed that before the stress achieves the yield stress σ_{yield}, the mechanical behavior is linear elastic, and it becomes nonlinear when the material comes to the plastic phase. Thus, the elastic–plastic response along the y direction can be expressed by the following:

$$\sigma(x, z) = \begin{cases} \frac{N(x)}{A(x)} + \frac{M_y(x)}{I_y(x)} z, & |z| \leq \frac{d}{2} + \delta_e,\ \sigma \leq \sigma_{yield} \\ \sigma_p(x, z), & |z| > \frac{d}{2} + \delta_e,\ \sigma > \sigma_{yield} \end{cases}, \tag{7}$$

where $I_y(x)$ denotes the moment of inertia with respect to the y axis, δ_e indicates the elastic zone with the z direction and d is the inner diameter. σ_p is the plastic stress to be determined by our proposed constitutive model in the following section.

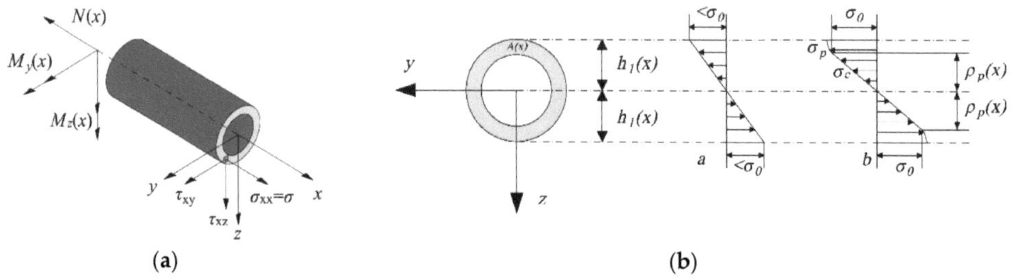

Figure 1. Elastic and elastic–plastic stress distribution in a circular steel tube bending problem. (**a**): A straight circular steel tube subjected to the bending moment $M_y(x)$, $M_z(x)$, and uniaxial force $N(x)$. (**b**): The stress distribution along the y and z directions.

2.1.2. The Constitutive Model

According to the research over past decades, people have generally used the conventional Prandtl–Reuss material model, wherein the elastic–perfectly plastic stress–strain relations are derived on the basis of the von Mises yield criterion and its associated flow rule [35]. The Prandtl–Reuss model is the simplest ideal elasticity model. The material yield function adopts the Mises yield function, and its expression is

$$\phi_{(\sigma_{ij})} = \sqrt{J_2} - k, \tag{8}$$

where J_2 is the second stress tensor invariant, and k is the hardening coefficient. Then, the partial derivative of the yield function Φ with respect to the deviatoric stress tensor S_{ij} is written as

$$\frac{\partial \Phi}{\partial S_{ij}} = \frac{S_{ij}}{2\sqrt{\frac{1}{2}S_{ij}S_{ij}}}. \tag{9}$$

Furthermore, the variation of strain $\delta\varepsilon_{ij}$ reads such that

$$\delta\varepsilon_{ij} = \frac{\delta S_{ij}}{2G} + \frac{\delta\sigma_m}{3K}\delta_{ij} + \frac{d\lambda S_{ij}}{2k}, \tag{10}$$

where G indicates the shear modulus, K represents the bulk modulus, $d\lambda$ is the plastic factor, and σ_m is the average of principal stress. Then, the variation of stress $\delta\sigma_{ij}$ is denoted as

$$\delta\sigma_{ij} = K\delta\varepsilon^p_{kk}\delta_{ij} + 2G\delta e^p_{ij} - G\frac{d\lambda}{k}S_{ij}, \tag{11}$$

where we have

$$d\varepsilon^p_{kk} = 3d\lambda\frac{\partial\Phi}{\partial\sigma_{kk}},\ de^p_{ij} = d\lambda\frac{\partial\Phi}{\partial S_{ij}}. \tag{12}$$

Thus, $d\lambda$, $\delta\varepsilon_{ij}$ and $\delta\sigma_{ij}$ can be written as follows:

$$d\lambda = \frac{\frac{S_{ij}}{2k}\delta e_{ij}}{\frac{1}{2}\frac{S_{ij}S_{ij}/2}{(\sqrt{J_2})^2}}, \tag{13}$$

$$\delta\varepsilon_{ij} = \frac{\delta S_{ij}}{2G} + \frac{\delta\sigma_m}{3K}\delta_{ij} + \frac{S_{mn}\delta e_{mn}}{2k^2}S_{ij}, \tag{14}$$

$$\delta\sigma_{ij} = K\delta\varepsilon^p_{kk}\delta_{ij} + 2G\delta e^p_{ij} - G\frac{S_{mn}\delta e_{mn}}{k^2}S_{ij}. \tag{15}$$

2.2. Methodology

The analysis process on the static stability of the reticulated shell using the stochastic defect mode method is illustrated in Figure 2. The simulation is carried out based on the software ANSYS. It can be seen in Figure 2 that the finite element model of structure is established firstly. Then a reasonable initial geometric defect distribution pattern and magnitude are applied, and the appropriate numerical iterative calculation coefficients are set. The main factors to determine the convergence numerical method are the quality of grids, the reference arc length radius factor and the load steps. Since the grids are fixed once the finite element model is generated, if the numerical iteration is not convergent, the calculation coefficients involving the reference arc length radius factor and the load steps need to be adjusted until the solution satisfies the criterion. Finally, output the load-displacement result.

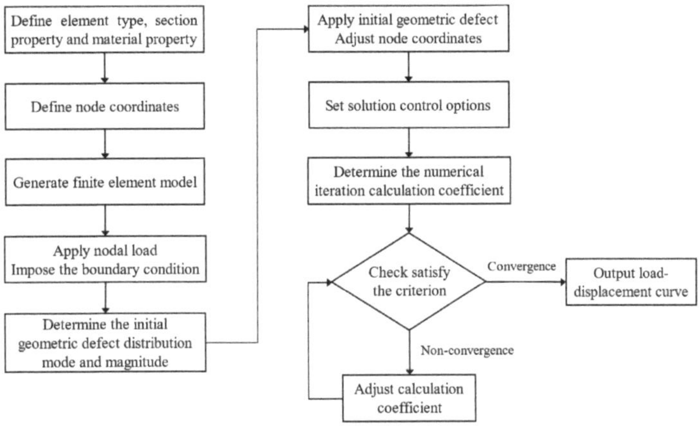

Figure 2. The schematic diagram of static stability analysis for reticulated shell using stochastic defect mode method.

2.2.1. Analysis Model

For the analysis method, the complicated point is how to propose the hypothesis to impose the IGDM using the random defect mode method. In this work, two basic assumptions are adopted in the following:

1. The installation deviation of each node in three directions of the coordinate axis conforms to the normal probability density function within the two-fold mean variance range, that is, the random variable of the installation deviation of each node is $\delta X/2$ (δ is the maximum allowable installation deviation, namely, the maximum calculated value of the initial geometric defect), and the random variable X obeys the standard normal distribution. The range of the error random variable is $[-\delta, \delta]$;
2. The random variable of each node installation error for the structure is mutually independent.

Based on the above assumptions, the installation deviation of each node of the structure is one multidimensional independent random variable, and each space sample point corresponds to one initial defect distribution pattern of the structure. Therefore, n samples can be taken out for nonlinear stability analysis, and the corresponding n SCL can be obtained.

In order to obtain the reasonable IGDM, it is essential to determine the SCL under the each IGDM. The calculation steps of the SCL of the reticulated shell within the random defect mode method are as follows:

1. The installation deviation with random variable $RW/2$ is introduced for each node of calculation model (R is the maximum allowable installation deviation, and W is a random variable, which obeys the standard normal distribution). For each nonlinear buckling analysis of the model, a SCL can be obtained.
2. After repeated calculation for n times, n sample space for SCL is generated.
3. The SCL sample space follows the standard normal distribution, and hence the SCL of each model is determined by the "3σ" principle [36].

The SCL sample space capacity is set as n = 100.

2.2.2. Calculation Model

The difficulty of the calculation model is in computing the structural nonlinear stability during the development from the stable state to unstable state. It is required to study the equilibrium routine for the whole computation process. In this work, the arc-length method [37–39] is used to track the equilibrium path. The method is briefly introduced as follows. The linear finite element increment equation based on the energy variation principle can be expressed as

$$\boldsymbol{K}_{\mathrm{t}}\Delta\boldsymbol{U}^{(\mathrm{i})} = \boldsymbol{F}_{\mathrm{t+\Delta t}} - \boldsymbol{N}_{\mathrm{t+\Delta t}}^{(\mathrm{i-1})}, \tag{16}$$

and employing the incremental displacement strategy proposed by Batoz and Dhatt [40], we can rewrite the Equation (16) by following

$$\boldsymbol{K}_{\mathrm{t}}\Delta\overline{\boldsymbol{U}}^{(\mathrm{i})} = \lambda_{\mathrm{t+\Delta t}}^{(\mathrm{i})}\boldsymbol{F} - \boldsymbol{N}_{\mathrm{t+\Delta t}}^{(\mathrm{i-1})}, \tag{17}$$

in which

$$\Delta\boldsymbol{U}^{(\mathrm{i})} = \Delta\overline{\boldsymbol{U}}^{(\mathrm{i})} + \lambda_{\mathrm{t+\Delta t}}\Delta\overline{\boldsymbol{U}}^{*(\mathrm{i})}, \tag{18}$$

$$\Delta\boldsymbol{U}_{\mathrm{t+\Delta t}}^{(\mathrm{i})} = \Delta\boldsymbol{U}_{\mathrm{t+\Delta t}}^{(\mathrm{i-1})} + \Delta\boldsymbol{U}^{(\mathrm{i})}, \tag{19}$$

$$\Delta\lambda_{\mathrm{t+\Delta t}}^{(\mathrm{i})} = \Delta\lambda_{\mathrm{t+\Delta t}}^{(\mathrm{i-1})} + \Delta\lambda^{(\mathrm{i})}, \tag{20}$$

where $\boldsymbol{K}_{\mathrm{t}}$ is the structural tangential stiffness matrix at time t, $\Delta\boldsymbol{U}^{(\mathrm{i})}$ is the iteration displacement increment at current time step, $\boldsymbol{F}$ is the load vector, and $\lambda_{\mathrm{t+\Delta t}}^{(\mathrm{i})}$ is the proportional coefficient of load at the i-th iteration. There are N + 1 unknown $\Delta\boldsymbol{U}^{(\mathrm{i})}$ and $\Delta\lambda^{(\mathrm{i})}$ but only N linear equations above. Therefore, a constraint equation is demanded. For different types of arc-length methods, the constraint equations are different. Here, two kinds of constraint equations are presented, which read as follows:

(1) Spherical arc-length method

$$\left(\Delta\lambda_{\mathrm{t+\Delta t}}^{(\mathrm{i})}\right)^2 + \left(\boldsymbol{U}^{(\mathrm{i})}\right)^{\mathrm{T}}\boldsymbol{U}^{(\mathrm{i})} = \Delta l^2, \tag{21}$$

(2) Cylindrical arc-length method

$$\left(\boldsymbol{U}^{(\mathrm{i})}\right)^{\mathrm{T}}\boldsymbol{U}^{(\mathrm{i})} = \Delta l^2, \tag{22}$$

where Δl is the increment of the arc length at each iteration.

The arc-length method uses an arc-length increment to determine the loading step, and the calculation is proceeded along the arc direction of the curve so that it is more adaptable than other methods. Regarding two arc-length methods referred above, the load increment method is used at the first calculation, and the structure displacement vector $\boldsymbol{U}_{\Delta\mathrm{t}}$ is obtained after iteration until it is convergent. Starting from the second step of the calculation, $\boldsymbol{U}_{\Delta\mathrm{t}}$ is substituted into Equations (21) or (22) to calculate the arc-length increment $\Delta l'$. The arc-length increment needs to be calculated before each computation, which yields the following:

$$\Delta l = \sqrt[4]{(N_1/N_2)^3} \cdot \Delta l', \tag{23}$$

where Δl is the arc length increment of the current computation step, N_1 is the assumed optimal number of iterations at each step, and N_2 is the number of iterations in the previous calculation step. Although the arc-length method is quite adaptable, it still has some limitations to some extent. It is suggested that adopting the iterative strategy flexibly and combining multiple methods reasonably is a good way to achieve an optimal balanced-path tracking.

2.2.3. Numerical Model

In the present paper, for the four types of commonly used, single-layer reticulated shells (as shown in Figure 3) are investigated through geometric and material nonlinear analysis within the random defect mode method, with the magnitude of $\delta = L/1000$, $L/900$, $L/800$, $L/700$, $L/600$, $L/500$, $L/400$, $L/300$, $L/200$, and $L/100$ (δ is the IGDM). Parameters of single-layer reticulated shells are shown in Table 1. Steel with the elastic modulus $E = 2.06 \times 10^6$ MPa, the Poisson ratio $\nu = 0.3$ and the yield strength $f_y = 240$ MPa were determined as the material properties of calculation models. The roof static load is 0.8 kN/m^2, and the roof live load, which is considered the whole span arrangement, is 0.5 kN/m^2.

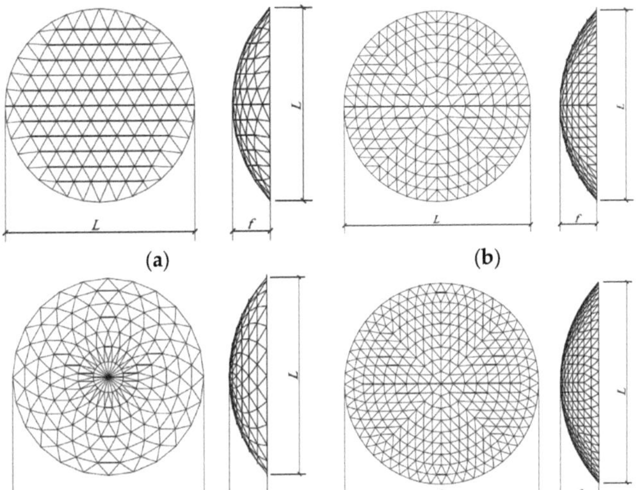

Figure 3. The single-layer reticulated shell models. (**a**) Geodesic; (**b**) Kiewitt; (**c**) Sunflower; (**d**) Kiewitt 8-Sunflower.

Table 1. Parameters of single-layer reticulated shells.

Type	L(m)	f/L	Radial and Hoop Members	Oblique Members
			Outer Diameter × Wall Thickness (mm)	
Geodesic	40	1/7, 1/6, 1/5, 1/4	Φ121 × 5	Φ121 × 5
Kiewitt 8	50		Φ168 × 6	Φ152 × 5
Sunflower	60		Φ180 × 6	Φ168 × 5
Kiewitt 8-Sunflower	70		Φ219 × 7/Φ152 × 6	Φ180 × 6/Φ203 × 6

3. Results

In this paper, 4000 cases of elastic–plastic stability analysis were conducted. The numerical results of SCL were based on 100 cases and are presented in Table 2.

Table 2. Maximum SCL under different IGDMs.

δ	Geodesic (f/L = 1/7)			Kiewitt 8 (f/L = 1/5)			Sunflower (f/L = 1/6)			Kiewitt 8-Sunflower (f/L = 1/4)		
	Δ (mm)	F_δ (kN/m^2)	β_δ	Δ (mm)	F_δ (kN/m^2)	β_δ	Δ (mm)	F_δ (kN/m^2)	β_δ	Δ (mm)	F_δ (kN/m^2)	β_δ
0	0	5422	—	0	10.237	—	0	5158	—	0	9.810	—
$L/1000$	40	3815	0.70	50	6953	0.68	60	3603	0.70	70	6918	0.71
$L/900$	44	3498	0.65	56	6616	0.65	67	3387	0.66	78	6566	0.67
$L/800$	50	3222	0.59	63	6255	0.61	75	3154	0.61	88	6123	0.62
$L/700$	57	3007	0.55	71	5846	0.57	86	3001	0.58	100	5699	0.58
$L/600$	67	2794	0.52	83	5482	0.54	100	2788	0.54	117	5284	0.54
$L/500$	80	2540	0.47	100	4869	0.48	120	2497	0.48	140	4886	0.50
$L/400$	100	2432	0.45	125	4206	0.41	150	2196	0.43	175	4412	0.45
$L/300$	133	2262	0.42	167	3979	0.39	200	1998	0.39	233	3953	0.40
$L/200$	200	1958	0.36	250	3532	0.35	300	1656	0.32	350	3217	0.33
$L/100$	400	1488	0.27	500	2911	0.28	600	0963	0.19	700	2112	0.22

Note: Δ, F_δ, and β_δ denote IGD, the maximum SCL in 100 samples with the same IGDM, and the F_δ ratio of the imperfect model to the perfect model, respectively.

It is documented in the literature [32] that when the IGDM reaches $L/300$, the SCL falls to the minimum value. Then, with the increase in IGDM, the SCL performs an increase trend. Therefore, it is suggested in ref. [32] that the $L/300$ be regarded as the IGDM of the reticulated shell. Nevertheless, as it can be seen in Table 2, for all the models, with the increase in IGDM, the SCL always falls down, even until the end of calculation ($\delta = L/100$). It is clear that the obtained result in our work is different from that in ref. [32]. The reasons can be attributed to two points by the following. (1) The random imperfections modal method is adopted in our work, while the consistent mode imperfection method is employed in ref. [32]. (2) Only the geometric nonlinearity of the structure is considered in ref. [32], whereas this paper takes into account the geometric and material nonlinearity. According to previous research, due to the large span of the reticulated shell, the plastic response of the member quite affects the mechanical behavior of the structure. He et al. [16] shows that the elastic–plastic ultimate strength of the structure could be computed by taking into account a plastic influence coefficient of 0.60 on the basis of the elastic result. Liu et al. [27] demonstrated that the ultimate bearing capacity of the latticed shell that considers the influence of material nonlinearity can be determined by multiplying the result of the ultimate bearing capacity of the latticed shell that does not consider the material nonlinearity by a reduction factor of 0.742. Hence, in order to find out a qualified IGDM, the double nonlinearity of the structure is used to analyze the single-layer reticulated shell structure within the random defect mode method in this manuscript.

4. Discussion

4.1. Stress Analysis

It is well known that the latticed shell structure is developed from a thin shell structure. In the early stability analysis of the latticed shell structure, the "method of simulated shell" is used to transform the reticulated shell structure into a continuous shell structure, and then some approximate nonlinear methods are used to solve the SCL of the reticulated shell structure. Although the finite element method is widely used, the analysis method of the thin shell theory can clearly understand the force characteristics of the reticulated shell structure. The internal forces generated by external forces in the reticulated shell structure are divided into "film internal force" and "bending internal force", which correspond to the "no-moment theory" and "moment theory" of the thin shell structure, respectively. It can be seen from the force analysis of the reticulated shell by adopting the bending moment theory and the non-bending moment theory that the film internal force mainly drives the axial stress, while the bending internal force mainly triggers the bending stress and shear stress (including the torsional shear stress and transverse shear stress). In the stability analysis of the single-layer reticulated shell, the increase in IGDM not only reduces the SCL of structure, but also changes the SFS at SCS. Although the accepted reticulated shell structure in actual engineering inevitably has initial geometric defects, it is controlled in a small range, and hence the SFS is still dominated by the film internal force at SCS. If excessive IGD is applied in the stability analysis of latticed shell, the SFS may not conform to its actual working state, and the structure stability cannot be truly reflected. Therefore, this section puts forward the calculation formula to judge the SFS at SCS.

4.1.1. Analysis of the Member Force State

The identification of the SFS at SCS needs to be based on the analysis of the member force state. In order to study the member force state at SCS, the ratio of normal stress produced by the bending moment and axial force (RBA) is defined as

$$\alpha_{\mathrm{i}} = |\sigma_{\mathrm{Mi}}/\sigma_{\mathrm{Ni}}|, \tag{24}$$

where α_{i}, σ_{Mi} and σ_{Ni} are the RBA, maximum bending stress and axial stress of the *i*-th member, respectively. It should be pointed out that since the torsional and transverse shear stress of the member are rather smaller than the axial and bending stress during stability analysis of reticulated shell, the RBA can be used to represent the member force state directly. In this manuscript, the member with RBA less than 0.50 is defined as the member in which the force caused by axial stress dominates (MAS); otherwise, it is the member with the force led by both axial stress and bending stress (MAB).

4.1.2. Analysis of the SFS

Based on the classification of the member force state, the SFS at SCS can be examined subsequently. As discussed above, the member of the latticed shell structure is either MAS or MAB. This paper defined the following variables to represent the percentage of MAS and MAB among all the effective members, which read

$$\gamma_{\mathrm{N}} = \left(\sum_{s=1}^{j} \alpha_{\mathrm{Ns}}\right) \Big/ j, \tag{25}$$

$$\eta_{\mathrm{N}} = j/m, \tag{26}$$

$$\gamma_{\overline{\mathrm{N}}} = \left(\sum_{s=1}^{k} \alpha_{\overline{\mathrm{N}}\mathrm{s}}\right) \Big/ k, \tag{27}$$

$$\eta_{\overline{\mathrm{N}}} = k/m, \tag{28}$$

where γ_{N}, $\sum_{s=1}^{j} \alpha_{\mathrm{Ns}}$ and j are the average of the RBA of the MAS, sum of the RBA of the MAS, and number of the MAS, respectively, η_{N} represents the percentage of MAS to effective member, $\gamma_{\overline{\mathrm{N}}}$, $\sum_{s=1}^{k} \alpha_{\overline{\mathrm{N}}\mathrm{s}}$ and k are the mean value of the RBA of the MAB, sum of the RBA of the MAB and number of the MAB, respectively, $\eta_{\overline{\mathrm{N}}}$ denotes the percentage of MAB to the effective member, and m is the number of effective members. As the bending stress and axial stress of the outermost ring member connected with the fixed hinge support are small, the value of RBA appears abnormal. Moreover, the outermost ring rod connected with the fixed hinge support has little influence on the overall buckling behavior of the structure. Therefore, when calculating the percentages of two kinds of members, the ring members at the support are not taken into account, which means the number of effective members. Meanwhile, m is computed by the difference between the number of all members and the number of outermost ring members.

The random defect mode method requires multiple calculations, and each calculation can produce a percentage of MAS and RBA of MAS. Therefore, the average of multiple calculations is applied by the following definitions:

$$\varphi_{\mathrm{N}} = \left(\sum_{s=1}^{n} \gamma_{\mathrm{Ns}} \right) \Big/ n, \quad (29)$$

$$\theta_{\mathrm{N}} = \left(\sum_{s=1}^{n} \eta_{\mathrm{Ns}} \right) \Big/ n, \quad (30)$$

$$\varphi_{\overline{\mathrm{N}}} = \left(\sum_{s=1}^{n} \gamma_{\overline{\mathrm{N}}\mathrm{s}} \right) \Big/ n, \quad (31)$$

$$\theta_{\overline{\mathrm{N}}} = \left(\sum_{s=1}^{n} \eta_{\overline{\mathrm{N}}\mathrm{s}} \right) \Big/ n, \quad (32)$$

where φ_{N} indicates the average of the MAS with RBA, θ_{N} represents the average percentage of MAS to effective member, $\varphi_{\overline{\mathrm{N}}}$ expresses the average of the MAB with RBA, and $\theta_{\overline{\mathrm{N}}}$ denotes the average of percentage of MAB to effective member. Hence, the SFS is determined by the average of percentage of MAS to effective member and the average of RBA.

4.2. Selection of IGDM Based on the SFS

4.2.1. Determination Criteria for SFS

This section uses the same calculation model (four types of commonly used single-layer reticulated shells) as those in Section 2.2.2 to analyze the SFS under the SCS. The SFS under different IGDMs are shown in Table 3.

Table 3. The SFS under different IGDMs.

δ	Geodesic (f/L = 1/7)			Kiewitt 8 (f/L = 1/5)			Sunflower (f/L = 1/6)			Kiewitt 8-Sunflower (f/L = 1/4)		
	Δ (mm)	φ_{N}	θ_{N} (%)	Δ (mm)	φ_{N}	θ_{N} (%)	Δ (mm)	φ_{N}	θ_{N} (%)	Δ (mm)	φ_{N}	θ_{N} (%)
0	0	0.169	82.8	0	0.141	92.9	0	0.175	89.0	0	0.183	89.3
$L/1000$	40	0.186	82.1	50	0.165	83.8	60	0.198	80.2	70	0.195	83.6
$L/900$	44	0.195	80.5	56	0.177	82.2	67	0.204	78.5	78	0.201	82.3
$L/800$	50	0.206	79.5	63	0.190	79.9	75	0.215	77.2	88	0.211	81.1
$L/700$	57	0.218	78.3	71	0.203	77.6	86	0.223	75.5	100	0.226	79.8
$L/600$	67	0.224	75.0	83	0.225	75.3	100	0.234	73.4	117	0.235	78.3
$L/500$	80	0.235	71.7	100	0.249	72.0	120	0.240	70.5	140	0.244	76.0
$L/400$	100	0.281	65.3	125	0.279	63.9	150	0.282	65.0	175	0.293	68.5
$L/300$	133	0.325	54.4	167	0.319	55.2	200	0.306	55.4	233	0.323	54.5
$L/200$	200	0.411	42.9	250	0.402	41.0	300	0.376	40.2	350	0.409	40.0
$L/100$	400	0.505	22.2	500	0.498	21.2	600	0.437	20.9	700	0.499	20.5

It is clear in Table 3 that when the IGDM is less than $L/500$, with the increase in the IGDM, the decrease in the MAS with percentage and the increase in the MAS with RBA are slow. While after the IGDM is greater than $L/500$, with the increase in the IGDM, the drop of the MAS with percentage and the rise in the MAS with RBA are sharp. It indicates that with the rise in the IGDM, the SFS at SCS changes from the film internal force to the bending internal force. In this paper, the dual-control principle is proposed to judge the SFS at the SCS. To be specific, when the average of the percentage of MAS to the effective member is larger than 0.70, and meanwhile, in these members, the average of RBA is not larger than 0.25, and the SFS is dominated by the film internal force. As it is mentioned in Section 4.1, the SFS of the reticulated shell can be interpreted to be dominated by either the internal film force or bending moment, depending on the percentage of the MAS member (namely, the value of θ_N in Equation (30)). When considering the allowed IGDM for the reticulated shell in practical engineering, it is safe to control the SFS of the structure to perform as an internal film force. As it is illustrated in Table 3, before the IGDM δ reaches the $L/500$, the θ_N for all the four types of reticulated shell structures (Geodesic, Kiewitt 8, Sunflower, and K8-Sunflower) decreases steadily and the value of it remains beyond 70%. After that, the value of θ_N drops remarkably with the rise in IGDM δ, and its value falls beneath 50%, and to 20% finally. This means that the SFS of the shell structure changes to be dominated by the bending moment.

4.2.2. Determination of the IGDM

In this section, the value of IGDM is allocated as $L/500$ and $L/300$, respectively, which aims to figure out the effect of IGDM on the SFS at the critical situation. Four types of single-layer reticulated shells (Geodesic L = 40 m, Kiewitt 8 L = 50 m, Sunflower L = 60 m and Kiewitt 8-Sunflower L = 70 m) with a rise–span ratio (f/L = 1/4, 1/5, 1/6 and 1/7) are investigated within the random defect mode method and the 100 samples (n = 100). The SFS under the IGDM with $L/500$ and $L/300$ are reported in Table 4. The average of the percentage of MAS and the average of RBA of MAS are presented in Figures 4 and 5, respectively.

Table 4. The SFS under the IGDM with $L/500$ and $L/300$.

Type	L(m)	f/L	$\delta = L/500$		$\delta = L/300$	
			φ_N	θ_N(%)	φ_N	θ_N(%)
Geodesic	40	1/4	0.244	79.0	0.326	48.6
		1/5	0.240	73.0	0.326	58.9
		1/6	0.232	74.3	0.325	56.4
		1/7	0.235	71.7	0.325	54.4
Kiewitt 8	50	1/4	0.248	75.8	0.314	51.2
		1/5	0.249	72.0	0.319	55.2
		1/6	0.236	71.1	0.326	48.5
		1/7	0.240	75.1	0.328	43.4
Sunflower	60	1/4	0.236	74.0	0.303	51.5
		1/5	0.246	76.4	0.306	42.2
		1/6	0.240	70.5	0.306	55.4
		1/7	0.238	77.6	0.308	43.6
Kiewitt 8-Sunflower	70	1/4	0.244	76.0	0.323	54.5
		1/5	0.246	72.8	0.325	42.7
		1/6	0.242	70.5	0.326	49.5
		1/7	0.248	75.6	0.327	45.9

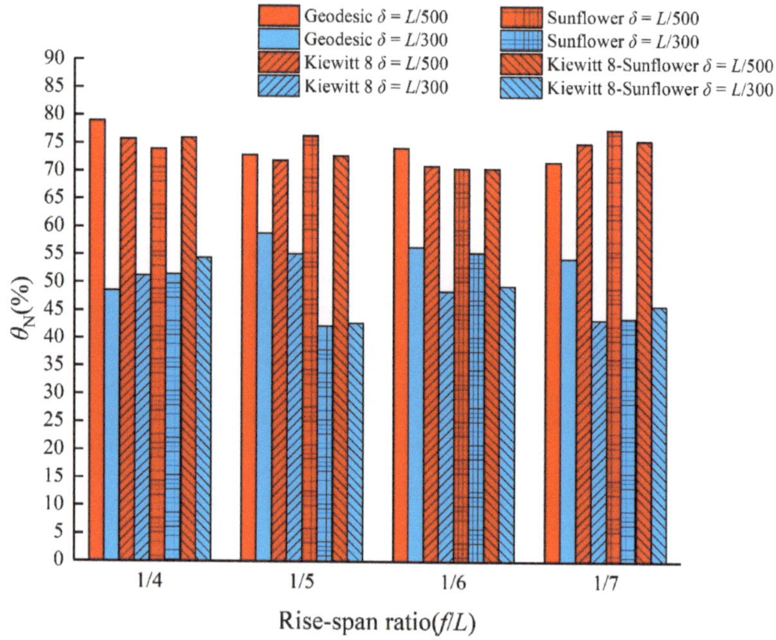

Figure 4. The average of percentage of MAS with $L/300$ and $L/500$.

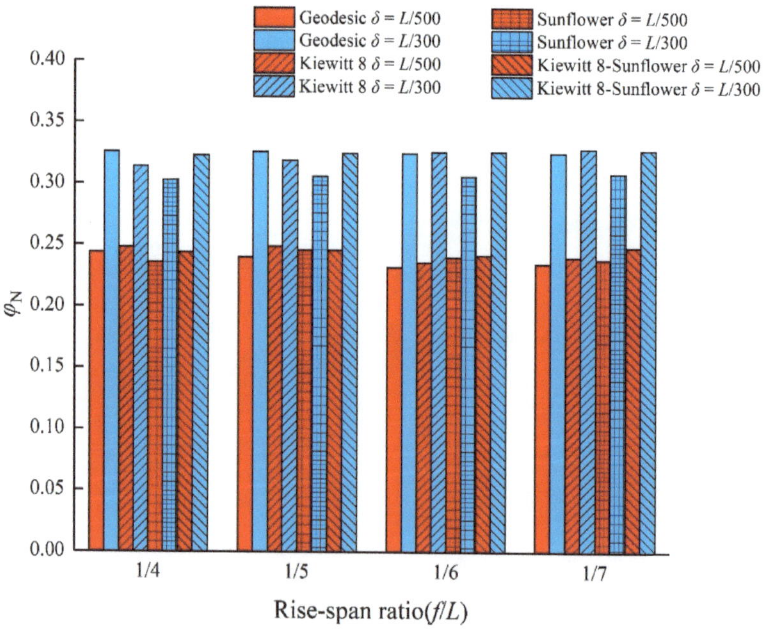

Figure 5. The average of RBA of MAS with $L/300$ and $L/500$.

For each model, when the IGDM is $L/500$, the average of the percentage of MAS is over 70%, and the average of RBA is less than 0.25. It demonstrates that the film internal force is the main SFS. For each model, when the IGDM rises to $L/300$, the average of the percentage of MAS is less than 50%, and the average of RBA is greater than 0.30. It denotes that the SFS at SCS changes from the film internal force to the bending internal force (from Figures 4 and 5). Therefore, it can be concluded that it is more reasonable to take the IGDM as $L/500$ in the stability analysis of the single-layer reticulated shell.

4.3. Selection of the IGDM Based on the Defect Coefficient

4.3.1. Choosing of the Defect Coefficient

To some extent, the IGD determines the SCL of the structure. In order to quantify this effect, this paper introduced a defect factor β_δ to represent the decreasing amplitude of SCL on the latticed shell structure, which is expressed as follows:

$$\beta_\delta = F_\delta / F_0, \tag{33}$$

where δ is the IGDM, and F_δ and F_0 are the SCL of imperfected and perfect structures, respectively. In this paper, the defect coefficient β_δ is utilized to help to identify the IGDM. As we can see in Figure 6, for each structure, β_δ can clearly reflect the influence of IGD on the SCL. It was suggested by Shen and Chen [32] and Fan et al. [9] that 50% of the SCL of the perfect lattice shell should be defined as the SCL of the defected latticed shell. It was proposed in ref. [16] that the ultimate bearing capacity of the structure with considering the geometric imperfection influence can be obtained through the perfect structure results multiplied by the reduction coefficient of 0.46. Therefore, it is appropriate to take the defect coefficient around 0.50. Note that the defect coefficient β_δ defined in Equation 33 relies on the IGDM δ. As it is referred in Sections 4.1 and 4.2.2, the primary factor to identify IGDM is to guarantee that the SFS of the shell mainly behaves as the internal film force, which is determined by the key parameter θ_N defined in Equation (30). Recall the definition of θ_N; we find that it derives from the coefficient α_i (RBA) in Equation (24). Look at the expression of α_i in Equation (24); we recognize that it involves the performance of bending stress σ_{Mi} and axial stress σ_{Ni}, which is dependent on the material property. Hence, we assume that our defect model can be extended to different materials.

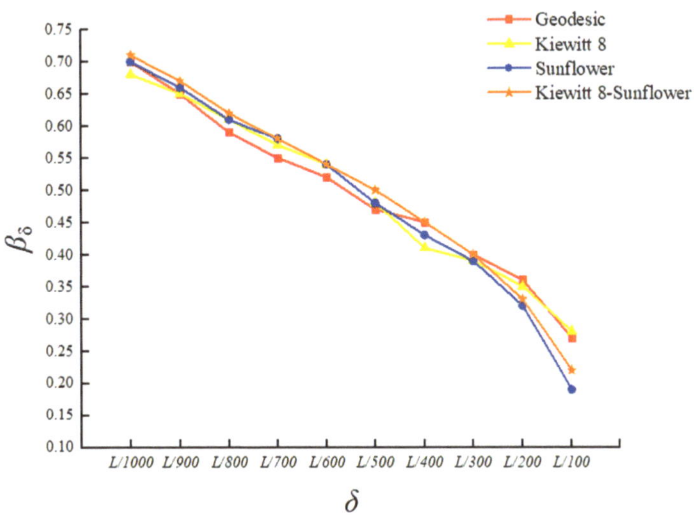

Figure 6. The relationship between defect factor (β_δ) and IGDM (δ).

As it is plotted in Figure 6, when the IGDM is $L/1000$, the defect coefficient β_δ is about 70% for four kinds of reticulated shell structures, which means that the stability critical load (SCL) drops around 30%. With the increase in IGDM, the values of β_δ decrease gradually with the nearly linear trend until the IGDM arrives $L/300$. Afterwards, the β_δ falls sharply to a small magnitude finally (around 17% for the lowest value). This is because, as we mentioned in Section 4.2, as long as the IGDM is over $L/300$, the SFS transfer from the state is dominated by the internal film force to the situation controlled by the bending moment, which indicates that the shell structure is not so stable anymore, leading to the tremendous reduction in SCL. Therefore, in a word, it is assumed that the defect coefficient β_δ is sensitive to IGDM δ in this work.

4.3.2. Identification of the IGDM

The same calculation model as those in Section 4.2.2 was used to figure out the feasible IGDM under the defect coefficient. The defect coefficient under the IGDM with $L/300$ and $L/500$ is shown in Table 5; when the IGDM is $L/500$, the defect coefficient of each model is around 0.48. As the IGDM increases to $L/300$, nearly all defect coefficients of the calculation model are less than 0.40. Thus, the selected IGDM is equal to $L/500$, which is more reasonable for the single-layer reticulated shell; meanwhile, it demonstrates that 0.50 as the defect coefficient is reasonable.

Table 5. Defect coefficient under the IGDM with $L/300$ and $L/500$.

Type	L (m)	f/L	F_0 (kN/m^2)	$\delta = L/500$		$\delta = L/300$	
				F_δ (kN/m^2)	β_δ	F_δ (kN/m^2)	β_δ
Geodesic	40	1/4	9598	4650	0.48	3793	0.40
		1/5	8829	4527	0.51	3314	0.38
		1/6	6802	3396	0.50	2722	0.40
		1/7	5422	2540	0.47	2262	0.42
Kiewitt 8	50	1/4	12,077	5997	0.50	4795	0.40
		1/5	10,237	4869	0.48	3979	0.39
		1/6	9154	4438	0.48	3418	0.37
		1/7	8163	3861	0.47	3059	0.37
Sunflower	60	1/4	8179	4012	0.49	3196	0.39
		1/5	6050	2972	0.49	2419	0.40
		1/6	5158	2497	0.48	1998	0.39
		1/7	4545	1996	0.44	1817	0.40
Kiewitt 8-Sunflower	70	1/4	9810	4886	0.50	3953	0.40
		1/5	8646	4229	0.49	3371	0.39
		1/6	7538	3722	0.49	2963	0.39
		1/7	5940	2780	0.47	2484	0.42

5. Conclusions

Based on four types of commonly used single-layer reticulated shells (Geodesic, Kiewitt 8, Sunflower, and Kiewitt 8-Sunflower), more than 5200 numerical cases of the elastic–plastic load-displacement of single-layer reticulated shells were investigated within the random defect mode method. Afterwards, an updated criterion to identify the structure force state at the stability critical state (SCS) was developed, and the reasonable initial geometric defect magnitude (IGDM) in the stability analysis of the single-layer reticulated shell was discussed. The main conclusions can be drawn as follows:

1. Increasing the initial geometric defect magnitude (IGDM) within a rational range always leads to a fall down of the stability critical load (SCL), and it does not perform a smooth or increase trend as did the earlier research [32].
2. It is feasible to select 0.50 as the defect coefficient, which is better to describe the influence of the initial geometric defect (IGD) on the structural stability.
3. The structure force state at the stability critical state (SCS) could be estimated by the ratio of normal stress produced by the bending moment and axial force. Briefly speaking, when the average of percentage of the member in which the force caused by the axial stress dominates (MAS) and is larger than 0.70 and meanwhile in these

members, the average of the ratio of normal stress produced by the bending moment and axial force (RBA) is less than 0.25, the structural force state is dominated by the film internal force.

4. When the initial geometric defect magnitude (IGDM) is adopted as $L/300$, the structure force state (SFS) is not the film internal force, and hence in this case, the initial geometric defect (IGD) on the structural stability is overestimated. Therefore, considering the structure force state (SFS) and the influence of the initial geometric defect (IGD) on the structural stability, it is more proper to select $L/500$ as the initial geometric defect magnitude (IGDM).

6. Recommendation for Future Research and Engineering Practice

Future research in static stability analysis for reticulated shells is required to consider the influence of more parameters, including the initial geometric defect of the bar member, the semi-rigid connection, etc.

It is more reasonable to appropriately relax the requirement of the construction acceptance in the maximum coefficient of the initial geometric defect. The defect coefficient with 0.50 is recommended.

Author Contributions: Conceptualization, S.H. and P.Y.; methodology, S.H.; software, X.H. and W.Y.; validation, S.H., X.H. and W.Y.; formal analysis, S.H.; investigation, X.H. and W.Y.; resources, S.H.; data curation, X.H. and W.Y.; writing—original draft preparation, S.H.; writing—review and editing, P.Y.; visualization, X.H.; supervision, P.Y.; project administration, P.Y.; funding acquisition, S.H. and P.Y. All authors have read and agreed to the published version of the manuscript.

Funding: This study was funded by Guangxi Nature Science Foundation (2018JJB160052), Application of Key Technology in Building Construction of Prefabricated Steel Structure (BB30300105), Research Grant for 100 Talents of Guangxi Plan, Guangxi Key Research and Development Program (AB22036007), Guangxi Ba-Gui Scholars Program (2019A33), Guangxi Major Research Program (AB19259013).

Data Availability Statement: Some or all data, models, or code generated or used during the study are proprietary or confidential in nature and may only be provided with restrictions.

Conflicts of Interest: The authors declare that they have no conflict of interest.

Nomenclature

IGD (Δ)	Initial Geometric Defect	R	installation deviation Maximum allowable
IGDM (δ)	Initial Geometric Defect Magnitude	W	Random variable
SFS	Structure Force State	K_t	Structural tangential stiffness matrix at time t
SCS	Stability Critical State	$\Delta u^{(i)}$	Iteration displacement increment at current time step
SCL	Stability Critical Load	F	Load vector
RBA (α_i)	Ratio of normal stress produced by Bending moment and Axial force	$\lambda_{t+\Delta t}^{(i)}$	Proportional coefficient of load at the i-th iteration
MAS	Member in which the force caused by Axial Stress dominates	Δl	Increment of arc length at each iteration
MAB	Member with the force led by both Axial stress and Bending stress	N_1	Assumed optimal number of iterations at each step
L	Structural span	N_2	Number of iterations in the previous calculation step
A	Area of cross section of straight circular steel tube	E	Elastic modulus
M_y	Bending moment with y direction	ν	Poisson ratio

M_z	Bending moment with z direction	f_y	Yield strength
N	Uniaxial force	f	Structural rise
l	Length of the tube	β_δ	SCL ratio of the imperfect model to the perfect model
D	Outer diameter for cross section	σ_{Mi}	Maximum bending stress
τ_{xy}	Shear stress with y direction	σ_{Ni}	Axial stress of the i-th member
τ_{xz}	Shear stress with z direction	γ_N	Average of the RBA of the MAS
σ_{xx}	Normal stress	$\sum_{s=1}^{j} \alpha_{Ns}$	Sum of the RBA of the MAS
σ_{yield}	Yield stress	j	Number of the MAS
I_y	Moment of inertia with respect to the y axis	η_N	Percentage of MAS to effective member
δ_e	Elastic zone with z direction	$\gamma_{\overline{N}}$	Mean value of the RBA of the MAB
d	Inner diameter	$\sum_{s=1}^{k} \alpha_{\overline{N}s}$	Sum of the RBA of the MAB
σ_p	Plastic stress	k	Number of the MAB
J_2	Second stress tensor invariant	$\eta_{\overline{N}}$	Percentage of MAB to effective member
k	Hardening coefficient	m	Number of effective members
Φ	Mises yield function	φ_N	Average of the MAS with RBA
S_{ij}	Deviatoric stress tensor	θ_N	Average percentage of MAS to effective member
G	Shear modulus	$\varphi_{\overline{N}}$	Average of the MAB with RBA
K	Bulk modulus	$\theta_{\overline{N}}$	Average of percentage of MAB to effective member
$d\lambda$	Plastic factor	F_δ	SCL of imperfected structure
σ_m	Average of principal stress	F_0	SCL of perfect structure
$\delta\upsilon_{ij}$	Variation of stress		

References

1. Ma, J.; Fan, F.; Zhang, L.; Wu, C.; Zhi, X. Failure modes and failure mechanisms of single-layer reticulated domes subjected to interior blasts. *Thin Wall Struct.* **2018**, *132*, 208–216. [CrossRef]
2. Fan, F.; Wang, D.; Zhi, X.; Shen, S. Failure modes of reticulated domes subjected to impact and the judgment. *Thin Wall Struct.* **2010**, *48*, 143–149. [CrossRef]
3. Yu, P.; Yun, W.; Bordas, S.; He, S.; Zhou, Y. Static Stability Analysis of Single-Layer Reticulated Spherical Shell with Kiewitt-Sunflower Type. *Int. J. Steel Struct.* **2021**, *21*, 1859–1877. [CrossRef]
4. He, S.; Wang, H.; Bordas, S.P.A.; Yu, P. A Developed Damage Constitutive Model for Circular Steel Tubes of Reticulated Shells. *Int. J. Struct. Stab. Dyn.* **2020**, *20*, 2050106. [CrossRef]
5. He, S.; Nie, Y.; Bordas, S.P.A.; Yu, P. Damage-Plastic Constitutive Model of Thin-Walled Circular Steel Tubes for Space Structures. *J. Eng. Mech.* **2020**, *146*, 4020131. [CrossRef]
6. Fan, F.; Yan, J.; Cao, Z. Elasto-plastic stability of single-layer reticulated domes with initial curvature of members. *Thin Wall Struct.* **2012**, *60*, 239–246. [CrossRef]
7. Fan, F.; Yan, J.; Cao, Z. Stability of reticulated shells considering member buckling. *J. Constr. Steel Res.* **2012**, *77*, 32–42. [CrossRef]
8. Zhao, Z.W.; Liu, H.Q.; Liang, B.; Yan, R.Z. Influence of random geometrical imperfection on the stability of single-layer reticulated domes with semi-rigid. *Adv. Steel Constr.* **2019**, *15*, 93–99. [CrossRef]
9. Fan, F.; Cao, Z.; Shen, S. Elasto-plastic stability of single-layer reticulated shells. *Thin Wall Struct.* **2010**, *48*, 827–836. [CrossRef]
10. Elishakoff, I. Uncertain buckling: Its past, present and future. *Int. J. Solids Struct.* **2000**, *37*, 6869–6889. [CrossRef]
11. Bruno, L.; Sassone, M.; Venuti, F. Effects of the Equivalent Geometric Nodal Imperfections on the stability of single layer grid shells. *Eng. Struct.* **2016**, *112*, 184–199. [CrossRef]
12. Chen, G.; Zhang, H.; Rasmussen, K.J.R.; Fan, F. Modeling geometric imperfections for reticulated shell structures using random field theory. *Eng. Struct.* **2016**, *126*, 481–489. [CrossRef]
13. Kashani, M.; Croll, J.G.A. Lower bounds for overall buckling of spherical space domes. *J. Eng. Mech.* **1994**, *120*, 949–970. [CrossRef]
14. Gioncu, V. Buckling of Reticulated Shells: State-of-the-Art. *Int. J. Space Struct.* **1995**, *10*, 1–46. [CrossRef]
15. Kato, S.; Mutoh, I.; Shomura, M. Collapse of semi-rigidly jointed reticulated domes with initial geometric imperfections. *J. Constr. Steel Res.* **1998**, *48*, 145–168. [CrossRef]

16. He, Y.; Zhou, X.; Liu, D. Research on stability of single-layer inverted catenary cylindrical reticulated shells. *Thin Wall Struct.* **2014**, *82*, 233–244. [CrossRef]
17. Zhu, S.; Ohsaki, M.; Guo, X.; Zeng, Q. Shape optimization for non-linear buckling load of aluminum alloy reticulated shells with gusset joints. *Thin Wall Struct.* **2020**, *154*, 106830. [CrossRef]
18. Shan, Z.; Ma, H.; Yu, Z.; Fan, F. Dynamic failure mechanism of single-layer reticulated (SLR) shells with bolt-column (BC) joint. *J. Constr. Steel Res.* **2020**, *169*, 106042. [CrossRef]
19. Chen, X.; Shen, S.Z. Complete Load-Deflection Response and Initial Imperfection Analysis of Single-Layer Lattice Dome. *Int. J. Space Struct.* **1993**, *8*, 271–278. [CrossRef]
20. Yamada, S.; Takeuchi, A.; Tada, Y.; Tsutsumi, K. Imperfection-sensitive overall buckling of single-layer lattice domes. *J. Eng. Mech.* **2001**, *127*, 382–386. [CrossRef]
21. Zhao, J.; Zhang, Y.; Lin, Y. Study on mid-height horizontal bracing forces considering random initial geometric imperfections. *J. Constr. Steel Res.* **2014**, *92*, 55–66. [CrossRef]
22. Liu, H.; Zhang, W.; Yuan, H. Structural stability analysis of single-layer reticulated shells with stochastic imperfections. *Eng. Struct.* **2016**, *124*, 473–479. [CrossRef]
23. See, T.; McConnel, R.E. Large displacement elastic buckling of space structures. *J. Struct. Eng.* **1986**, *112*, 1052–1069. [CrossRef]
24. Borri, C.; Spinelli, P. Buckling and post-buckling behaviour of single layer reticulated shells affected by random Buckling and post-buckling behaviour of single layer reticulated shells affected by random imperfections. *Comput. Struct.* **1988**, *4*, 937–943. [CrossRef]
25. Morris, N.F. Effect of imperfections on lattice shells. *J. Struct. Eng.* **1991**, *117*, 1796–1814. [CrossRef]
26. Ministry of Housing and Urban-Rural Development of the People's Republic of China. *Technical Specification for Space Frame Structure. (JGJ7–2010)*; China Architecture Industry Press: Beijing, China, 2010.
27. Liu, H.; Ding, Y.; Chen, Z. Static stability behavior of aluminum alloy single-layer spherical latticed shell structure with Temcor joints. *Thin Wall Struct.* **2017**, *120*, 355–365. [CrossRef]
28. Cui, X.Z.; Li, Y.G.; Hong, H.P. Effect of spatially correlated initial geometric imperfection on reliability of spherical latticed shell considering global instability. *Struct. Saf.* **2020**, *82*, 101895. [CrossRef]
29. Xiong, Z.; Guo, X.; Luo, Y.; Zhu, S. Elasto-plastic stability of single-layer reticulated shells with aluminium alloy gusset joints. *Thin Wall Struct.* **2017**, *115*, 163–175. [CrossRef]
30. Guo, J.M. Research on distribution and magnitude of initial geometrical imperfection affecting stability of suspen-dome. *Adv. Steel Constr.* **2011**, *7*, 344–358.
31. Shu, Z.; Li, Z.; He, M.; Chen, F.; Hu, C.; Liang, F. Bolted joints for small and medium reticulated timber domes: Experimental study, numerical simulation, and design strength estimation. *Arch. Civ. Mech. Eng.* **2020**, *20*, 1–24. [CrossRef]
32. Shen, S.Z.; Chen, X. *Stability of Lattice Shell Structures*; Science Press of China: Beijing, China, 1999.
33. Štok, B.; Halilovič, M. Analytical solutions in elasto-plastic bending of beams with rectangular cross section. *Appl. Math. Model.* **2009**, *33*, 1749–1760. [CrossRef]
34. Dai, H.H.; Huo, Y. Asymptotically approximate model equations for nonlinear dispersive waves in incompressible elastic rods. *Acta Mech.* **2002**, *157*, 97–112. [CrossRef]
35. Chen, W.-F.; Han, D.-J. *Plasticity for Structural Engineers*; Springer Science & Business Media: Berlin, Germany, 2012.
36. Rice, J.A. *Mathematical Statistics and Data Analysis*; Cengage Learning: Boston, MA, USA, 2006.
37. Crisfield, M.A. A faster modified newton-raphson iteration. *Comput. Methods Appl. Mech. Eng.* **1979**, *20*, 267–278. [CrossRef]
38. Crisfield, M.A. An arc-length method including line searches and accelerations. *Int. J. Numer. Methods Eng.* **1983**, *19*, 1269–1289. [CrossRef]
39. Riks, E. An incremental approach to the solution of snapping and buckling problems. *Int. J. Solids Struct.* **1979**, *15*, 529–551. [CrossRef]
40. Batoz, J.; Dhatt, G. Incremental displacement algorithms for nonlinear problems. *Int. J. Numer. Methods Eng.* **1979**, *14*, 1262–1267. [CrossRef]

buildings

MDPI

Article

Mesoscale Study on Dilation Behavior of Plain Concrete under Axial Compression

Peng Chen [1,2,*], Xiaomeng Cui [1], Huijun Zheng [3] and Shengpu Si [1]

1 Department of Engineering Mechanics, Shijiazhuang Tiedao University, Shijiazhuang 050043, China; cxm971226@163.com (X.C.); sishengpu@163.com (S.S.)
2 Hebei Key Laboratory of Mechanics of Intelligent Materials and Structures, Shijiazhuang Tiedao University, Shijiazhuang 050043, China
3 Department of Construction Engineering, Hebei Petroleum University of Technology, Chengde 067000, China; cdpczhj@126.com
* Correspondence: chenpeng@stdu.edu.cn

Abstract: The dilation of concrete in the radial direction is crucial in understanding the failure process and the key to predicting the confining level of passively confined concrete. To better understand this problem, we established a mesoscale model of concrete by considering the random distribution of coarse aggregate and the different properties between mortar and concrete. The model's validity was demonstrated by comparing with the stress–strain curves in code and the lateral–axial strain curves in test. The simulation results show that the lateral dilation is non-uniformly distributed along the specimen height and the circumferential direction of sections. Moreover, the deformation mainly occurs in the middle part of the specimen ranging from 3/8 to 5/8. The strength of concrete influences the stress ratio at maximum compressive strain, while it slightly influences the stress ratio at zero volumetric strain. The secant strain ratio is about 0.5 as the compressive stress reaches the strength of concrete. Compared with the simulation, the relationship between lateral strain and axial strain proposed by Teng and Binici shows excellent performance on the dilation trend prediction of plain concrete.

Keywords: dilation of concrete; lateral–axial strain relationship; secant strain ratio; mesoscale model; finite element analysis

Citation: Chen, P.; Cui, X.; Zheng, H.; Si, S. Mesoscale Study on Dilation Behavior of Plain Concrete under Axial Compression. *Buildings* **2022**, *12*, 908. https://doi.org/10.3390/buildings12070908

Academic Editors: Shan Gao, Jingxuan Wang, Dewen Kong and Yong Liu

Received: 21 May 2022
Accepted: 23 June 2022
Published: 27 June 2022

1. Introduction

As a composite material, concrete is widely used in buildings, bridges, tunnels, water conservancies, and other engineering projects due to its advantages, such as economy, convenience, high strength, and excellent durability. The economic and safe design of such structures cannot be realized until a deep understanding of the mechanical behavior of concrete is available. At present, the research mainly focuses on the analysis of concrete in the longitudinal direction [1–10], while research on the dilation behavior of concrete in the radial direction is relatively limited. Relevant research is crucial in understanding the failure process of concrete. Moreover, it is also the key to predicting the mechanical behavior of passively confined concrete, because the performance of the developed model depends on how well it captures the dilation tendency of concrete.

The dilation of plain concrete under axial load has attracted the interest of several researchers. Klink studied the distribution of dilation at the section of a concrete specimen experimentally [11], and Allos investigated the influences of compressive strength and age on the dilation of concrete [12]. The volumetric strain of concrete was analyzed in Ferretti's study [13], which was used to describe the biaxial behavior of concrete using damage mechanics [14]. Although relevant studies have been carried out, no theoretical formula for the dilation of plain concrete has been put forward in the above literature. In FRP-confined concrete, the confinement enhances the strength and ductility of concrete [15–19]. The performance of the developed model depends on how well it captures the dilation tendency of concrete. Thus, the dilation property of FRP-confined concrete has been investigated by many scholars [20–27], and can be used to predict its mechanical behavior through an iterative procedure. Harries et al. [24] developed a dilation ratio formula for FRP-confined concrete, in which the ratio kept constant at the value of Poisson's ratio before reaching 66% peak strain. After that, the ratio increased linearly until it reached a maximum value at two times the peak strain. Beyond this point, the increase in this ratio stopped and kept constant. A fractional equation was proposed by Mirmiran to describe the dilation rate of a concrete specimen confined by FRP jacket [25], in which the rate decreased to an asymptotic value after reaching a peak value. A similar conclusion was also obtained in the studies of Pimanmas and Nguyen [26,27] because the increasing confining pressure will inhibit the dilation of concrete. The authors believe that the increasing pressure makes the dilation of concrete specimens confined by FRP jacket differ from that of plain concrete, and thus, these formulas are not suitable for plain concrete. Based on the test data of plain concrete and actively confined concrete, researchers [28–30] proposed the lateral–axial strain relationship formula for concrete specimens confined by constant pressure. The dilation behavior of plain concrete can be obtained when the confining pressure is set as zero. However, the derived dilation curves of plain concrete are different in form and values, because the test data of plain concrete in literature are limited. Moreover, the dilation of plain concrete after reaching the peak stress has a large discreteness due to the local failure and non-uniform distribution of strain. Therefore, dilation analysis of plain concrete from limited experimental data is far from sufficient. The improvement in computing power promotes the development of the mesoscale simulation in concrete, in which the heterogeneity of concrete and the interaction among phases are considered [31–42]. As an essential supplement to experimental research, numerical simulation of dilation behavior will deepen our understanding of the concrete failure process.

The purpose of this study was to propose a new method to predict the dilation behavior of concrete. To achieve this goal, a mesoscale model was first established based on random distribution theory, in which concrete was divided into the mortar, aggregate, and interfacial transition zone (ITZ). After validation of the proposed model, the dilation of plain concrete subjected to axial compression was analyzed. Finally, we compared the predictions of the existing theoretical formulas with the simulation results.

2. Mesoscale Model of Concrete

2.1. Generation of the Coarse Aggregate

In this paper, the Fuller curve [31] was adopted to describe the particle gradation of coarse aggregate, which was proposed based on the theory of maximum density. The cumulative percentage P of the coarse aggregate that passes through the sieve is shown by the following equation.

$$P = 100 \times (d/d_{\max})^n \quad (1)$$

where d is the sieve size; $d_{\max}$ is the maximum diameter of coarse aggregate; n is a constant value ranging from 0.45 to 0.70, and 0.5 was used herein. Therefore, the aggregate volume V_p in size range of d_{s+1} to d_s is shown by Equation (2).

$$V_p = \frac{P(d_s) - P(d_{s+1})}{P(d_{\max}) - P(d_{\min})} \times v_p \times V \quad (2)$$

where V is the total volume of the specimen and v_p is the volume ratio of coarse aggregate.

Similar to the studies in the literature [32–34], the spherical aggregate particles were used in the simulation. The random distribution of coarse aggregate was fulfilled by adopting the Monte Carlo method in MATLAB (version 2016, MathWorks, Natick, MA, USA) until the aggregate content in each size range had been determined. To improve the computational efficiency, coarse aggregate less than 5 mm was assumed as the mortar phase, and a relatively lower aggregate percentage of 30% was used, which was also demonstrated by other studies [35–37]. In the mesoscale model, 10 mm and 20 mm were used as the equivalent diameter representing the aggregate in size range of 5–15 mm and 15–25 mm, respectively. Moreover, all coarse aggregates are in the range of specimen volume and do not intersect with each other.

2.2. Determination of the Three Phases

The specimen was divided into hexahedra elements of the same size using the mapping mesh method, and the attribute of each element can be determined according to the coordinates of element nodes and the center of spherical aggregate. The classification principle of each phase is shown in Figure 1.

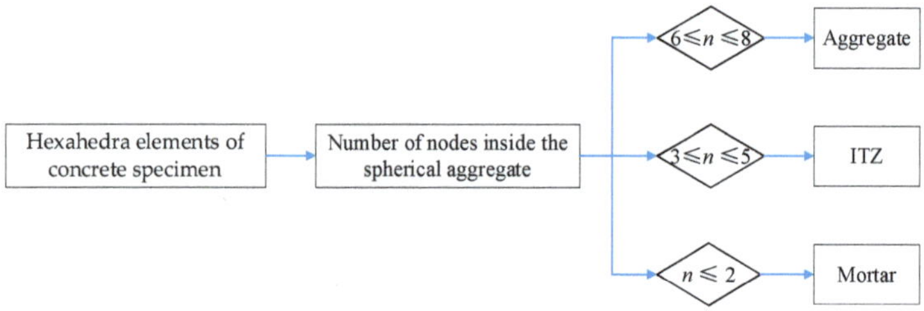

Figure 1. Classification principle of each phase.

Figure 1 shows the classification principle of each phase, and the principle of classification is as follows. (1) If 0–2 nodes of the hexahedra element are inside the spherical aggregate, the property is defined as the mortar phase. (2) If 3–5 nodes are inside the spherical aggregate, the property is defined as the interface phase. (3) If 6–8 nodes are inside the spherical aggregate, the attribute is defined as the aggregate phase.

Based on the studies in the literature [38–40], the actual size of ITZ is about 10–50 μm, and the variation in element size in the range of 0.5–2 mm only has a slight influence on the stress–strain curve of concrete. Thus, 2 mm was used as the element size in the mesoscale simulation to reduce the computational cost, and the same method was also adopted by previous scholars [41,42]. The numerical model is shown in Figure 2.

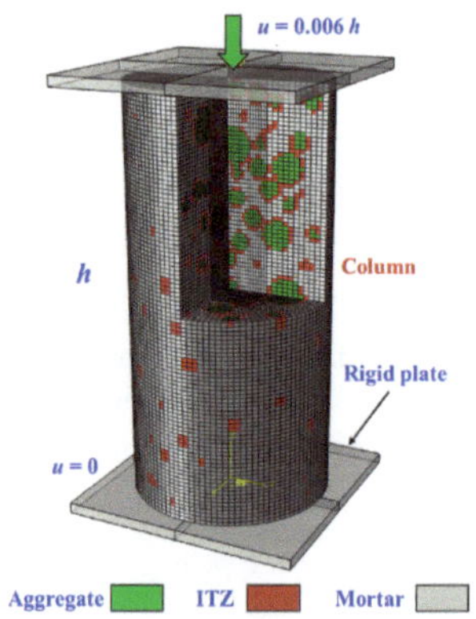

Figure 2. Mesoscale model of concrete cylinder.

In Figure 2, the numerical concrete column is subjected to axial compression, where the coarse aggregate, the mortar, and interfacial transition zone (ITZ) are represented by green, gray, and red colors, respectively.

2.3. Damage Plasticity Model of Concrete

The damage plasticity model established by Lubiliner and Lee et al. [43,44] has been widely used to describe the property of concrete, which accounts for the tension cracking and compression crushing of concrete. The stress–strain relationship of concrete under axial load can be defined by Equations (3) and (4), in which d_t, ε_t, and $\tilde{\varepsilon}_t^{pl}$ represent the damage coefficient, the total strain, and plastic strain under axial tension, respectively. The subscript c represents the variables under axial compression.

$$\sigma_c = (1 - d_c)E_0\left(\varepsilon_c - \tilde{\varepsilon}_c^{pl}\right) \tag{3}$$

$$\sigma_t = (1 - d_t)E_0\left(\varepsilon_t - \tilde{\varepsilon}_t^{pl}\right) \tag{4}$$

In the software of ABAQUS (version 6.14, Dassault Simulia, France), the plastic strain of concrete under compressive and tensile load depends on the definition of inelastic strain $\tilde{\varepsilon}_c^{in}$ and cracking strain $\tilde{\varepsilon}_t^{ck}$ of concrete. The relationship can be described by Equations (5) and (6). Meanwhile, the inelastic strain and cracking strain can be obtained until the constitutive model of the material is available, as shown by Equations (7) and (8). The characteristic length l_e is set as 2 mm, and the symbol w represents the cracking displacement.

$$\tilde{\varepsilon}_c^{pl} = \tilde{\varepsilon}_c^{in} - \frac{d_c}{1 - d_c}\frac{\sigma_c}{E_0} \tag{5}$$

$$\tilde{\varepsilon}_t^{pl} = \tilde{\varepsilon}_t^{ck} - \frac{d_t}{1 - d_t}\frac{\sigma_t}{E_0} \tag{6}$$

$$\tilde{\varepsilon}_c^{in} = \varepsilon_c - \frac{\sigma_c}{E_0} \tag{7}$$

$$\tilde{\varepsilon}_t^{ck} = \frac{w}{l_e} \tag{8}$$

Based on the theory of energy equivalence in the literature [45–47], the damage coefficient of concrete under axial load can be expressed by Equation (9), which reflects the stiffness degradation of the material.

$$d = 1 - \sqrt{\frac{\sigma}{E_0\varepsilon}} \tag{9}$$

2.4. Constitutive Model of Each Phase

Compared with the mortar and ITZ phase, the aggregate particles have a higher strength, and the fracture of aggregate in normal concrete is usually ignored. Therefore, it was assumed as an elastic body in the mesoscale simulation. The difference between aggregate and mortar in mechanical and thermal properties results in higher porosity and water–cement ratio in ITZ. Nevertheless, the ITZ can be regarded as the mortar with weakened mechanical properties, considering that they have a similar material composition [48]. Thus, the key to the mesoscale model lies in the definition of the constitutive model of the mortar.

Based on the test data and analysis in the literature [49–51], it was found that there existed great differences between the mortar and concrete in aspects of peak strain, elastic modulus, and decreasing index. The authors [52] conducted a parameter analysis to investigate the influence of these parameters on the properties of concrete, and then proposed three correction coefficients to consider the difference. Finally, a constitutive model of the mortar under axial compression and tension was proposed, as shown by Equations (10) and (11) and Equations (12)–(14), respectively. The stress-cracking relationship rather than the stress–strain relationship was used in the tension definition, which can reduce the mesh size dependency in simulation [53,54].

$$y = \begin{cases} 2x - x^2 & x \leq 1 \\ \frac{xr}{r-1+x^r} & x > 1 \end{cases} \tag{10}$$

$$\varepsilon_{m0} = 1260 + 310\sqrt{f_m} \tag{11}$$

where $y = \sigma_m / f_m$, $x = \varepsilon_m / \varepsilon_{m0}$, $r = E_m/(E_m - E_p)$, $E_m = 4000\sqrt{f_m}$, and $E_p = 1.44 f_m / \varepsilon_{m0}$.

$$\sigma_t = \begin{cases} f_t\left(1 - \frac{w}{w_0}\right) & w \leq w_0 \\ 0 & w > w_0 \end{cases} \tag{12}$$

$$f_t = 0.26(1.25 f_c)^{2/3} \tag{13}$$

$$G_f = (1.25 d_{max} + 10)(f_c/10)^{0.7} \times 10^{-3} \quad (\mathrm{N/mm}) \tag{14}$$

2.5. Validation of the Mesoscale Model

The concrete studied in this work is made up of conventional cement and additives, and the compressive strength of the mortar instead of the water–cement ratio was taken as the basic parameter in the mesoscale simulation. This is because the mesoscale model is not able to describe the interaction between water and cement. Figure 3a depicts the stress–strain relationship of the concrete specimen with varying mortar strengths. To reflect the accuracy of the simulation results, the stress–strain curves derived from code GB50010-2010 are also added to this figure. The comparison shows that the mesoscale model can make a satisfactory estimation of the properties of concrete under axial compression. Figure 3b shows the relationship between the mortar strength and peak values of concrete. It can be noticed that the peak stress and peak strain increase linearly with the increase in mortar strength, and the strength relationship between mortar and concrete is shown by Equation (15), which can be utilized to simulate the concrete with different strengths in the following sections.

$$f_c = 2.7 + 0.67 f_m \tag{15}$$

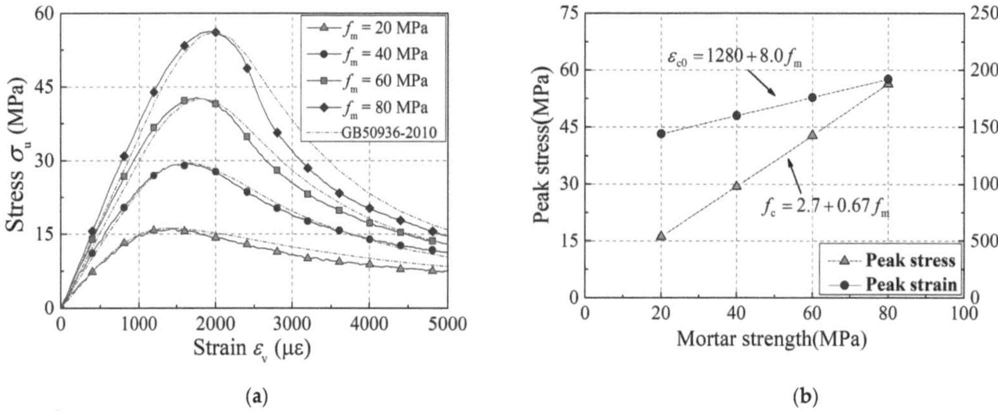

Figure 3. Mechanical properties of concrete specimens with varying mortar strengths: (**a**) stress–strain curve; (**b**) peak values.

To further illustrate the validity of the established mesoscale model, the lateral–axial strain curves of plain concrete with six strengths were collected from the existing literature [55–59], as shown in Figure 4. The strength of concrete is 26 MPa, 35.8 MPa, 40 MPa, 50 MPa, 60 MPa, and 68 MPa, respectively, while the corresponding strength of the mortar in the mesoscale simulation is 34.8 MPa, 49.4 MPa, 55.7 MPa, 70.6 MPa, 85.5 MPa, and 97.5 MPa, according to the strength relationship shown by Equation (15).

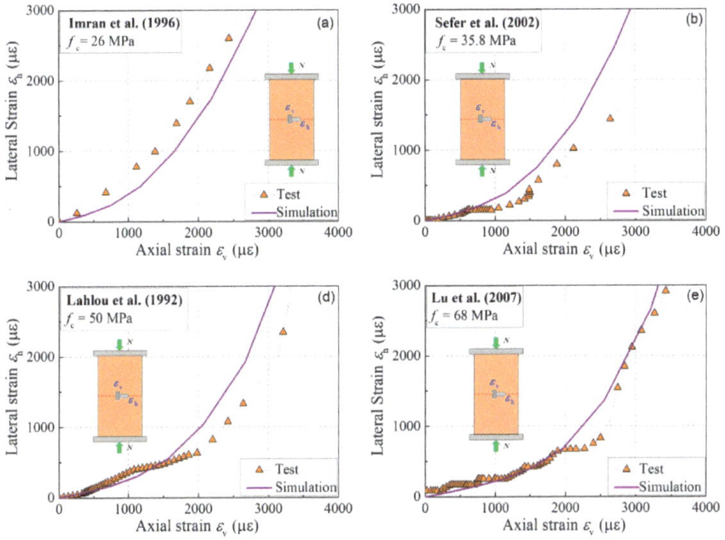

Figure 4. Comparison of the lateral–axial strain relationship: (**a**) f_c = 26 Mpa; (**b**) f_c = 35.8 Mpa; (**c**) f_c = 40/60 Mpa; (**d**) f_c = 50 Mpa; (**e**) f_c = 68 MPa.

Figure 4 shows the comparison of the lateral–axial strain relationship between the simulation and test results in the middle section of the specimen. The established model can make an adequate prediction of the development trend of the lateral–axial strain curve. Although there are some differences in specific values, the model is still deemed effective considering the sensitivity and discreteness of the strain measurement.

In sum, a mesoscale model of concrete was established in this section, which can be utilized to investigate the dilation properties of plain concrete. Based on this, the dilation law of concrete under axial compression is discussed in the following section.

3. Results

3.1. Failure Process of the Concrete Specimen

The properties of concrete specimens subjected to axial compression were studied through parameter analysis, and it was found that they had a similar failure process. Figure 5 shows the schematic diagram of axial strain and lateral strain. In the simulation, the axial displacement Δh and radial displacement Δr were extracted and divided by the height h and diameter r of the specimen, respectively, as shown in Equations (16) and (17), and the axial strain ε_v of the specimen and lateral strain ε_h at the middle section could be obtained. The secant strain ratio is the ratio between lateral strain and axial strain, as shown in Equation (18).

$$\varepsilon_v = \frac{\Delta h}{h} \quad (16)$$

$$\varepsilon_h = \frac{\Delta r}{r} \quad (17)$$

$$\mu_s = \frac{\varepsilon_h}{\varepsilon_v} \quad (18)$$

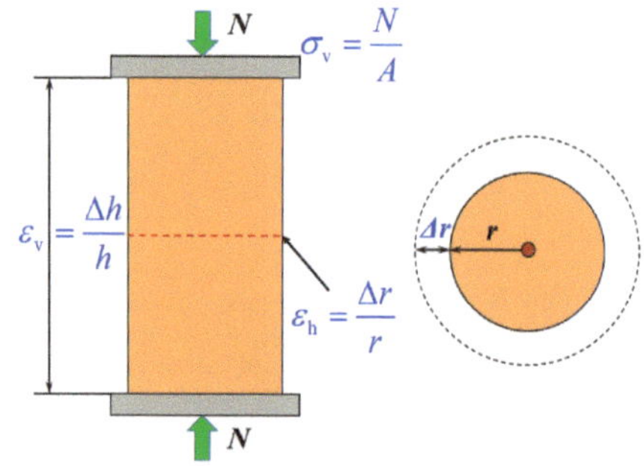

Figure 5. The axial and lateral strain of the specimen.

Figure 6 depicts the development of the stress–strain curve and secant strain ratio–strain curve. It shows that the secant strain ratio increases slowly when the axial strain is less than 1000 με, because the cracks in the specimen grow slowly in the elastic stage. The result is consistent with the study conducted by Spoelstra and Marques [60,61]. With the increase in axial strain, the secant strain ratio increases rapidly.

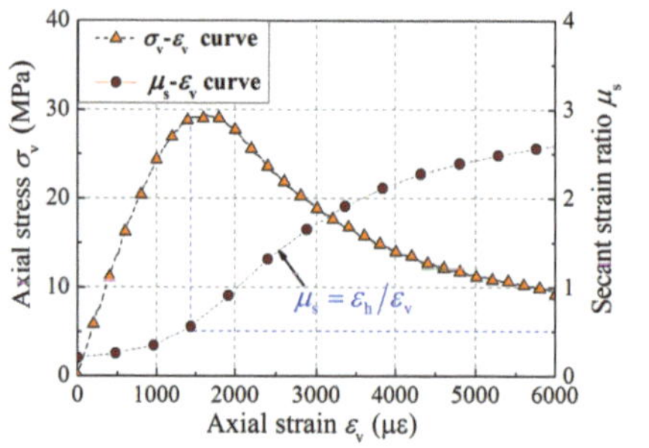

Figure 6. Secant strain ratio of concrete under axial load.

The axial stress is close to concrete strength as the secant strain ratio reaches 0.5. The secant strain ratio increases slowly after the axial strain reaches 4000 με, which may be related to the larger value of the current axial strain. The residual stress is about 1/3 of the concrete strength as axial strain reaches 6000 με. At this time, the cracks in concrete have been fully developed and the secant strain ratio is about 2.5, which is close to the statistical results in the literature [28].

3.2. *Distribution of Lateral Strain in the Specimen*

Seven sections were selected on the specimen to study the distribution of lateral strain along the specimen height, as shown in Figure 7a. The specimen height is designed as 200 mm, and the spacing of each section is 25 mm. Based on Equation (17), the lateral strain of each section can be obtained by extracting the radial displacement. Figure 7b reflects the distribution of lateral strain along the specimen height. It shows that the lateral strain is relatively uniform along the specimen height when the axial strain is less than 1000 με. The lateral strain at the middle section increases sharply as the axial strain increases. The main reason is that the middle region of the specimen is less affected by the end constraint and the crack development is intensive. This phenomenon is also observed in Figure 8, where the radial displacement field at axial strains of 1000, 2000, and 4000 με is depicted. The figure shows that the dilation of concrete mainly occurs in the middle part of the specimen ranging from 3/8 to 5/8, where the stiffness degradation of the material is the most serious, as shown in Figure 8d.

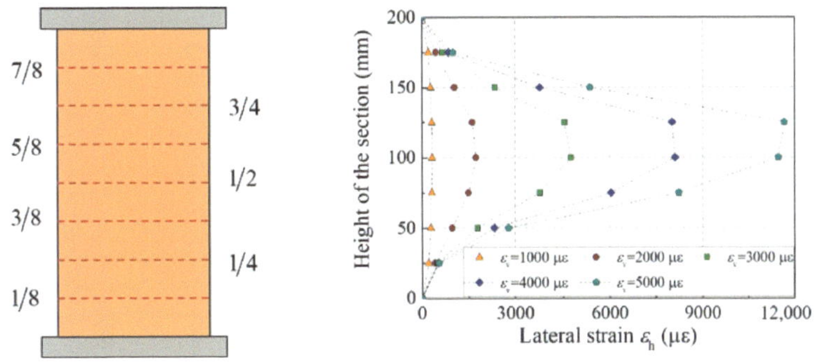

(**a**) (**b**)

Figure 7. Distribution of the lateral strain along the specimen height: (**a**) location of the measured section; (**b**) distribution of the lateral strain.

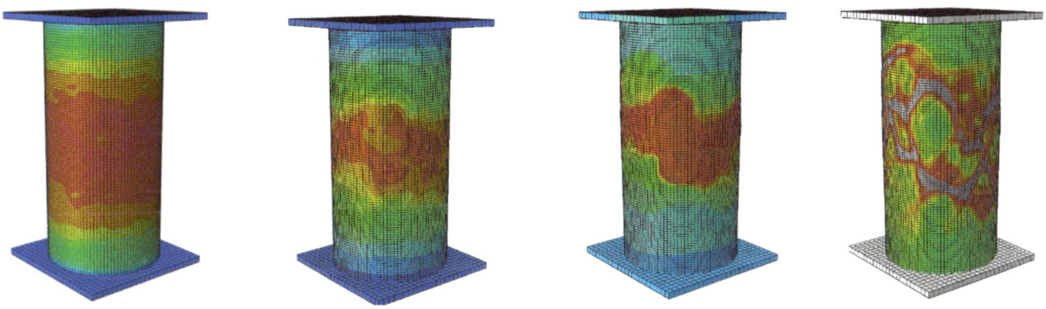

(**a**) (**b**) (**c**) (**d**)

Figure 8. Field of radial displacement and stiffness damage in concrete specimen: (**a**) Ur at 1000 με; (**b**) Ur at 2000 με; (**c**) Ur at 4000 με; (**d**) SDEG at 2000 με.

To study the distribution of lateral strain along the circumferential direction, the lateral strains of 50 points along the middle section of the specimen were extracted, as shown in Figure 9.

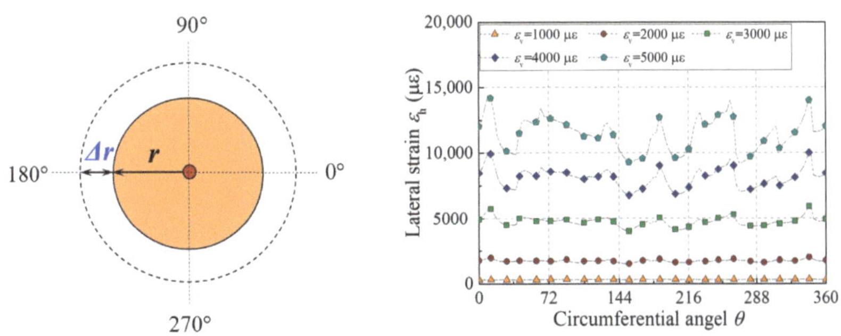

(**a**) (**b**)

Figure 9. Distribution of the lateral strain along circumferential direction: (**a**) middle section of the specimen; (**b**) distribution of the lateral strain.

One can see from Figure 9 that the lateral strain at the middle section is evenly distributed along the circumferential direction when the axial strain is less than 2000 με.

As the axial strain reaches 3000 με, the lateral strain along the circumferential direction begins to fluctuate, which shows that the strain starts to distribute non-uniformly. This is because the extended cracks pass through different positions at the middle section in this stage, resulting in differences in radial displacement.

3.3. Lateral–Axial Strain Relationship for Plain Concrete

The cracks mainly concentrate in the middle part of the specimen, and the lateral strain develops quickly in this area. Thus, the average strain at the sections of 3/8, 1/2, and 5/8 is taken as the lateral strain of the specimen, and the ratio between lateral strain and axial strain is defined as the secant strain ratio. Figure 10 shows the lateral strain–axial strain curves and secant strain ratio–axial strain curves of concrete with different strengths under axial compression. It shows that the development of lateral strain and secant strain ratio can be divided into three stages. The lateral strain is small and the secant strain ratio increases slowly from 0.2 in the first stage, because the fracture development is less in this stage. When the axial strain exceeds 1000 με, the internal cracks continue to accumulate, leading to a rapid increase in lateral strain and secant strain ratio. Moreover, the lateral strain and the secant strain ratio of concrete with higher strength are smaller than those with lower strength. The lateral–axial strain relationship is approximately linear in the last stage as the axial strain exceeds 4000 με. At the same time, the secant strain ratio grows slowly and approaches its critical value with the increase in axial strain.

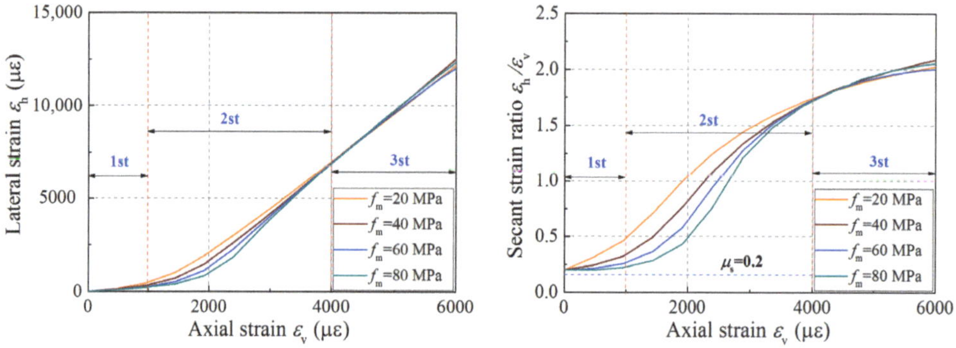

(a) (b)

Figure 10. Dilation behavior of concrete in radial direction: (**a**) ε_h–ε_v curve; (**b**) μ_s–ε_v curve.

The volumetric strain is an important parameter to illustrate the failure process of concrete, and its definition is shown by Equation (19). Figure 11a depicts the stress–volumetric strain relationship of the specimens with different strengths under axial compression. One can see that the volume of the specimen decreases slowly due to the compression, and then expands rapidly with the development of internal cracks.

Figure 11b depicts the stress ratio at maximum compressive strain and zero volumetric strain, where stress ratio represents the ratio between stress at a certain strain and the maximum stress. It shows that the stress ratio at maximum compressive strain increases from 0.64 to 0.82 when the mortar strength varies from 20 MPa to 80 MPa, illustrating that the initiation of cracks starts later in concrete specimens with higher strength. The volume of concrete changes from compression to dilation when the volumetric strain reaches zero, and the stress ratio at this point is approximately equal to 1.0. That is to say, the secant strain ratio is about 0.5 when the axial stress reaches the concrete strength, which is consistent with the literature [28,60,61].

$$\varepsilon_{vol} = \varepsilon_v + 2\varepsilon_h = (1 - 2\mu_s)\varepsilon_v \tag{19}$$

In sum, the development trend of secant strain ratio, the distribution of lateral strain, and the lateral–axial strain relationship of plain concrete under axial compression were studied in this section, which deepens our understanding of the dilation behavior of concrete.

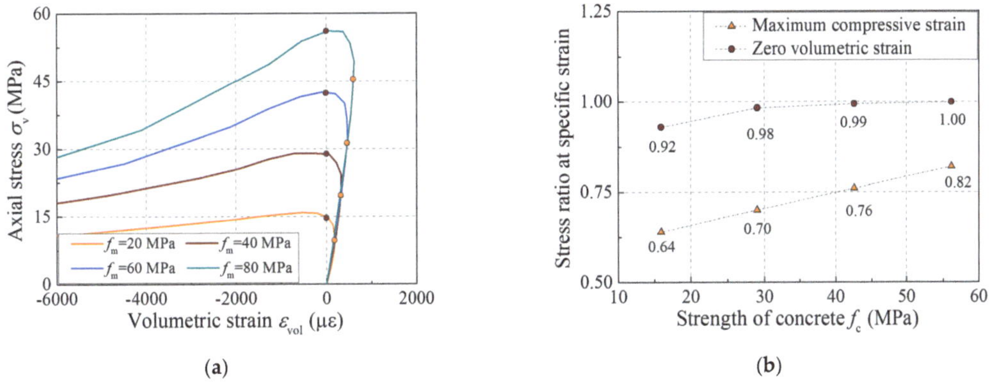

Figure 11. Volumetric dilation behavior of the concrete specimen: (**a**) σ_v–ε_{vol} curve; (**b**) stress ratio at maximum compressive strain and zero volumetric strain.

4. Comparison with Current Formulas

Currently, several formulas are available for determining the lateral–axial strain relationship of concrete specimen. To illustrate the accuracy of the formulas for different strengths of concrete, three formulas, namely, the Binici model [28], Teng model [29], and Lim model [30], are utilized to make a comparison.

Equation (20) was proposed by Binici [28] to describe the secant strain ratio of concrete under axial compression. In the elastic stage, the secant strain ratio keeps constant at the value of Poisson's ratio, which is between 0.15 and 0.2. With the increase in axial strain, the secant strain ratio increases nonlinearly. The authors believe an ultimate strain ratio exists as the axial strain becomes much larger, and the critical value can be calculated by Equation (21). The diameter ϕ is the ratio between the confinement pressure and concrete strength, and the value is set as zero for plain concrete.

$$\mu_s = \begin{cases} \mu_0 & \varepsilon \le \varepsilon_e \\ \mu_l - (\mu_l - \mu_0)\exp\left[-\left(\frac{\varepsilon_v - \varepsilon_e}{\Delta}\right)^2\right] & \varepsilon > \varepsilon_e \end{cases} \tag{20}$$

$$\Delta = \frac{\varepsilon_{co} - \varepsilon_e}{\sqrt{-\ln\beta}} \quad \beta = \frac{\mu_l - \mu_p}{\mu_l - \mu_0} \quad \mu_l = \mu_p + \frac{1}{(\phi + 0.85)^4} \tag{21}$$

where μ_0 represents Poisson's ratio of concrete, and 0.2 is adopted here; μ_p represents the secant strain ratio as axial strain reaches peak strain, and 0.5 is set as the value; μ_l represents the critical secant strain ratio when the axial strain is much larger.

Based on the test data of plain concrete and confined concrete, Teng et al. [29] proposed a lateral–axial strain curve for confined concrete, which can be used to study the properties of FRP confined concrete. The equation is shown as follows:

$$\frac{\varepsilon_v}{\varepsilon_{co}} = 0.85\left(1 + 8\frac{p}{f_{co}}\right)\left\{\left[1 + 0.75\left(\frac{\varepsilon_h}{\varepsilon_{co}}\right)\right]^{0.7} - \exp\left[-7\left(\frac{\varepsilon_h}{\varepsilon_{co}}\right)\right]\right\} \tag{22}$$

where p is the confining pressure; its value is zero for plain concrete.

After that, Lim et al. [30] also proposed a lateral–axial strain curve for concrete based on statistical test data shown by Equation (23). The strain relationship for plain concrete can be obtained when the confining pressure is set as zero. A parameter n was introduced in this formula to consider the influence of concrete strength on initial turning point of the curves, and its value is defined by Equation (24).

$$\varepsilon_{\mathrm{v}} = \frac{\varepsilon_{\mathrm{h}}}{\mu_0\left[1+\left(\frac{\varepsilon_{\mathrm{h}}}{\mu_0\varepsilon_{\mathrm{co}}}\right)^n\right]^{1/n}} + 0.04{\varepsilon_{\mathrm{h}}}^{0.7}\left[1+21\left(\frac{p}{f_{\mathrm{co}}}\right)^{0.8}\right] \tag{23}$$

$$n = 1 + 0.03 f_{\mathrm{co}} \tag{24}$$

Figure 12 compares the secant strain ratio–axial strain relationship between simulation and formula predictions. The comparison illustrates that the simulation results are consistent with those predicted by formulas of Teng et al. and Binici et al. [28,29] when the axial strain is less than 2000 με. At this stage, the stress–strain curve of concrete is in the ascending branch. However, the formula proposed by Lim et al. [30] makes a relatively more significant difference due to the longer constant stage. The stress–strain curve of concrete enters the descending branch as the axial strain exceeds 2000 με, and a certain degree of difference occurs between the simulation and formulas proposed by Teng et al. and Binici et al. [28,29]. Nevertheless, the simulation result is in the range of the theoretical results, which demonstrates the validation of the mesoscale model in predicting the dilation of plain concrete. Therefore, the mesoscale simulation is an efficient tool to investigate the dilation behavior of concrete.

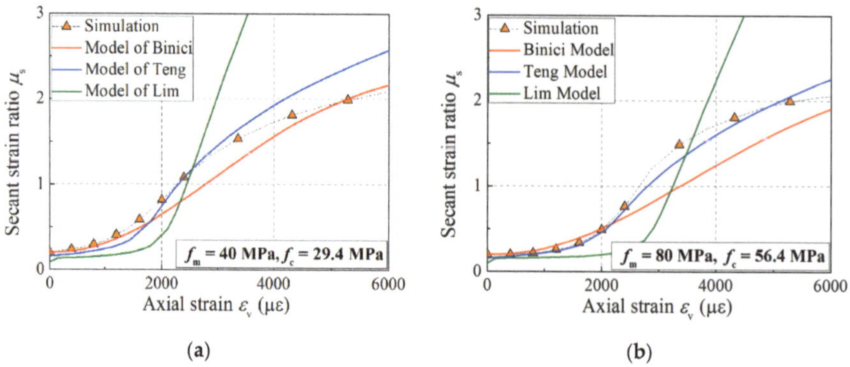

Figure 12. The secant strain ratio–strain relationship: (**a**) f_{m} = 40 MPa; (**b**) f_{m} = 80 MPa.

5. Conclusions

In this work, we established a mesoscale model of concrete by considering the random distribution of coarse aggregate and the different properties between mortar and concrete. The validity of the proposed model was demonstrated by code curves and test data; then, the failure process of the concrete, the distribution of lateral strain, and the lateral–axial strain relationship of concrete were analyzed. The main conclusions are summarized as follows:

(1) The lateral strain is non-uniformly distributed along the specimen height, and the lateral deformation mainly occurs in the middle part of the specimen ranging from 3/8 to 5/8, where the stiffness degradation is the most serious. Moreover, the lateral strain along the circumferential direction becomes non-uniform as axial strain reaches 3000 με.

(2) The development of lateral strain and secant strain ratio can be divided into three stages. In the first stage, the lateral strain is small and the secant strain ratio increases slowly from 0.2. In the second stage, the lateral strain and the secant strain ratio increase rapidly as the internal cracks continue to accumulate. In the third stage, the lateral–axial strain curve is approximately linear, and the secant strain ratio grows slowly and approaches its critical value.

(3) The strength of concrete influences the stress ratio at maximum compressive strain, and the stress ratio varies from 0.62 to 0.82 when the mortar strength varies from 20 MPa to 80 MPa, illustrating that the initiation of cracks starts later in concrete with higher strength. Moreover, the secant strain ratio is about 0.5 as the stress reaches the concrete strength.

(4) The secant strain ratio–axial strain curves in the simulation are consistent with the results predicted by the formulas of Teng et al. and Binici et al. in the ascending branch of the stress–strain curve of concrete. When the stress–strain curve of concrete enters the descending branch, the simulation of the secant strain ratio is in the range of the theoretical results, which demonstrates the validation of our mesoscale model in predicting the dilation of plain concrete.

We have proposed a new method to predict the dilation behavior of concrete, and the relevant analysis will deepen our understanding of the failure process of concrete. The content in this manuscript is part of our current work, based on which we are trying to investigate the dilation of concrete columns under varying confining pressures, and then propose a formula that can be applied to both unconfined and confined concrete.

Author Contributions: Conceptualization, P.C.; methodology, P.C.; software, X.C.; validation, X.C.; formal analysis, P.C. and H.Z.; investigation, H.Z.; writing—original draft preparation, S.S.; writing—review and editing, H.Z.; visualization, S.S.; supervision, H.Z.; funding acquisition, P.C. All authors have read and agreed to the published version of the manuscript.

Funding: This research was funded by the National Natural Science Foundation of China (grant number 51908381).

Institutional Review Board Statement: Not applicable.

Informed Consent Statement: Not applicable.

Data Availability Statement: Not applicable.

Conflicts of Interest: The authors declare no conflict of interest.

References

1. Mander, J.B.; Priestley, M.J.N.; Park, R. Theoretical stress-strain model for confined concrete. *J. Struct. Eng.* **1988**, *114*, 1804–1826. [CrossRef]
2. Teng, J.G.; Lam, L. Compressive behavior of carbon fiber reinforced polymer-confined concrete in elliptical columns. *J. Struct. Eng.* **2002**, *128*, 1535–1543. [CrossRef]
3. Han, L.H.; An, Y.F. Performance of concrete-encased CFST short columns under axial compression. *J. Constr. Steel Res.* **2014**, *93*, 62–76. [CrossRef]
4. Tang, Y.; Xiao, J.; Zhang, H.; Duan, Z.; Xia, B. Mechanical properties and uniaxial compressive stress-strain behavior of fully recycled aggregate concrete. *Constr. Build. Mater.* **2022**, *323*, 126546. [CrossRef]
5. Zheng, D.; Song, W.; Fu, J.; Xue, G.; Li, J.; Cao, S. Research on mechanical characteristics, fractal dimension and internal structure of fiber reinforced concrete under uniaxial compression. *Constr. Build. Mater.* **2020**, *258*, 120351. [CrossRef]
6. Ma, L.; Li, Z.; Liu, J.; Duan, L.; Wu, J. Mechanical properties of coral concrete subjected to uniaxial dynamic compression. *Constr. Build. Mater.* **2019**, *199*, 244–255. [CrossRef]
7. Shi, X.; Park, P.; Rew, Y.; Huang, K.; Sim, C. Constitutive behaviors of steel fiber reinforced concrete under uniaxial compression and tension. *Constr. Build. Mater.* **2020**, *233*, 117316. [CrossRef]
8. Zhou, W.; Feng, P.; Lin, H. Constitutive relations of coral aggregate concrete under uniaxial and triaxial compression. *Constr. Build. Mater.* **2020**, *251*, 118957. [CrossRef]
9. Yang, L.; Xie, H.; Fang, S.; Huang, C.; Chao, Y.J. Experimental study on mechanical properties and damage mechanism of basalt fiber reinforced concrete under uniaxial compression. *Structures* **2021**, *31*, 330–340. [CrossRef]
10. Zhu, H.; Yu, H.; Ma, H.; Yang, S. Uniaxial compressive stress-strain curves of magnesium oxysulfate cement concrete. *Constr. Build. Mater.* **2020**, *232*, 117244. [CrossRef]
11. Klink, S.A. Poisson's ratio variations in concrete. *Exp. Mech.* **1975**, *15*, 139–141. [CrossRef]
12. Allos, A.E.; Martin, L.H. Factors affecting Poisson's ratio for concrete. *Build. Environ.* **1981**, *16*, 1–9. [CrossRef]
13. Ferretti, E. On Poisson's ratio and volumetric strain in concrete. *Int. J. Fract.* **2004**, *126*, 49–55. [CrossRef]
14. Smahi, R.; Bouafia, Y.; Kachi, M.S. Modeling the biaxial behavior of concrete by damage mechanics with Poisson's ratio variable. *Appl. Mech. Mater.* **2015**, *749*, 391–397. [CrossRef]

15. Raza, A.; Ahmad, A. Prediction of axial compressive strength for FRP-confined concrete compression members. *KSCE J. Civ. Eng.* **2020**, *24*, 2099–2109. [CrossRef]
16. Moran, D.A.; Pantelides, C.P.; Reaveley, L.D. Mohr-coulomb model for rectangular and square FRP-confined concrete. *Compos. Struct.* **2019**, *209*, 889–904. [CrossRef]
17. Pour, A.F.; Nguyen, G.D.; Vincent, T.; Ozbakkaloglu, T. Investigation of the compressive behavior and failure modes of unconfined and FRP-confined concrete using digital image correlation. *Compos. Struct.* **2020**, *252*, 112642. [CrossRef]
18. Ghorbel, E.; Limaiem, M.; Wardeh, G. Mechanical performance of bio-based FRP-confined recycled aggregate concrete under uniaxial compression. *Materials* **2021**, *14*, 1778. [CrossRef]
19. Ribeiro, F.; Sena, J.; Branco, F.G. Júlio, E. 3D finite element model for hybrid FRP-confined concrete in compression using modified CDPM. *Eng. Struct.* **2019**, *190*, 459–479. [CrossRef]
20. Shayanfar, J.; Rezazadeh, M.; Barros, J.A. Analytical model to predict dilation behavior of FRP confined circular concrete columns subjected to axial compressive loading. *J. Compos. Constr.* **2020**, *24*, 04020071. [CrossRef]
21. Shayanfar, J.; Rezazadeh, M.; Barros, J.; Ramezansefat, H. A new dilation model for FRP fully/partially confined concrete column under axial loading. In Proceedings of the 3rd RILEM Spring Convention 2020 Ambitioning a Sustainable Future for Built Environment: Comprehensive Strategies for Unprecedented Challenges, Guimarães, Portugal, 10–14 March 2020; pp. 435–446. [CrossRef]
22. Chang, W.; Zheng, W.; Hao, M. Lateral dilation and limited value of volumetric ratio of stirrups for ultra-high strength concrete confined with spiral stirrups. *Mater. Struct.* **2021**, *54*, 125. [CrossRef]
23. Dong, C.; Kwan, A.; Ho, J. Effects of confining stiffness and rupture strain on the performance of FRP confined concrete. *Eng. Struct.* **2015**, *97*, 1–14. [CrossRef]
24. Harries, K.A.; Kharel, G. Behavior and modeling of concrete subjected to variable confining pressure. *ACI Mater.* **2002**, *99*, 180–189.
25. Mirmiran, A.; Shahawy, M. Dilation characteristics of confined concrete. *Mech. Cohesive Frict. Mater.* **1997**, *2*, 237–249. [CrossRef]
26. Pimanmas, A.; Saleem, S. Dilation characteristics of PET FRP-confined concrete. *J. Compos. Constr.* **2018**, *22*, 04018006. [CrossRef]
27. Nguyen, H.D.; Choi, E.; Park, K. Dilation behavior of normal strength concrete confined by FRP wire jackets. *Constr. Build. Mater.* **2018**, *190*, 728–739. [CrossRef]
28. Binici, B. An analytical model for stress-strain behavior of confined concrete. *Eng. Struct.* **2005**, *27*, 1040–1051. [CrossRef]
29. Teng, J.G.; Huang, Y.L.; Lam, L. Theoretical model for fiber-reinforced polymer-confined concrete. *J. Compos. Constr.* **2007**, *11*, 201–211. [CrossRef]
30. Lim, J.C.; Ozbakkaloglu, T. Lateral strain-to-axial strain relationship of confined concrete. *J. Struct. Eng.* **2015**, *141*, 04014141. [CrossRef]
31. Fuller, W.B.; Thompson, S.E. The laws of proportioning concrete. *Trans. Am. Soc. Civ. Eng.* **1907**, *59*, 7–143. [CrossRef]
32. Wriggers, P.; Moftah, S.O. Mesoscale models for concrete: Homogenization and damage behavior. *Finite Elem. Anal. Des.* **2006**, *42*, 623–636. [CrossRef]
33. Gal, E.; Ganz, A.; Hadad, L. Development of a concrete unit cell. *Int. J. Multiscale Comp.* **2008**, *6*, 499–510. [CrossRef]
34. Shahbeyk, S.; Hosseini, M.; Yaghoobi, M. Mesoscale finite element prediction of concrete failure. *Comput. Mater. Sci.* **2011**, *50*, 1973–1990. [CrossRef]
35. Jin, L.; Xia, H.; Jiang, X. Size effect on shear failure of CFRP-strengthened concrete beams without web reinforcement: Meso-scale simulation and formulation. *Compos. Struct.* **2020**, *236*, 11895. [CrossRef]
36. Donza, H.; Cabrera, O.; Irassar, E.F. High-strength concrete with different fine aggregate. *Cem. Concr. Res.* **2002**, *32*, 1755–1761. [CrossRef]
37. Meddah, M.S.; Zitouni, S.; Belaabes, S. Effect of content and particle size distribution of coarse aggregate on the compressive strength of concrete. *Constr. Build. Mater.* **2010**, *24*, 505–512. [CrossRef]
38. Jin, L.; Yu, W.; Du, X. Meso-scale simulation of size effect on concrete dynamic splitting tensile strength: Influence of aggregate content and maximum aggregate size. *Eng. Fract. Mech.* **2020**, *230*, 106979. [CrossRef]
39. Kim, S.M.; Al-Rub, R.K.A. Meso-scale computational modeling of the plastic-damage response of cementitious composites. *Cem. Concr. Res.* **2011**, *41*, 339–358. [CrossRef]
40. Song, Z.; Lu, Y. Mesoscopic analysis of concrete under excessively high strain rate compression and implications on the interpretation of test data. *Int. J. Impact Eng.* **2012**, *46*, 41–55. [CrossRef]
41. Jin, L.; Yu, W.; Du, X. Mesoscopic numerical simulation of dynamic size effect on the splitting tensile strength of concrete. *Eng. Frac. Mech.* **2019**, *209*, 317–332. [CrossRef]
42. Jin, L.; Chen, H.; Wang, Z. Size effect on axial compressive failure of CFRP-wrapped square concrete columns: Tests and simulations. *Compos. Struct.* **2020**, *254*, 112843. [CrossRef]
43. Lubliner, J.; Oliver, J.; Oller, S. A plastic-damage model for concrete. *Int. J. Solids Struct.* **1989**, *25*, 299–326. [CrossRef]
44. Lee, J.; Fenves, G.L. Plastic-damage model for cyclic loading of concrete structures. *J. Eng. Mech.* **1998**, *124*, 892–900. [CrossRef]
45. Cao, M. Research on damage plastic calculation method of ABAQUS concrete damaged plasticity model. *Transp. Stand.* **2012**, *2*, 51–54.
46. Tian, L.; Hou, J. Reasonable plastic damaged factor of concrete damaged plastic model of ABAQUS. *J. Hubei Univ. (Nat. Sci.)* **2015**, *37*, 340–347.

47. Yao, F.; Guan, Q.; Wang, P. Simulation analysis of damage factor based on ABAQUS damage plasticity model. *Struct. Eng.* **2019**, *35*, 76–81.
48. Grote, D.L.; Park, S.W.; Zhou, M. Dynamic behavior of concrete at high strain rates and pressures: Experimental characterization. *Int. J. Impact Eng.* **2001**, *25*, 869–886. [CrossRef]
49. Su, J. The Research on the Size Effect of Concrete Behavior in Compression and Tension. Hunan: Hunan University 2013. Available online: https://kns.cnki.net/kcms/detail/detail.aspx?dbcode=CDFD&dbname=CDFD1214&filename=1013348096.nh&uniplatform=NZKPT&v=SIF9m1fDwexpnbE4w2N6m5P5aeKE8AWXCZifpfHgEbdPdOzGZlGihSf10XAXu1v3 (accessed on 20 March 2022). (In Chinese).
50. Xiao, J.; Li, W.; Corr, D.J. Effects of interfacial transition zones on the stress-strain behavior of modeled recycled aggregate concrete. *Cem. Concr. Res.* **2013**, *52*, 82–99. [CrossRef]
51. Guo, Z.H.; Zhang, X.Q.; Zhang, D.C. Experimental investigation of the complete stress-strain curve of concrete. *J. Build. Struct.* **1982**, *1*, 1–12.
52. Chen, P.; Liu, J.X.; Cui, X.M.; Si, S.P. Mesoscale analysis of concrete under axial compression. *Constr. Build. Mater.* **2022**, *337*, 127580. [CrossRef]
53. Huang, Y.J.; Yang, Z.J.; Chen, X.W. Monte Carlo simulations of mesoscale dynamic compressive behavior of concrete based on X-ray computed tomography images. *Int. J. Impact Eng.* **2016**, *97*, 102–115. [CrossRef]
54. Jin, L.; Du, M.; Li, D. Effects of cross-section size and transverse rebar on the behavior of short squared RC columns under axial compression. *Eng. Struct.* **2017**, *142*, 223–239. [CrossRef]
55. Imran, I.; Pantazopoulou, S.J. Experimental study of plain concrete under triaxial stress. *ACI Struct. J.* **1996**, *93*, 589–601. [CrossRef]
56. Sefer, D.; Carol, I.; Gettu, R.; Etse, G. Study of the behavior of concrete under triaxial compression. *J. Eng. Mech.* **2002**, *128*, 156–163. [CrossRef]
57. Candappa, D.C.; Sanjayan, J.G.; Setunge, S. Complete triaxial stress-strain curves of high-strength concrete. *J. Mater. Civ. Eng.* **2001**, *13*, 209–215. [CrossRef]
58. Lahlou, K.; Aitcin, P.C.; Chaallal, O. Behavior of high-strength concrete under confined stress. *Cem. Concr. Compos.* **1992**, *14*, 185–193. [CrossRef]
59. Lu, X.; Hsu, C.T.T. Stress-strain relations of high-strength concrete under triaxial compression. *J. Mater. Civil Eng.* **2007**, *19*, 261–268. [CrossRef]
60. Spoelstra, M.R.; Monti, G. FRP-confined concrete model. *J. Compos. Constr.* **1999**, *3*, 143–150. [CrossRef]
61. Marques, S.P.; Marques, D.C.; Siva, J.; Cavalcante, M.A. Model for analysis of short columns of concrete confined by fiber-reinforced polymer. *J. Compos. Constr.* **2004**, *8*, 332–340. [CrossRef]

buildings

MDPI

Article

Associated Effects of Sodium Chloride and Dihydrate Gypsum on the Mechanical Performance and Hydration Properties of Slag-Based Geopolymer

Quan Shen [1,2], Benxiao Li [2], Wei He [2,*], Xia Meng [3] and Yinlan Shen [4]

1 Shanghai Interlink Road & Bridge Engineering Co., Ltd., Shanghai 201213, China; shenquansh@163.com
2 School of Civil and Transportation Engineering, Beijing University of Civil Engineering and Architecture, Beijing 100044, China; libxhi@foxmail.com
3 Architectural Design and Research Institute of Tsinghua University Co., Ltd., Beijing 100084, China; mengxia@thad.com.cn
4 Faculty of Architecture, Civil and Transportation Engineering, Beijing University of Technology, Beijing 100124, China; shenyinlan@bjut.edu.cn
* Correspondence: hewei@bucea.edu.cn

Abstract: The associated effect of sodium chloride and dihydrate gypsum on the mechanical performance of a slag-based geopolymer activated by quicklime was investigated by compressive strength, shrinkage, and square circle anti-cracking tests of mortar with a 0.5 water–binder ratio and a 1:3 binder–sand ratio, as well as paste soundness, powder X-ray diffraction (XRD), thermogravimetric analysis (TGA), scanning electron microscopy with energy dispersive spectroscopy (SEM-EDS), and mercury intrusion porosimetry (MIP) of the paste. The results indicate that (1) when dihydrate gypsum is used alone, it combines with calcium aluminate hydrate (C-A-H) to form calcium sulfoaluminate hydrate (AFt), which encourages the hydration process of slag. A 7.5% addition can result in an increase of 97.33% and 36.92% in 3-day and 28-day compressive strengths, respectively. When NaCl is used by itself, it facilitates the condensation of the aluminum silicate tetrahedron unit and generates zeolite. A 2% dosage can lead to a 66.67% increase in the 3-day compressive strength, while causing a 15.89% reduction in the 28-day compressive strength. (2) The combined effect of 2% NaCl and 7.5% gypsum results in the formation of needle-like and rod-shaped AFt, Friedel's salt, and plate-like Kuzel's salt in the geopolymer. This leads to an increase in 3-day and 28-day compressive strengths by 148% and 37.85%, respectively. Furthermore, it reduces the porosity by 18.7%. (3) Both NaCl and gypsum enhance the paste soundness of the slag-based geopolymer, and they do no harm to the crack resistance of the geopolymer. The drying shrinkage of the geopolymer at 28 days is just 0.48×10^{-3}, which is only 66.7% of OPC. This slag-based geopolymer has a simple preparation process, good volume stability, low raw material cost, low energy consumption, and low carbon emissions. It can be used instead of 32.5 slag Portland cement in plain concrete applications, and has high engineering, economic, and environmental values.

Keywords: sodium chloride; dihydrate gypsum; slag-based geopolymer; mechanical performance; hydration mechanism; Friedel's salt; Kuzel's salt

Citation: Shen, Q.; Li, B.; He, W.; Meng, X.; Shen, Y. Associated Effects of Sodium Chloride and Dihydrate Gypsum on the Mechanical Performance and Hydration Properties of Slag-Based Geopolymer. *Buildings* **2023**, *13*, 1285. https://doi.org/10.3390/buildings13051285

Academic Editors: Shan Gao, Jingxuan Wang, Dewen Kong and Yong Liu

Received: 10 April 2023
Revised: 30 April 2023
Accepted: 11 May 2023
Published: 15 May 2023

1. Introduction

The worldwide construction sector bears the onus of generating 39% of CO_2 emissions and depleting over 50% natural resources extracted [1], and cement manufacturing is directly responsible for approximately 5% of artificial CO2 emissions, with 0.8–1.1 tons of CO2 released against each ton of cement production [2]. Geopolymers are one of the several alternative cementitious materials to Portland cement being investigated. They are formed by the alkaline activation of industrial solid wastes, such as slag, steel slag, fly ash [3], red mud [4], as well as natural pozzolan [5]. Their preparation process has low energy consumption and low emissions, so the advantages of geopolymers fit well with China's strategic goals of "carbon peak and carbon neutrality" [6]. Additionally, geopolymers have exceptional characteristics, such as fast setting and hardening, high strength [7,8], resistance to high temperature [9] and acid corrosion [10–13], excellent impermeability [14], and good freeze–thaw resistance [15–17]. They can recycle industrial solid waste and have great potential for use in various applications.

Traditional slag-based geopolymers usually require a strong alkaline activator. Although they have a remarkable activation effect, they have several drawbacks, including over-rapid early-stage strength formation, low later-stage strength potential, and severe shrinkage. The shrinkage of alkali-activated slag concrete is generally at least twice that of ordinary Portland cement concrete, and the shrinkage of alkali-activated slag mortar activated by sodium hydroxide and water glass is three and six times that of ordinary Portland cement [18–22], respectively. Currently, it is widely believed that the high gel content of geopolymers and the excessively high proportion of pores smaller than 10 nm cause significant negative pore pressure during dehydration, resulting in excessive shrinkage [23].

Cheng et al. [24] found that the addition of NaCl at a dosage between 0% and 5% resulted in a greater improvement in the strength of slag-based geopolymers. However, when the NaCl content exceeded 5%, the strength increase in slag-based geopolymers becomes lower. Lv et al. [25] observed that the addition of 1% NaCl increased the strength of fly-ash-based geopolymer, but when the NaCl content is 3~11%, the strength of fly-ash-based geopolymers drops. Cao et al. [26] also found that adding 1% NaCl accelerated the hydration process of cement. He et al. [27] discovered that increasing the NaCl level to 2% increased the strength of slag-based geopolymers. Liu et al. [28] discovered that the geopolymer strength is greatest when the gypsum component is 10%. Zhang et al. [29] found that the strength of geopolymer mortar increased when adding less than 2% gypsum and decreased significantly when adding 5% gypsum. Xing et al. [30] found that the compressive strength of the geopolymer test block is the greatest when the gypsum component is 10%.

Previous studies [25,31–33] used a composite activator consisting of light calcined magnesia (MgO), sodium chloride (NaCl), building gypsum, and quicklime (CaO) to activate slag. During the hydration process, abundant crystalline hydration products, such as Fredel's salt, Kuzel's salt, and Aft, filled the gel pores, while MgO compensated for the shrinkage. The geopolymer met the strength requirements of 32.5 slag Portland cement. However, MgO is costly and limits its application potential. So, in this study, we removed the MgO component and used more gypsum and quicklime to compensate for shrinkage. By adjusting, it can further reduce energy consumption and emissions and be cheaper than OPC and strong alkali-activated slag-based geopolymer. The mechanical properties of this geopolymer were investigated by the compressive strength, square circle anti-cracking tests of mortar, and soundness of paste, and its microscopic mechanism was studied using XRD, SEM, TGA, and MIP. The differences in shrinkage between this geopolymer and P.O. 42.5 cement were also compared to comprehensively evaluate the performance of the slag-based geopolymer and provide theoretical support for its application in concrete engineering, especially in plain concrete engineering. The research of the composite-activated slag-based geopolymer has positive significance for reusing granulated blast-furnace slag and reducing carbon emissions caused by cement production.

2. Materials and Methods

2.1. Materials

The slag utilized in this paper is S95 slag powder manufactured by Hebei JINTAICHENG Environmental Resources Co., Ltd., Xingtai, China, with a density of 2.87 g/cm^3. Chemical compositions were analyzed using XRF, and the instrument used was Japanese Rigaku Supermini 200. The specific surface area and particle size distribution were analyzed using a Laser Diffraction Particle Size Analyzer. The instrument used was Mastersize 3000-F made by Malvern Panalytical. The results are shown in Table 1, Table 2, and Figure 1, respectively.

The slag's chemical module was K = 2.17; alkalinity modules were Mo = 1.14; and the observed 28-days activity coefficient was 95. The XRD spectrum shown in Figure 2 clearly exhibits a 'steamed bread peak', suggesting the presence of amorphous phase minerals in the slag, primarily active SiO_2 and active Al_2O_3 [34].

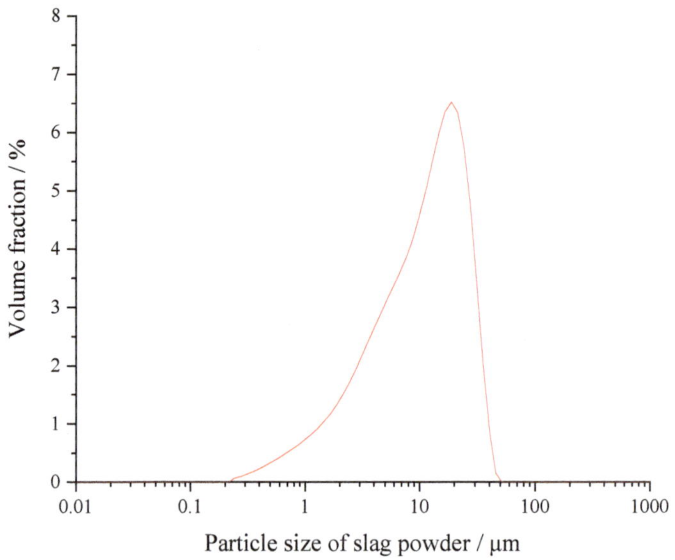

Figure 1. Particle size distribution curve.

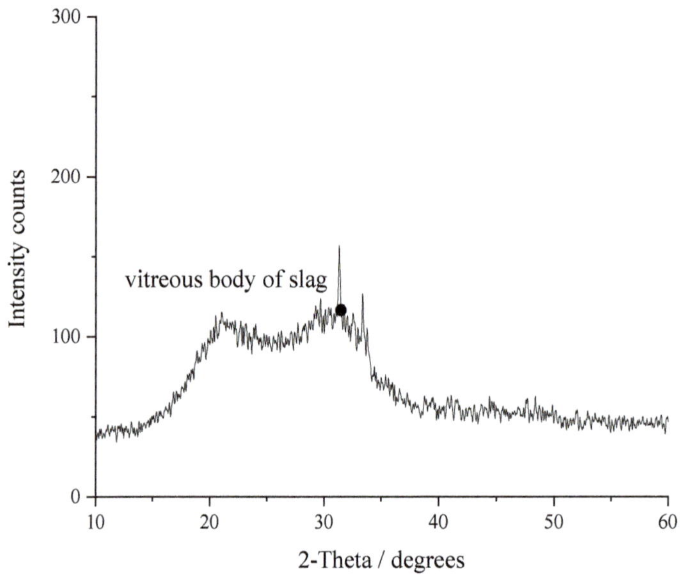

Figure 2. X-ray diffraction spectra of the slag.

Table 1. Chemical composition of slag (mass %).

SiO_2	Al_2O_3	SO_3	MgO	CaO	Fe_2O_3	Na_2O	K_2O
27.56	15.80	2.89	7.52	42.19	0.31	0.45	0.45
Cl	P_2O_5	TiO_2	MnO	SrO_2	Y_2O_3	Others	
0.09	0.05	1.97	0.61	0.09	0.01	0.01	

Table 2. Particle size distribution.

Specific Surface Area	Dx (10)	Dx (50)	Dx (90)
432.9 m^2/kg	2.23 μm	12.0 μm	28.3 μm

Tianjin BEICHEN, Tianjin BEILIAN, and Tianjin ZHIYUAN Chemical Industry provided the analytical pure NaCl (>99.9%), dihydrate gypsum ($CaSO_4{\cdot}2H_2O$) (>99.9%), and quicklime (>99.9%) employed in the experiment, respectively. Quicklime reacts with water to form calcium hydroxide, which increases the alkalinity of the gelling system and plays the role of an alkaline activator. Xiamen AISOU standard sand Co., Ltd., Xiamen, China manufactured the ISO standard sand. A comparison experiment was carried out with 42.5 ordinary Portland cement produced by Beijing JINYU Group Co., Ltd., Beijing, China, with a specific surface area of 349.8 m^2/kg, to assess the volume stability difference between the compound slag-based geopolymer and ordinary Portland cement.

The flowchart of this study is shown in Figure 3.

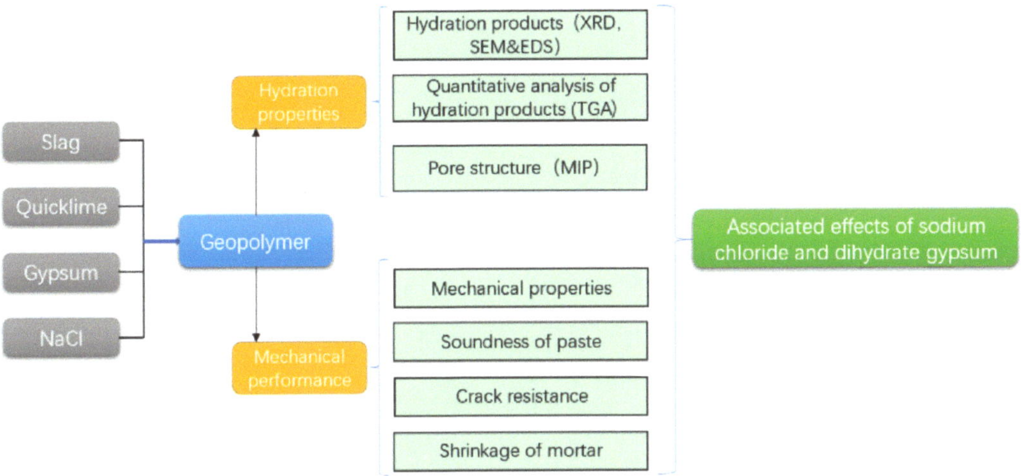

Figure 3. Flowchart of the research.

2.2. Experiment Methods

Although NaCl and gypsum have overlapping reactants in slag-based geopolymers activated by NaCl, gypsum, and quicklime, they are also distinct, and the composite application can give superior results. Based on previous experiments, the strength is stronger when the composite activator content is 25%. Therefore, under the condition of 25% activator content, the NaCl dosage was set at 2%, 4%, and 6%, marked as L, M, and H, respectively; the gypsum dosage was set at 2.5%, 5%, 7.5%, and 10%, represented by A, B, C, and D, respectively. The effects of NaCl and gypsum on slag-based geopolymer characteristics were investigated. The test group LC with the highest strength was chosen as the control group without NaCl (0C), without gypsum (L0), and neither NaCl nor gypsum

(OQ, only quicklime) by replacing the NaCl and gypsum components with quicklime in order to study the activating mechanism of NaCl and gypsum. The amount of activators in the LC group was adjusted to 20%, 30%, and 35%, and the optimum activator dose was tested. Because conventional geopolymer has a far higher dry shrinkage than Portland cement, it can cause interior micro fractures and strength degradation [18].

The calcium oxide level was relatively low, and the alkaline environment that may be created was restricted. Previous research had also shown that the excitation process takes longer with this excitation combination, so the 28-day strength of geopolymer under compound activators may be insufficient; therefore, a 56-day strength test was performed for the LC group. The volume stability of the geopolymer was compared with a P.O. 42.5 cement control group (OPC). Mixture proportions are detailed in Table 3. The reaction mechanism of the compound slag-based geopolymer activated by quicklime was analyzed, the law of mechanical properties was studied, and the volume stability was compared and evaluated by studying the products, pore structure, and mortar strength of the composite system of NaCl, gypsum, quicklime, and slag with different ratios and comparing the stability, shrinkage, and crack resistance of the paste. The reaction mechanism is comparable for the test groups with the same composition and little change in dose, and the test group with high strength as well as good volumetric stability (or not) has research value. As a result, the conventional test group and the control group are primarily investigated. In Table 3, a checkmark (√) indicates that this set of tests was performed, whereas a slash (/) indicates that it is not within the scope of the research.

Table 3. Mixture proportions (SC, NaCl; DG, dihydrate gypsum; QL, quicklime; AT, activator/%).

No.	Ratio	SC	DG	QL	AT	Slag	Paste			Mortar	
							Micro-Test	Soundness	Strength	Shrinkage	Cracking
1	LA	2	2.5	20.5	25	75	/	√	√	/	/
2	LB	2	5	18	25	75	/	√	√	/	/
3	LC	2	7.5	15.5	25	75	√	√	√	√	√
4	LD	2	10	13	25	75	/	/	√	/	/
5	MA	4	2.5	18.5	25	75	/	/	√	/	/
6	MB	4	5	16	25	75	/	/	√	/	/
7	MC	4	7.5	13.5	25	75	/	/	√	/	/
8	MD	4	10	11	25	75	/	/	√	/	/
9	HA	6	2.5	16.5	25	75	/	/	√	/	/
10	HB	6	5	14	25	75	/	/	√	/	/
11	HC	6	7.5	11.5	25	75	/	/	√	/	/
12	HD	6	10	9	25	75	/	/	√	/	/
13	0C	0	8	17	25	75	√	√	√	/	√
14	L0	3	0	22	25	75	√	√	√	/	√
15	OQ	0	0	25	25	75	√	√	√	/	√
16	LC 20%	1.5	6	12.5	20	80	/	/	√	/	/
17	LC 30%	2.5	9	18.5	30	70	/	/	√	/	/
18	LC 35%	3	11	21	35	65	/	/	√	/	/

(1) Micro-tests of Hardened paste

A pure paste sample was made with a 0.35 water–binder ratio and placed in a test tube for standard curing. The sample was taken out at the age of 3 and 28 days and split into little pieces before being sealed in a test tube containing anhydrous ethanol to stop hydration. The samples were collected from a sealed anhydrous ethanol test tube and vacuum-dried for 24 h at 60 °C for tests in the LC test group as well as the 0C, L0, and OQ control groups.

a. X-ray diffraction (XRD)

The paste sample was ground using a mortar until there was no particle sensation and then passed through a 45 μm sieve. The groove on the diffractometer sample table was filled with the sieved paste powder. To scan the phase, a Cu target was used with a tube voltage of 40 kV, a tube current of 40 mA, a scanning speed of 6 °/min, and a test range of 3 to 80°.

b. Thermogravimetric analysis (TGA)

The AFt, $Ca(OH)_2$, and chemical bonding water of the samples were evaluated by mass changes with temperature after being heated from 30 °C to 1000 °C at 10 °C/min in a nitrogen environment.

c. Scanning electron microscopy with energy dispersive spectroscopy (SEM-EDS)

The microstructure of the shattered sample was examined using a ZEISS Gemini 300 scanning electron microscope, manufactured by Zeiss, Germany, after a portion of it was sprayed with carbon.

d. Mercury intrusion porosimetry (MIP)

The mercury injection experiment technique involves calculating the diameter and volume of various-sized holes based on the functional connection between the amount of mercury injected into the hardened paste and the applied pressure. The fractal dimension of pore volume may be determined directly from the experimental data of mercury intrusion porosimetry based on the variable features of pore volume and pore diameter. The curves are displayed after calculating the logarithms of dV/dr and dr, and the fractal dimension of the pore volume is estimated using the slope of the curve.

(2) Soundness of paste

The SO_3 content of Portland cement specified in GB 175–2007 "Universal Portland Cement" standard is not more than 3.5%, mainly because of the potential risk of soundness caused by AFt. So, the pat paste was prepared in accordance with the measured standard consistency water consumption for study of soundness under standard curing conditions. Considering the mechanical properties and volume stability test results, LA, LB, LC, 0C, L0, and OQ test groups were selected for key research, and the cracking was observed and recorded regularly. The boiling test method was not used in the soundness test. AFt plays an important role in compensating for shrinkage and preventing cracking at the early stage of hardening, while AFt decomposes above 70 °C.

(3) Rupture and compressive strength tests of mortar

According to GB/T 17671-1999 test method for the strength of cement mortar (ISO method), 40 mm × 40 mm × 160 mm cement mortar samples were made with a 0.5 water–binder ratio and a 1:3 binder–sand ratio in line with Table 3. The rupture strength and compressive strength were measured after 3 and 28 days of curing.

(4) Shrinkage tests of mortar

The LC group with the highest strength was compared with grade 42.5 ordinary Portland cement. The mortar test blocks were prepared according to a binder–sand ratio of 1:3 and a water–binder ratio of 0.5. The drying shrinkage mortar samples were made in a 40 mm × 40 mm × 160 mm triple steel mold, and the plates at both ends of the mold have three 6 mm blind holes. A 25 mm long, 6 mm round head copper rod probe was placed before shaping, the surface was covered with a layer of the fresh-keeping film after forming, and it was placed in a standard curing room (temperature (20 ± 2) °C, relative humidity above 90%). The mold was removed after curing for 1 day, the initial value of the samples' length (accurate to 0.001 mm) was measured using a screw micrometer, the samples were placed on the confined frame, moved into the constant temperature and humidity room for curing (temperature (20 ± 2) °C, relative humidity 60% ± 5%), and the dial indicator readings of the samples at the age of 1 to 28 days were read. Paraffin wax was applied to the three removed mortar cubes, the interface was isolated with water in the air, and the autogenous shrinkage value was measured for 14 days under the same circumstances, as shown in Figure 4.

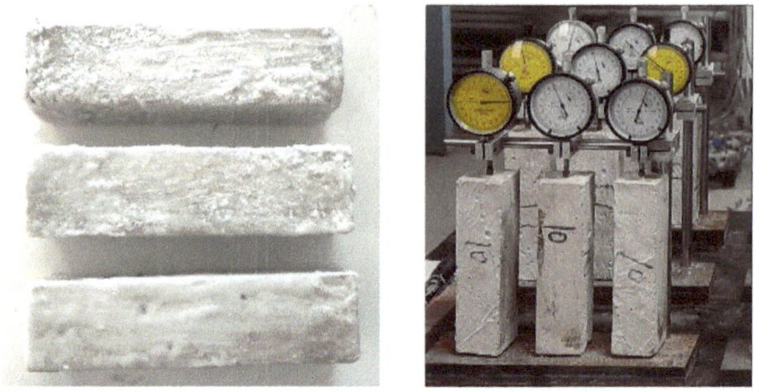

(a) (b)

Figure 4. (**a**) Coated specimens with paraffin wax; (**b**) shrinkage test frame.

The autogenous and dry shrinkage strain values of each age was calculated according to the following formula:

$$\varepsilon = (l_0 - l_t)/l_0$$

where ε represents the value of shrinkage strain; l_0 is the initial length of the specimen; and l_t is the length of the specimen at the age of t days. The average value of the shrinkage strain of three specimens was taken for each ratio.

(5) Square circle anti-cracking tests of mortar

The traditional ring method is widely used in the early crack resistance test of concrete. The earliest ring-cracking method was proposed and applied by Carlson [35]. The ring device is composed of two steel rings. The uniform constraint on the specimen is realized by limiting the shrinkage of cement-based materials by the inner ring. However, the commonly used uniaxial constraint method is complex and cannot be compared in large quantities; the plate constraint method has some problems, such as uneven constraint. Therefore, this study adopts a new cracking evaluation model—square and circular anti-cracking mold—which can not only uniformly restrict the concrete sample but also crack the surface of the concrete sample within a certain time, so as to measure the crack length and width conveniently and quickly. The center of the square round anti-crack test mold used in this experiment is square, and the location of the crack is guided by the cube edges and corners, so as to observe the process of crack generation and propagation [36]. The outer ring diameter of the ring test mold used in the test is 210 mm, the inner ring diameter is 200 mm, the height is 100 mm, and the side length of the central square is 106.05 mm [37], as shown in Figure 5. The test samples were placed indoors, observed for 28 days, and the number and width of cracks were recorded.

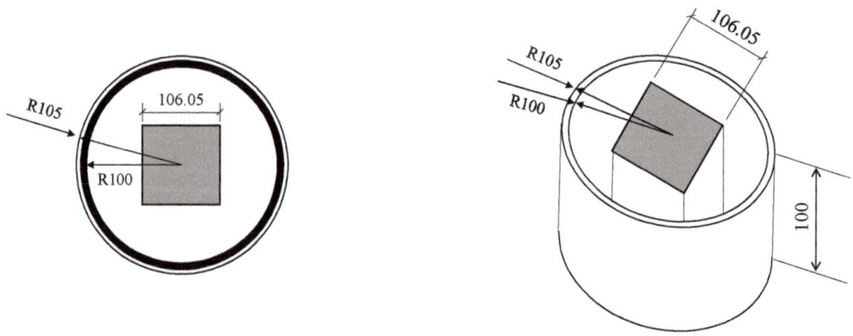

Figure 5. Square circle anti-cracking test mold.

3. Results and Discussion

3.1. Influence of NaCl and Gypsum on the Properties of Hardened Paste

To explore the hydration products and reaction process, the experimental groups LC, 0C, and L0 were principally investigated and compared to the control group OQ.

3.1.1. XRD

The hydration products of the geopolymer activated by quicklime, gypsum, and NaCl were analyzed using XRD, and the types of hydration products were obtained by qualitative analysis of the diffractogram, as shown in Figure 6.

In Figure 6, the LC group exhibits obvious diffraction peaks of $Ca(OH)_2$, AFt ($3CaO \cdot Al_2O_3 \cdot CaSO_4 \cdot 12H_2O$), Friedel's salt ($3CaO \cdot A1_2O_3 \cdot CaCl_2 \cdot 10H_2O$), and Kuzel's salt ($3CaO \cdot A1_2O_3 \cdot 0.5CaCl_2 \cdot 0.5CaSO_4 \cdot 12H_2O$), as well as unknown zeolite. Friedel's salt and Kuzel's salt are both double-layer metal hydroxides with positive charge on their laminates. To maintain overall electrical neutrality, a specific number of anions such as Cl^- and SO_4^{2-} are present between the laminates; at the same time, a particular number of water molecules are also present between layers [38,39]. The hydration of CaO generates $Ca(OH)_2$, which subsequently reacts with slag to produce C-A-H. It then reacts with gypsum to form AFt. With time, the combined effect of NaCl and gypsum promotes the continued formation of Friedel's salt and Kuzel's salt, resulting in the enhancement of their diffraction peaks at 28 days. The diffraction peaks of Friedel's salt are found in the L0 group without adding gypsum because the slag includes a trace of SO_3. In contrast, no diffraction peaks of Friedel's salt or Kuzel's salt were detected in the 0C group.

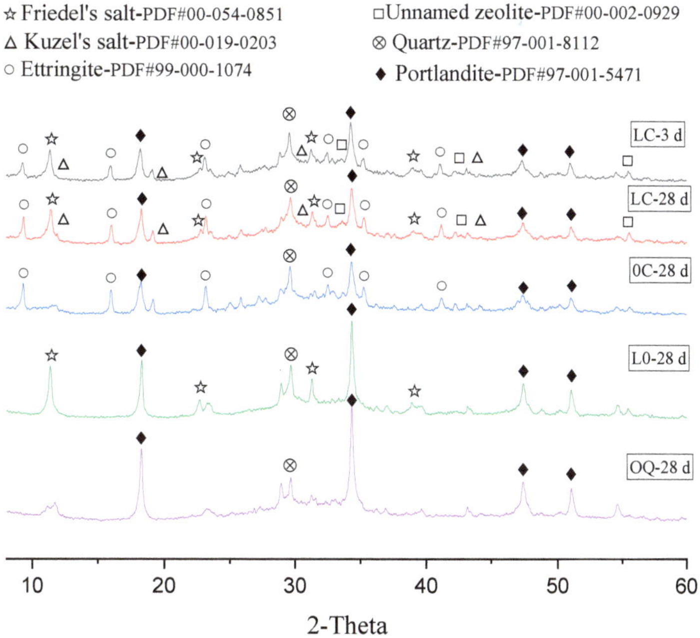

Figure 6. XRD pattern of geopolymer paste.

3.1.2. TGA

To quantify the influence of NaCl and gypsum on the hydration of slag, thermogravimetric analysis was carried out on the LC, 0C, L0, and OQ groups, and the TG-DTG curves are shown in Figure 7.

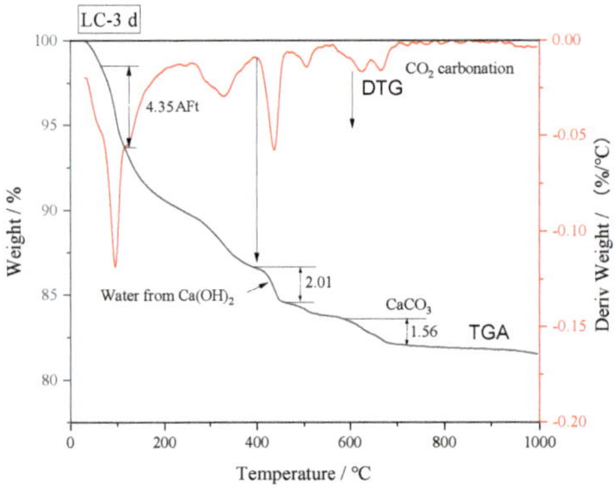

(**a**) LC-3 days

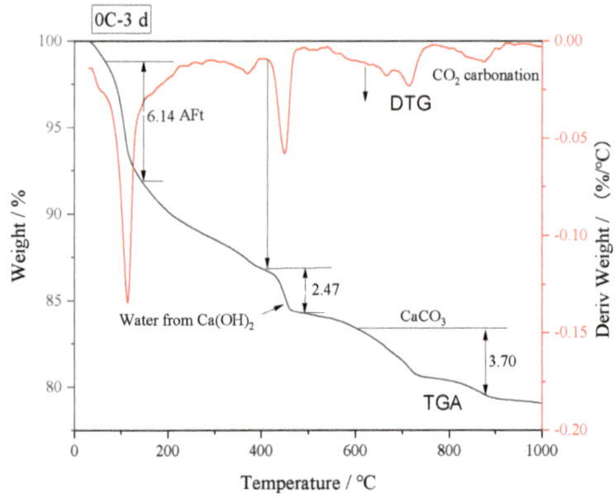

(**b**) 0C-3 days

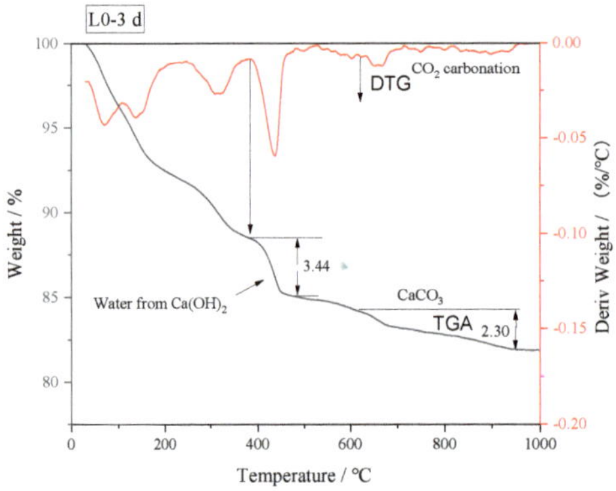

(**c**) L0-3 days

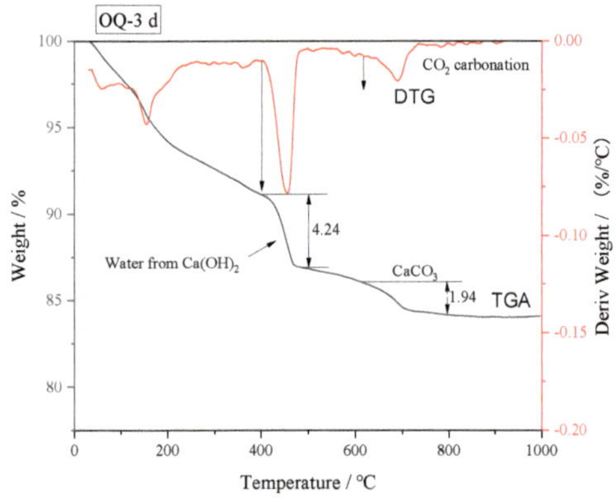

(**d**) OQ-3 days

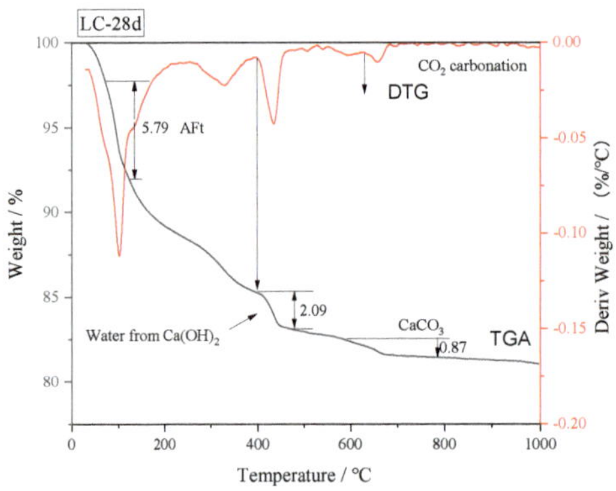

(**e**) LC-28 days

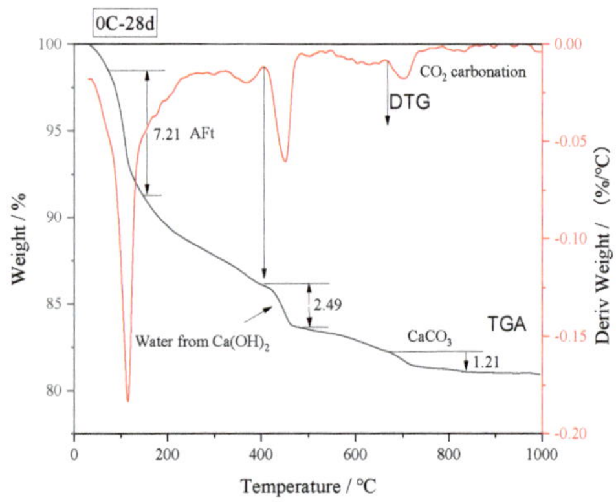

(**f**) 0C-28 days

Figure 7. *Cont.*

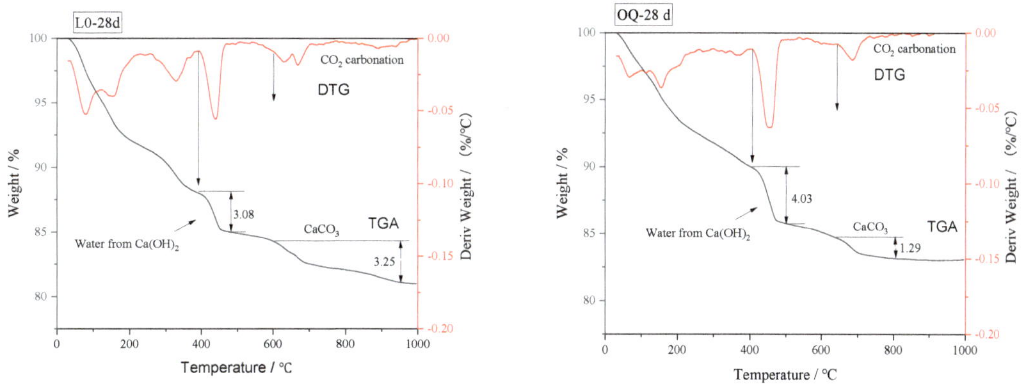

(**g**) L0-28 days (**h**) OQ-28 days

Figure 7. TG-DTG curves of geopolymer paste.

As shown in Figure 7, obvious AFt decomposition peaks occurring at 70~150 °C [40] can be seen in the 0C and LC groups. Hydration of CaO produced $Ca(OH)_2$, which can stimulate the activity of slag and promote hydration. As $Ca(OH)_2$ acts as an activator in this process, a lower residual amount of $Ca(OH)_2$ leads to a better slag hydration.

TG analysis can be used to determine the residual amount of $Ca(OH)_2$, as it decomposes and releases water at 350 °C to 550 °C. Although the thermogravimetric loss of C-S-H can occur throughout the entire temperature range, certain inaccuracies in the predicted quantities of AFt and $Ca(OH)_2$ cannot be measured. However, under the same conditions, measuring the relative concentration of $Ca(OH)_2$ in each test group might still represent the degree of slag hydration. As the chemical bonding water content includes the water released by the thermogravimetric loss of $Ca(OH)_2$, the hydration degree of slag must be subtracted from the quantity of water released by residual $Ca(OH)_2$.

The amounts of AFt, $Ca(OH)_2$, and chemical bonding water content are calculated and shown in Figure 8.

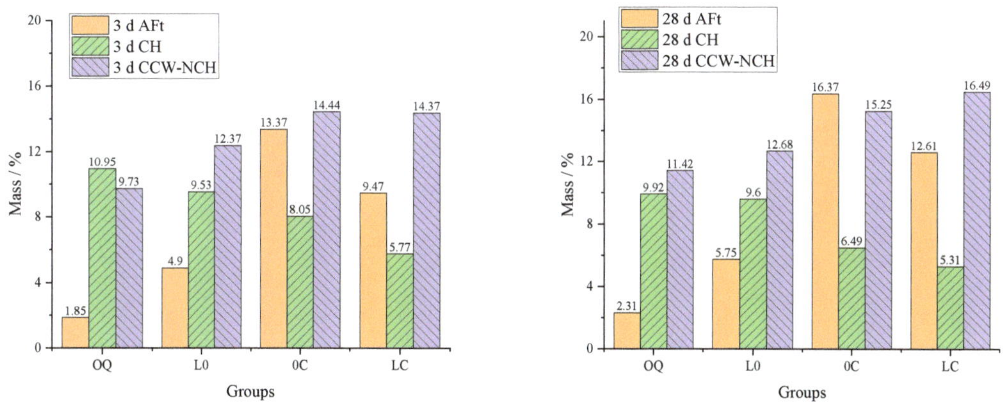

Figure 8. Percentage of AFt/CH/CCW-NCH (CCW-NCH deducts the amount of water released by the decomposition of $Ca(OH)_2$ (CH)).

As shown in Figure 8, the CCW-NCH of the L0 and 0C groups are significantly higher than that of the 00 group at both 3 and 28 days, implying that sodium chloride and gypsum can effectively promote hydration individually. Furthermore, the CCW-NCH value of the

LC group is higher than both groups, demonstrating that the combined use of sodium chloride and gypsum results in a more pronounced promotion of hydration. Notably, the AFt contribution to CCW-NCH is relatively high in the 0C and L0 groups. AFt content in the LC group is 3.9% lower than 0C, suggesting that NaCl serves to boost slag hydration and reduce AFt formation. Moreover, the 28-day CCW-NCH increase is greater in the LC group than in the 0C group, thereby indicating that the addition of NaCl facilitates subsequent hydration enhancement. Within a few hours of mixing with water, AFt was produced, thereby improving early strength, while the formation of Freidel's salt and Kuzel's salt typically occurred later than that of AFt. The formation of Freidel's salt and Kuzel's salt consumed chloride ions, sulfate ions, and C-A-H, thereby reducing the formation of AFt in the hardened paste. There is no noticeable AFt decomposition peak observed in the OQ test group's DTG curve, which is attributed to the absence of gypsum. Furthermore, the $Ca(OH)_2$ content of the 0C and L0 groups is lower than that of OQ, and the CCW-NCH content rose concomitantly, indicating that both gypsum and NaCl serve to enhance the reaction degree of slag. Finally, the LC group exhibits the greatest 28-day CCW-NCH content, indicating the highest reaction degree of slag and the most effective activation when NaCl and gypsum are used together.

NaCl reacts with the hydration products of slag to release NaOH, which further promotes the hydration of slag. The amount of chemically combined water in each group is increased after 28 days, compared to 3 days, indicating that the slag undergoes enhanced hydration over time, while the amount of $Ca(OH)_2$ decreases. At 28 days, the $Ca(OH)_2$ content of the LC group is 5.31%, meaning that 65.74% of the quicklime has engaged in the hydration process, thereby indicating that the amount of quicklime was sufficient.

3.1.3. SEM and EDS

The hydration products of the geopolymer were meticulously analyzed using SEM and EDS. The identification of the hydration products was based on their visual appearance and elemental composition, which are presented in Figures 9 and 10, as well as in Table 4.

Table 4. EDS analysis results of dots (atomic percentage).

Elements	Ca	Si	O	Al	S	Mg	Na	Cl
dot 1	20.90	6.16	58.23	2.34	3.38	1.49	1.57	1.33
dot 2	22.34	1.79	53.49	10.48	0.22	0.33	1.3	4.94
dot 3	13.49	4.64	59.27	3.22	1.08	0.75	3.43	1.10
dot 4	15.51	3.62	58.61	4.19	1.24	0.85	2.93	1.29

(**a**) OQ (×5000) (**b**) LC (×5000)

Figure 9. *Cont.*

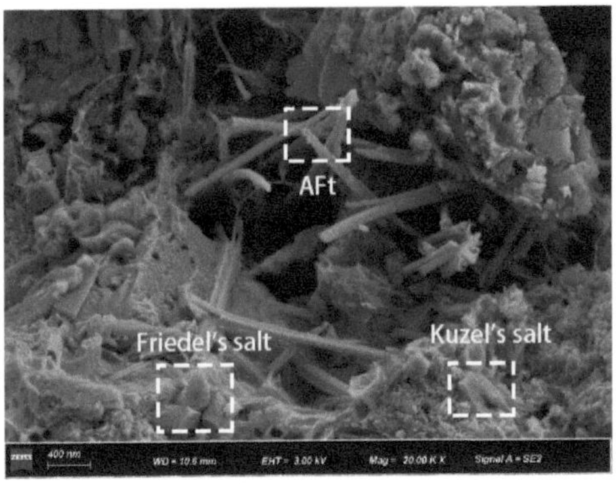

(**c**) LC (×20,000)

Figure 9. SEM images of the paste samples.

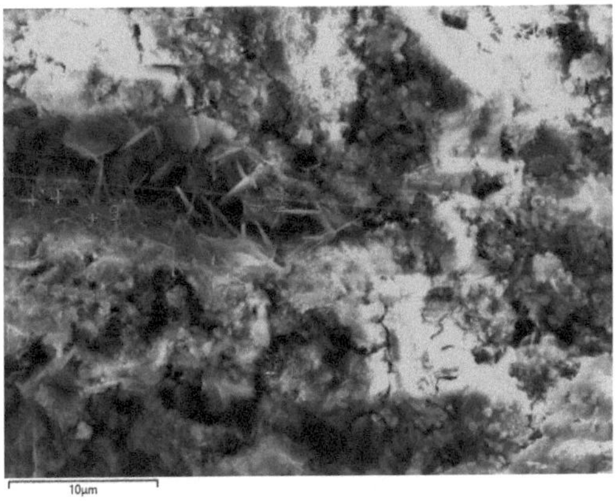

(**a**) Micrographs indicating the dot 1-4 locations of energy spectrum analysis.

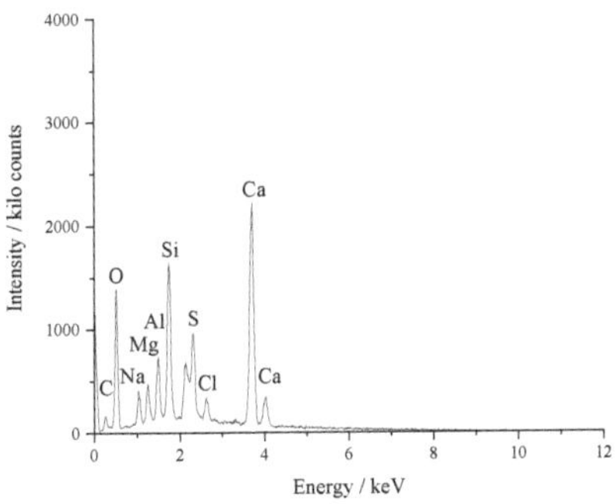

(**b**) dot 1

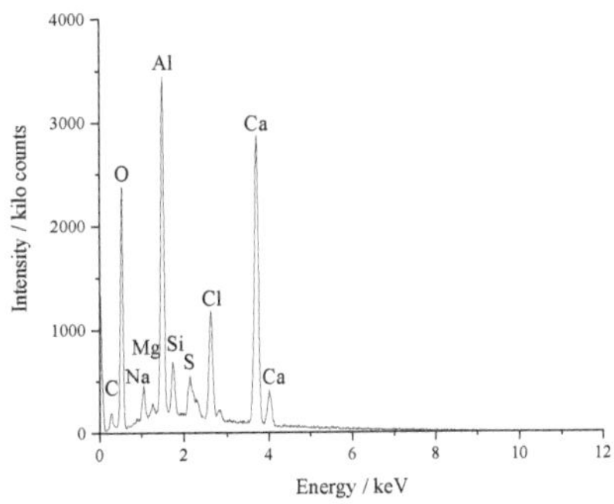

(**c**) dot 2

Figure 10. *Cont.*

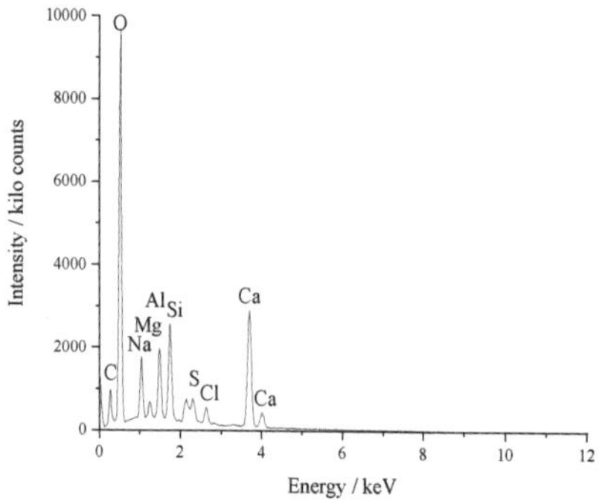

(**d**) dot 3

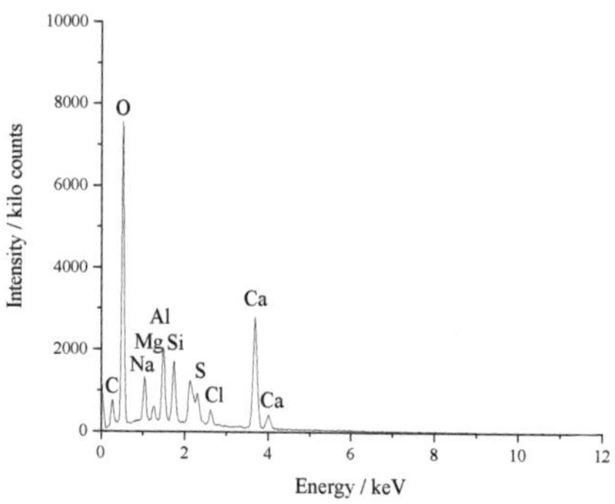

(**e**) dot 4

Figure 10. EDS spectra of geopolymer paste (the LC group).

It can be seen from Figure 10 and Table 4 that since the diameter of the determination point of the energy spectrum analysis may be larger than the size of the chlorine-containing compounds, the determination areas are mixed with other compounds, such as C-S-H gel, which may cause errors, so the influence of this part should be eliminated as much as possible. C-S-H gel has a calcium–silicon ratio that ranges between 0.8 and 2. It is assumed that the calcium–silicon ratio is constant and that all Si is present in the hydrated silica gel. After subtracting the calcium in the hydrated calcium silicate in Table 4, the Ca-Al-S-Cl ratio is recalculated to yield Table 5.

Table 5. Atom ratio.

Elements	Ca	Si	O	Al	S	Mg	Cl	Content
dot 1	4.95	0	24.2	2.34	0.67	0	1.33	KS
dot 2	12.16	0	31.24	5.48	0.00	0	4.94	FS
dot 3	6.53	0	40.78	3.22	1.08	0	1.10	AFt + FS
dot 4	8.08	0	39.39	4.19	1.24	0	1.29	AFt + FS

The combination of AFt, gypsum, portlandite, Friedel's salt, and Kuzel's salt with zeolite products (including Na and forsterite) can cause changes in the relative content of elements. The atom ratio of Ca-Al-Cl in Friedel's salt ($3CaO{\cdot}Al_2O_3{\cdot}CaCl_2{\cdot}10H_2O$) is 4:2:2, while the atom ratio of Ca-Al-S in AFt ($3CaO{\cdot}Al_2O_3{\cdot}CaSO_4{\cdot}12H_2O$) is 4:2:1. Table 5 shows that dot 1 represents Kuzel's salt ($3CaO{\cdot}Al_2O_3{\cdot}0.5CaCl_2{\cdot}0.5CaSO_4{\cdot}12H_2O$) with a Ca-Al-S-Cl ratio of 4.2:2:1.1:0.57; dot 2 mainly consists of Friedel's salt with a Ca-Al-S-Cl ratio of 4.4:2:0:1.8. Since a mixture of 2/3 AFt and 1/3 Friedel's salt was mixed, the Ca-Al-S-Cl ratio in dot 3 is 4:2:1:0 and 4.1:2:0:1.8, and dot 4 has ratios of 4:2:1:0 and 3.6:2:0:1.5, respectively. The concentration of calcium is slightly higher due to the effects of zeolite minerals and unreacted $Ca(OH)_2$. Overall, there were AFt, Friedel's salt, and Kuzel's salt generated in the geopolymer according to the XRD, SEM, and EDS results.

Quantities of flake crystals can be seen in the OQ group in Figure 9a because $Ca(OH)_2$ was generated when quicklime was combined with slag alone. The $Ca(OH)_2$ content remaining in this group is still as high as 9.92% at 28 days, and the consumption of $Ca(OH)_2$ is also lower, suggesting that the hydration degree is low when quicklime is employed alone and the enrichment of $Ca(OH)_2$ crystals will also have a negative influence on the strength. Because gypsum interacts with C-A-H and produces a lot of AFt, quantities of acicular crystals may be visible in the LC group after adding 7.5% gypsum (as seen in Figure 9b LC $\times$ 5000). Excessive AFt volume growth in the hardened paste may result in poor volume stability. Needle-rod and flake crystals are readily visible in Figure 9c LC $\times$ 20,000. The reaction is completely dominated by NaCl, yielding coarse needle-rod

Friedel's salt and flake Kuzel's salt. The XRD pattern also shows that the diffraction peak intensity of Friedel's salt in the LC group was lower than that of L0, indicating that the concentration of Friedel's salt declines, while the amount of Kuzel's salt grows.

NaCl functions as an auxiliary activator in the hydration reaction, releasing the alkali metal cation Na^+ and generating NaOH; NaOH offers an alkaline environment for slag hydration [41]. The hydration of slag in an alkaline environment experienced the dissolution and diffusion of aluminosilicate minerals in alkaline solution, the polymerization of aluminosilicate solutions and complexes caused by the gel phase ($Mx(AlO_2)y(SiO_2)z{\cdot}nMOH{\cdot}mH_2O$), and the consolidation and hardening process of the gel phase [42], in which the polymerization refers to the chemical reaction between various aluminosilicates and strong alkaline silicate solutions [43,44]. In addition to the aforementioned process, activated silicon oxide and activated alumina in slag react with $Ca(OH)_2$ in an alkaline environment to form hydrated calcium silicate and hydrated calcium aluminate. Simultaneously, sodium hydroxide can generate a geopolymer by reacting with active silica and active alumina [45]. In addition, slag can also reduce the diffusion rate of Cl^- [46] and produce Friedel's salt. One of the AFm compounds is Friedel's salt. There are several different forms of AFm compounds, and their chemical makeup varies widely. The majority of them are lamellar crystals, with the most common being $3CaO{\cdot}Al_2O_3{\cdot}CaSO_4{\cdot}xH_2O$, $3CaO{\cdot}Al_2O_3{\cdot}CaCO_3{\cdot}xH_2O$, and $3CaO{\cdot}Al_2O_3{\cdot}CaCl_2{\cdot}xH_2O$ [47,48]. Because the addition of NaCl and gypsum causes Cl^-, $SO_4{}^{2-}$, $CO_3{}^{2-}$, and other ions to exist in the hydration reaction, it may simultaneously produce a variety of compounds, such as hydrated calcium chloroaluminate, hydrated calcium sulphoaluminate, hydrated calcium carboaluminate, and more complex compounds produced after the substitution of each acid radical ion. The hydration products containing Cl^- are $3CaO{\bullet}{\bullet}Al_2O_3{\bullet}CaCl_2{\bullet}xH_2O$, C-S-H gels, and zeolite containing $SO_4{}^{2-}$, Mg^{2+}, and Na^+. The molecular formula of the lamellar crystal and needle-rod crystal components is $3CaO{\cdot}Al_2O_3{\cdot}(0.5CaCl_2{\cdot}0.5CaSO_4){\cdot}12H_2O$, which is extremely close to the insoluble Kuzel's salt.

It has been established that NaCl participates in the hydration process and generates chlorine-containing molecules. The released Na^+ generates NaOH, which raises the PH of the geopolymer and promotes the hydration of slag, hence increasing strength. The addition of NaCl can promote the dissolution of gypsum, thus accelerating $SO_4{}^{2-}$ to form AFt [30,49], providing early strength and compensating for shrinkage. AFt is produced before Friedel's salt when NaCl is added. With the extension of hydration age, NaCl is easier to react with C-A-H and AFt to produce Friedel's salt [22]. Meanwhile, chloride and sulfate react with C-A-H and AFt to form Kuzel's salt [50–53]; the creation of Friedel's salt and Kuzel's salt consumes Cl^-, and the released OH^- combines with Na^+ to generate NaOH, enhancing the alkalinity and accelerating slag hydration. Simultaneously, the creation of Kuzel's salt consumes $SO_4{}^{2-}$ and decreases the quantity of AFt in the hardened paste.

The hydration reaction equation of the slag-based geopolymer activated by NaCl, gypsum, and quicklime are as follows:

(1) CaO in the compound activator reacts with water:

$$CaO + H_2O = Ca(OH)_2 \tag{1}$$

(2) In an alkaline environment, OH^- depolymerizes Ca-O, Si-O, and Al-O bonds in the vitreous body and accelerates the slag reaction, which results in hydration products such as C-S-H gel and calcium aluminate hydrate [54–56]:

$$xCa(OH)_2 + SiO_2 + m\,H_2O = xCaO{\cdot}SiO_2{\cdot}(m + x)H_2O \tag{2}$$

$$yCa(OH)_2 + Al_2O_3 + nH_2O = yCaO{\cdot}Al_2O_3{\cdot}(y + n)H_2O \tag{3}$$

$$yCaO{\cdot}Al_2O_3{\cdot}(y + n)H_2O + CaSO_4 + H_2O \rightarrow Ca_6Al_2(SO_4)_3(OH)_{12}{\cdot}26H_2O \tag{4}$$

(3) In addition to AFt, partially hydrated calcium aluminate ($yCaO{\cdot}Al_2O_3{\cdot}(y+n)H_2O$) reacted with $Ca(OH)_2$ and NaCl to form Friedel's salt and NaOH:

$$yCaO{\cdot}Al_2O_3{\cdot}nH_2O + NaCl + Ca(OH)_2 \rightarrow 3CaO{\cdot}A1_2O_3{\cdot}CaCl_2{\cdot}10H_2O + NaOH \quad (5)$$

(4) The addition of gypsum provided $SO_4{}^{2-}$, and part of Friedel's salt continued to react with gypsum, replacing part of the Cl^- and generating Kuzel's salt ($3CaO{\cdot}A1_2O_3{\cdot}0.5CaCl_2{\cdot}0.5CaSO_4{\cdot}12H_2O$) [37]:

$$\begin{gathered}3CaO{\cdot}A1_2O_3{\cdot}CaCl_2{\cdot}10H_2O + CaSO_4{\cdot}2H_2O + H_2O \rightarrow \\ 3CaO{\cdot}A1_2O_3{\cdot}0.5CaCl_2{\cdot}0.5CaSO_4{\cdot}12H_2O + CaCl_2\end{gathered} \quad (6)$$

(5) The NaOH generated in reaction (5) can be employed as an activator in the formation of zeolite ($0.8CaO{\cdot}0.2Na_2O{\cdot}A1_2O_3{\cdot}3SiO_2{\cdot}6H_2O$) by reacting with slag:

$$Ca(OH)_2 + SiO_2 + Al_2O_3 + NaOH + H_2O \rightarrow 0.8CaO{\cdot}0.2Na_2O{\cdot}A1_2O_3{\cdot}3SiO_2{\cdot}6H_2O \quad (7)$$

The hydration reaction mechanism in the cementitious system is shown in Figure 11. And the reaction Equations (1)–(7) are labeled.

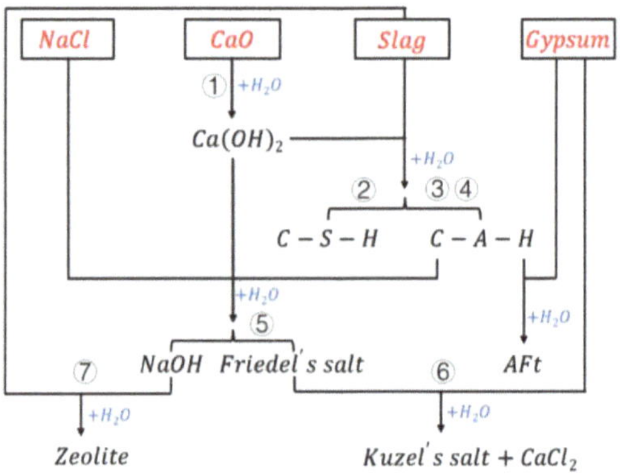

Figure 11. Hydration reaction diagram.

3.1.4. MIP

Based on the study of the hydration mechanism and product composition of the slag-based geopolymer, the pore structure of hardened paste was analyzed using MIP.

According to Figure 12, the most likely pore size of the LC group is 30 nm, and the pore size distribution curve is largely below that of the 0C group in the 10 to 100 nm range, indicating that the pore volume of the LC group is smaller than that of the 0C group. This is due to the crystals of Friedel's salt and Kuzel's salt filling the pores between gels and increasing the density of hardened paste. At 28 days, the CCW-NCH content of the LC group is greater than that of the 0C, L0, and OQ groups. NaCl and gypsum boosts slag hydration and the formation of additional gel phases, which raises the compactness of hardened paste and decreased porosity. Pores in the 0C group are spread in the 100–130 nm range, but those in the LC group vanish, and the pore size distribution curve of the LC group between 50,000 and 400,000 nm was much smaller than those of the 0C, L0, and OQ groups. The damaging pores cause a decrease in strength. As a result, the LC group has fewer damaging pores and greater strength. The L0 group has the most pores in the 10–30 nm and 20,000–40,000 nm ranges, and its porosity is the highest among the four groups.

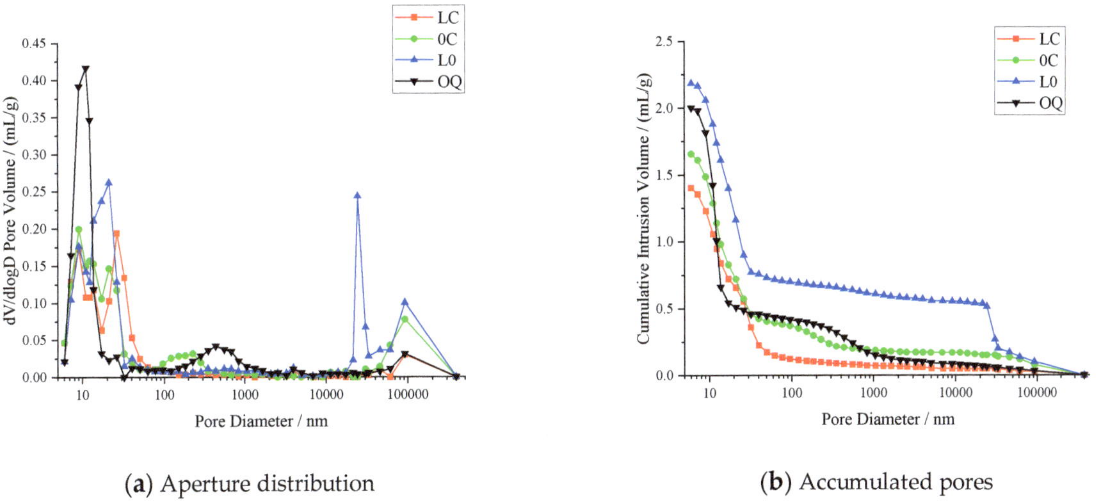

Figure 12. Pore structure analysis curve of different mixture proportions.

The D value of pore fractal dimension describes the pore section's complexity. The more intricate the spatial distribution of the pores, the higher the score. The greater the fractal dimension of pore volume, the more complex the spatial geometric properties of pores can be interpreted from fractal theory and pore fractal model, i.e., the more complicated the spatial distribution of pores in the material and the stronger the spatial filling ability. It also implies that when the pores are reduced, the average pore size lowers, the number of macropores reduces, the number of tiny holes grows, and the specific surface area of the pores increases, implying that the pore structure is refined and optimized to some extent. Therefore, the fractal dimension of pore volume may be regarded as an all-encompassing characteristic of pore form and spatial distribution [57]. As shown in Figure 13, using NaCl and quicklime (L0) as activators enhanced porosity by 46.7% when compared to quicklime alone. Porosity rose by 9.9% when gypsum and lime (0C) were employed as activators instead of quicklime alone. The porosity was reduced by 18.7% when NaCl, gypsum, and quicklime (LC) were employed simultaneously. The fractal dimension shrunk from 2.9 to 2.0. Porosity rose with single additions of NaCl or gypsum but decreased with compound additions of NaCl and gypsum.

XRD results show that there are no Friedel's salt and Kuzel's salt diffraction peaks in the 0C group, and no Friedel's salt and Kuzel's salt crystals are found in SEM photos. Friedel's salt and Kuzel's salt could not be produced without NaCl or gypsum, and they could not play the role of filling pores, so the porosity of the 0C group is higher. Crystal products cross-link in the pores and compact the paste [58], which is beneficial to the strength and volume stability. Friedel's salt and Kuzel's salt diffraction peaks were not discovered in the 0C group, and no Friedel's salt and Kuzel's salt crystals were found in SEM pictures. Friedel's salt and Kuzel's salt cannot be made without NaCl or gypsum, and they cannot fill pores; therefore, the porosity of the 0C group is larger. Crystal products form cross-links in the pores and compact the paste, which can improve strength and volume stability.

3.2. Strength of Mortar

The compressive strength of geopolymer mortar at 3 days and 28 days is shown in Figure 14.

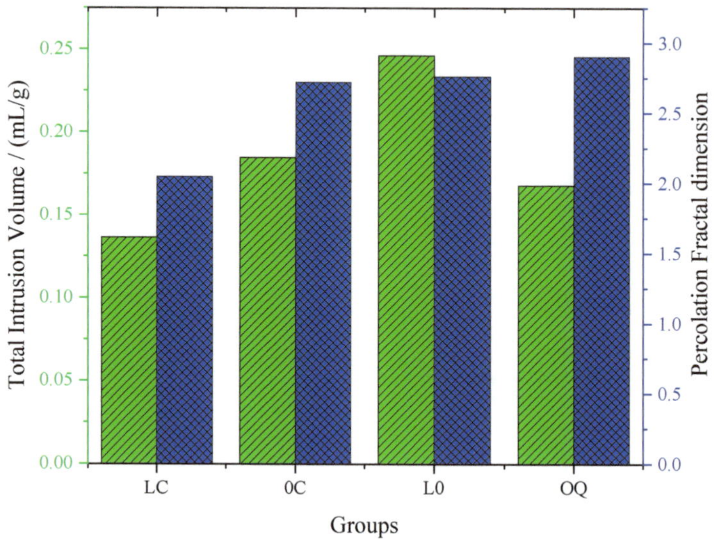

Figure 13. Porosity and fractal dimension.

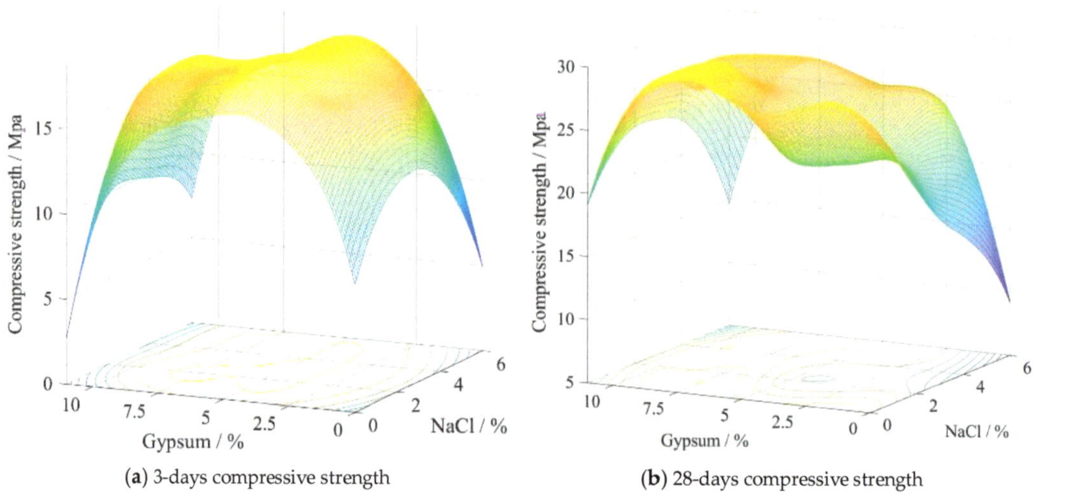

Figure 14. Fitting surface of compressive strength of mortar.

According to Figure 14a, the inclusion of a combination of 2% NaCl and 5% gypsum can significantly enhance the 3-day strength of the material, and the maximum 3-day compressive strength achieved is 18.8 MPa. Specifically, when the NaCl dosage is maintained at 2%, the 3-day compressive strength initially experiences a gradual increase as the amount of gypsum is increased from 0 to 5%, but then undergoes a sharp decline upon further increase in the gypsum content from 7.5% to 10%. Furthermore, the addition of 7.5% gypsum leads to a noteworthy increase of 48.8% in the 3-day strength, from an initial value of 12.5 MPa to 18.6 MPa. However, the strength subsequently declines as the gypsum content is increased to 10%. The 3-day compressive strength exhibits a 25.6% increase from 14.8 MPa to 18.6 MPa as NaCl increases from 0 to 2% when the gypsum content is 7.5%; however, the strength decreases by 15.1% when the NaCl content is further increased from 2 to 6%.

Based on Figure 14b, incorporating a mixture of 2% NaCl and 7.5% gypsum can enhance the 28-day compressive strength of the material. Specifically, when the gypsum content is 7.5%, the 28-day compressive strength initially experiences a slight improvement as the NaCl content increases from 0 to 2%, but then undergoes a decline when the NaCl content is increased from 4 to 6%. When the gypsum content is 10%, the 28-day compressive strength decreases monotonically with an increase in NaCl content. Moreover, when the NaCl content is maintained at 2%, the 28-day compressive strength initially increases gradually as the gypsum content is increased from 0 to 7.5%, but then experiences a decline as the gypsum content is further increased from 7.5 to 10%. Significantly, the addition of 7.5% gypsum results in a notable 63.8% increase in the 28-day compressive strength, from an initial value of 18.0 MPa to 29.5 MPa, which represents the maximum value achieved.

The LC with a 25% activator displayed the highest strength when compared to LC with 20%, 30%, and 35% activators. The optimal activating effect is achieved by utilizing a mixture of NaCl, gypsum, quicklime, and slag in the proportions of 2:7.5:15.5:75 (LC), which resulted in a 148% increase in 3-day compressive strength and a 37.85% increase in 28-day compressive strength, as compared to the OQ group. The 28-day rupture strength of the LC group is found to be 9.9 MPa; additionally, the 28-day compressive strength of the LC group reached 29.5 MPa, and the compressive strength at 56 days reached 34.2 MPa.

The addition of gypsum has a significant impact on the strength of the geopolymer. A low gypsum dosage (2.5%) results in insufficient sulfate content, which limits the activation of the slag. Conversely, a high gypsum dosage (10%) decreases the quicklime content, which is necessary for the hydration reaction of the slag. Additionally, excessive gypsum continues to react in the hardened paste and generate Aft, causing volume expansion that may damage the hardened paste structure. At a 7.5% gypsum dosage, adding 2% NaCl can significantly improve 3-day compressive strength; however, further increases in NaCl content can result in decreased strength.

The experimental results demonstrate that the geopolymer achieves its highest strength when quicklime content is approximately 15.5%, mainly because the CaO is the source of $Ca(OH)_2$ in the system. Insufficient quicklime content leads to a depletion of $Ca(OH)_2$ content in the geopolymer, resulting in incomplete reactions of slag and inadequate hydration products that are unable to form an effective space network structure in the geopolymer. Sufficient quicklime content can result in a high pH value, creating a highly alkaline environment, promoting the disintegration of the slag, releasing active components such as silica tetrahedrons and alumina tetrahedrons, accelerating the hydration process of slag, generating more C-S-H, and enhancing the density of the geopolymer. However, excessive amounts of quicklime not only reduce the proportion of slag in the geopolymer, but the excess Ca^{2+} and OH^- causes the formation of hexagonal flake $Ca(OH)_2$ crystals, which have poorer microscopic properties than other hydration products, such as C-S-H, resulting in lower strength.

When slag is activated by quicklime alone, its hydration degree is limited. As a result, the OQ group has the lowest CCW-NCH content and a 3-day compressive strength of only 7.5 MPa, despite having a $Ca(OH)_2$ content of 10.95% after 3 days. After adding 2% NaCl and 7.5 gypsum, compared to the OQ group, CCW-NCH content at 3 and 28 days increased by 4.64% and 5.07%. This results in an increase in compressive strength of 148% and 37.85% at 3 days and 28 days, respectively.

The LC group achieved the highest 28-d compressive strength. In this group, the combination of NaCl and gypsum leads to the generation of NaOH through the interaction with C-A-H, thereby promoting slag hydration and increasing its strength. Gypsum can stimulate slag activity and enhance its hydration, leading to increased strength. However, excessive amounts of gypsum can lead to the production of excessive amounts of AFt [59], resulting in volume expansion and the breaking of hardened paste. Previous studies have shown that the optimal proportion of gypsum in slag-based geopolymer is between 6% and 10% [60]. Additionally, the reactivity of slag powder increases significantly when the concentration of gypsum and quicklime is within the range of 5–10% and 10–20%, respectively [61]. The appropriate dosage of gypsum and sodium chloride in this study is consistent with these findings.

The LC group exhibits higher rupture strength due to the formation of more crystalline materials such as AFt and Kuzel's salt. The AFt content is as high as 12.61%, which improves the compactness of the paste by filling the pores and increases rupture strength by withstanding tensile stress better than the gel phase. The LC group achieved the highest 28-day compressive strength of 29.5 MPa due to the gradual depolymerization of the vitreous network structure in slag resulting from the synergistic activation of quicklime and gypsum. AFt and C-S-H gel are the primary hydration products in the early stage of hardened paste. As the hydration process continues, crystal products pack into the paste's pores and are tightly bonded to the whole, enhancing the compactness of the paste structure and improving early mortar strength [62]. C-A-H interacts with Cl^- and SO_4^{2-} to form Friedel's salt and Kuzel's salt as it ages. The addition of NaCl and gypsum provides Cl^- and SO_4^{2-} for the creation of Kuzel's salt. Kuzel's salt is an insoluble salt that fills the paste's pores and contributes to strength, while the development of zeolite minerals also improves strength.

3.3. Soundness, Crack Resistance, and Shrinkage

Visible cracks in the geopolymer paste are shown in Figure 15.

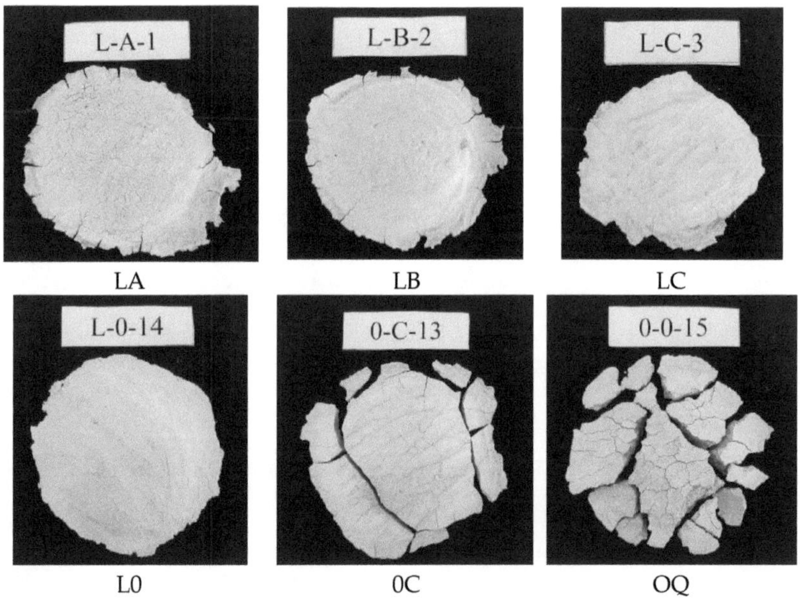

Figure 15. Cracking test of geopolymer paste.

As shown in Figure 15, gypsum has a positive effect on the stability of slag geopolymers. The LA group exhibits numerous micro-cracks on the surface, accompanied by multiple piercing fractures along the edge. As the amount of gypsum increases, the LB group exhibits fewer surface fissures and no visible fractures, while the LC group shows no discernible cracks. Furthermore, gypsum also leads to an improvement in the 28-day compressive strength. Appropriate NaCl content also confers a significant advantage in enhancing stability. During the hardening process, the OQ group experiences severe cracking within an hour, resulting in fractured surfaces broken into small pieces. X-ray diffraction (XRD) patterns show significant peaks for $Ca(OH)_2$, thermogravimetric (TG) analysis indicates that over 60% of $Ca(OH)_2$ remains unreacted with slag after three days, and scanning electron microscopy (SEM) reveals the presence of many hexagonal flake $Ca(OH)_2$ crystals. In contrast, the L0 group treated with 2% NaCl does not exhibit any cracks. The addition of NaCl and gypsum to the LC group almost completely eliminates the cracks. Both NaCl and gypsum enhance the stability of slag-based geopolymers, with NaCl having the most significant impact.

According to Figure 16a, the 14-day autogenous shrinkage of the LC group is 0.48×10^{-3}, which is similar to the OPC group's autogenous shrinkage of 0.45×10^{-3}. In addition, as seen in Figure 16b, the early drying shrinkage growth trend of the LC group is similar to that of the OPC group, but the LC group begins to gradually stabilize at 7 days, whereas the OPC group begins to gradually stabilize at 21 days. The drying shrinkage of the LC group at 28 days is just 0.48×10^{-3}, which is only 66.7% of OPC. The addition of gypsum supplies SO_4^{2-} to form AFt in the hydration products, leading to volume expansion and compensating for shrinkage [63]. MIP results indicate that the porosity of the LC group is low, as the crystal products of Friedel's salt and Kuzel's salt fill the pores, resulting in the lowest porosity of the LC group. Consequently, the capillary pore negative pressure is low, and the autogenous shrinkage is also small.

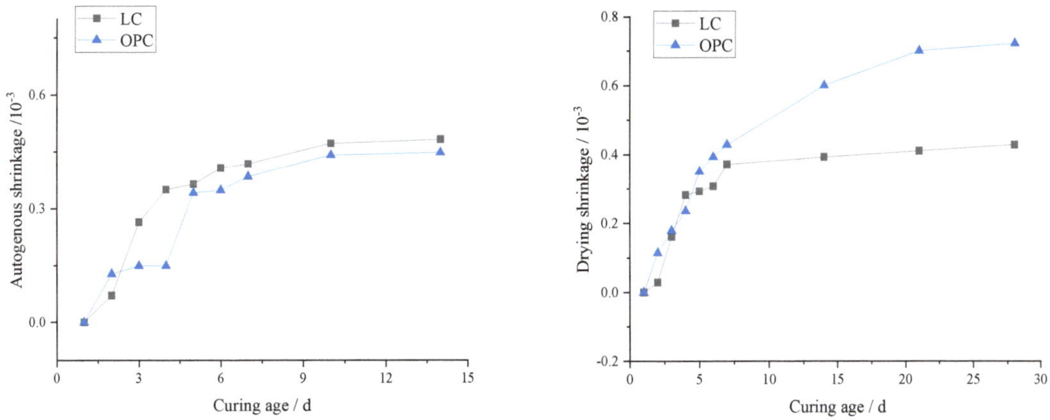

(**a**) Autogenous shrinkage (**b**) Drying shrinkage

Figure 16. Shrinkage comparison.

Both drying shrinkage and autogenous shrinkage are caused by a decrease in water in the pores of hardened cement paste, resulting in certain pores being unable to be saturated by water. The hydrophilic capillary wall causes the water surface in the pores to form a meniscus, which brings the particles around the pores closer together, resulting in a reduction of the paste's macro volume [64,65]. The gel has a more porous structure in concrete, making it more prone to shrinkage and cracking, whereas the total drying shrinkage strain of typical slag-based geopolymers can reach as high as 1700 $\mu\varepsilon$ [66,67], which is four times that of ordinary concrete [68,69]. The drying shrinkage of slag-based geopolymers activated by sodium hydroxide and water glass is much greater than that of Portland cement [70,71]. According to Cartwright's research, the shrinkage value of slag-based geopolymer is six times more than that of Portland cement [19]. Atis et al. compares the drying shrinkage of slag-based geopolymers to the impacts of liquid sodium silicate, sodium hydroxide, and sodium carbonate, and determined that the shrinkage of slag-based geopolymer activated by liquid sodium silicate and sodium hydroxide is greater [20]. When sodium hydroxide or sodium silicate is employed as an activator, the drying shrinkage can be three to six times that of Portland cement mortar [18], which is a major cause for the failure of slag-based geopolymer applications on a wide scale. However, slag-based geopolymers activated by NaCl, gypsum, and quicklime exhibit lower shrinkage than the OPC group and demonstrate a promising potential for broad application.

As shown in Figure 17, all test groups did not crack in 28 days, showing good crack resistance, the same as the OPC group. The addition of NaCl and gypsum, as well as their combination, does no harm to the crack resistance of the geopolymer. According to research, slag-based geopolymer mortar activated by sodium silicate and sodium hydroxide breaks along the radial direction in the ring cracking test, and the frequency of cracks near the limit column rises [72]; this is due to the substantial shrinkage of slag-based geopolymer activated by sodium silicate, which causes stress concentration at the limiting prism angle, resulting in cracking. Differently, the volume expansion of AFt compensates for shrinkage, and the volume stress induced by lesser volume shrinkage is smaller; therefore, the compound geopolymer does not crack. Obviously, when compared to the slag-based geopolymer activated by water glass and caustic soda, the slag-based geopolymer activated by NaCl, gypsum, and quicklime has better stability, shrinkage, and fracture resistance, which is important for engineering applications.

Figure 17. Square circle anti-cracking test of the geopolymer.

4. Conclusions

In this study, the mechanical properties and hydration mechanism of slag-based geopolymer activated by sodium chloride, gypsum, and quicklime were investigated. The following conclusions can be drawn based on the results presented:

(1) Needle-like crystal AFt, Friedel's salt, and flake Kuzel's salt are generated in the slag-based geopolymer activated by NaCl compounding gypsum. Na+ is released when Friedel's salt is formed, thus promoting the hydration of slag. Under the combined effect of NaCl and gypsum, the amount of AFt as well as $Ca(OH)_2$ is decreased, while the chemical combined water is increased by 4.64% and 5.07% at 3 days and 28 days. In addition, the porosity is reduced by 18.7% for the filling effect of those crystal products.

(2) There is a superimposed effect on compressive strength when 2% NaCl and 7.5% gypsum are combined. By using a NaCl: gypsum: quicklime: slag ratio of 2:7.5:15.5:75 (the LC group), the compressive strengths were increased by 148% and 37.8%, respectively, compared to the 00 group. The LC's compressive strength at 3 days, 28 days, and 56 days were 18.6 MPa, 29.5 MPa, and 34.2 MPa, respectively. Under appropriate dosage, the formation of AFt, Friedel's salt, and Kuzel's salt through the addition of NaCl and gypsum is beneficial for enhancing the strength of the geopolymer. The combined employment of both sodium chloride and gypsum effectively addresses the issue of 28-day strength reduction that arises when sodium chloride is used in isolation and the problem of poor paste soundness that arises when gypsum is used alone.

(3) According to the square circle anti-crack tests, the co-presence of NaCl and gypsum in the slag-based geopolymer does not have a detrimental effect on crack resistance. In addition, the autogenous shrinkage of the LC group was found to be similar to the OPC group, and the drying shrinkage was only 66.75% of that of the OPC group. Generally, the shrinkage of slag-based geopolymer activated by strong alkali can reach several times that of ordinary Portland cement, so the shrinkage of the slag-based geopolymer activated by the composite activator is much smaller than that of the geopolymer activated by strong alkali.

(4) This geopolymer is made from cheap and easily available industrial products, such as sodium chloride, gypsum, quicklime, and slag, and does not use strong alkali activators; therefore, this geopolymer is more economical and energy-efficient and has lower carbon emissions than geopolymers activated by strong alkali, rendering it an environmentally friendly and cost-effective cementitious material in plain concrete applications. More research, such as frost resistance, carbonization resistance, etc., will be conducted in the future. The application of this geopolymer to solidify coastal saline soils is also ongoing.

Author Contributions: Conceptualization, W.H. and X.M.; methodology, W.H. and Q.S.; validation, W.H. and Q.S.; formal analysis, B.L. and Q.S.; investigation, B.L. and Q.S.; data curation, Y.S. and Q.S.; writing—original draft preparation, B.L. and X.M.; funding acquisition, W.H. and Y.S. All authors have read and agreed to the published version of the manuscript.

Funding: This work is sponsored by the National Key R&D Program of China (2022YFC3803400), National Natural Science Foundation of China (No. 52208413 and 52008011), Beijing Municipal Natural Science Foundation (No. 8204058), BUCEA Post Graduate Innovation Project (No. 02081022003), and the R&D Program of Beijing Municipal Education Commission (No. KM202210016011).

Data Availability Statement: The data used to support the findings of this study are available from the corresponding author upon request.

Conflicts of Interest: The authors declare no conflict of interest.

References

1. Jesus, C.F.; Arruda Junior, E.S.; Braga, N.T.S.; Silva Junior, J.A.; Barata, M.S. Coloured concrete produced from low-carbon cements: Mechanical properties, chromatic stability and sustainability. *J. Build. Eng.* **2023**, *67*, 106018. [CrossRef]
2. Ijaz, N.; Ye, W.; Rehman, Z.U.; Ijaz, Z. Novel application of low carbon limestone calcined clay cement (LC3) in expansive soil stabilization: An eco-efficient approach. *J. Clean. Prod.* **2022**, *371*, 133492. [CrossRef]
3. Yang, T.; He, X. Research status and prospect of blast furnace slag resource utilization. *China Environ. Prot. Ind.* **2020**, *3*, 4.
4. Ahmadi, H.; Khalaj, G.; Najafi, A.; Abbasi, S.M.; Safari, M. Metakaolin-red mud/carbon nanotubes geopolymer nanocomposite: Mechanical properties and structural studies. *Mater. Res. Express* **2022**, *9*, 025011. [CrossRef]
5. Salami, B.A.; Ibrahim, M.; Al-Osta, M.A.; Nasir, M.; Ali, M.R.; Bahraq, A.A.; Wasiu, A. Engineered and green natural pozzolan-nano silica-based alkali-activated concrete: Shrinkage characteristics and life cycle assessment. *Environ. Sci. Pollut. Res.* **2023**, *30*, 17840–17853. [CrossRef]
6. Cui, K.; Liang, K.; Chang, J.; Lau, D. Investigation of the macro performance, mechanism, and durability of multiscale steel fiber reinforced low-carbon ecological UHPC. *Constr. Build. Mater.* **2022**, *327*, 126921. [CrossRef]
7. Provis, J.L.; Bernal, S.A. Geopolymers and Related Alkali-Activated Materials. *Annu. Rev. Mater. Res.* **2014**, *44*, 299–327. [CrossRef]
8. Nasir, M.; Johari, M.A.M.; Maslehuddin, M.; Yusuf, M.O. Sodium sulfate resistance of alkali/slag activated silico-manganese fume-based composites. *Struct. Concr.* **2021**, *22* (Suppl. S1), E415–E429. [CrossRef]
9. Sofi, M.; Deventer, J.S.J.V.; Mendis, P.A.; Lukey, G.C. Engineering properties of inorganic polymer concretes (IPCs). *Cem. Concr. Res.* **2007**, *37*, 251–257. [CrossRef]
10. Rashad, A.M.; Zeedan, S.R.; Hassan, A.A. Influence of the activator concentration of sodium silicate on the thermal properties of alkali-activated slag pastes. *Constr. Build. Mater.* **2016**, *102*, 811–820. [CrossRef]
11. Bakharev, T. Resistance of geopolymer materials to acid attack. *Cem. Concr. Res.* **2005**, *35*, 658–670. [CrossRef]
12. Bakharev, T. Durability of geopolymer materials in sodium and magnesium sulfate solutions. *Cem. Concr. Res.* **2005**, *35*, 1233–1246. [CrossRef]
13. Yang, N.R. Physicochemical basis of formation of alkali cementitious materials. *J. Chin. Ceram. Soc.* **1996**, *24*, 209–215. [CrossRef]
14. Roy, D.M. Alkali-activated cements Opportunities and challenges. *Cem. Concr. Res.* **1999**, *29*, 249–254. [CrossRef]
15. Shi, C.J. Strength, pore structure and permeability of alkali-activated slag mortars. *Cem. Concr. Res.* **1996**, *26*, 1789–1799. [CrossRef]
16. Fu, Y.W.; Cai, L.C.; Wu, Y.J. Freeze-thaw cycle test and damage mechanics models of alkali-activated slag concrete. *Constr. Build. Mater.* **2011**, *25*, 3144–3148. [CrossRef]
17. Fu, Y.W.; Cai, L.C.; Cao, D.G.; Wu, Y.G. Freeze-thaw durability and damage mechanics model of high performance alkali-slag concrete. *Eng. Mech.* **2012**, *23*, 103–109.
18. Duxson, P.; Provis, J.L.; Lukey, G.C.; Deventer, J.S.J.V. The role of inorganic polymer technology in the development of 'green concrete'. *Cem. Concr. Res.* **2007**, *37*, 1590–1597. [CrossRef]
19. Collins, F.; Sanjayan, J.G. Strength and Shrinkage Properties of Alkali-Activated Slag Concrete Containing Porous Coarse Aggregate. *Cem. Concr. Res.* **1999**, *29*, 607–610. [CrossRef]
20. Zhang, L.F. *Alkali Activated Slag Cement and Concrete*, 1st ed.; Southwest Jiaotong University Press: Chengdu, China, 2018; pp. 132–140. ISBN 978-7-5643-6401-4.

21. Cartwright, C.; Rajabipour, F.; Radlinska, A. Shrinkage Characteristics of Alkali-Activated Slag Cements. *J. Mater. Civ. Eng.* **2014**, *27*, B4014007. [CrossRef]
22. Atiş, C.D.; Bilim, C.; Çelik, Ö.; Karahan, O. Influence of activator on the strength and drying shrinkage of alkali-activated slag mortar. *Constr. Build. Mater.* **2009**, *23*, 548–555. [CrossRef]
23. Ibrahim, M.; Nasir, M.; Hussaini, S.R.; Najamuddin, S.K.; Ewebajo, A. Performance of structurally viable green concrete derived from natural pozzolan and nanosilica. *Mag. Civ. Eng.* **2021**, *7*, 10710.
24. Cheng, Y.; Huang, X. Effect of chloride on strength of alkali activated slag paste. *J. Beijing Univ. Aeronaut. Astronaut.* **2015**, *4*, 693–700. [CrossRef]
25. Lv, Q.F.; Wang, Z.S.; Cheng, Y. Effect mechanism of sodium chloride on strength of alkali-activated geopolymer. *J. Funct. Mater./Gongneng Cailiao* **2020**, *51*, 73–98+7. [CrossRef]
26. Cao, Y.Z.; Guo, L.P.; Xue, X.L. Effects of sodium chloride and sodium sulfate on hydration process. *J. Southeast Univ. (Nat. Sci. Ed.)* **2019**, *49*, 8. [CrossRef]
27. He, W. Effects of sodium chloride on the mechanical properties of slag composite matrix geopolymer. *Adv. Cem. Res.* **2019**, *31*, 389–398. [CrossRef]
28. Liu, C.B.; Gao, X.Q.; Zhou, Y.C. Research on hydration mechanism of slag composite cementitious material mixed with NaCl. *J. Chem. Pharm. Res.* **2014**, *6*, 845–849.
29. Zhang, L.F.; Song, S.S.; Liang, Q.S. Effect of gypsum content on properties of alkali-activated slag cement mortar. *Appl. Chem. Ind.* **2020**, *49*, 6. [CrossRef]
30. Xing, J.; Hu, J.W.; Li, C.; Qiu, J.P.; Sun, X.G. The effect of gypsum on the cementitious performance of blast furnace slag stimulated by calcium oxide. *China Min. Mag.* **2019**, *28*, 166–171. [CrossRef]
31. Provis, J.L.; Lukey, G.C.; Deventer, J.V. Do Geopolymers Actually Contain Nanocrystalline Zeolites? A Reexamination of Existing Results. *Chem. Mater.* **2005**, *17*, 3075–3085. [CrossRef]
32. Zhang, Y.X.; Yang, Z.H.; Guo, D.; Geng, H.; Dong, C. Effect of chloride salts and bicarbonate on solubility of $CaSO_4$ in aqueous solutions at 37 °C. *Procedia Environ. Sci.* **2013**, *18*, P84–P91. [CrossRef]
33. Liu, C.B.; Yuan, Y.K.; He, W.; Zhang, L. Durability analysis of seashore saline soil bound with a slag compound binder. *Soils Found.* **2019**, *59*, 1456–1467. [CrossRef]
34. Xiong, L.C.; Wan, Z.J.; Zhang, Y.; Wang, F.T.; Kang, Y.L. Fly ash particle size effect on pore structure and strength of fly ash foamed geopolymer. *Adv. Polym. Technol.* **2019**, *2019*, 1098027. [CrossRef]
35. Carlson, R.W.; Reading, T.J. Model Study of Shrinkage Cracking in Concrete Building Walls. *Struct. J.* **1988**, *85*, 395–404.
36. Luo, H.; Lv, L.Y.; Cui, H.Z. Research progress of cracking resistance evaluation methods for early age concrete. In Proceedings of the Industrial Architecture Academic Exchange Conference, Xi'an, China, 11–13 December 2020.
37. Song, S.M.; Shi, H.X.; Lian, H.Z.; Zhang, H.Y.; Zhang, L.Y. The utility model relates to a cracking sensitivity testing device for cement-based materials. U.S. Patent CN207198158U, 2018.
38. Ma, J.Y.; Li, Z.B. Preparation and desilication of Ca-Al hydrotalcite in sodium aluminate solution. *Sci. Sin. Chim.* **2010**, *40*, 577–584.
39. Mesbah, A.; François, M.; Cau-dit-Coumes, C.; Frizon, F.; Filinchuk, Y.; Leroux, F.; Ravaux, J.; Renaudin, G. Crystal structure of Kuzel's salt $3CaO{\cdot}Al_2O_3{\cdot}1/2CaSO_4{\cdot}1/2CaCl_2{\cdot}11H2O$ determined by synchrotron powder diffraction. *Cem. Concr. Res.* **2011**, *41*, 504–509. [CrossRef]
40. Durdzinski, P.T. *Hydration of Multi-Component Cements Containing Cement Clinker, Slag, Calcareous Fly Ash and Limestone*; EPFL: Lausanne, Switzerland, 2016.
41. Ji, H.G.; Liu, C.B.; Liu, J.H. Characteristics of slag fine-powder composite cementitious material-cured coastal saline soil. *Emerg. Mater. Res.* **2014**, *3*, 282–291. [CrossRef]
42. Van Jaarsveld, J.; Van Deventer, J.; Schwartzman, A. The potential use of geopolymeric materials to immobilise toxic metals: Part II. Material and leaching characteristics. *Miner. Eng.* **1999**, *12*, 75–91. [CrossRef]
43. Jaarsveld, J.G.S.V.; Deventer, J.S.J.V.; Lukey, G.C. The characterisation of source materials in fly ash-based geopolymers. *Mater. Lett.* **2003**, *57*, 1272–1280. [CrossRef]
44. Xu, H.; Deventer, J.S.J.V. The geopolymerisation of alumino-silicate minerals. International Journal of Mineral Processing. *Int. J. Miner. Process.* **2000**, *59*, 247–266. [CrossRef]
45. Liu, C.B. The research of slag composite cementitious material strength and hydration mechanism. *Sichuan Build. Sci.* **2012**, *38*, 169–172. [CrossRef]
46. Roy, D.M.; Jiang, W.M.; Silsbee, M.R. Chloride diffusion in ordinary, blended, and alkali-activated cement pastes and its relation to other properties. *Cem. Concr. Res.* **2000**, *30*, 1879–1884. [CrossRef]
47. Matschei, T.; Lothenbach, B.; Glasser, F.P. The AFm phase in Portland cement. *Cem. Concr. Res.* **2007**, *37*, 118–130. [CrossRef]
48. Wang, S.D.; Huang, Y.B.; Wang, Z. Concrete resistance to chloride ingress: Effect of cement composition. *J. Chin. Ceram. Soc.* **2000**, *6*, 570–574. [CrossRef]
49. Haghtalab, A.; Badizad, M.H. Solubility of gypsum in aqueous NaCl + K_2SO_4 solution using calcium ion selective electrode-investigation of ionic interactions. *Fluid Phase Equilibria* **2016**, *409*, 341–353. [CrossRef]
50. Hirao, H.; Yamada, K.; Takahashi, H.; Zibara, H. Chloride Binding of Cement Estimated by Binding Isotherms of Hydrates. *J. Adv. Concr. Technol.* **2005**, *3*, 77–84. [CrossRef]

51. Ye, H.; Jin, X.Y.; Chen, W.; Fu, C.Q.; Jin, N.G. Prediction of chloride binding isotherms for blended cements. *Comput. Concr.* **2016**, *17*, 665–682. [CrossRef]
52. Birnin-Yauri, U.A.; Glasser, F.P. Friedel's salt, Ca2Al(OH)6(Cl,OH)·2H2O: Its solid solutions and their role in chloride binding. *Cem. Concr. Res.* **1998**, *28*, 1713–1723. [CrossRef]
53. Thomas, M.D.A.; Hooton, R.D.; Scott, A.; Zibara, H. The effect of supplementary cementitious materials on chloride binding in hardened cement paste. *Cem. Concr. Res.* **2012**, *42*, 1–7. [CrossRef]
54. Walther, J.V. Experimental determination of portlandite and brucite solubilities in supercritical H_2O. *Geochim. Et Cosmochim. Acta* **1986**, *50*, 733–739. [CrossRef]
55. Xing, G.F.; Xu, C.; Ye, G.B.; Yang, X.M. Mechanism Analysis of Influence of Soluble Salt Ions on Strength of Salt-rich Cement-soil. *China J. Highw. Transp.* **2008**, *21*, 26–30. [CrossRef]
56. Yi, Y.L.; Liska, M.; Jin, F.; Al-Tabbaa, A. Mechanism of reactive magnesia—Ground granulated blastfurnace slag (GGBS) soil stabilization. *Can. Geotech. J.* **2016**, *53*, 773–782. [CrossRef]
57. Hou, Y.F.; Liu, J.T.; Peng, X.D.; Zhao, S.R. Effect of Iron Tailing Powder Fineness on the Pore Structure of Cement Mortar. *Fly Ash Compr. Util.* **2017**, *5*, 23–26+35.
58. Cui, K.; Chang, J. Hydration, reinforcing mechanism, and macro performance of multi-layer graphene-modified cement composites. *J. Build. Eng.* **2022**, *57*, 104880. [CrossRef]
59. Ma, H.Z.; Li, Z.Q. Ettringite Formation in Concrete. *Build. Sci.* **2007**, *11*, 105–110. [CrossRef]
60. Wang, F.S.; Zhu, Y.N.; Ma, J.L.; Sun, R.L. Experimental Research of Sodium Chloride on Activation and Binding Mode of Slag Portland Blend Cement. *Bull. Chin. Ceram. Soc.* **2009**, *28*, 784–791. [CrossRef]
61. Lu, Q.B.; Yang, Q.B. Influencing factors to activate the potential activity of slag. In Proceedings of the 7th Council Meeting and Academic Exchange Conference of Concrete Cement Products Branch of China Silicate Society, Beijing, China, 19–21 October 2005; Volume 6, pp. 131–134.
62. Liu, S.L.; Li, G.C.; Liu, G.L.; Wang, F.G.; Wang, J. Early Hydration Mechanism of Gypsum Slag-Lime Composite Cementitious System. *Nonferrous Met. Eng.* **2021**, *11*, 102–109. [CrossRef]
63. Yu, Z.Y.; Zhao, Y.D.; Ba, H.J.; Liu, M.H. Synergistic effects of ettringite-based expansive agent and polypropylene fiber on early-age anti-shrinkage and anti-cracking properties of mortars. *J. Build. Eng.* **2021**, *39*, 102275. [CrossRef]
64. Hou, D.W. *Integrative Studies on Autogenous and Drying Shrinkage of Concrete and Related Issues*; Tsinghua University: Beijing, China, 2010.
65. Yan, H.D.; Sun, W. Study on the relationship between autogenous shrinkage and drying shrinkage of fly ash mortar. *J. Chin. Ceram. Soc.* **2003**, *31*, 428–433. [CrossRef]
66. Yuan, B.; Yu, Q.L.; Dainese, E.; Brouwers, H.J.H. Autogenous and drying shrinkage of sodium carbonate activated slag altered by limestone powder incorporation. *Constr. Build. Mater.* **2017**, *153*, 459–468. [CrossRef]
67. Liu, H.H. *Test Research on Mechanical Performance for Structural Member of Inorganic Polymer Concrete*; Wuhan University of Technology: Wuhan, China, 2014. [CrossRef]
68. Palacios, M.; Puertas, F. Effect of shrinkage-reducing admixtures on the properties of alkali-activated slag mortars and pastes. *Cem. Concr. Res.* **2007**, *37*, 691–702. [CrossRef]
69. Taghvayi, H.; Behfarnia, K.; Khalili, M. The Effect of Alkali Concentration and Sodium Silicate Modulus on the Properties of Alkali-Activated Slag Concrete. *J. Adv. Concr. Technol.* **2018**, *16*, 293–305. [CrossRef]
70. Zhang, B.; Zhu, H.; Cheng, Y.; Ghasan, F.H.; Kwok, W.S. Shrinkage mechanisms and shrinkage-mitigating strategies of alkali-activated slag composites: A critical review. *Constr. Build. Mater.* **2022**, *318*, 125993. [CrossRef]
71. Zhang, B.; Zhu, H.; Feng, P.; Zhang, P. A review on shrinkage-reducing methods and mechanisms of alkali-activated/geopolymer systems: Effects of chemical additives. *J. Build. Eng.* **2022**, *49*, 104056. [CrossRef]
72. Yao, J.B.; Yuan, J.; Song, X.F.; Ju, J.T. Influence of Super-absorbent Polymer on the Drying Shrinkage and Cracking of Alkali Activated Slag. *Bull. Chin. Ceram. Soc.* **2019**, *38*, 2335–2339. [CrossRef]

buildings

MDPI

Article

Bond Performance of Anti-Corrosion Bar Embedded in Ceramsite Concrete in Freeze–Thaw Cycles and Corrosive Environments

Yan Liu [1,2], Qiang Zhu [2], Jinhua Teng [2], Peng Deng [1,2,*] and Yan Sun [2]

[1] Shandong Provincial Key Laboratory of Civil Engineering Disaster Prevention and Mitigation, Shandong University of Science and Technology, Qingdao 266590, China
[2] College of Civil Engineering and Architecture, Shandong University of Science and Technology, Qingdao 266590, China
* Correspondence: dengpeng1226@sdust.edu.cn

Abstract: At present, basalt fiber-reinforced polymer (BFRP) bars and epoxy-coated steel reinforcing bars (ECRs) are very promising in ocean engineering. In this study, the bond strength degradation characteristics of BFRP bars, ECR, and ordinary steel bars (OSBs) embedded in ceramsite concrete (CC) were compared in a single-corrosive environment (acid, salt, and alkaline salt, respectively) coupled with freeze–thaw (FT) cycles (0, 15, or 30); a total of 111 specimens were designed. In the three corrosive environments, the BFRP-bar-CC specimens and OSB-CC specimens all failed to pull out, while the ECR-CC specimens showed splitting failure. When corrosive and FT cycles acted together, the failure modes of BFRP-bar-CC specimens and ECR-CC specimens did not change. However, when the FT cycles increased from 15 to 30, the type of failure for the OSB-CC specimens changed from pullout failure to splitting failure. In addition, the bonding strength of the three kinds of bars decayed most rapidly in an acid environment. When 30 FT cycles were applied, the bond strength of ECR-CC specimens and OSB-CC specimens decreased most rapidly in the acid environment, by 9.12% and 18.62%, respectively. However, the bond strength of BFRP-bar-CC decreased most rapidly, by 17.2%, in an alkaline salt environment.

Keywords: basalt fiber-reinforced polymer bars; bond strength; ceramsite concrete; corrosive environment; epoxy-coated steel bars; freeze–thaw cycle

Citation: Liu, Y.; Zhu, Q.; Teng, J.; Deng, P.; Sun, Y. Bond Performance of Anti-Corrosion Bar Embedded in Ceramsite Concrete in Freeze–Thaw Cycles and Corrosive Environments. *Buildings* **2023**, *13*, 884. https://doi.org/10.3390/buildings13040884

Academic Editors: Shan Gao, Jingxuan Wang, Dewen Kong and Yong Liu

Received: 9 March 2023
Revised: 26 March 2023
Accepted: 26 March 2023
Published: 28 March 2023

1. Introduction

In the main normative documents, such as ACI 318 [1], EN 1992 [2], and GB 50010-2010 [3], the properties of the bonding between the reinforcement and the surrounding concrete are crucial for the bearing capacity or safety performance of the reinforced concrete (RC) structure. Lightweight aggregate concrete (LWAC), especially ceramsite concrete (CC), can reduce the self-weight of reinforced concrete members. In addition, it possesses relatively good freeze–thaw (FT) resistance and realizes good heat insulation. Therefore, it holds promising prospects for application, particularly in cold or intensely corrosive ocean areas [4–6]. Although CC has good corrosion resistance, the weak corrosion resistance of steel bars affects bond performance [7]. As a result, numerous studies have focused on developing reasonable alternative materials for OSBs in order to elevate the bond performance of LWAC structures.

The use of stainless steel, fiber-reinforced polymers (FRPs), and epoxy-coated reinforcements (ECRs) is considered to be a good approach to addressing corrosion. However, the use of stainless-steel reinforcement is limited by its high cost [8]. Epoxy-coated reinforcement (ECR) has the advantages of low cost, corrosion resistance, and good ductility and durability [9–12]. However, the surface of ECR is easily damaged during construction [13]. Despite this defect, the engineering application of ECR has been gradually popularized;

thus, its bonding with CC requires our attention. In order to achieve better corrosion resistance, increasing attention has been paid to FRP in recent years.

Basalt FRP (BFRP) bars can be used as a suitable substitute for steel bars in reinforced concrete structures [14], due to their good chemical corrosion resistance, high fatigue resistance, thermal stability, and low water absorption [15–19]. The function of these excellent characteristics on the bond properties of FRP-reinforced concrete structures under FT conditions, cold regions, and ocean environments requires further study. Green et al. [20] introduced FT cycles to investigate the effect of FRP plates and concrete bonding properties, revealing that, with failure, the surface turned into an adhesive interface with the FRP plates with an increasing number of cycles. In contrast, in the absence of FT cycles, failure occurs on concrete substrates, which makes the bond strength of FRP bars controlled not by concrete strength but by interlaminar shear strength beneath the resin layer of reinforcement [21,22]. Shi et al. [23] reported that the durability of BFRP composites was largely maintained in an FT environment. Firas et al. [24] conducted a supplementary experiment to simulate the marine environment by soaking samples in brine; the results showed that when the bearing capacity decreased by 48%, the resin–concrete interface was damaged. FRP is exposed to one or more aging media during its service life, such as FT cycles, acid rain, or marine erosion. Therefore, the durability of FRP in corrosive environments remains a problem [25]. For instance, Hassan M et al. [26], Yan et al. [27], Wu et al. [28], and Mohamed Al et al. [29] confirmed that an alkaline solution (attempting to simulate concrete-pore solutions) was the main influencing condition for BFRP bars. Altalmas et al. [30] studied the bond durability of BFRP in different corrosive environments (acid, salt, or alkali). It was found that the bond strength of BFRP specimens decreased by 25% after 90 days of immersion in seawater and alkaline solutions, and decreased by 14% when they were immersed in an acidic solution for 90 days.

Many previous studies only investigated the bonding properties between bars (BFRP bars, ECRs) and ordinary Portland cement (OPC), but few of them studied reinforcement with CC, and even fewer studied the bonding properties in harsh environments. This study evaluated the decay of the bond strength of BFRP bars under corrosive conditions. The combined effects of FT cycling and corrosion conditions were evaluated and compared with their effects on ECRs and OSBs. The failure mechanism of the specimens was analyzed and the bond–slip relationship was proposed.

2. Materials and Methods

2.1. Specimen Size and Structure

A total of 111 specimens were prepared, including 81 central pullout cube specimens with a side length of 150 mm and 30 compressive strength cube specimens with a side length of 100 mm. The 150 mm concrete cubes were concentrically reinforced with 330 mm of BFRP, ECR, and ordinary steel bar (OSB), and the BFRP and ECR were compared with the OSB. According to ACI440.3R-04 [31], the embedding length was 5 d [32] (Figure 1a), where d is the diameter of the ECR. In order to increase the required bonding length, the PVC sleeve is set at the loading end position of each specimen, and the gap between the PVC sleeve and the steel bar is filled and sealed with a foaming agent to prevent its grouting from affecting the test results. The samples were cured in a curing chamber with 95% relative humidity at ambient temperature (20 $\pm$ 2 °C) for 28 days. (Figure 1b). The cured specimens were subjected to the FT cycle test and corrosion solution test, respectively.

2.2. Raw Materials

2.2.1. Concrete

Lightweight aggregate (LWA) uses shale ceramic products produced by Yichang Huiteng Ceramic Products Trading Co., Ltd., Yichang, China. Table 1 shows the characteristic parameters of shale ceramsite (SC). Table 2 shows the proportions of CC used to cast the specimens. The cement is made of 42.5 OPC, the fine aggregate is made of natural medium sand with a fineness coefficient of 2.6, and the coarse aggregate is made of SC with a maximum particle size of 20 mm. Tap water was used. In order to improve the fluidity of the cement mixture, Grey Ba brand high-efficiency water-reducing agent was added to the test at a dosage of 0.1~0.3%. The concrete compressive strengths were determined using 100 × 100 × 100 mm cube specimens. The 28-day compressive strengths are listed in Table 3.

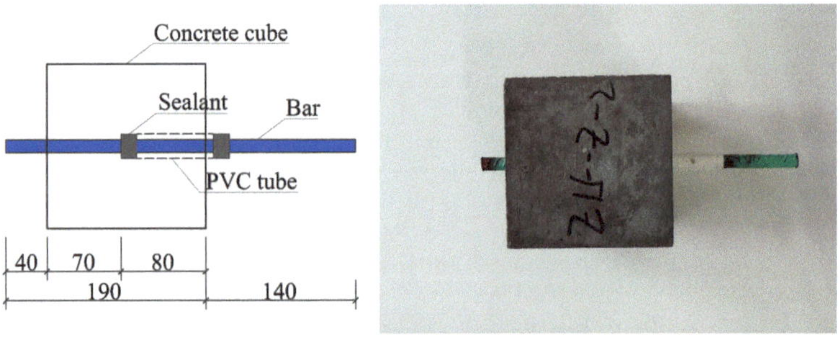

(a) (b)

Figure 1. Pullout samples. (**a**) Schematic diagram of the bond test sample (mm); (**b**) sample of concrete after curing.

Table 1. Characteristic parameters of shale ceramsite (SC).

Ceramsite Type	Density Grad (kg/m^3)	Nominal Diameter (mm)	Packing Density (kg/m^3)	Water Absorption 1 h (%)	Compressive Strength (MPa)
SC	500	5–20	470	7.6	2.6

Table 2. Mixture ratio of CC.

Compressive Strength	Cement (kg/m^3)	Water (kg/m^3)	Fine Aggregate (kg/m^3)	Coarse Aggregate (kg/m^3)
LC30 *	400	200	680	490

* LC30, the compressive strength of CC after 28 days of curing is close to 30 MPa.

Table 3. The measured values of concrete strength.

Compressive Strength	Measured Compressive Strength (MPa)			Mean (MPa)
	1	2	3	
LC30	34.56	33.58	31.48	33.21

2.2.2. Bars

The diameter and length of the three reinforcing bars were 14 mm and 330 mm, respectively (Figure 2). BFRP bars were customized by Jingdong Construction Technology Co., Ltd., Shanghai, China. A pultrusion process was used to combine basalt fibers with epoxy resins. The fiber-to-epoxy volumetric ratios were 65% and 35%, respectively, and the BFRP bars met the minimum requirements of ACI440.3R-04 [31] and CSAS807 [33]. The epoxy coated steel bar was customized by Deyang Wanteng Metal Products Co., Ltd., Deyang, China, the coating thickness was 0.17 mm, the main raw materials of epoxy coating included epoxy resin, curing agent, and additives. The tensile properties of the reinforcement test are shown in Table 4. BFRP is not suitable for reinforcing flexural members due to its low elastic modulus, as it will lead to excessive deflection and cracking of the reinforced members.

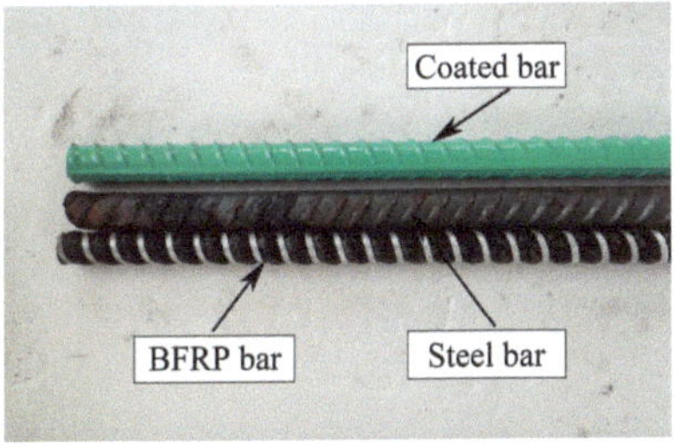

Figure 2. Bar types.

Table 4. Mechanical properties of bars.

Bar Type	Diameter (mm)	Ultimate Tensile Strength (MPa)	Yield Strength (MPa)	Elastic Modulus (GPa)
BFRP bars	14	703	-	40
ECRs	14	660	470	208
OSBs	14	655	566	202

2.3. Experimental Scheme

2.3.1. Bond Specimens and Concrete Test Cube

In this study, a series of experiments were designed to evaluate the influence of different factors on bond strength. The specific test groups are shown in Table 5. These variables were the type of reinforcement, FT cycles, the corrosive environment, and CC compressive strength. All bond specimens were labeled in the following order: bar type (basalt, B; epoxy-coated rebar, E; or ordinary steel bar, D), middle numbers indicating FT environment (0, 15, or 30) experienced by the specimen, and corrosive exposure (S for acid, Z for salt, and J for alkaline salt). For example, B-0-S is a specimen of BFRP bars and CC under a single acidic condition. The compressive strength samples were labeled in the following order: number of FT (0, 15, 30), conditional exposure (S for acid, Z for salt, and J for alkaline salt). For instance, 15-S is the specimen of CC under the combined action of 15 FT cycles and an acid environment.

Table 5. Testing plan.

Grouping	Specimens	Bar Type	FT Cycles	Corrosive Environment	Number of Specimens
Group 1 *	B-0-S/J/Z	BFRP	0	pH = 2/10/7	3/3/3
Group 2	E-0-S/J/Z	ECR	0	pH = 2/10/7	3/3/3
Group 3	D-0-S/J/Z	OSB	0	pH = 2/10/7	3/3/3
Group 4	B-15-S/J/Z	BFRP	15	pH = 2/10/7	3/3/3
Group 5	E-15-S/J/Z	ECR	15	pH = 2/10/7	3/3/3
Group 6	D-15-S/J/Z	OSB	15	pH = 2/10/7	3/3/3
Group 7	B-30-S/J/Z	BFRP	30	pH = 2/10/7	3/3/3
Group 8	E-30-S/J/Z	ECR	30	pH = 2/10/7	3/3/3
Group 9	D-30-S/J/Z	OSB	30	pH = 2/10/7	3/3/3
Group 10	-	-	Untreated specimen		3
Group11	0-S/J/Z	-	0	pH = 2/10/7	3/3/3
Group12	15-S/J/Z	-	15	pH = 2/10/7	3/3/3
Group13	30-S/J/Z	-	30	pH = 2/10/7	3/3/3

* Three corrosive environments were set up for each group of specimens: S/J/Z represent acid/alkaline salt/salt and pH of 2/10/7, respectively.

2.3.2. FT Cycles

The FT cabinet (NJW-HDK-5) was used for the accelerated aging test of the samples. Temperature monitoring was carried out by temperature-sensitive elements placed in the center and diagonal of the instrument.

FT cycle test scheme was based on ASTMC666/C666M [34]. The highest temperature of the FT test was set at 6 ± 2 °C, and the lowest temperature was -18 ± 2 °C. Before the FT test, the specimens, after curing for 24 days, were immersed in water at 20 ± 2 °C for 4 days. After soaking, the specimens were taken out and exposed to sunlight for 2 days to remove excess moisture.

2.3.3. Three Different Corrosive Solutions

Three artificial corrosive environments were used in this study: (1) an alkaline salt environment of pH 10, using a $NaHCO_3$ solution to simulate the alkaline environment in ordinary reinforced concrete; (2) a salt environment of pH 7, using a NaCl solution with a chloride concentration of 3.5% to simulate the climate of coastal areas; (3) an acid solution of pH 2.0, using an oxalic acid solution to simulate a polluted industrial area. After the FT cycles test, the specimens were immersed in the three solutions, respectively. There were three specimens in each solution, all of which were soaked for 60 days. The pH value of the three corrosive environments was monitored in real-time by a pH meter, and the corresponding solution was added in time to ensure that the pH was unchanged.

2.4. Pullout Test Scheme

The center pullout test was performed according to RILEM-FIP-CEB [35]. MTS-SANS series universal testing machine. A range of 400 kN was used for the single-end drawing of the specimen (Figure 3a,b). The displacement meter used epoxy resin glue to paste an angle steel bar horizontally on the free end of the steel bar, and the concrete surface, ensuring that the tie rod of the displacement meter and the angle steel surface were always at 90 degrees. LVDT1 was used to record the slip of the unloaded end rod, whereas LVDT2 recorded the slippage of the free end on the concrete surface. The displacement control was loaded at a rate of 0.2 mm/min. When the free-end slip reached 10 mm, all tests were halted unless the specimen failed before then (split failure).

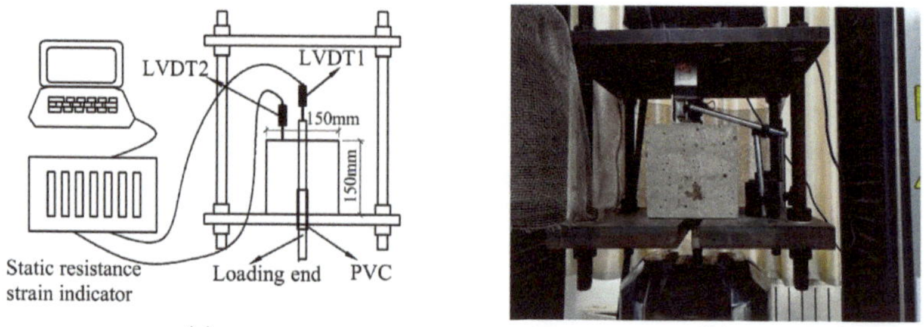

(**a**) (**b**)

Figure 3. Pullout test setup. (**a**) Pullout test equipment; (**b**) universal tensile-compression testing machine.

3. Test Results and Discussion

3.1. The Bond Stress

Due to the low elastic modulus of BFRP bars, their elastic deformation cannot be ignored. The elastic elongation of BFRP bars can be obtained using Equation (1) [36]:

$$\delta_e = \frac{P}{EA} L \tag{1}$$

where δ_e is the elastic elongation of unbonded BFRP bars under drawing force, E is the elastic modulus of BFRP bars, A is the section area of BFRP bars, and L is the length of the bonding section.

The slip is then subtracted from the elastic elongation of the BFRP bar to obtain the true slip. This can be expressed as Equation (2) [37,38]:

$$s = s_m - \delta_e \quad (2)$$

where s is the real slip, and s_m is the slip amount of the BFRP bar measured by the displacement sensor.

The average bond stress between the bar and CC can be expressed as Equation (3) [39]:

$$\tau = \frac{P}{\pi d l_a} \quad (3)$$

where τ is the average bond stress (MPa), P is the pullout force of the bar (N), d is the diameter of the bar (mm), and l_a is the bond length of the bar (mm).

Table 6 shows the results of the drawing test. To ensure the reliability of the test results, three samples were tested in each group, and the final test result was the average value of the data of the three specimens.

Table 6. Bond test results of BFRP, ECR, and OSB specimens exposed to different environments.

Specimen	Failure Mode	P (kN)			$\overline{P}$ (kN) [3]	τ_{max} (MPa) [4]	S (mm)			$\overline{S}$ (mm) [5]
		1	2	3			1	2	3	
B-0-S	PO [1]	29.01	28.52	27.49	28.34	9.24	1.33	1.30	1.24	1.29
B-0-J	PO	28.43	28.02	29.68	28.71	9.45	1.20	1.16	1.27	1.21
B-0-Z	PO	28.34	28.90	/ [6]	28.62	9.42	1.13	1.25	/	1.19
E-0-S	S [2]	37.86	35.53	36.94	36.77	11.95	1.08	1.15	0.93	1.05
E-0-J	S	38.79	36.43	39.53	38.25	12.59	1.04	0.88	0.99	0.97
E-0-Z	S	35.23	40.49	39.30	38.34	12.52	0.79	1.12	0.85	0.92
D-0-S	PO	/	36.32	39.68	38.00	12.35	/	0.77	0.95	0.86
D-0-J	PO	39.69	40.39	39.95	40.01	13.17	0.86	0.90	0.73	0.83
D-0-Z	PO	37.60	42.60	39.38	39.86	13.12	0.70	0.94	0.79	0.81
B-15-S	PO	25.52	27.77	30.02	27.77	9.03	1.21	1.31	1.53	1.35
B-15-J	PO	25.76	26.05	29.01	26.94	8.76	1.30	1.27	1.36	1.31
B-15-Z	PO	28.54	29.62	27.67	28.61	9.30	1.20	1.38	1.17	1.25
E-15-S	S	33.27	37.64	32.59	34.50	11.55	1.11	1.17	0.99	1.09
E-15-J	S	40.61	37.26	35.26	37.71	12.26	1.03	0.78	0.74	0.85
E-15-Z	S	36.91	38.51	/	37.71	12.26	0.89	1.05	/	0.97
D-15-S	PO	31.60	33.82	37.51	34.31	11.15	0.80	0.80	0.95	0.85
D-15-J	PO	40.35	39.50	37.63	39.16	12.73	0.88	0.67	0.67	0.74
D-15-Z	PO	39.75	35.32	41.39	38.82	12.62	0.54	0.66	0.57	0.59
B-30-S	PO	25.67	26.03	28.73	26.81	8.71	1.79	1.87	2.01	1.89
B-30-J	PO	23.38	/	24.76	24.07	7.82	1.63	/	1.55	1.59
B-30-Z	PO	27.19	26.59	29.32	27.70	9.00	1.49	1.34	1.61	1.48
E-30-S	S	34.81	31.59	33.83	33.41	10.86	1.39	1.09	1.27	1.25
E-30-J	S	38.77	33.83	36.96	36.52	11.87	1.04	0.86	1.07	0.99
E-30-Z	S	37.96	33.12	38.36	36.48	11.86	0.80	0.67	0.90	0.79
D-30-S	S	29.86	31.96	/	30.91	10.05	1.15	1.01	/	1.08
D-30-J	S	38.40	36.24	39.12	37.92	12.32	1.02	0.79	0.92	0.91
D-30-Z	S	39.37	38.21	36.09	37.89	12.31	0.81	0.62	0.52	0.65

[1] PO, pullout failure; [2] S, splitting failure; [3] $\overline{P}$, mean value of ultimate load; [4] τ_{max}, maximum bond strength; [5] $\overline{S}$, mean free end slip; [6] /, the sample value exceeds the value specified in the specification and is not accepted.

3.2. Concrete Compressive Strength

Table 7 shows the change in compressive strength of untreated specimens and specimens in three corrosive environments. The compressive strength in alkaline and salt environments was similar to that of non-corroded specimens, which was similar to previous studies [30]. However, the compressive strength of concrete immersed in an acid solution decreased significantly due to acid erosion. Compared with the alkaline salt environment and the salt environment, the acid environment had the greatest influence on the compressive strength of the CC specimen.

Table 7. Compressive strength of concrete after corrosion and FT.

Grouping	Specimen	$f_{c,pH}$ (MPa) [1]			$\overline{f}_{c,pH}$ (MPa) [2]
		1	2	3	
Group 10	Untreated specimen	34.56	33.58	31.48	33.21
Group 11	0-S	28.41	28.72	28.22	28.45
	0-J	34.22	33.41	32.62	33.42
	0-Z	31.51	32.56	34.63	32.90
Group 12	15-S	22.02	21.76	21.09	21.62
	15-J	16.90	26.75	28.90	26.75
	15-Z	28.04	26.53	25.35	26.64
Group 13	30-S	18.26	18.32	17.62	18.07
	30-J	25.17	20.92	22.43	22.84
	30-Z	21.96	21.53	22.92	22.14

[1] $f_{c,pH}$, compressive strength (MPa); [2] $\overline{f}_{c,pH}$, the average compressive strength (MPa).

3.3. Failure Mode

Table 6 shows the failure models of all the specimens. It can be seen that the BFRP-bar-CC specimens all experienced pullout failure, but all the ECR-CC specimens failed in splitting. The failure mode of the OSB-CC specimens was the most notable. Under three single-corrosive environments, it experienced pullout failure. However, under the combined action of FT and corrosion, the pullout failure became a splitting failure as the FT number increased from 15 to 30. The failure mode of this specimen was mainly divided into two types: steel bar pullout failure and concrete splitting failure. In the pullout failure, when the loading starts, the displacement at both ends of the sample was very small. When the tension gradually rose to 70–90% of the peak load, a significant displacement of the steel bar could be observed at the free end. Although the splitting failure of concrete is a result of brittleness, there is no obvious sign before the splitting failure occurs. When the drawing force reached the ultimate load, the damage occurred suddenly, accompanied by a loud sound, and the specimen was broken into two or three segments.

In order to analyze the failure mode of specimens more carefully, we studied typical specimens in three corrosive environments (e.g., B-0-S, D-0-Z, E-0-S, E-0-Z, E-0-J) and typical specimens under the combined action of corrosive environments and FT cycles (e.g., E-30-J, B-30-J, D-15-J, D-30-J).

3.3.1. Specimens in Corrosive Environments

Figure 4 shows the failure phenomena of typical specimens in three corrosive environments. After the test, the samples were segmented to observe the condition of the reinforcement and concrete in the embedded area. The BFRP-bar-CC specimens all failed to pull out in three kinds of corrosive environments, which was due to the corrosion resistance of epoxy resin-based grease on the surface of BFRP bars. We were able to observe that the surface of BFRP bars was smooth and did not change significantly in the corrosive environment (Figure 4a). The OSB-CC specimens also failed to pull out in three kinds of corrosive environments. When we cut the specimen open, we were able to observe that the surface of the rib material was slightly corroded (Figure 4b). However, all the ECR-CC specimens experience splitting failure because the outer protective layer effectively resisted the erosion of the acid, alkali, and salt environments. In the acid environment, a layer of white mesh material was attached to the inner reinforcement surface of CC, and the rib surface fell off (Figure 4c). In the salt environment, the surface of ECR was slightly peeled off (Figure 4d). In the alkaline salt environment, the surface coating clearly fell off and adhered to the CC (Figure 4e).

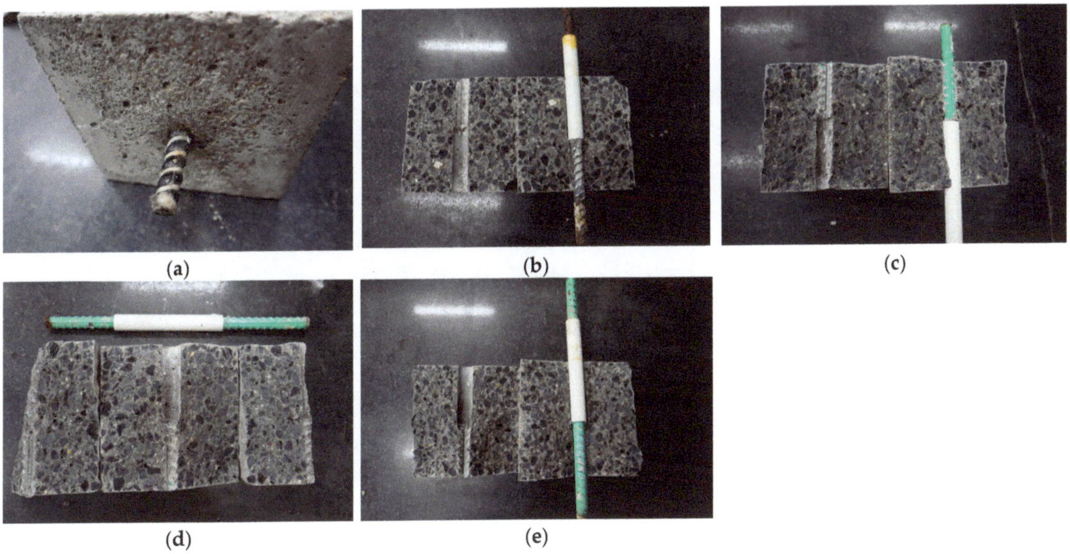

Figure 4. Typical failure modes for samples after a corrosive environment. (a) B-0-S; (b) D-0-S; (c) E-0-S; (d) E-0-Z; (e) E-0-J.

3.3.2. Specimens in FT Cycles and Corrosive Environments

It can be seen from Table 6 that the combination of FT cycles and corrosive environments had a more significant effect on the bond strength of specimens than the single corrosive environment. The failure mode of part of the bar-CC bond specimen is shown in Figure 5. The ECR-CC experienced splitting failure due to the high force required to pull the coated rebar out of the CC. In the alkaline salt corrosion environment, the coating on the surface of the coated steel rib fell off (Figure 5a), white reactants were formed between the reinforcement and CC, and an epoxy coating was attached to the CC. Under the coupled action of FT cycling and a corrosive environment, the BFRP-bar-CC specimens all failed to pull out. After the interaction of 30 FT cycles (FT30) and an alkaline salt environment (Figure 5b), the surface strands of BFRP bars fell off, and the surface became significantly rough. Under the action of 15 FT cycles (FT15) and a corrosive environment, the OSB specimens were mostly pulled out, the steel bars were gradually pulled out from the loading end of the specimen, and no cracks were observed on the surface of the specimen (Figure 5c). However, under the coupling effect of FT30 and a corrosive environment, the common ribbed-steel specimen largely failed by splitting, and the concrete broke into two or three segments. Due to the sudden failure, the crack length and width were large, and a loud sound occurred during the failure (Figure 5d).

3.4. Weight Loss

Due to the FT cycles and the corrosive environments, concrete surface spalling occurred. Generally speaking, the quality of concrete has a certain relationship with its surface state. Especially in FT cycle environments, spalling usually starts locally and expands to a larger area or even the entire concrete surface. Finally, the surface of the sample will appear uneven or aggregate exposure may occur [27,40]. Equation (4) is used to analyze the weight loss of all specimens after erosion:

$$\Delta W_n = \frac{W_0 - W_r}{W_r} \times 100\% \quad (4)$$

where ΔW_n is the weight loss of CC specimens after environmental erosion, W_0 is the weight of the sample before it erodes the environment, and W_r is the weight of the corroded sample.

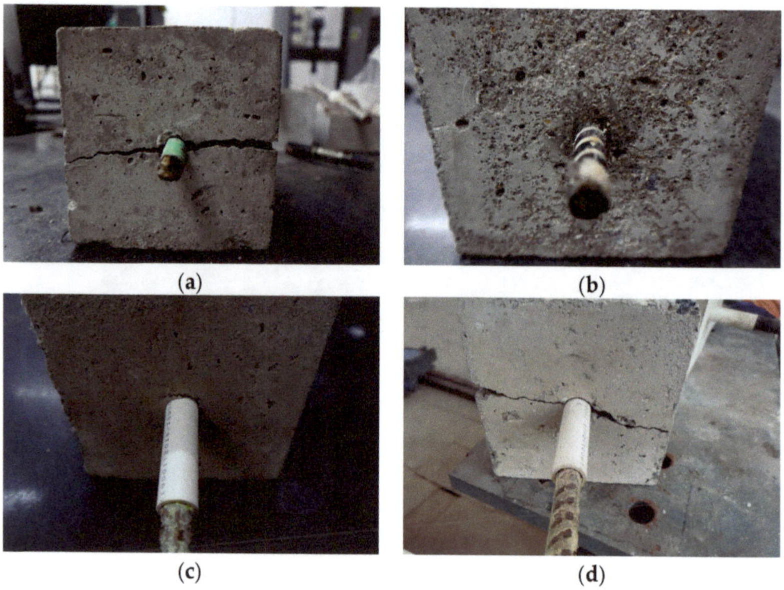

Figure 5. Typical failure modes for samples after the corrosive environment and FT cycles. (**a**) E-30-J; (**b**) B-30-J; (**c**) D-15-J; (**d**) D-30-J.

As can be seen from Table 8 and Figure 6, for the specimens that were exposed to single-corrosive environments, the weight-loss rate of CC specimens exposed to the alkaline salt environment and the salt environment was similar: 0.18% and 0.12%, respectively. The weight loss rate of CC exposed to acid was 0.24%. This shows that acid corrosive solution can peel off the surface cementitious material of concrete [41]. In addition, it can be seen from Table 8 that when FT cycles are not considered, the influence of the three corrosive environments on CC specimens was smaller than the coupling effect of the FT cycles and corrosive environments. However, when the specimens underwent the coupling effect of FT30 and a corrosive environment, the weight-loss rate of the specimens increased to 1.85% (acid environment), 1.64% (alkaline salt environment), and 1.65% (salt environment). It can be seen that the weight-loss rate of specimens subjected to the coupled effect of FT cycles and corrosive environments was significantly higher than that of specimens subjected to a single factor.

Table 8. Weight loss of CC samples undergoing FT environment and corrosive environment.

Specimen	W_0 (g)			W_r (g)			ΔW_n (%)			Mean (%)
	1	2	3	1	2	3	1	2	3	
0-S	1749	1698	1704	1745	1694	1700	0.23	0.24	0.24	0.24
0-J	1681	1704	1662	1678	1698	1654	0.18	0.19	0.18	0.18
0-Z	1648	1668	1645	1646	1668	1643	0.12	0.00	0.12	0.12
15-S	1664	1709	1737	1649	1693	1721	0.91	0.95	0.93	0.93
15-J	1632	1651	1671	1619	1638	1656	0.80	0.79	0.91	0.83
15-Z	1636	1623	1677	1620	1609	1665	0.83	0.87	0.72	0.81
30-S	1656	1657	1705	1626	1627	1674	1.85	1.84	1.85	1.85
30-J	1712	1640	1740	1685	1613	1712	1.60	1.67	1.64	1.64
30-Z	1695	1775	1647	1668	1746	1620	1.62	1.66	1.67	1.65

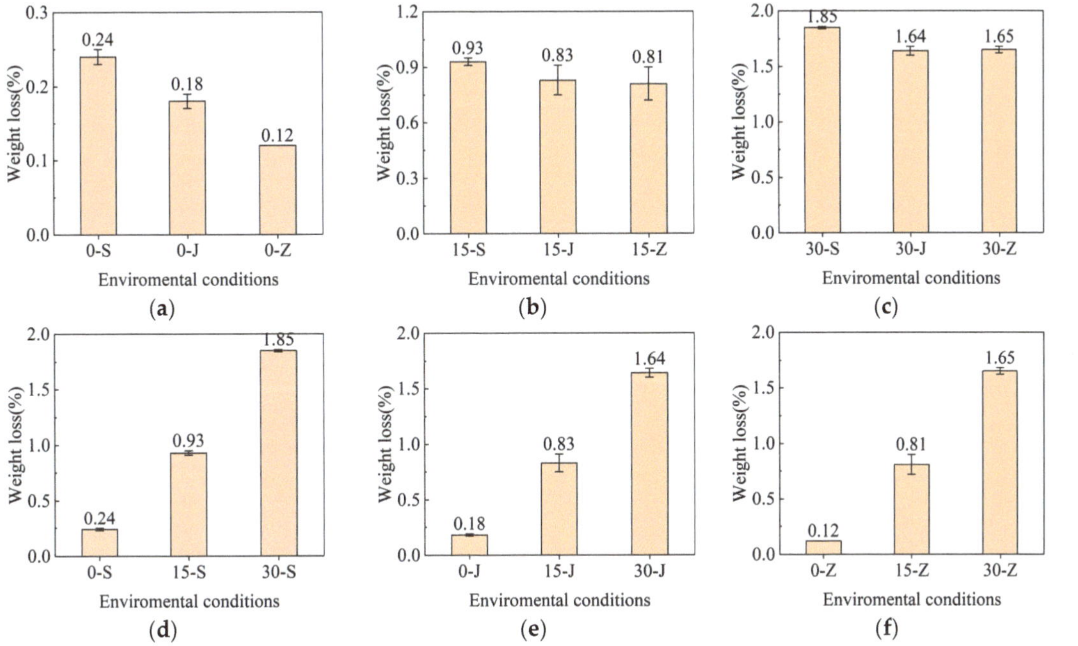

Figure 6. Mass loss in different corrosive environments and FT environments. (**a**) Different corrosive environments; (**b**) 15 FT cycles and different corrosive environments; (**c**) 30 FT cycles and different corrosive environments; (**d**) different FT cycles and the acid environment; (**e**) different FT cycles and the alkaline salt environment; (**f**) different FT cycles and the salt environment.

3.5. Bond Stress–Slip Curves

Figure 7 shows the bond stress–slip curves of three types of reinforcement in different corrosive environments. Each specimen's bond performance was slightly different in different corrosive environments. Among them, we can see that the τ_{max} is the lowest in the acid environment. Kong and Orbison [42] found that the strength of concrete decreased gradually under acidic conditions (pH 2), which was about 70% of the 28-day age strength of uncorroded concrete specimens. For BFRP bar-CC specimens, the τ_{max} (9.45 MPa) in an alkaline salt environment was only 0.3% higher than that under a salt environment (9.42 MPa). In an acid environment, the τ_{max} of BFRP-bar-CC specimens was 9.24 MPa, which is 2.2% lower than that in an alkaline salt environment. For OSB specimens, the τ_{max} of acid (12.35 MPa) was 6.2% lower than that of alkaline salt (13.17 MPa). In the acid environment, the τ_{max} of ECR–CC specimens was 11.95 MPa, which was 3.2% lower than that of OSB-CC specimens. To sum up, the τ_{max} of the same reinforced specimen in an alkaline salt environment, a salt environment, and the acid environment went from higher to lower in a single-corrosive environment.

As shown in Figure 8, the bond–slip curves of all the specimens present obvious upward trends before failure. However, the samples that experienced splitting failure (e.g., E-15-S) belong to the brittle failure category, and the bond–slip curve showed no downward trend. When exposed to 30 FT cycles under the corrosive conditions of an acid environment or salt environment, the bond strength loss of the BFRP bar-CC specimens were similar: 3.3% and 4.4%, respectively. When the corrosive environment was alkaline salt and the number of FT cycles was increased from 15 to 30 times, it could be observed that the τ_{max} of the BFRP-bar-CC specimen was reduced by about 10.66%, from 8.76 MPa to 7.82 MPa. The results show that in the same corrosive environment, the bond strength of the specimen decreased and the slip increased with an increase in the number of FT cycles.

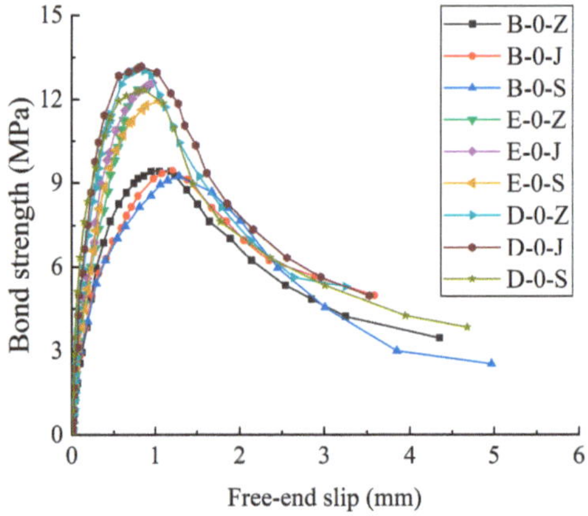

Figure 7. Bond stress–slip curves at the free end for BFRP bar specimens, ECR specimens, and OSB specimens in different corrosive environments.

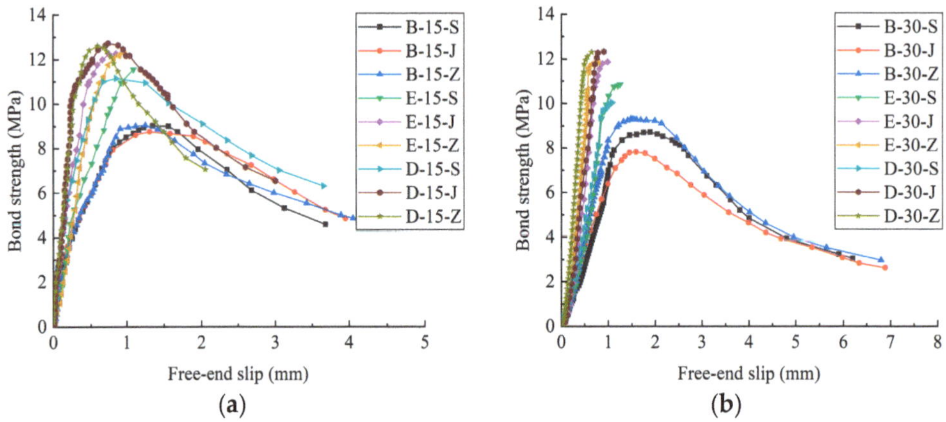

Figure 8. Bond stress–slip curves at the free end for BFRP bar specimens, ECR specimens, and OSB specimens under the combined action of FT cycles and corrosive environments. (**a**) 15 FT cycles and corrosive environments; (**b**) 30 FT cycles and corrosive environments.

In the same environmental conditions, the correspondence of τ_{max} with different reinforcement specimens was quite different. Taking the coupling effect of the acid environment and FT30 as an example, the $\overline{S}$ corresponding to the τ_{max} of the BFRP-bar–CC specimen was 1.89 mm, that of ECR-CC was 1.25 mm, and that of the OSB-CC specimen was the smallest, that is 1.08 mm. It can be seen that BFRP bars and ECRs showed a high slippage due to the smooth epoxy resin-based grease coating.

3.6. Analysis of Influence Factors of Bond Strength

Bank et al. [43] found a correlation between the degradation of fibrous materials and the reduction of bond strength. As can be seen in Figure 9c, under the combined action of FT30 and an alkaline salt environment, the τ_{max} of the three reinforcement materials, in descending order, were: OSB, ECR, and BFRP bar. We can see that the τ_{max} of BFRP bar specimen is always the lowest, and the reason is that there is a layer of epoxy resin on the surface of the bar. The analysis of the τ_{max} of these three kinds of steel bars shows that under the coupled effect of FT30 and an alkaline salt environment, the τ_{max} of ECR is 3.7% lower than that of OSB, and the τ_{max} of BFRP bar is 36.52% lower than that of OSB.

However, under the coupled effect of FT30 and an acid environment, the τ_{max} of OSB is 7.47% lower than that of ECR, and the τ_{max} of BFRP bar is 19.47% lower than that of ECR.

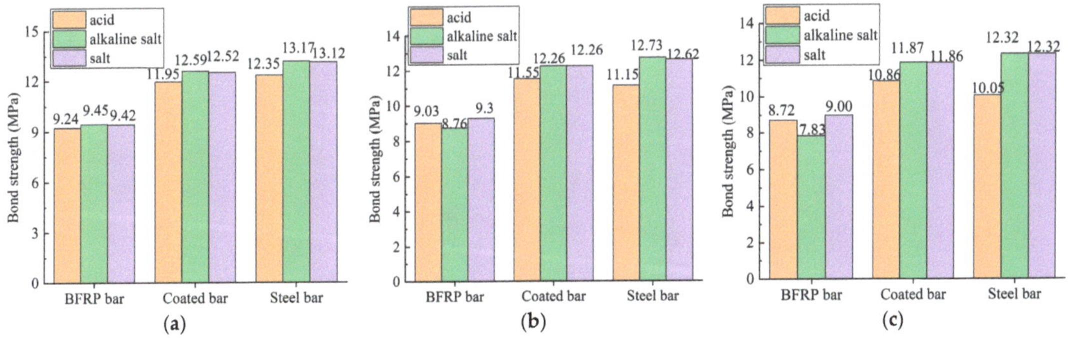

Figure 9. The influence of different factors on bonding performance. (**a**) Single corrosive environment; (**b**) FT15 + corrosive environment; (**c**) FT30 + corrosive environment.

It can be seen from Figure 9c that the τ_{max} of ECR in an alkaline salt environment is similar to that in a salt environment: both were about 11.80 MPa. The τ_{max} of the specimen in the acid environment was the lowest (10.86 MPa), which was 8.43% lower than that in the salt environment. In addition, it can be seen that the bond strength of ECRs and OSBs immersed in the alkaline and salt environments was greater than that of the specimens exposed to an acid environment. This is due to the high alkalinity and moisture of the internal environment of the concrete, with a pH of 10.5 to 13.5 [44]. The chloride in the water erodes the concrete, forming chemicals that can be absorbed and expanded, thus damaging the concrete [45–47]. In this case, while the strength of concrete decreases, the bond strength of the specimen in the acid environment decreases.

4. Calibration of Bonding Models to FT Cycles and Corrosive Environments

In this study, the bond properties of reinforced concrete under drawing loads were analyzed by using existing models. Xu et al. [48] studied the influence of various factors on the bond strength of ordinary crescent steel. However, they did not consider the influence of FT cycle times (N) and corrosive environments (pH).

$$\tau_{max} = \left(0.82 + 0.9\frac{d}{l_a}\right)\left(1.6 + 0.7\frac{c}{d}\right)f_t \quad (5)$$

where d is the diameter of reinforcement material, l_a is the bond strength, c is the thickness of CC protective layer, f_t is the splitting tensile strength of CC, and N is the FT cycle number and the pH of the corrosive solution.

Wang Peng et al. [49] studied the relationship between splitting tensile strength and compressive strength and proposed Equation (6).

$$f_{t,0} = 0.33{f_{c,0}}^{\frac{2}{3}} \quad (6)$$

Based on Equations (5) and (6), the type of reinforcement material, diameter of reinforcement material d, bond length L_a, thickness of protective layer c, splitting tensile strength f_t, number of FT cycles N, and the pH of the corrosive solution, the ultimate bond strength of different reinforcement materials and CC under the combined action of FT cycle and corrosive environment is established, as shown in Equation (7):

$$\tau_{max} = \left(a_0 + a_1 \times N + a_2 \times pH + a_3 \times \frac{d}{L_a}\right)\left(a_4 \times N + a_5 \times pH + a_6 \times \frac{c}{d}\right) \times f_t \quad (7)$$

Origin 2018 was subsequently used to fit Formula (7) and to obtain Formulas (8)–(10), and each parameter was obtained. For BFRP-reinforced concrete calibration, the parameters a_0, a_1, a_2, a_3, a_4, a_5, and a_6 were 0.5, −0.0126, −0.0565, 7.2611, 0.0342, 0.0417, and 0.3162, respectively. Similarly, ECR concrete calibration parameters a_0, a_1, a_2, a_3, a_4, a_5, and a_6 were 0.7, −0.0297, 0.00182, −12.19, 0.0138, 0.0155, and −0.4676, respectively. OSB concrete calibration parameters a_0, a_1, a_2, a_3, a_4, a_5, and a_6 were 0.65, −0.0257, 0.0486, −14.3947, 0.0110, −0.0391, and −0.3702, respectively. These parameters are closely related to the bond stress and the corresponding slip.

According to Equations (6) and (7), we can calculate the ultimate bond strength of BFRP-bar-CC specimens, ECR-CC specimens, and OSB-CC specimens using Equations (8)–(10):

$$\tau_{\max,B} = \left(0.5 - 0.0126N - 0.0565\text{pH} + 7.2611\frac{d}{l_a}\right)\left(0.0342N + 0.0417\text{pH} + 0.3162\frac{c}{d}\right) \times f_t \quad (8)$$

$$\tau_{\max,E} = \left(0.7 - 0.0297N + 0.00182\text{pH} + 12.19\frac{d}{l_a}\right)\left(0.0138N + 0.0155\text{pH} - 0.4676\frac{c}{d}\right) \times f_t \quad (9)$$

$$\tau_{\max,D} = \left(0.65 - 0.0257N + 0.0486\text{pH} - 14.3947\frac{d}{l_a}\right)\left(0.011N - 0.0391\text{pH} - 0.3702\frac{c}{d}\right) \times f_t \quad (10)$$

In this paper, the ultimate bond strength calculated according to the formula of the bond strength is compared with the ultimate bond strength obtained by a test, and the results are shown in Figure 10. In order to further analyze the degree of coincidence between the calculated value and the test value, the root-mean-square error was used. The closer the root mean square error value is to zero, the higher the degree of coincidence. For the BFRP bar, ECR, and OSB, the root-mean-square error formula produced values of 0.148, 0.124, and 0.137, respectively. As can be seen from the figure, the fitting results are in good agreement with the experimental results.

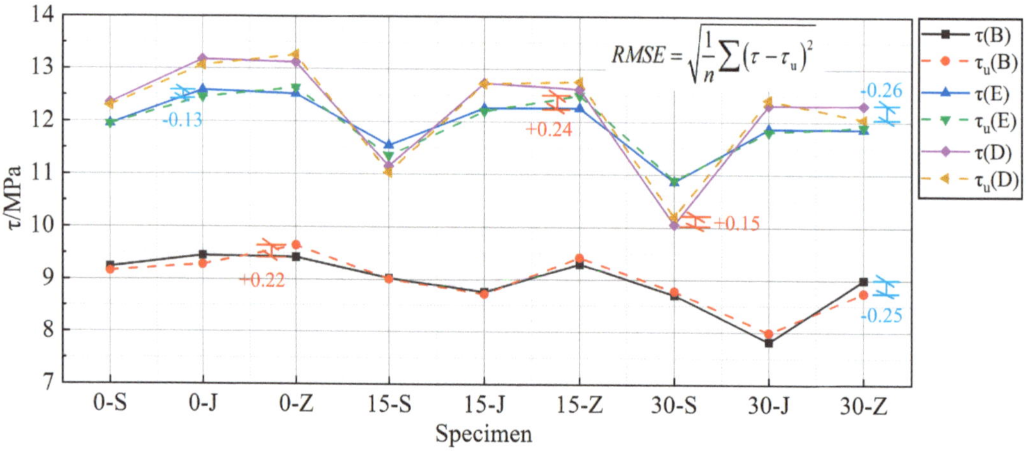

Figure 10. Comparison between the calculated values τ_u and experimental values τ of ultimate bond stress.

5. Conclusions

In this paper, by analyzing the influence of single corrosion environment (acid, alkali, alkaline salt) and the coupling effect of corrosion and FT cycle (0, 15, 30) on the bond strength attenuation of three bar-CC specimens (BFRP-bar-CC, OSB-CC, ECR-CC), the following conclusions were reached:

1. In three corrosive environments, the failure mode of BFRP-bar-CC specimens and OSB-CC specimens was a pullout failure, while the failure mode of ECR-CC specimens

was a split failure. When FT and corrosion acted together, the failure mode of BFRP-bar-CC specimens and ECR-CC specimens did not change, while the failure mode of OSB-CC specimens changed from a pullout failure to a splitting failure as the number of FT cycles increased from 15 to 30;
2. For the same bar-CC specimens, different corrosive environments had a slight influence on the bonding property of the specimens. For BFRP-bar-CC specimens, the most rapid decline of bond strength was observed in an acid environment. Compared with an acid environment, the bond strength in an alkaline salt environment and a salt environment increased by 2.2% and 1.9%, respectively;
3. When exposed to the coupled effect of an acid environment and 30 FT cycles, the bond strength of BFRP-bar-CC specimens, OSB-CC specimens, and ECR-CC specimens were decreased 5.7%, 18.6%, and 9.12%, respectively. This indicates that the BFRP-bar-CC specimens possessed good acid corrosion resistance, which can effectively prevent bond strength reduction;
4. When exposed to the coupled effect of 30 FT cycles and a corrosive environment, BFRP-CC specimens showed the most rapid decline (17.2%) in an alkaline environment, while OSB-CC specimens and ECR-CC specimens showed the most rapid degradation (18.6% and 9.12%, respectively) in an acid environment;
5. The formulas for calculating the ultimate bond strength of BFRP bars, ECRS, and OSBS with CC under the combined action of FT cycles and corrosive environments are provided. The obtained results are in good agreement with the experimental results.

Author Contributions: Conceptualization, Y.L. and Q.Z.; methodology, Q.Z.; formal analysis, Q.Z. and J.T.; investigation, J.T. and Y.S.; resources, Y.L. data curation, Y.L. and J.T.; writing—original draft preparation, Q.Z.; writing—review and editing, Y.L. and J.T.; visualization, Q.Z. and J.T.; supervision, Y.L. and P.D.; project administration, P.D. All authors have read and agreed to the published version of the manuscript.

Funding: This research received no external funding.

Data Availability Statement: The data presented in this study are available on request from the corresponding author.

Conflicts of Interest: The authors declare no conflict of interest.

References

1. *ACI 318*; Building Code Requirements for Structural Concrete (ACI 318-08) and Commentary (ACI 318R-08). American Concrete Institute: Farmington Hills, MI, USA, 2009.
2. *EN 1992*; Eurocode 2: Design of Concrete Structures—Part 1-1: General Rules and Rules for Buildings. European Committee for Standardization (CEN): Brussels, Belgium, 2014.
3. *GB 50010-2010*; Code for Design of Concrete Structures. China Architecture & Building Press: Beijing, China, 2015.
4. Mehany, S.; Mohamed, H.M.; Benmokrane, B. Contribution of lightweight self-consolidated concrete (LWSCC) to shear strength of beams reinforced with basalt FRP bars. *Eng. Struct.* **2021**, *231*, 111758. [CrossRef]
5. Ji, T.; Zheng, D.; Chen, X.; Lin, X.; Wu, H. Effect of prewetting degree of ceramsite on the early-age autogenous shrinkage of lightweight aggregate concrete. *Constr. Build. Mater.* **2015**, *98*, 102–111. [CrossRef]
6. Tanyildizi, H.; Coskun, A. The effect of high temperature on compressive strength and splitting tensile strength of structural lightweight concrete containing fly ash. *Constr. Build. Mater.* **2008**, *22*, 2269–2275. [CrossRef]
7. Fu, X.; Chung, D. Effect of corrosion on the bond between concrete and steel rebar. *Cem. Concr. Res.* **1997**, *27*, 1811–1815. [CrossRef]
8. Bautista, A.; Blanco, G.; Velasco, F. Corrosion behaviour of low-nickel austenitic stainless steels reinforcements: A comparative study in simulated pore solutions. *Cem. Concr. Res.* **2006**, *36*, 1922–1930. [CrossRef]
9. Nie, R.; Huang, Y.; Li, X.; Sun, H.; Li, D. Bond of epoxy-coated reinforcement to seawater coral aggregate concrete. *Ocean Eng.* **2020**, *208*, 107350. [CrossRef]
10. McDonald, D.B.; Pfeifer, D.W.; Sherman, M.R. Corrosion evaluation of epoxy-coated, metallic-clad and solid metallic reinforcing bars in concrete. *Aust. Civ. Eng. Trans.* **2002**, *44*, 103–107.
11. Huang, Y.; Qi, X.; Li, C.; Gao, P.; Wang, Z.; Ying, J. Seismic behaviour of seawater coral aggregate concrete columns reinforced with epoxy-coated bars. *Structures* **2022**, *36*, 822–836. [CrossRef]

12. Gerardo, G.C. *Resistances of a Stainless Steel-Clad Reinforcing Bar to Chloride-Induce Corrosion in Concrete*; Virginia Transportation Research Council: Charlottesville, VA, USA, 2004.
13. Wang, J.; Gong, J.; Gong, Z.; Yang, X.; Wang, B. Effect of curing agent polarity on water absorption and free volume in epoxy resin studied by PALS. *Nucl. Instrum. Methods Phys. Res. Sect. B Beam Interact. Mater. At.* **2010**, *268*, 2355–2361. [CrossRef]
14. Sagar, B.; Sivakumar, M.V.N. Performance evaluation of basalt fibre-reinforced polymer rebars in structural concrete members—A review. *Innov. Infrastruct. Solut.* **2021**, *6*, 75. [CrossRef]
15. Lopresto, V.; Leone, C.; De Iorio, I. Mechanical characterisation of basalt fibre reinforced plastic. *Compos. Part B Eng.* **2011**, *4*, 717–723. [CrossRef]
16. Lapko, A.; Urbanski, M. Experimental and theoretical analysis of deflections of concrete beams reinforced with basalt rebar. *Arch. Civ. Mech. Eng.* **2015**, *15*, 223–230. [CrossRef]
17. Wang, X.; Wu, G.; Wu, Z.; Dong, Z.; Xie, Q. Evaluation of prestressed basalt fiber and hybrid fiber reinforced polymer tendons under marine environment. *Mater. Des.* **2014**, *64*, 721–728. [CrossRef]
18. Garcia-Espinel, J.D.; Castro-Fresno, D.; Gayo, P.P.; Ballester-Munoz, F. Effects of sea water environment on glass fiber reinforced plastic materials used for marine civil engineering constructions. *Mater. Des.* **2015**, *66*, 46–50. [CrossRef]
19. Borrie, D.; Liu, H.B.; Zhao, X.L.; Singh, R.R.K.; Bai, Y. Bond durability of fatigued CFRP-steel double-lap joints preexposed to marine environment. *Compos. Struct.* **2015**, *131*, 799–809. [CrossRef]
20. Mark, F.G.; Luke, A.B.; Yves, B. Effect of freeze–thaw cycles on the bond durability between fibre reinforced polymer plate reinforcement and concrete. *Can. J. Civ. Eng.* **2000**, *27*, 949–959. [CrossRef]
21. Achillides, Z.; Pilakoutas, K. Bond Behavior of Fiber Reinforced Polymer Bars under Direct Pullout Conditions. *J. Compos. Constr.* **2005**, *8*, 173–181. [CrossRef]
22. Refai, A.; Ammar, M. Bond Performance of Basalt Fiber-Reinforced Polymer Bars to Concrete. *J. Compos. Constr.* **2015**, *19*, 1–12. [CrossRef]
23. Shi, J.; Zhu, H.; Wu, Z. Bond Behavior between Basalt Fiber–Reinforced Polymer Sheet and Concrete Substrate under the Coupled Effects of Freeze-Thaw Cycling and Sustained Load. *J. Compos. Constr.* **2013**, *17*, 530–542. [CrossRef]
24. Firas, A.; Jean-Michel, M.; Mohamed, S. Bond strength of different strengthening systems-Concrete elements under freeze-thaw cycles and salt water immersion exposure. *Constr. Build. Mater.* **2014**, *70*, 399–409. [CrossRef]
25. Dong, Z.Q.; Wu, G. Research progress on durability of FRP bars reinforced concrete structures. *China Civ. Eng. J.* **2019**, *52*, 1–19. [CrossRef]
26. Hassan, M.; Benmokrane, B.; ElSafty, A. Bond durability of basalt-fiber-reinforced-polymer (BFRP) bars embedded in concrete in aggressive environments. *Compos. Part B Eng.* **2016**, *106*, 262–272. [CrossRef]
27. Fei, Y.; Zhibin, L.; Dalu, Z. Experimental study on bond durability of glass fiber reinforced polymer bars in concrete exposed to harsh environmental agents: Freeze-thaw cycles and alkaline-saline solution. *Compos. Part B Eng.* **2017**, *116*, 406–421. [CrossRef]
28. Wu, G.; Dong, Z.Q.; Wang, X.; Zhu, Y. Prediction of long-term performance and durability of BFRP bars under the combined effect of sustained load and corrosive solutions. *J. Compos. Constr.* **2015**, *19*, 04014058. [CrossRef]
29. Rifai, M.A.; El-Hassan, H.; El-Maaddawy, T. Durability of basalt FRP reinforcing bars in alkaline solution and moist concrete environments. *Constr. Build. Mater.* **2020**, *243*, 118258. [CrossRef]
30. Altalmas, A.; Refai, A.; Abed, F. Bond degradation of basalt fiber-reinforced polymer (BFRP) bars exposed to accelerated aging conditions. *Constr. Build. Mater.* **2015**, *81*, 162–171. [CrossRef]
31. *ACI 440.6M-08*; Guide Test Methods for Fiber-Reinforced Polymer (FRP) Composites for Reinforcing or Strengthening Concrete and Masonry Structures. American Concrete Institute: Farmington Hills, MI, USA, 2004.
32. Lu, Z.; Su, L.; Lai, J.; Xie, J.; Yuan, B. Bond durability of BFRP bars embedded in concrete with fly ash in aggressive environments. *Compos. Struct.* **2021**, *271*, 114121. [CrossRef]
33. *CAN/CSA S807e10 (R2015)*; Specification for Fiber Reinforced Polymers. Canadian Standards Association: Rexdale, ON, Canada, 2015.
34. *ASTM C666/C666M*; Standard Test Method for Resistance of Concrete to Rapid Freezing and Thawing. American Society for Testing and Materials: West Conshohocken, PA, USA, 2015.
35. Rilem, F.C. Tentative recommendations, recommendations for reinforcing steel, bond test for reinforcing steel: 1-Beam test (7-II-28 d) 2-pull-out test (7-II-128). *Mater. Struct.* **1973**, *6*, 79–118.
36. Amini, K.; Jalalpour, M.; Delatte, N. Advancing concrete strength prediction using non-destructive testing: Development and verification of a generalizable model. Constr. *Build. Mater.* **2016**, *102*, 762–768. [CrossRef]
37. Institute of Building Structure and Chinese Academy of Building Sciences. *Research and Application of Lightweight Aggregate Concrete*; China Architecture & Building Press: Beijing, China, 1981.
38. Deng, P.; Cong, Z.R.; Liu, Y.; Huang, Y.Y.; Zhu, Q. Effect of Dry-Wet Cycles on BFRP Bars and Modified Ceramsite Concrete in Marine Environments. *J. Mater. Civ. Eng.* **2022**, *34*, 040221251–0402212512. [CrossRef]
39. Shang, H.S.; Zhao, T.J.; Cao, W.Q. Bond behavior between steel bar and recycled aggregate concrete after freeze–thaw cycles. *Cold Reg. Sci. Technol.* **2015**, *118*, 38–44. [CrossRef]
40. Deng, P.; Wang, Y.J.; Sun, Y.; Liu, Y.; Guo, W.H. Bond durability of basalt-fiber-reinforced-polymer bars embedded in lightweight aggregate concrete subjected to freeze–thaw cycles. *Struct. Concr.* **2021**, *22*, 2829–2848. [CrossRef]

41. Wang, H.Z. An Experimental Study on the Deterioration of Properties of Fiber–Reinforced Concrete in Acid Corrosive Environments. Master's Thesis, Zhejiang University of Technology, Hangzhou, China, 2018.
42. Kong, H.L.; Orbison, J.G. Concrete deterioration due to acid precipitation. *ACI Mater. J.* **1987**, *84*, 110–115. [CrossRef]
43. Bank, L.C.; Puterman, M.; Katz, A. Effect of material degradation on bond properties of fiber reinforced plastic reinforcing bars in concrete. *ACI Mater. J.* **1998**, *95*, 232–243.
44. Barneyback, R.S.; Diamond, S. Expression and analysis of pore fluids from hardened cement pastes and mortars. *Cem. Concr. Res.* **1981**, *11*, 279–285. [CrossRef]
45. Fang, C.Q.; Karin, L.G.; Chen, L.G. Corrosion influence on bond in reinforced concrete. *Cem. Concr. Res.* **2004**, *10*, 2159–2167. [CrossRef]
46. Lee, S.T. Performance Deterioration of Portland Cement Matrix due to Magnesium Sulfate Attack. *KSCE J. Civ. Eng.* **2007**, *11*, 157–163. [CrossRef]
47. Lee, S.T.; Kim, D.G. Sulfate Attack of Cement Matrix Containing Inorganic Alkali-free Accelerator. *KSCE J. Civ. Eng.* **2009**, *13*, 49–54. [CrossRef]
48. Xun, Y.L.; Liu, L.X.; Guan, P.W.; Zeng, D.G. Anchorage Performance and Design Method of Epoxy Resin Coated Steel Bar. *Port Waterw. Eng.* **1999**, *8*, 33–37. [CrossRef]
49. Wang, P. Experimental Research on the Fundamental Mechanical Behavior of Ceramisite Concrete. Master's Thesis, Changsha University of Science & Technology, Changsha, China, 2008. [CrossRef]

buildings

MDPI

Article

Analytical and Numerical Study on the Performance of the Curved Surface of a Circular Tunnel Reinforced with CFRP

Fan Yang [1], Gan Qin [2,*], Kang Liu [1], Feng Xiong [1] and Wu Liu [1,*]

1 School of Civil Engineering, Hefei University of Technology, Hefei 230009, China
2 Shenzhen Urban Public Safety and Technology Institute, Shenzhen 518046, China
* Correspondence: qingan@szsti.org (G.Q.); liuwu168@hfut.edu.cn (W.L.)

Abstract: Pasting carbon fiber reinforced polymer (CFRP) has become an effective method to reinforce the circular tunnel. For this reinforcement method, the mechanical performance of the curved substrate is important to keep the coordinated deformation of CFRP and the lining concrete. To investigate the effect of interface curvature on the stresses of the reinforced interface, an analytical model is proposed for the curved reinforced interface with the consideration of the interface bond–slip relationship. Additionally, a 3D numerical model is established to further investigate the effects of some important parameters (CFRP's layer, length, elastic modulus, thickness and the adhesive's elastic modulus, thickness) on the reinforced interface stresses. The results reveal that the stress state of the curved reinforced interface is more complex than that of the plane reinforced interface. With decreasing the radius of the curved reinforced interface, the interface radial stresses are increased significantly, while the circumferential stresses hardly change. For the adhesive, decreasing the elastic modulus and thickness of the adhesive layer can significantly improve the stress state of the reinforced interface. For the CFRP, decreasing the thickness, elastic modulus and layer number of CFRP is conducive to full utilization of materials and long-term combined work of the concrete and CFRP.

Keywords: tunnel reinforcement; CFRP; curved interface; interface stress; finite element model

Citation: Yang, F.; Qin, G.; Liu, K.; Xiong, F.; Liu, W. Analytical and Numerical Study on the Performance of the Curved Surface of a Circular Tunnel Reinforced with CFRP. *Buildings* **2022**, *12*, 2042. https://doi.org/10.3390/buildings12112042

Academic Editors: Shan Gao, Jingxuan Wang, Dewen Kong and Yong Liu

Received: 19 October 2022
Accepted: 18 November 2022
Published: 21 November 2022

1. Introduction

Circular tunnels are usually buried deep in a complex environment. After a period of operation, quality defects such as lining cracks and water leakage are prone to appear. The existence of the above defects will inevitably deteriorate the stress of the lining structure and lead to unfavorable conditions, such as stress concentration and excessive deformation, which will have a negative impact on the safety of the whole tunnel project in some serious cases [1]. At the same time, with the rapid development of urban construction, the external conditions of the tunnels are also constantly changing, including the construction of upper roads, high-speed railways, high-rise buildings, upper ponds, and various new infrastructures. These changes in the external conditions have a potential impact on the tunnel, which may make the lining structures treacherous [2]. Once the external conditions are changed and cannot meet the requirements for the tunnel safety and stability, it is necessary to adopt some corresponding reinforcement schemes for the lining structures.

The technology of carbon fiber reinforced polymer (CFRP) reinforcement in concrete structures was first studied by the Swiss Federal Laboratories for Materials, which used the method of applying CFRP in the reinforcement of the Ibach bridge and achieved good reinforcement results [3]. Subsequently, this CFRP reinforcement technology was widely applied in various structural reinforcements because of the good mechanical properties of CFRP and high construction efficiencies, such as compressive or seismic reinforcement of concrete columns [4–6] and the reinforcement of various beam and slab structures [7]. The successful application of CFRP reinforcement technology in the field of civil building reinforcement has also inspired engineers and technicians to continuously explore the possibility and effectiveness of CFRP reinforcement technology in the field of hydraulic tunnel lining. Since CFRP is characterized by high strength, stable performance, corrosion resistance and light weight [8,9], it can effectively control the continual expansion of lining cracks and prevent the lining from falling off and spalling, which was well applied in the circular hydraulic tunnels of Taohua River Reservoir, Qingshan Reservoir and Dahuofang Reservoir [10]. When the concrete is reinforced with CFRP, the CFRP contributes to the structural strength, mainly by bearing the stress transferred from the reinforced interface. Therefore, many researchers focus on the mechanical properties of the interface between CFRP and concrete by experimental and analytical methods. For the experimental methods, some single shear tests [11–13] and double shear tests [14] were performed to obtain the stress–slip relationship of the reinforced interface. Additionally, in-plane shear tests are also important methods to obtain the bond–slip constitutive models of the reinforced interface, and based on these tests, the Neubauer and Rostasy model [15], Monti model [16], Nakaba model [17] and Savioa model [18] are proposed. For analytical methods, the elastic mechanic method is a common method to calculate the interface stress between CFRP and concrete, such as the stage analytical method [19,20], considering the deformation compatibility method [21–24]. The above analytical methods are mainly based on two assumptions: the reinforcement materials are linear elastic and the interfacial stresses are constants along the direction of the adhesive thickness. Once the above two assumptions are not considered, the interfacial stresses are difficult to solve or to express explicitly [25,26].

In addition, the existing research mainly focuses on concrete beams reinforced with CFRP to improve their bending or shear resistance, the main contents are related to the shear behavior and peel failure of the reinforced interface. For these structures, the reinforced interfaces are mostly planes, and the bond stresses of the interface are mainly shear stresses [27]. However, for the circular tunnel reinforced with CFRP, the reinforced interfaces are mostly a concave curved surface, such as the reinforced concrete (RC) arch in Figure 1. Due to the existence of the interface curvature, this reinforced structure will be more prone to peel failure [28]. Therefore, the above research cannot be directly used to analyze the mechanical properties of the curved interface for circular tunnels reinforced with CFRP.

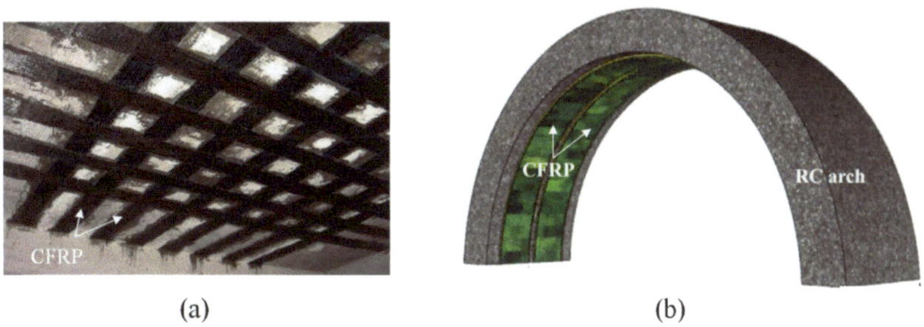

Figure 1. Reinforced interface: (**a**) plane surface; (**b**) concave curved surface.

Therefore, to investigate the reinforced interface stress of the circular tunnel reinforced with CFRP, an analytical model is proposed for the interaction between CFRP, adhesive and the lining concrete, and the analytical expression of the interface stress is derived with the consideration of the bond–slip at the reinforced interface. According to the analytical model, the influence of the curvature on the interface peel failure is analyzed, and the key parameters affecting the stresses of the curved reinforced interface are obtained. Subsequently, a detailed three-dimensional numerical model, which is verified by the experimental results, is established for a circular tunnel reinforced with CFRP. Based on the numerical model, the influence of the key parameters on the stresses of the curved reinforced interface is further analyzed.

2. Analytical Study on the Stresses of the Curved Reinforced Interface

2.1. Analytical Model

As shown in Figure 2, an analytical model is proposed for the curved reinforced interface of a circular tunnel reinforced with CFRP. The thickness of CFRP is t_c, the elastic modulus of CFRP is E_c, and the pasted length of CFRP is L, the thickness of the adhesive layer is t_a, the elastic modulus of the adhesive layer is E_a, the radius of the curved interface is r. A force (F) is applied to the CFRP along the tangential direction. The stress condition of an infinitesimal section is presented in Figure 2b.

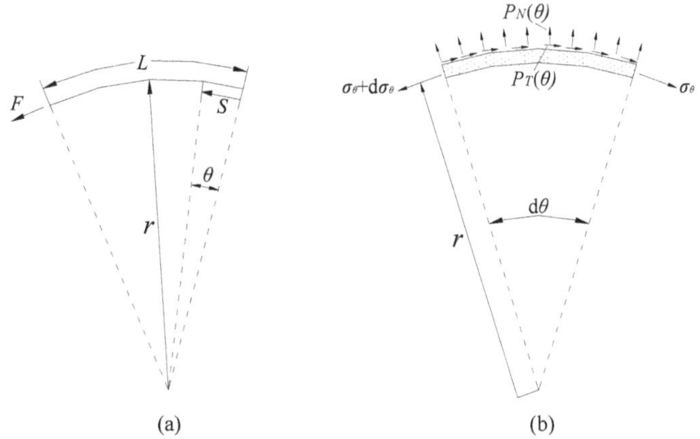

Figure 2. Analytical model: (**a**) curved reinforced interface dimensions; (**b**) infinitesimal section.

To simplify the calculation, the CFRP and adhesive can be simplified as a layer of reinforcement material [29], and the equivalent thickness is:

$$t = t_c + t_a \tag{1}$$

The equivalent elastic modulus of the reinforcement material can be calculated by:

$$E = \frac{E_c E_a}{E_c + E_a} \tag{2}$$

The circumferential stress of the reinforcement material is σ_θ, the distance from the calculated point to the free end is s, and the radial and tangential stresses of the reinforced interface are P_N and P_T, respectively. As shown in Figure 2b, the equilibrium equations of the infinitesimal section can be expressed as:

$$P_T \times \mathrm{d}s = t \times \mathrm{d}\sigma_\theta \tag{3}$$

$$P_N \times r \times \mathrm{d}\sigma_\theta = (\sigma_\theta + \mathrm{d}\sigma_\theta + \sigma_\theta) \times t \times \sin\frac{\mathrm{d}\sigma_\theta}{2} \tag{4}$$

According to Hook's law, the stress–strain relationship of the reinforcement material can be obtained by:

$$\sigma_\theta = E\varepsilon_\theta \tag{5}$$

Substitute Equation (5) into Equations (3) and (4), we obtain:

$$P_T = Et\frac{\mathrm{d}\varepsilon_\theta}{\mathrm{d}s} \tag{6}$$

$$P_N = \frac{Et}{r}\varepsilon_\theta \tag{7}$$

The relative tangential slip (g_T) and radial slip (g_N) at the interface between the adhesive and concrete can be defined by the following equations.

$$g_T = u_\theta - u_{\theta,s} \tag{8}$$

$$g_N = u_r - u_{r,s} \tag{9}$$

where u_θ is the tangential displacement of the adhesive at the interface between the adhesive and the concrete, $u_{\theta,s}$ is the tangential displacement of the concrete at the interface between the adhesive and the concrete; u_r is the radial displacement of the adhesive at the interface between the adhesive and the concrete, $u_{r,s}$ is the radial displacement of the concrete at the interface between the adhesive and the concrete. Since the deformation of the concrete is much smaller than that of the adhesive the values of $u_{\theta,s}$ and $u_{r,s}$ are much smaller than those of u_θ and u_r, and they can be neglected in the analytical model [30].

The relationship between the interface stress and the interface slip can be obtained by:

$$P_T = k_T g_T \tag{10}$$

$$P_N = k_N g_N \tag{11}$$

where k_T is the tangential stiffness of the interface, and k_N is the radial stiffness of the interface. According to Equations (6) to (11), we obtain:

$$k_T u_\theta = Et\frac{\mathrm{d}\varepsilon_\theta}{\mathrm{d}s} \tag{12}$$

$$k_N u_r = \frac{Et}{r}\varepsilon_\theta \tag{13}$$

The geometric equation in the polar coordinate is:

$$\varepsilon_\theta = \frac{\mathrm{d}u_\theta}{\mathrm{d}s} + \frac{u_r}{r} \tag{14}$$

Substituting Equation (14) into Equations (12) and (13), we obtain:

$$k_T u_\theta = Et\left(\frac{\mathrm{d}^2 u_\theta}{\mathrm{d}s^2} + \frac{1}{r}\frac{\mathrm{d}u_r}{\mathrm{d}s}\right) \tag{15}$$

$$k_N u_r = \frac{Et}{r}\left(\frac{\mathrm{d}u_\theta}{\mathrm{d}s} + \frac{u_r}{r}\right) \tag{16}$$

According to Equation (16),

$$u_r = \frac{Etr}{k_N r^2 - Et}\frac{\mathrm{d}u_\theta}{\mathrm{d}s} \tag{17}$$

Substituting Equation (17) into Equation (15), we obtain:

$$\frac{\mathrm{d}^2 u_\theta}{\mathrm{d}s^2} - \frac{k_T}{Et}\left(1 - \frac{Et}{k_N r^2}\right)u_\theta = 0 \tag{18}$$

We define a variable by:

$$\lambda^2 = \frac{k_T}{Et}\left(1 - \frac{Et}{k_N r^2}\right) \tag{19}$$

Substituting Equation (19) into Equation (18), we obtain:

$$\frac{\mathrm{d}^2 u_\theta}{\mathrm{d}s^2} - \lambda^2 u_\theta = 0 \tag{20}$$

According to Equations (20) and (17), u_θ and u_r can be expressed as:

$$u_\theta = A_1 \sin\text{h}(\lambda s) + A_2 \cos\text{h}(\lambda s) \tag{21}$$

$$u_r = \frac{Etr}{k_N r^2 - Et}(A_1 \cos\text{h}(\lambda s) + A_2 \sin\text{h}(\lambda s)) \tag{22}$$

where A_1 and A_2 are the constants to be determined by the boundary conditions. Substituting Equations (21) and (22) into Equation (14), ε_θ can be calculated by:

$$\varepsilon_\theta = \frac{k_N r^2}{k_N r^2 - Et}\lambda(A_1 \cos\text{h}(\lambda s) + A_2 \sin\text{h}(\lambda s)) \tag{23}$$

As the analytical model shows in Figure 2a, the boundary conditions are:

$$\varepsilon_\theta(0) = 0 \tag{24}$$

$$\varepsilon_\theta(L) = \frac{F}{Et} \tag{25}$$

According to Equations (23) to (25), the constants can be obtained:

$$A_1 = 0 \tag{26}$$

$$A_2 = \frac{F\lambda}{k_T \sin\text{h}(\lambda L)} \tag{27}$$

Substituting Equations (26) and (27) into Equations (21) and (22), u_θ and u_r can be calculated by:

$$u_\theta = \frac{F\lambda}{k_T \sin\text{h}(\lambda L)} \cos\text{h}(\lambda s) \tag{28}$$

$$u_r = \frac{F\lambda}{k_N r \sin\text{h}(\lambda L)} \sin\text{h}(\lambda s) \tag{29}$$

According to Equations (10), (11), (28) and (29), the interface radial stress ($P_T(s)$) and the interface circumferential stress ($P_N(s)$) can be obtained by Equations (30) and (31), separately.

$$P_T(s) = \frac{F\lambda}{\sin\text{h}(\lambda L)} \cos\text{h}(\lambda s) \tag{30}$$

$$P_N(s) = \frac{F}{r \sin\text{h}(\lambda L)} \sin\text{h}(\lambda s) \tag{31}$$

2.2. *Effect of the Curvature on Interface Stresses*

Based on Equations (30) and (31), the effects of the interface curvature on the reinforced interface stresses are investigated. The involved parameters and load in the analytical model are presented in Table 1, and the calculated interface stresses are presented in Figure 3.

Table 1. Parameters and load.

Parameter	E_a	t_a	E_c	t_c	L	F
Value	2 GPa	1 mm	250 GPa	0.111 mm	150 mm	50 N/m

Figure 3a presents that the interface curvature has little influence on the interface circumferential stresses. However, according to Figure 3b, the interface radial stresses are significantly influenced by the interface curvature. Obviously, with decreasing the interface curvature radius, the interface radial stresses tend to be increased, which is unfavorable for protecting the reinforced interface from peel failure.

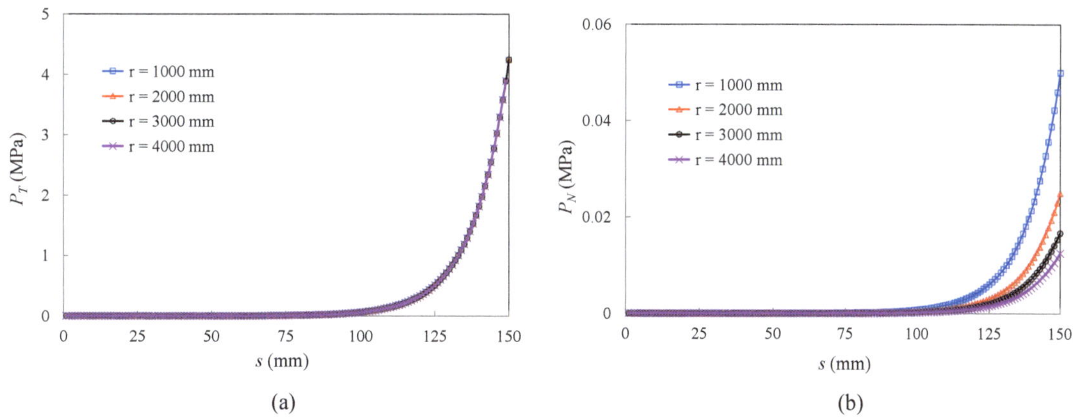

Figure 3. Analytical results of the interface stresses: (**a**) circumferential stress; (**b**) radial stress.

For the concrete beams or columns reinforced with CFRP, the reinforced interface is mostly flat and mainly in a pure shear state (State A), as shown in Figure 4. For the circular tunnel reinforced with CFRP on the internal surface, the reinforced interface is a curved surface. The analytical results show that the existence of interface curvature causes not only circumferential stresses (State A) but also radial stresses (State B) to exist on the curved reinforced interface. Therefore, the interface stress state (State C) of the curved reinforced interface is more complex than that of the plane interface.

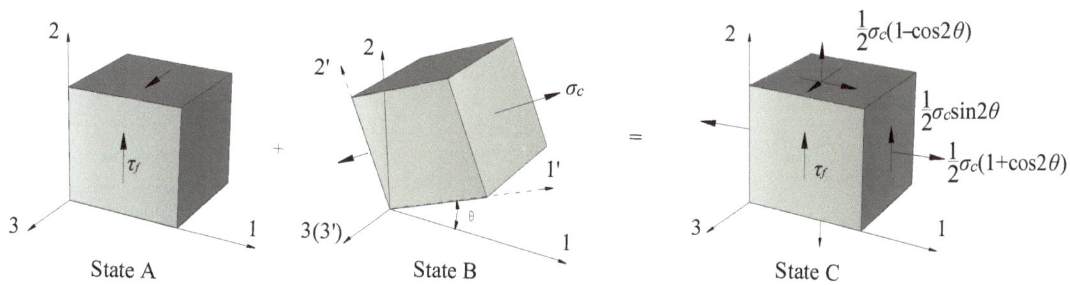

Figure 4. Stress state of the reinforced interface: State A pure shear state; State B radial stress state; State C interface stress state.

2.3. *Effect of the Stress State on Peel Failure*

When the principal stress of the concrete cover exceeds the tensile strength of the concrete, the reinforced interface exhibits peel failure. Therefore, the peel stress can be expressed as:

$$\sigma_b = \frac{P_N}{2} + \sqrt{\left(\frac{P_N}{2}\right)^2 + P_T^2} \quad (32)$$

For the interface radial stress, the positive value represents the curved reinforced interface between the adhesive and lining concrete, which is in tension, and the negative value represents the curved reinforced interface between the adhesive and lining concrete, which is in compression. For the interface circumferential stress, the positive value and negative value represent two opposite shear directions. For a circular tunnel reinforced with CFRP, the stress states of the concrete cover can be divided into two cases. (1) The interface of the concrete cover is in radial compression and circumferential shear. According to Equation (32), in this case, whether the peeling failure occurs is mainly determined by the circumferential shear stress P_T, and the existence of radial compressive stress P_N is beneficial to prevent the interface from experiencing peel failure. (2) The interface of the concrete cover is in radial tension and circumferential shear. According to Equation (32), this case is the most unfavorable condition, and the existence of the radial tensile stress will further increase the reinforced interface stresses, which will lead to earlier peel failure of the structure.

To decrease the radial tensile stresses of the reinforced interface that have a negative effect on this composite structure, some material parameters and construction schemes should be optimized. According to Equations (30) and (31), the interface stresses are mainly affected by several parameters such as the CFRP's length (L), layer (n), thickness (t_c), elastic modulus (E_c), and the adhesive's thickness (t_a), elastic modulus (E_a). For the CFRP, it is generally divided into high-strength CFRP and high-elastic CFRP, for which the thicknesses and the elastic moduli are quite different. In addition, since the types of the adhesive are diverse, the elastic moduli are quite different. Additionally, the thickness of the adhesive layer can also be affected by the brushing operation and circumstances. Therefore, it is necessary to further investigate the effects of L, n, t_c, E_c, t_a and E_a on the stress state of the curved reinforced interface, on the basis of a practical example.

Since the factors that affect the performance of the tunnel reinforced with CFRP are complex and diverse, the analytical model is mainly used for qualitative analysis, that is, to find out the main factors. However, the analytical model is somewhat simplified. For example, the bonding slip between the adhesive layer and the liner concrete is ignored. Therefore, based on the analytical results, detailed experimental or numerical studies should be conducted for quantitative analysis to reveal the influence of these factors on the performance of the tunnel reinforced with CFRP.

3. Experimental Study on a Tunnel Reinforced with CFRP

The effect of a CFRP reinforcement scheme for a circular tunnel with vertical loads is preliminarily investigated by comparing the experimental results of the specimen reinforced with CFRP and the specimen without any reinforcement scheme [31]. The dimensions of the specimen are presented in Figure 5, the internal diameter is 300 mm, the thickness is 30 mm, and the longitudinal length is 1000 mm. Six rebars with a diameter of 4 mm are erected in the longitudinal direction for the concrete, and the rebars are arranged at an interval of 60° along the circumferential direction. In addition, spiral stirrups with a diameter of 3 mm are used in the circumferential direction, and the spacing of the spiral stirrups is 74 mm.

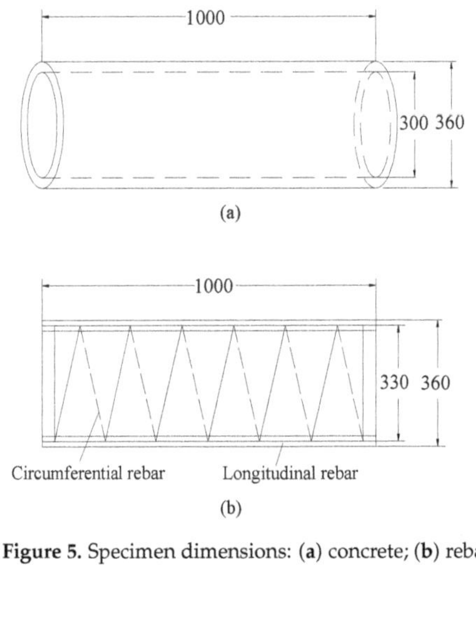

Figure 5. Specimen dimensions: (**a**) concrete; (**b**) rebars (mm).

The loading condition of the experimental test is shown in Figure 6. The involved materials are presented in Table 2. To investigate the stresses and strains of the concrete and CFRP, as shown in Figure 6b, several strain gauges are arranged on the internal surfaces of the concrete and CFRP.

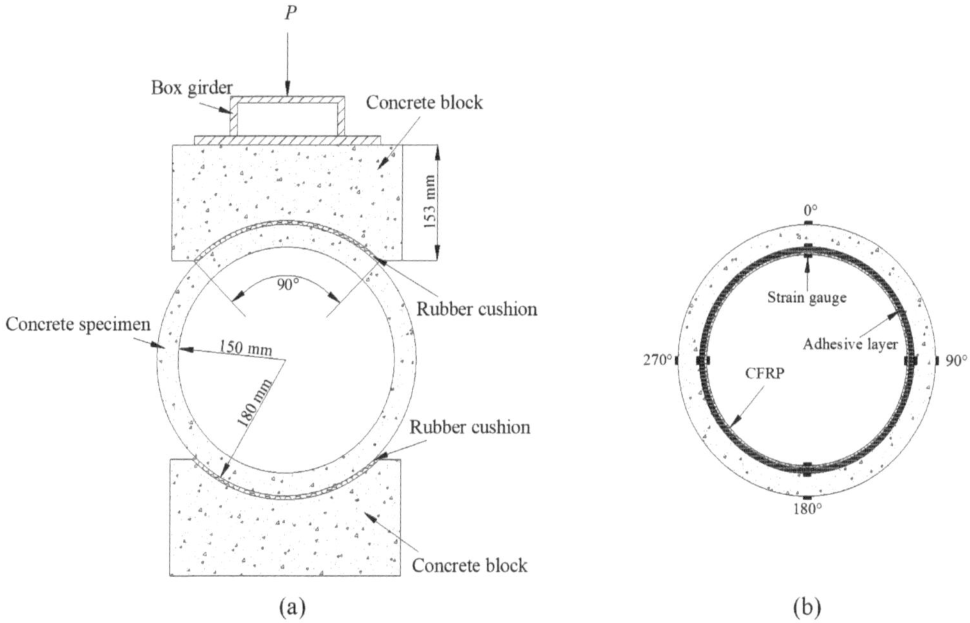

Figure 6. Details of the experimental test: (**a**) device; (**b**) strain gauge positions.

Table 2. Materials.

Material	Elastic Modulus (GPa)	Poisson's Ratio	Density (kg/m^3)
C30 concrete	30	0.167	2400
Longitudinal rebar	210	0.3	7850
Circumferential rebar	210	0.3	7850
Rubber cushion	7.8	0.48	1200
CFRP	235	0.28	—
Adhesive layer	2.5	0.35	—

4. Numerical Simulation and Validation

4.1. Numerical Model

The experimental test is reproduced by the numerical software ANSYS, and the 3D finite element model is presented in Figure 7. The concrete and the adhesive are simulated by Solid 65 elements and Solid 45 elements separately; the CFRP is simulated by Shell 63 elements, and the rebar is simulated by Link 8 elements. The multilinear isotropic hardening material model is used for the concrete, the bilinear kinematic hardening material model is used for the steel, and the linear elastic model is used for the CFRP and adhesive.

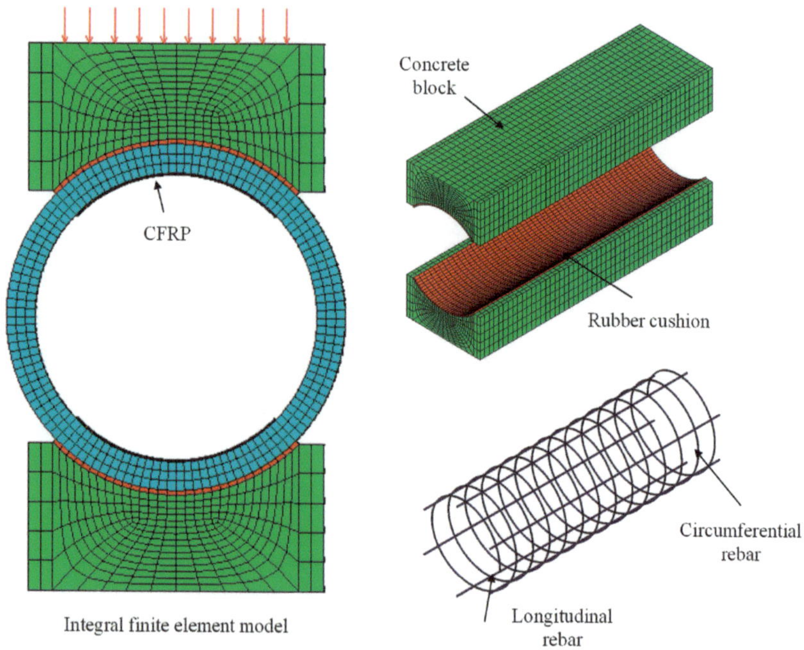

Figure 7. Numerical model.

The performance parameters of the adhesive layer are shown in Table 3. The spring elements (Combine 39) are used to simulate the bonding and slipping behavior of the contact interface between the adhesive layer and the concrete lining [32,33]. One end of the spring is connected to the node of the adhesive element, and the other end is connected to the node of the concrete element. The stiffness of each spring is the first-order inverse of the force-displacement relationship curve, which can be calculated as shown in Equations (33)–(36) [10,30].

$$K_{st} = \frac{\mathrm{d}F_T}{\mathrm{d}g_T} \tag{33}$$

$$F_T = P_T \times A_i \tag{34}$$

$$K_{sn} = \frac{\mathrm{d}F_N}{\mathrm{d}g_N} \tag{35}$$

$$F_N = P_N \times A_i \tag{36}$$

where K_{st} is the stiffness of the tangential spring; K_{sn} is the stiffness of the radial spring; A_i is the force area represented by each spring at the interface.

Table 3. Performance parameters of the adhesive layer.

<table>
<tr><th colspan="2" rowspan="2">Performance Items</th><th colspan="2">Performance Requirements</th></tr>
<tr><th>Grade A Adhesive</th><th>Grade B Adhesive</th></tr>
<tr><td rowspan="5">Adhesive layer performance</td><td>Tensile strength (MPa)</td><td>≥40</td><td>≥30</td></tr>
<tr><td>Modulus of elasticity under tension (MPa)</td><td>≥2500</td><td>≥1500</td></tr>
<tr><td>Elongation rate (%)</td><td colspan="2">≥1.5</td></tr>
<tr><td>Flexural strength (MPa)</td><td>≥50</td><td>≥40</td></tr>
<tr><td>Compressive strength (MPa)</td><td colspan="2">≥70</td></tr>
<tr><td>Bonding capacity</td><td>Positive tensile bond strength with concrete (MPa)</td><td colspan="2">≥2.5</td></tr>
</table>

Full constraints are applied to the bottom of the model, normal constraints are applied to the front and rear of the model, and the upper part of the model is free. The whole model contains 57,000 Solid 65 elements, 1758 Link 8 elements, 3000 Solid 45 elements and 18,360 Combin 39 elements. The main loads considered in the calculation include the top vertical load and the gravity. The sequence of the load application and calculation process is as follows. (1) Establish the finite element model; (2) "Kill" the adhesive elements and CFRP elements, apply the gravity and calculate; (3) Perform restart analysis, "alive" the adhesive elements and CFRP elements, apply the top vertical load and calculate.

4.2. Validation

As presented in Table 4, with the vertical load of 20 kN, the circumferential micro-strains of the concrete and CFRP at 90° are obtained by the numerical simulation and compared with the experimental results. For the concrete micro-strain, the average difference between the numerical results and experimental results is within 10%. For the CFRP micro-strain, since the adhesive layer and CFRP are simplified as linear elastic materials, the differences between the numerical results and experimental results are relatively large, and the average difference is about 17%. Since the numerical results generally agree well with the experimental results, the 3D numerical model is verified, and then we can use this verified 3D numerical model to perform the following extended parameter analysis.

Table 4. Comparison between the experimental results and analytical results.

Force (kN)	Concrete Strain			CFRP Strain		
	Experimental Results (10^{-6})	Numerical Results (10^{-6})	Differences	Experimental Results (10^{-6})	Numerical Results (10^{-6})	Differences
5	−22.20	−23.02	3.7%	−23.39	−23.37	−0.1%
10	−45.14	−42.96	−4.8%	−53.12	−46.76	−12.0%
15	−68.08	−62.02	−8.9%	−82.86	−70.15	−15.3%
20	−91.03	−82.84	−9.0%	−112.59	−93.54	−16.9%

5. Parametric Study

According to the analytical study in Section 2, the interface stresses are mainly affected by L, n, t_c, E_c, t_a and E_a. Based on the numerical model, the affecting degrees of these parameters on the interface stresses are further investigated. According to some engineering practices [9,34,35], the ranges of these parameters are listed in Table 5.

Table 5. Parameter values.

Parameter	Reference Value	Discussion Value
L	180°	30°, 60°, 90°, 120°
E_a	2 GPa	0.24, 2, 11.2
t_a	2 mm	1, 2, 4
E_c	235 GPa	100, 300, 500
t_c	0.1 mm	0.1, 0.15, 0.2
n	1	1, 2, 3, 4

5.1. Effect of L

The effects of the CFRP length (L), which is determined by the pasted angle, on the interface stresses are investigated. The distributions of the concrete tensile area with different pasted angles of 30°, 60°, 90° and 120° are presented in Figure 8. The gray areas represent when the concrete is compressed, and the other color areas represent when the concrete is tensioned. For the angles of 30, 60° and 90°, the CFRP cannot cover all the concrete tensile areas. According to Figure 9a, the radial tensile stress concentration appears at the edges of the reinforced areas in these three cases. However, for the angle of 120°, all the concrete tensile areas can be completely covered by CFRP, and the interface at the edges of the reinforced areas is radial compressed. The maximum radial tensile stress is 0.014 MPa, which is much smaller than those in the three previous cases. For the full-pasted case (180°), the interface radial tensile stresses mainly appear at the crown and bottom, and the maximum radial tensile stress is 0.013 MPa. It indicates that when the pasted angle is larger than 120°, the maximum radial tensile stress cannot be effectively decreased by extending the pasted length of CFRP.

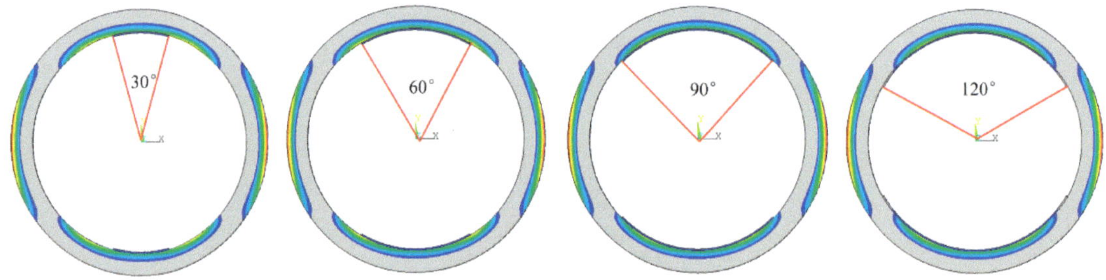

Figure 8. Concrete tensile areas and pasted angles.

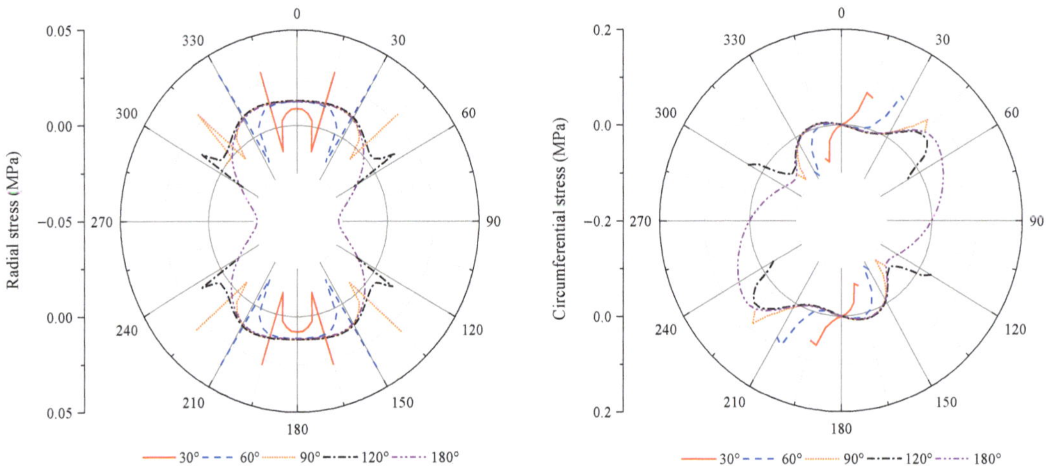

(a) (b)

Figure 9. Effect of L on the reinforced interface stresses: (**a**) radial stress; (**b**) circumferential stress.

According to Figure 9b, for the angles of 30, 60° and 90°, the maximum circumferential stresses all appear near the edges of the reinforced areas. Since these positions are radial tensioned in Figure 9a, the interface peel failure is most likely to occur there. For the angles of 120°, with the further extension of the CFRP pasted length, the distribution of the interface circumferential stress was effectively improved, and the maximum circumferential stress was significantly decreased. For the full pasted case (180°), the curve of the circumferential stress distribution was further smoothed, and the maximum circumferential stress was close to that in the case of 120°.

In total, for a circular tunnel partially reinforced with CFRP, the interface peel failure is most likely to occur near the edges of the reinforced areas. However, the interface radial stresses and circumferential stresses can be effectively decreased by extending the pasted length of CFRP to a certain extent. Therefore, to prevent the peeling failure, the pasted length of CFRP should be large enough to at least cover all the concrete tensile areas. In this studied case, to save CFRP materials, the pasted angle of 120 is more appropriate.

5.2. Effect of n

The effects of the CFRP layer (n) on the interface stresses are investigated. According to Figure 10, the interface stresses are significantly affected by n. With the increase in n from 1 to 3, the maximum radial tensile stress increases by 187%, and the maximum circumferential stress increases by 93%. In addition, pasting multiple layers of CFRP is not conducive to the full immersion of adhesive in CFRP. Therefore, for the partially pasted CFRP reinforcement scheme, decreasing the CFRP layers as much as possible is suggested when it can meet the condition of structural safety.

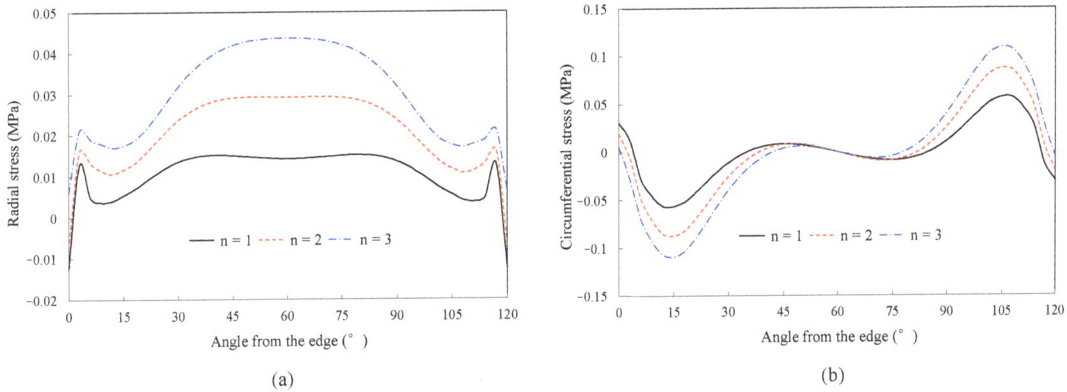

Figure 10. Effect of n on the reinforced interface stresses: (**a**) radial stress; (**b**) circumferential stress.

5.3. Effect of t_c

The effects of the CFRP thickness (t_c) on the interface stresses are investigated. According to Figure 11, the interface radial stresses and circumferential stresses are significantly affected by t_a. With the increase in t_c from 0.1 mm to 0.2 mm, the maximum radial tensile stress increases by 58%, and the maximum circumferential stress increases by 37%. Therefore, for the partially pasted CFRP reinforcement scheme, decreasing the CFRP thickness can improve the stress state of the interface.

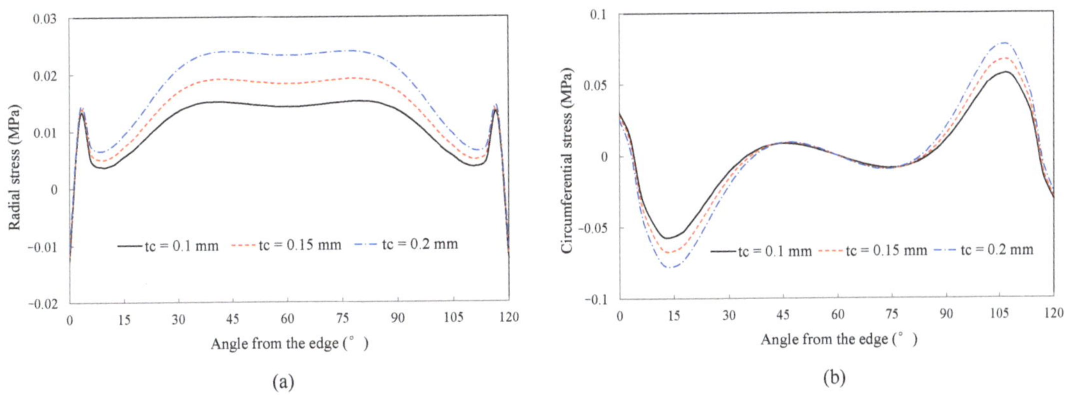

Figure 11. Effect of t_c on the reinforced interface stresses: (**a**) radial stress; (**b**) circumferential stress.

5.4. Effect of E_c

The effects of the CFRP elastic modulus (E_c) on the interface stresses are investigated. According to Figure 12, the interface radial stresses and circumferential stresses are significantly affected by E_c. With the increase in E_c from 100 GPa to 500 GPa, the maximum radial tensile stress increases by approximately 150%, and the maximum circumferential stress increases by approximately 131%. Normally, the elastic modulus of the high strength CFRP is approximately 235 GPa, and the elastic modulus of the high modulus CFRP ranges from 345 GPa to 690 GPa. Although the high modulus CFRP has the advantages of good bearing performance and preventing crack propagation, it is detrimental to the long-term combined work of the lining concrete and CFRP.

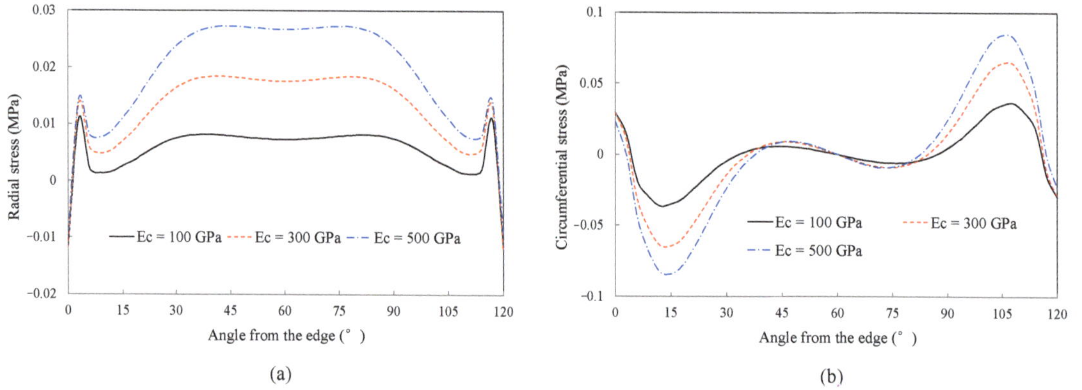

Figure 12. Effect of E_c on the reinforced interface stresses: (**a**) radial stress; (**b**) circumferential stress.

5.5. Effect of t_a

The effects of the adhesive layer thickness (t_a) on the interface stresses are investigated. According to Figure 13, the interface radial stresses are significantly affected by t_a. With the increase in t_a from 1 mm to 4 mm, the radial stresses near the central and marginal position of the interface are obviously increased, and the maximum radial tensile stress increases by 39%. However, according to Figure 13b, t_a has a negligible effect on the interface circumferential stresses. With increasing t_a, the maximum circumferential stress increases by only 18%. Therefore, for the partially pasted CFRP reinforcement scheme, decreasing the adhesive layer thickness is suggested to decrease the interface radial tensile stress and prevent peel failure.

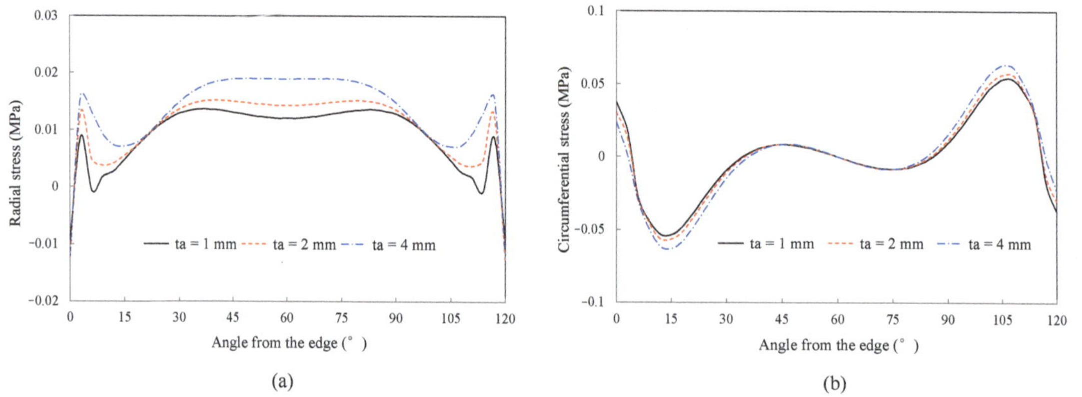

Figure 13. Effect of t_a on the reinforced interface stresses: (**a**) radial stress; (**b**) circumferential stress.

5.6. Effect of E_a

The effects of the adhesive elastic modulus (E_a) on the interface stresses are investigated. According to Figure 14, the interface radial stresses and circumferential stresses are significantly affected by E_a. When E_a is 0.24 GPa, the distributions of the interface radial stresses and circumferential stresses are relatively uniform. However, with increasing E_a, the interface stress concentration tends to be obvious. With the increase in E_a from 0.24 GPa to 11.2 GPa, the maximum radial tensile stress increases by 80%, and the maximum circumferential stress increases by 136%. Therefore, for the partially pasted CFRP reinforcement scheme, decreasing the adhesive elastic modulus has the advantage of preventing interface stress concentration.

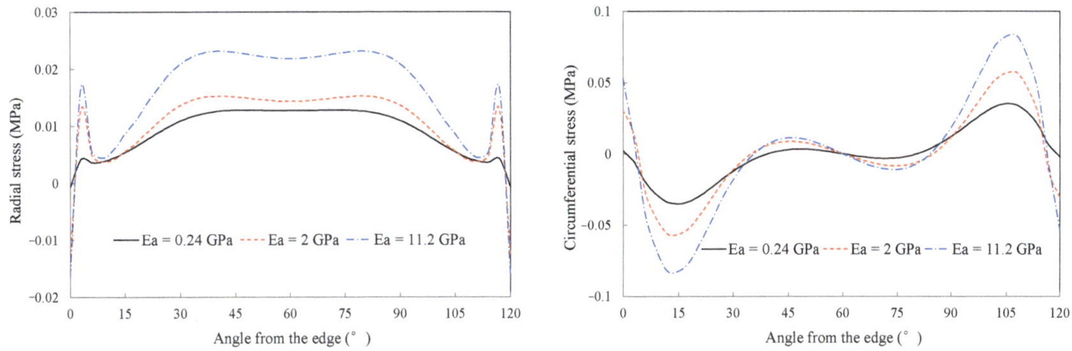

(a) (b)

Figure 14. Effect of E_a on the reinforced interface stresses: (**a**) radial stress; (**b**) circumferential stress.

6. Conclusions

An analytical model is proposed for the curved reinforced interface with the consideration of the interface bond–slip relationship, and a 3D numerical model is established to further investigate the effects of some important parameters. According to the analytical and numerical results, the following conclusions were obtained.

(1) The analytical results reveal that the interface curvature has a significant effect on the interface radial stresses. With decreasing the radius of the curvature, the interface radial stresses increase significantly.
(2) The numerical results reveal that the reinforced curved interface stresses are mainly affected by the CFRP's length, layer, thickness, elastic modulus, and the adhesive's thickness, elastic modulus. In engineering practice, these reinforcement parameters should be optimized to decrease the interface stresses and prevent premature peel failure of the structure.
(3) For the adhesive, decreasing the elastic modulus and thickness of the adhesive layer can significantly improve the stress state of the reinforced interface. For the CFRP, on the premise of meeting the structural bearing requirements, the pasted length of CFRP should be large enough to at least cover all the concrete tensile areas, and the CFRP with smaller thickness, smaller elastic modulus and fewer layers is conducive to the full utilization of materials and long-term combined work of the concrete and CFRP.

As the results of this study, the factors that affect the performance of the tunnel reinforced with CFRP are complex and diverse. This paper preliminarily reveals the influence law of the involved parameters. Multitudinous calculations and orthogonal test analysis should be conducted to further investigate the influence degrees and thresholds of these parameters under various conditions.

Author Contributions: Conceptualization, F.Y.; methodology, F.Y.; software, F.Y. and G.Q.; validation, F.Y. and G.Q.; formal analysis, F.Y. and G.Q.; investigation, F.Y. and G.Q.; resources, W.L.; data curation, F.Y. and G.Q.; writing—original draft preparation, F.Y., G.Q., K.L, F.X. and W.L.; writing—review and editing, F.Y., G.Q., K.L., F.X. and W.L.; visualization, F.Y.; supervision, W.L.; project administration, F.Y. and W.L.; funding acquisition, F.Y. and W.L. All authors have read and agreed to the published version of the manuscript.

Funding: This research was funded by the National Key Research and Development Program of China (Grant No. 2019YFC0810702), the Fundamental Research Funds for the Central Universities (Grant No. JZ2022HGTA0335) and the Natural Science Foundation of Anhui Province (Grant No. 2208085ME153).

Institutional Review Board Statement: Not applicable.

Informed Consent Statement: Not applicable.

Data Availability Statement: Not applicable.

Conflicts of Interest: The authors declare no conflict of interest.

References

1. Liu, D.; Huang, H.; Yue, Q.; Xue, Y.; Wang, M. Behaviour of tunnel lining strengthened by textile-reinforced concrete. *Struct. Infrastruct. Eng.* **2015**, *12*, 964–976. [CrossRef]
2. Wang, T.-T.; Lee, C.-H. Life-Cycle Design Considerations for Hydraulic Tunnels: Lessons Learned from Inspection and Maintenance Cases. *J. Perform. Constr. Facil.* **2013**, *27*, 796–806. [CrossRef]
3. Meier, U.; Deuring, M. *CFRP Bonded Sheets, FRP Reinforcement for Concrete Structures: Properties and Applications*; Elsevier Science Publishers: Amsterdam, The Netherlands, 1993.
4. Zeng, J.-J.; Liao, J.; Ye, Y.-Y.; Guo, Y.-C.; Zheng, Y.; Tan, L.-H. Behavior of FRP spiral strip-confined concrete under cyclic axial compression. *Constr. Build. Mater.* **2021**, *295*, 123544. [CrossRef]
5. Liao, J.J.; Zeng, J.J.; Gong, Q.M.; Quach, W.M.; Gao, W.Y.; Zhang, L. Design-oriented stress-strain model for FRP-confined ultra-high performance concrete (UHPC). *Constr. Build. Mater.* **2022**, *318*, 126200. [CrossRef]
6. Liao, J.J.; Zeng, J.J.; Chen, J.; Li, J.X.; Yuan, J.S. Stress-strain behavior and design-oriented model for FRP spiral strip-confined concrete. *Compos. Struct.* **2022**, *293*, 115747. [CrossRef]
7. Wang, Q.; Zhu, H.; Su, W.; Du, H.; Chen, D. Fatigue performance of CFRP reinforced pretensioned prestressed beams. *Constr. Build. Mater.* **2022**, *324*, 126509. [CrossRef]
8. Hollaway, L.C. A review of the present and future utilisation of FRP composites in the civil infrastructure with reference to their important in-service properties. *Constr. Build. Mater.* **2010**, *24*, 2419–2445. [CrossRef]
9. Mugahed Amran, Y.H.; Alyousef, R.; Rashid, R.S.M.; Alabduljabbar, H.; Hung, C.-C. Properties and applications of FRP in strengthening RC structures: A review. *Structures* **2018**, *16*, 208–238. [CrossRef]
10. Qin, G.; Yang, F.; Jin, D. Stress Transferring Mechanism of a Pressure Tunnel Lining Strengthened with CFRP. *Lat. Am. J. Solids Struct.* **2021**, *18*, 405–423. [CrossRef]
11. Yao, J.; Teng, J.; Chen, J. Experimental study on FRP-to-concrete bonded joints. *Compos. Part B Eng.* **2004**, *36*, 99–113. [CrossRef]
12. Dai, J.; Ueda, T.; Sato, Y. Development of the Nonlinear Bond Stress–Slip Model of Fiber Reinforced Plastics Sheet–Concrete Interfaces with a Simple Method. *J. Compos. Constr.* **2005**, *9*, 52–62. [CrossRef]
13. Cao, S.Y.; Chen, J.F.; Pan, J.W.; Sun, N. ESPI Measurement of Bond-Slip Relationships of FRP-Concrete Interface. *J. Compos. Constr.* **2007**, *11*, 149–160. [CrossRef]
14. Cui, E.; Jiang, S.; Wang, J.; Zeng, X. Bond behavior of CFRP-concrete bonding interface considering degradation of epoxy primer under wet-dry cycles. *Constr. Build. Mater.* **2021**, *292*, 123286. [CrossRef]
15. Neubauer, U.; Rostasy, F.S. Bond failure of concrete fiber reinforced polymer plates at inclined cracks-experiments and fracture mechanics model. In Proceedings of the 4th International Symposium on Fiber Reinforced Polymer Reinforcement for Reinforced Concrete Structures ACI, Farmington Hills, MI, USA, 1 August 1999; pp. 369–382.
16. Monti, M.; Renzelli, M.; Luciani, P. FRP adhesion in uncracked and cracked concrete zones. In *Fiber-Reinforced Polymer Reinforcement for Concrete Structures: Proceeding of the 6th International Symposium on FRP Reinforcement for Concrete Structures, Singapore, 8–10 July 2003*; World Scientific Publications: Singapore, 2003; pp. 183–192.
17. Nakaba, K.; Toshiyuki, K.; Tomoki, F.; Yoshizawa, H. Bond behavior between fiber-reinforced polymer laminates and concrete. *ACI Struct. J.* **2001**, *98*, 359–367.
18. Savioa, M.; Farracuti, B.; Mazzotti, C. Non-linear bond-slip law for FRP-concrete interface. In *Fiber-Reinforced Polymer Reinforcement for Concrete Structures: Proceeding of the 6th International Symposium on FRP Reinforcement for Concrete Structures, Singapore, 8–10 July 2003*; World Scientific Publications: Singapore, 2003; pp. 163–172.
19. Roberts, T.M. Approximate analysis of shear and normal stress concentrates in the adhesive layer of Plated RC Beams. *Struct. Eng.* **1989**, *67*, 222–233.

20. Roberts, T.M.; Haji-Kazemi, H. Theoretical study of the behavior of reinforced concrete beams strengthened by externally bonded steel plates. *Proc. Inst. Civ. Eng.* **1989**, *87*, 39–55.
21. Vilnay, O. The analysis of reinforced concrete beams strengthened by epoxy bonded steel plates. *Int. J. Cem. Compos. Light. Concr.* **1988**, *10*, 73–78. [CrossRef]
22. Liu, Z.H.; Zhu, B.L. Analytical solutions for R/C beams strengthened by externally bonded steel plates. *J. Tongji Univ.* **1994**, *22*, 21–26.
23. Täljsten, B. Strengthening of beams by plate bonding. *J. Mater. Civ. Eng.* **1997**, *9*, 206–212. [CrossRef]
24. Malek, A.M.; Saadatmanesh, H.; Ehsani, M.R. Prediction of failure load of R/C beams strengthened with FRP plate due to stress concentration at the plate end. *ACI Struct. J.* **1998**, *95*, 142–152.
25. Rabinovich, O.; Frostig, Y. Closed-Form High-Order Analysis of RC Beams Strengthened with FRP Strips. *J. Compos. Constr.* **2000**, *4*, 65–74. [CrossRef]
26. Shen, H.S.; Teng, J.G.; Yang, J. Interfacial stresses in beams and slabs bonded with thin plate. *J. Eng. Mech.* **2001**, *127*, 399–406.
27. Monti, G.; Liotta, M. Tests and design equations for FRP-strengthening in shear. *Constr. Build. Mater.* **2007**, *21*, 799–809. [CrossRef]
28. Wang, J.; Zhang, C. A three-parameter elastic foundation model for interface stresses in curved beams externally strengthened by a thin FRP plate. *Int. J. Solids Struct.* **2010**, *47*, 998–1006. [CrossRef]
29. Lu, X.Z. Study on FRP-Concrete Interface. Ph.D. Thesis, Tsinghua University, Beijing, China, 2004.
30. Lorenzis, L.D.; Zavarise, G. Interfacial stress analysis and prediction of debonding for a thin plate bonded to a curved substrate. *Int. J. Non-Linear Mech.* **2009**, *44*, 358–370. [CrossRef]
31. Wang, L.Y. The Experimental Study on Pressure Conduit of Ferroconcrete Reinforced by Fabric Reinforced Plastic. Master's Thesis, Wuhan University, Wuhan, China, 2004.
32. Hugo, C.B.; Carlos, C.; Manuel, A.G.S. linear and nonlinear analysis of bond slip models for interfaces between FRP composites and concrete. *Compos. Part B* **2013**, *45*, 1554–1568.
33. Obaidat, Y.T.; Heyden, S.; Dahlblom, O. Evaluation of Parameters of Bond Action between FRP and Concrete. *J. Compos. Constr.* **2013**, *17*, 626–635. [CrossRef]
34. Furtado, A.; Rodrigues, H.; Arêde, A.; Varum, H. Experimental tests on strengthening strategies for masonry infill walls: A literature review. *Constr. Build. Mater.* **2020**, *263*, 120520. [CrossRef]
35. Liu, Y.; Dong, A.; Zhao, S.; Zeng, Y.; Wang, Z. The effect of CFRP-shear strengthening on existing circular RC columns under impact loads. *Constr. Build. Mater.* **2021**, *302*, 124185. [CrossRef]

buildings

Article

Experimental Investigation and Numerical Analyses on Cyclic Behavior of the Prefabricated Concrete Frame Infilled with CFS-CLPM Composite Walls

Peifang Hu [1], Yong Liu [1,2,*], Jingfeng Wang [1,2], Wanqian Wang [1,2] and Guangdong Pan [1]

1 School of Civil Engineering, Hefei University of Technology, Hefei 230009, China
2 Anhui Key Laboratory of Civil Engineering Structures and Materials, Hefei 230009, China
* Correspondence: yongliu@hfut.edu.cn; Tel.: +86-5516-2901-434

Abstract: A novel CFS composite wall filled with cement-based lightweight polymer material (CFS-CLPM composite wall) has been proposed and proven to have excellent architectural and mechanical performance. To promote its application in prefabricated concrete (PC) frame structures, two full-scale specimens were designed and tested under cyclic loading to investigate the failure mode, hysteretic response and energy dissipation of the PC frame infilled with the CFS-CLPM composite wall. The experimental results indicated that CFS-CLPM composite walls can significantly improve the lateral behavior of the PC frame in terms of load capacity, elastic stiffness and energy dissipation capacity, while slightly reducing its ductility because of the infill-frame interaction. Subsequently, finite element (FE) analyses for the PC frame infilled with CFS-CLPM composite walls were developed and verified against the experimental results. The force-transferring mechanisms between the PC frame and the CFS-CLPM composite walls were revealed by analyzing the stress distributions. The parametric analyses demonstrated that the influential parameters for lateral resistances of the PC frame structure infilled with CFS-CLPM composite walls were the strength of CLPM, the span-to-height ratio and the thickness of CFS-CLPM composite walls. Finally, a formula considering the mechanical contribution of the CFS-CLPM composite wall was proposed to predict the elastic lateral stiffness of the structures. The results of this study could provide a basis for the application of CFS-CLPM composite walls in PC frame structures.

Keywords: CFS-CLPM composite walls; PC frames; cyclic behavior; numerical analyses; elastic stiffness

Citation: Hu, P.; Liu, Y.; Wang, J.; Wang, W.; Pan, G. Experimental Investigation and Numerical Analyses on Cyclic Behavior of the Prefabricated Concrete Frame Infilled with CFS-CLPM Composite Walls. *Buildings* **2022**, *12*, 1991. https://doi.org/10.3390/buildings12111991

Academic Editor: Abdelhafid Khelidj

Received: 24 October 2022
Accepted: 12 November 2022
Published: 16 November 2022

1. Introduction

Cold formed steel (CFS) walls filled with lightweight materials are novel composite walls developed from CFS structures, and have advantages of uniform quality, low weight, easy installation and convenient transportation [1–6]. Moreover, the lightweight fillers are usually made of industrial waste [7–9], which just meets the demand of green buildings and low-carbon society. Thus, the development and investigation of CFS composite walls filled with lightweight materials have been a hot topic. Many studies [4,10–12] investigating the structural performance of CFS composite walls filled with different lightweight materials indicated that the lightweight fillers could not only restrain the local buckling of the CFS frame effectively, but also improve the axial compressive capacity, cyclic performance and ductility.

Among these novel composite walls, CFS composite walls filled with cement-based lightweight polymer material (CFS-CLPM composite walls in abbreviation) was proposed by Wang et al. [12] as shown in Figure 1. The investigations of the CFS-CLPM composite walls have confirmed that such novel composite walls exhibited excellent axial compressive and cyclic performance [12,13]. Accordingly, industrial production lines have been established to prefabricate the CFS-CLPM composite walls, and they have been widely used as bearing walls or shear walls in low- and multi-rise buildings in China (Figure 2).

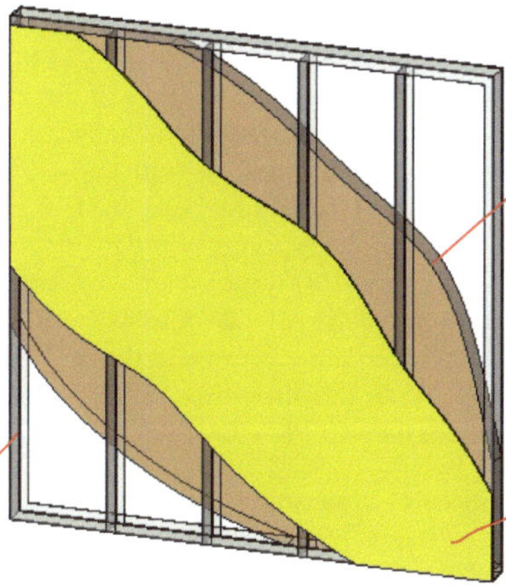

Figure 1. Configurations of CFS-CLPM composite walls.

Figure 2. Buildings with CFS-CLPM composite walls.

Furthermore, owing to the contribution of lightweight polymer material, CFS-CLPM composite walls are 800 kg/m^3 in density, which is much lighter than concrete, and have superior performance in sound insulation, thermal insulation and fire resistance than conventional CFS composite walls [13]. The CFS frame in the CFS-CLPM composite walls also made it easier to install using steel connectors. Combined with the excellent mechanical performance, CFS-CLPM composite wall is also a good option for walls in prefabricated frame structures. Prefabricated building is one of the important initiatives to achieve building industrialization and green construction, which have been vigorously promoted by the government in China. Owing to the popularity of concrete structures, prefabricated concrete (PC) buildings have gained great attention and are widely used in residential buildings and office buildings. Therefore, it is of great significance to apply CFS-CLPM composite walls into PC frame structures.

Recently, CFS walls have been introduced into frame structures by scholars to promote favorable CFS walls in high-rise buildings. Nevertheless, research on frame structure with CFS walls is still in its infancy so far. Wang et al. [14] conducted cyclic tests on the steel frame infilled with CFS composite walls. The experimental results showed that the steel frame cooperated well with the CFS composite walls and the lateral load capacity of this structure was determined by superposing shear strengths of the steel frame and the CFS composite walls. Wang et al. [15] reported that the sheathed CFS walls improved the progressive collapse capacity of steel frames but decreased the ductility. Kildashti et al. [16] proposed a hot-rolled steel knee-braced frames with CFS stud walls system. The nonlinear analyses indicated that CFS contribution to lateral loads can improve the seismic performance of the proposed system. However, studies on PC frames cooperating with CFS walls, especially CFS-CLPM composite walls, have not been reported.

To promote the application of CFS-CLPM composite walls in PC frame structures, it is significant to study the performance of the PC frames with CFS-CLPM composite walls. In this paper, two full-scale specimens were tested under cyclic loading to investigate the performance of PC frame infilled with CFS-CLPM composite walls. The effect of CFS-CLPM composite walls on the failure mode, hysteretic response, energy dissipation and horizontal displacement-strain response of the PC frame were studied. In addition, the numerical simulation using ABAQUS software was performed to explore the working mechanism of the PC frame infilled with CFS-CLPM composite walls under horizontal loads. Finally, a superposition formula with a good degree of accuracy was proposed to predict the elastic lateral stiffness of the structures. The results of this study could provide a basis for the application of CFS-CLPM composite walls in PC frame structures.

2. Experimental Program

2.1. Test Specimens

Two specimens, including one bare prefabricated concrete frame (PCF) and one prefabricated concrete frame infilled with CFS-CLPM composite walls (PCFW), were designed and examined under cyclic loads. The PC frames of two specimens were designed with identical configurations. Figure 3a illustrated the dimensional details of the PC frame, which is 3340 mm in span and 3030 mm in height. The PC frame contained precast columns with a dimension of 300 mm $\times$ 300 mm ($b \times h$), precast beams with a dimension of 200 mm $\times$ 350 mm ($b \times h$), and foundations with a dimension of 500 mm $\times$ 570 mm ($b \times h$), as shown in Figure 3b. These PC members adopted concrete with a nominal compressive strength of 40 Mpa and hot-rolled ribbed bars with a yield strength of 400 Mpa. In the assembly process of the PC frames (Figure 4), firstly, the precast foundation, columns and beams were installed to produce a frame. Next, the precast columns were fixed on the foundation by filling sleeves with high-strength concrete grouts [17]. The depth of reinforcements of columns inserted into sleeves with a length of 340 mm was 150 mm. Finally, the concrete with a nominal compressive strength of 50 Mpa was poured into beam-column joints [18,19].

For the specimen PCFW, the wall was assembled with prefabricated separated CFS-CLPM composite walls. In accordance with Ref. [13], configurations of separated CFS-CLPM composite walls and splicing connections of the walls were determined. Single galvanized C-section steel (89 mm $\times$ 41 mm $\times$ 11 mm $\times$ 0.9 mm) with a nominal strength of 550 Mpa was selected as studs and tracks in the CFS composite walls, as illustrated in Figure 5c. The tracks were connected to studs using ST4.8-grade self-drilling screws (d = 4.8 mm) to form the CFS frame. Then cement-based lightweight polymer materials (CLPM) were poured into the space of the CFS frame. Simultaneously, sheathing layers were also poured with CLPM. The mixed proportion of CLPM is presented in Table 1. Figure 5a illustrates the configuration and frame-wall joints of the specimen PCFW. Three separated CFS-CLPM composite walls were spliced together using steel strips and ST4.8-grade self-drilling screws (d = 4.8 mm), as depicted in Figure 5b. Steel angles were used as connectors by welding to both built-in fittings and 8.8-grade hook bolts (d = 10 mm). The fittings were embedded in the PC beams and the foundations while the hook bolts were embedded in the CFS composite walls.

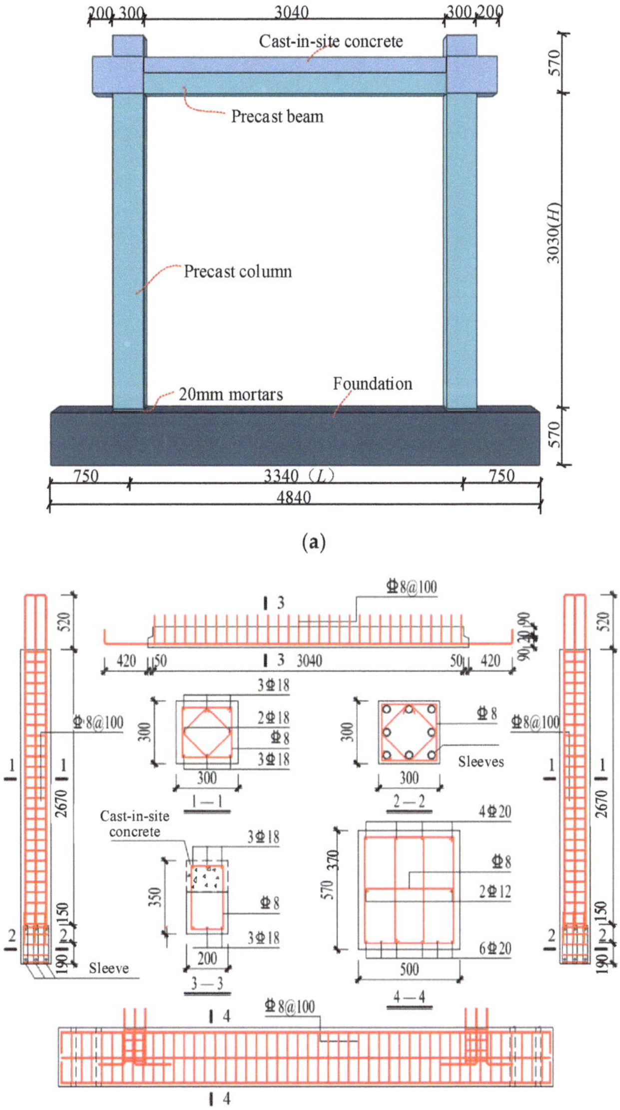

Figure 3. Details of PC frame: (**a**) Configurations of PC frame; (**b**) Reinforcements of precast concrete members.

Figure 4. Assembly procedure of PC frame.

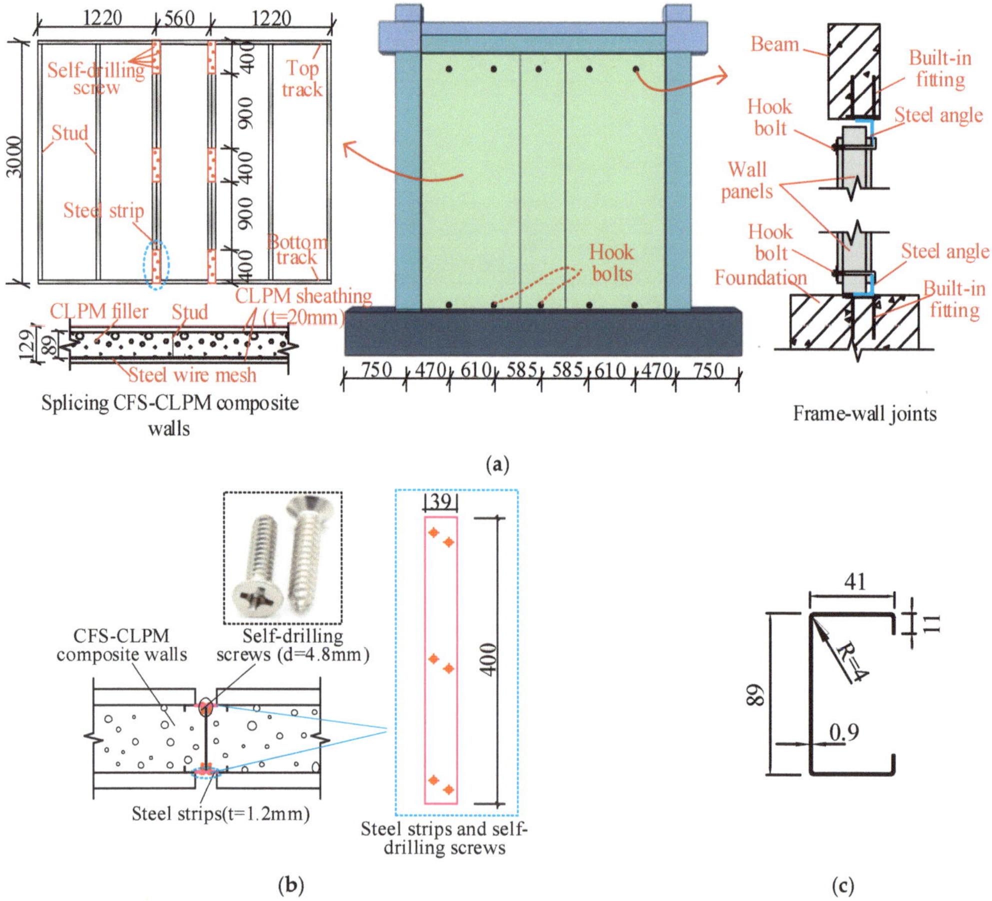

Figure 5. Details of specimen PCFW: (**a**) Configurations of specimen PCFW; (**b**) Splicing connection; (**c**) Size of C-section steel.

Table 1. Mix proportion of CLPM.

Cement (kg/m^3)	Fly Ash (kg/m^3)	Expansive Agent (kg/m^3)	Water-Reducing Agent (kg/m^3)	EPS (kg/m^3)	Water (kg/m^3)
300	65	125	2	0.6	225

2.2. Material Properties

In accordance with the Chinese Standard [20], the material properties of concrete are determined by cube coupons with the dimension of 150 mm. The measured compressive strengths of precast and cast-in-site concrete were respectively 45.2 MPa and 56.8 MPa, and the elastic modulus were 33,676.8 MPa and 35,575.6 MPa, respectively. The measured compressive strength and elastic modulus of the high-strength concrete grouts were 86.7 MPa and 34,000 MPa, respectively.

As per the Specification [21], the compressive strength and elastic modulus of CLPM were tested as 0.92 MPa and 250 MPa, respectively. The tensile coupons of the reinforcement and C-section steel were tested following the relevant provisions [22]. The mechanical characterizations of the reinforcement and C-section steel are summarized in Table 2.

Table 2. Mechanical properties of steel.

Steel Item	Thickness/ Diameter (mm)	Yield Strength (MPa)	Ultimate Strength (MPa)	Yield Strain	Elastic Modulus (MPa)
reinforcement	18	491	663	2338×10^{-6}	2.1×10^5
reinforcement	8	476	768	2268×10^{-6}	2.1×10^5
C-section steel	0.9	696	1006	3314×10^{-6}	2.05×10^5
C-section steel	1.2	695	1017	3310×10^{-6}	2.06×10^5

2.3. Test Setup and Measurements

The test setup is shown in Figure 6. The foundation was fixed on the rigid ground by four anchor bolts. Four 36 mm-diameter steel bars were used to connect the thick steel plates attached to the beam ends. The steel tube constraint device was employed to reduce hazards caused by out-of-plane collapse of specimens. The MTS hydraulic machine supplied the horizontal loads. Two vertical hydraulic jacks acting on the reaction steel frame were applied to exert axial loads at the top of each column. Two Polytetrafluoroethylene plates with high lubrication were installed between the reaction frame and the vertical hydraulic jacks to ensure that the jacks adapted to the specimens' lateral movements. In this study, twelve strain gauges numbered S1–S12 were arranged to record the longitudinal strains of reinforcements at the column ends as well as beam ends, as shown in Figure 7.

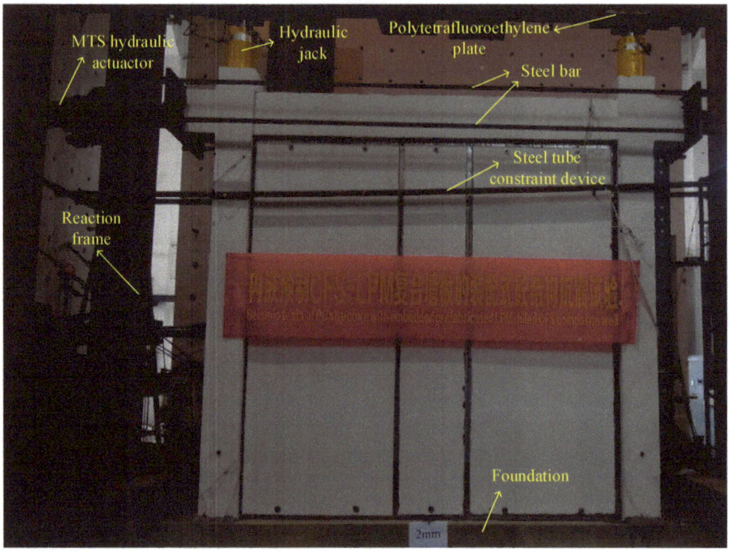

Figure 6. Test set up.

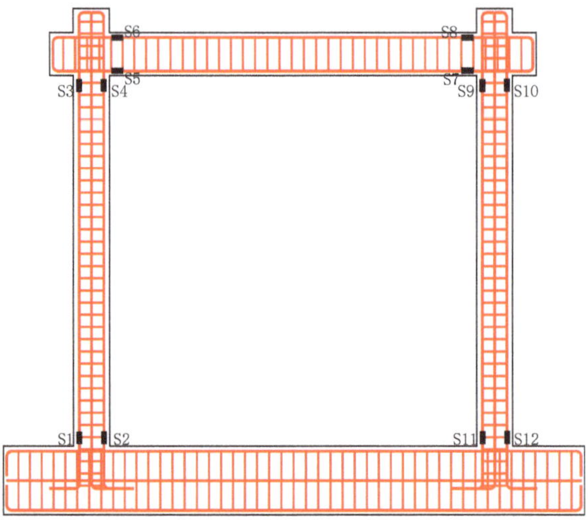

Figure 7. Strain gauges arrangement on the PC frame.

2.4. Loading Protocol

The axial load of 500 kN (calculated by an axial compression ratio of 0.3) at each top of the column remained constant during loading process. The horizontal loading was displacement-controlled as displayed in Figure 8. In the beginning, horizontal displacements of 2 mm, 3 mm, 4 mm, 5 mm, 6 mm and 7 mm were performed in one cycle per level. Afterwards, horizontal displacements of 8 mm, 16 mm, 24 mm, 32 mm, 40 mm, etc. (corresponding to the inter-story drift of 0.25%, 0.50%, 0.75%, 1.00%, 1.25% etc., respectively) were cycled three times per level. The loading procedure was terminated when the lateral load was down to 85% of the maximum lateral load or severe damage occurred on specimens.

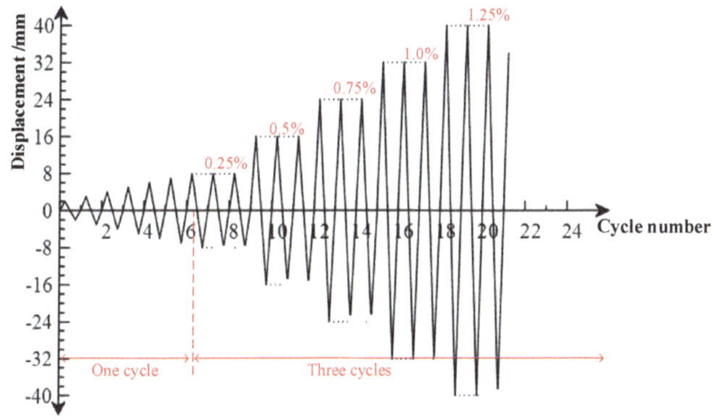

Figure 8. Loading history.

3. Experimental Results and Discussions

3.1. Failure Modes

3.1.1. Specimen PCF

Figure 9 exhibits the typical failure characteristics of specimen PCF. During the loading process with horizontal displacement of 2–7 mm, there were no damage phenomena on the specimen, indicating that the specimen was in elastic stage. When the inter-story drift increased to 0.25%, slight vertical cracks were first detected at beam ends. At the drift ratio of 0.50%, horizontal cracks occurred at the column ends near the top height of the sleeves. It showed that the strength and stiffness of columns around the length of sleeves were higher with the help of the additional steel sleeves than the rest part of the column without sleeves. Oblique cracks appeared on the beam-column joints at the drift ratio of 1.25%. As the horizontal drift increased, vertical cracks of the beam and horizontal cracks of the columns gradually widened and deepened, and eventually penetrated. When the inter-story drift was up to 4.0%, severe concrete crushing occurred at the bottom of the columns.

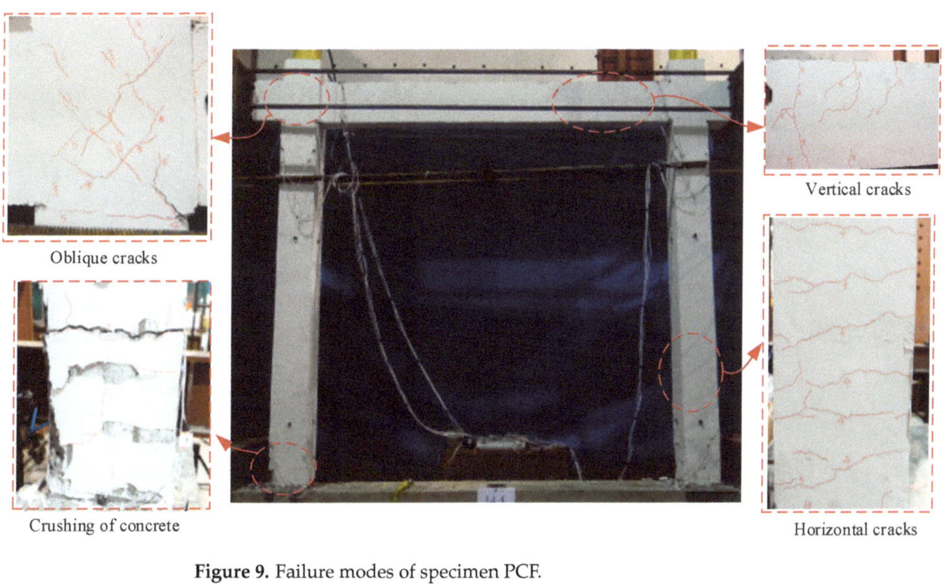

Figure 9. Failure modes of specimen PCF.

3.1.2. Specimen PCFW

In terms of the failure mode of the PC frame, specimen PCFW had similar phenomena to the specimen PCF during the cyclic loading. Concerning the failures of the CFS-CLPM composite walls, the diagonal cracks occurred and developed when the inter-story drift increased to 1.0%, as shown in Figure 10. It indicated that part of the horizontal loads was transferred to the CFS-CLPM composite walls. Moreover, the self-drilling screws on the steel strips tilted and loosened at the drift ratio of 2.25%. At the end of the loading, the drift reached 4.0% and severe concrete crushing occurred at the bottom of the columns. It was shown that the proposed frame-wall joints could inhibit out-of-plane movement of the infill wall. The CFS-CLPM composite walls remained quite intact with few spalling of CLPM at failure and the overall damage of CFS-CLPM composite walls in this study was less severe, compared with masonry infill walls in RC frames [23].

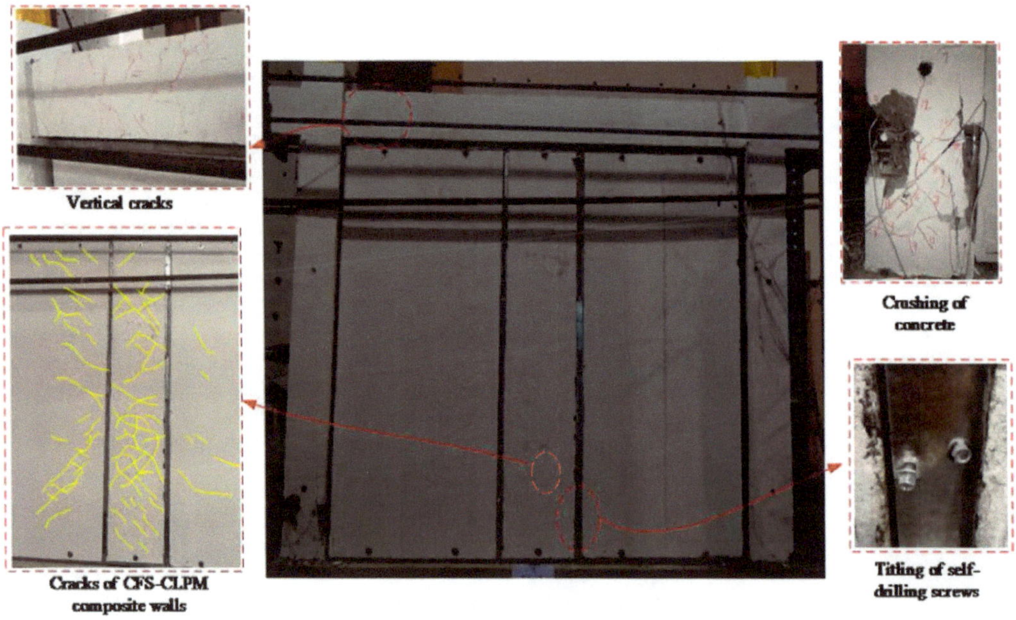

Figure 10. Failure modes of specimen PCFW.

3.2. Hysteretic Response

The load(P)-displacement(Δ) hysteresis curves are plotted in Figure 11. In the initial loading phase, hysteresis curves of all specimens were approximately linear due to the elastic deformation. With the increase of the horizontal displacement, the hysteresis loops exhibited shuttle-shaped when the test specimens reached the elastic–plastic stage. Beyond the peak point, the hysteresis loops of the specimen PCFW transformed to reverse S-shaped owing to the shear deformation including the shear cracks on CFS-CLPM composite walls and the tilting of self-drilling screws along steel trips. In general, the hysteresis loops of PCFW were plumper than that of specimen PCF, which demonstrated that the CFS-CLPM composite infill walls could also dissipate some energy in earthquakes.

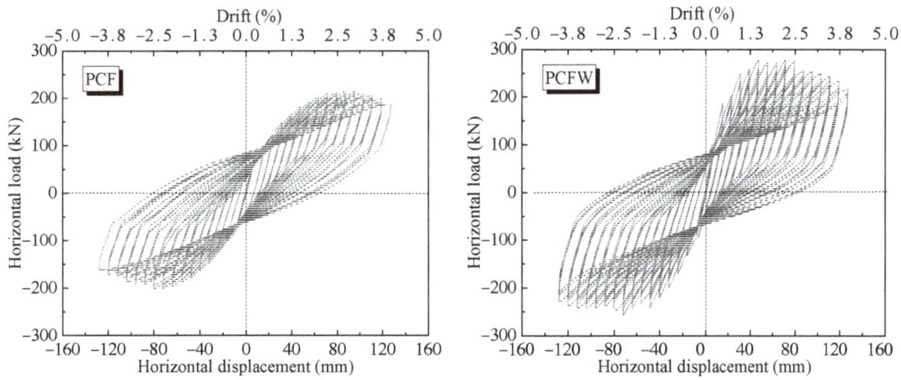

Figure 11. Force–displacement hysteresis curves of specimens.

3.3. Feature Values and Ductility

In this study, envelope curves (Figure 12) were utilized to determine the feature values. The characteristic loads are the yield load P_y, the maximum load P_m, and the failure load P_f, corresponding to the lateral displacement Δ_y, Δ_m, and Δ_f, respectively. The yield point was determined based on the 'graphing method' [24]. The elastic stiffness of the specimens is determined as the secant stiffness at 8 mm displacement in the P-Δ curves. Furthermore, the ductility coefficient is defined as $\mu_\Delta = \Delta_f/\Delta_y$. To minimize the errors of μ_Δ, the average μ_Δ was adopted in this study. The feature values of two specimens are listed in Table 3.

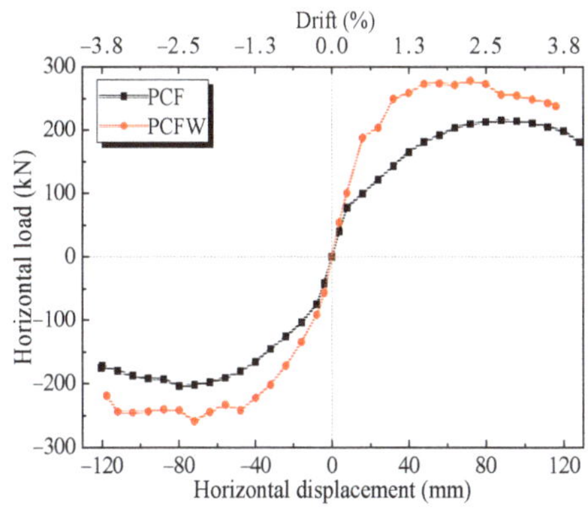

Figure 12. Envelope curves.

Table 3. Feature values of specimens.

Specimen	Δ_y/mm	P_y/kN	Δ_m/mm	P_m/kN	Δ_u/mm	P_f/kN	Elastic Stiffness kN/mm	μ_Δ
PCF (+)	40.9	166.2	88.0	212.58	127.3	180.7	10.22	3.26
PCF (−)	35.4	154.2	80.4	206.62	120.5	175.6	10.50	
PCFW (+)	37.7	248.3	72.0	279.8	116.1	237.8	14.53	3.16
PCFW (−)	36.5	223.2	72.0	258.0	117.7	219.3	14.87	

The effects of the CFS-CLPM composite walls on the cyclic performance of the PC frame were evaluated comprehensively. Specifically, the P_m of specimen PCFW was 24.9~31.6% higher than that of specimen PCF, and the elastic stiffness increased by 34.1~38.4%. The

Δ_y, Δ_m, and Δ_f of PCFW were close to those of PCF and the ductility coefficient of specimen PCFW was 3% lower than that of specimen PCF. However, the failure drifts of PCF and PCFW reached 3.87% and 3.65%, respectively, which can both meet the elastic–plastic drift of 2% required in the Chinese Code for Seismic Design of Buildings [25]. The comparison indicated that CFS-LPM composite walls can improve significantly the load capacity, elastic stiffness of the PC frame, while slightly reducing its ductility because of the infill-frame interaction.

3.4. Energy Dissipation

The energy dissipation capacity of specimens is assessed by calculating the areas of hysteretic loops. Figure 13 illustrates the energy consumption of the 1st loading cycle at each displacement level and cumulative energy dissipation, respectively. It was seen from Figure 13a that the energy consumption of each specimen was less before the displacement level up to 8 mm (drift of 0.25%) owing to the recoverable elastic deformation. Thereafter, it gradually increased as loading displacement increased. Comparisons of the 1st cycle energy consumption between specimens PCF and PCFW demonstrated that additional energy was dissipated by the CFS-CLPM composite walls even at high displacement levels. Furthermore, the cumulative energy of specimen PCFW was 1.14 times higher than that of specimen PCF at the horizontal displacement of 120 mm (drift of 3.75%). The PC frame structure infilled with CFS-CLPM composite walls exhibited better energy dissipation capacity.

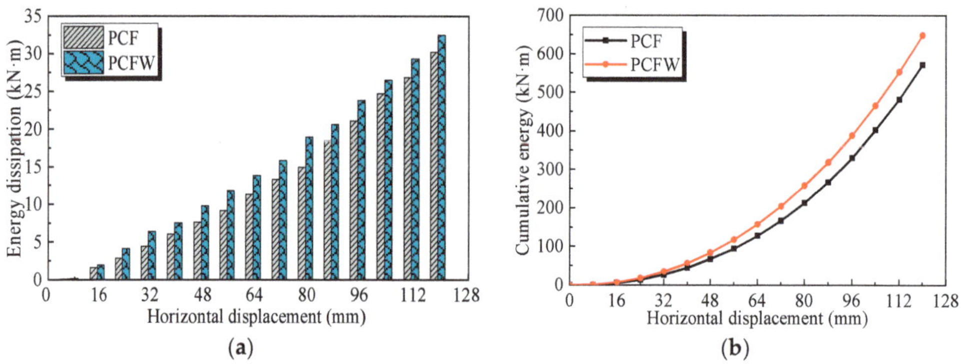

Figure 13. Energy dissipation: (**a**) 1-st cycle energy consumption at each displacement level; (**b**) Cumulative dissipated energy.

3.5. Strain Analyses

Key strains of reinforcement (S1–S6 in Figure 7) were plotted to analyze the influence of the CFS-CLPM composite walls on the strain development of the PC frame. Positive and negative values of the longitude strain represented tension and compression stress, respectively. The yield strain of longitude reinforcements is 2338με according to the material properties results in Section 2.3. As shown in Figure 14, the longitude reinforcements in both specimens yielded by tension earlier than by compression. Owing to the development of typical diagonal strut mechanisms in the infill wall, the higher horizontal load increased the strain of reinforcements around the upper beam–column joint. The strain of reinforcements at column bases was lower since part of the horizontal load was transferred to the infill wall through the compression between columns and infill walls.

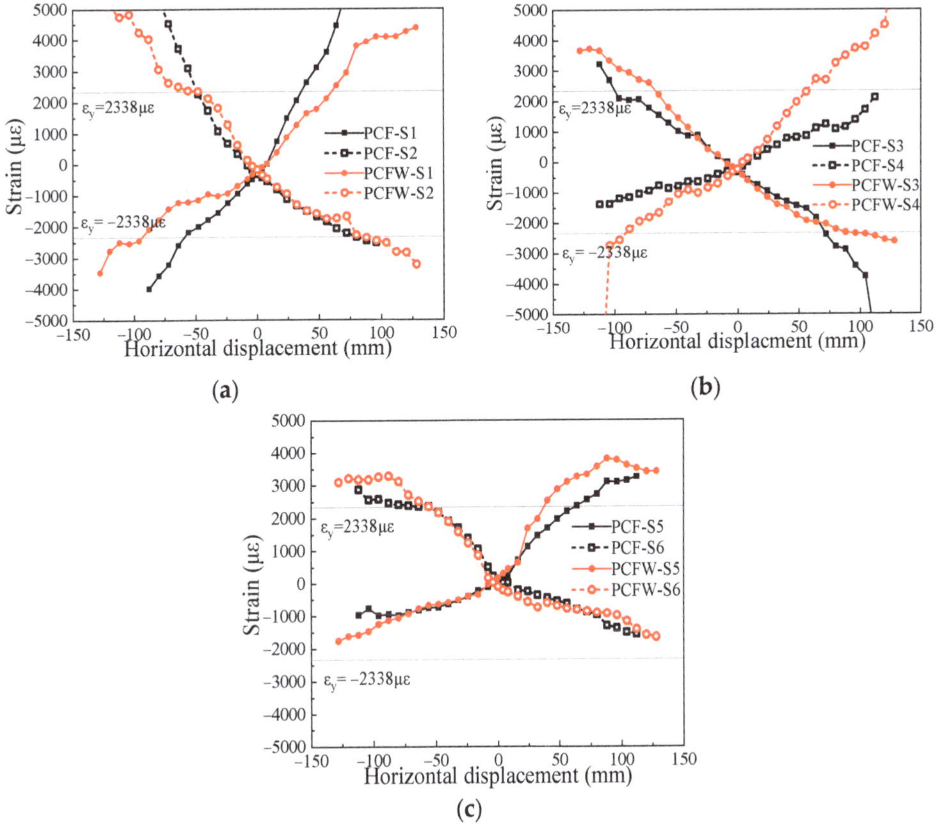

Figure 14. Strain of reinforcements: (**a**) Column feet; (**b**) Column upper ends; (**c**) Beam ends.

4. Numerical Analyses

The specimens examined in Section 2 were modeled by ABAQUS to analyze their lateral performance further in this section. Based on the validated FE models, parametric analyses were conducted to study the influence of dominant parameters, including material, geometric and load cases, on the lateral performance of PC frame structure infilled with CFS-CLPM composite walls.

4.1. Model Characteristics

4.1.1. Material Laws

In the FE modeling, the concrete damaged plasticity (CDP) model was applied to evaluate plastic behaviors of the precast and cast-in-site concrete. The stress–strain behavior of the CDP model was specified following the literature [26]. The bilinear kinematic hardening model (shown in Figure 15) was adopted to describe the stress–strain behavior of steel units of specimens. The f_y and E_s are respectively the yield strength and elastic modulus of the steel, which could be obtained from the preceding coupon tests.

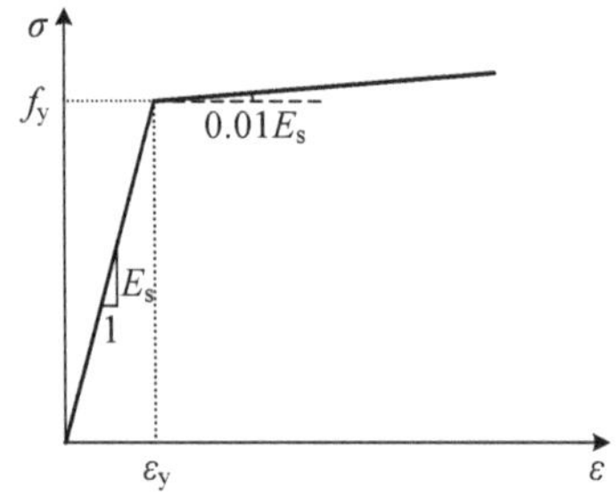

Figure 15. Steel constitutive model.

The CLPM material is essentially similar to the expanded polystyrene (EPS) concrete. Since the constitutive model of CLPM remains obscure at present, the stress–strain model for EPS concrete [27] was employed to simulate the compressive behavior of CLPM, as shown in Equation (1). The tensile strength of CLPM equals to 10% of the maximum compressive strength in this paper.

$$\sigma = \begin{cases} \frac{a(\varepsilon/\varepsilon_{c,m})+(b-1)(\varepsilon/\varepsilon_{c,m})^2}{1+(a-2)(\varepsilon/\varepsilon_{c,m})+b(\varepsilon/\varepsilon_{c,m})^2}\sigma_{c,m} & (0 \le \varepsilon \le \varepsilon_{c,m}) \\ \frac{1+c}{(\varepsilon/\varepsilon_{c,m})^d+c}\sigma_{c,m} & (\varepsilon_{c,m} < \varepsilon \le \varepsilon_{c,u}) \end{cases} \tag{1}$$

in which the parameters a, b, c and d vary with the dry density of CLPM; $\sigma_{c,m}$ is the maximum compressive strength; $\varepsilon_{c,m}$ is the strain corresponding to $\sigma_{c,m}$; $\varepsilon_{c,m}$ is the ultimate compressive strain.

4.1.2. Element Types and Cell Meshing

C3D8R solid elements were applied to concrete and CLPM fillers. Reinforcements were modeled with the T3D2 truss elements. The sleeves and C-section steel were modeled with S4R shell elements. Owing to the few deformations of the foundation during tests, the foundation was considered to be a discrete rigid body. After many attempts, the mesh size of 50 mm was applied for concrete, reinforcements, C-section steel as well as CLPM fillers.

4.1.3. Interaction and Boundary Conditions

In this FE modeling, the reinforcements were rigidly embedded within the concrete neglecting the bond-slip effects at the rebar-concrete interface. The contact surfaces between the precast and cast-in-site concrete were simulated with tie constraints. Based on the experimental phenomena, self-drilling screws used to assemble CFS frames rarely failed and could guarantee reliable connections between wall studs and tracks. Thus, the wall studs and tracks were merged into a whole instance in the FE models. For the CFS-CLPM composite walls, the CFS frame was embedded in the CLPM fillers to improve the convergence of FE models. In addition, the interaction between the beam and steel angles, as well as the foundation and steel angles were defined as tie constraints. Full adherence properties between hook bolts and CFS-CLPM composite walls were assumed by embedding the bolts into CLPM fillers. The behavior of frame-wall joints was simulated by linear springs that were used to describe the welding connections between the hook bolts and the steel angles. The hard contact and penalty with a friction coefficient of 0.3 were adopted to simulate the normal and tangential behavior in the contact surface between the columns and CFS-CLPM composite walls.

In terms of the boundary conditions of the FE models, all translations and rotations of the foundation were restricted to provide fixed properties. Out-of-plane deformation of the FE models was prevented by restricting the displacement and rotational freedoms (U_z, r_{xz}, r_{yz}) of all nodes on the side surfaces of the beam. According to the experimental testing program, axial loads were exerted at the top of each column, and a lateral load was supplied by the monotonic drifts to simplify the FE models and increase their convergence. Figure 16 shows the FE model details of specimen PCFW as an example.

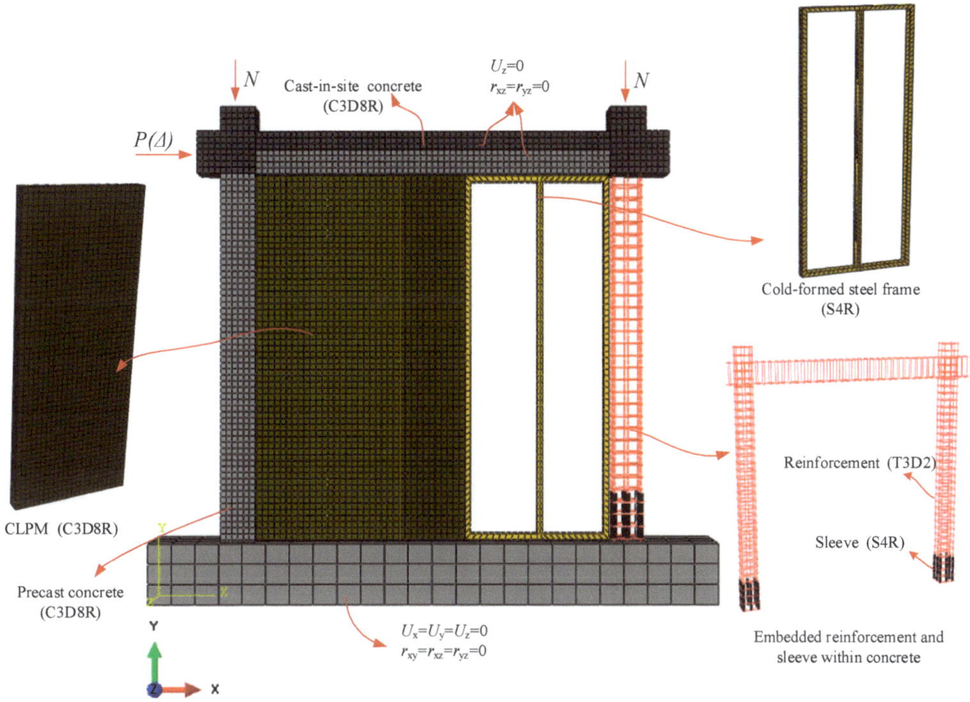

Figure 16. FE model of specimen PCFW.

4.2. Model Validation and Analyses

4.2.1. Model Validation

To verify the accuracy of the FE models, experimental and numerical results of specimens were compared comprehensively, displayed in Table 4 and Figures 17 and 18. The comparison results demonstrated that the FE modeling technique was reasonable to simulate the lateral performance of the PC frame structure.

Table 4. Comparison of experimental and numerical results.

Specimen	Test		Simulation		$K_{e,s}/K_{e,t}$	$P_{m,s}/P_{m,t}$
	$K_{e,t}$ (kN/mm)	$P_{m,t}$ (kN)	$K_{e,s}$ (kN/mm)	$P_{m,s}$ (kN)		
PCF (+)	10.22	212.58	8.53	193.70	0.83	0.91
PCFW (+)	14.53	279.80	15.34	293.31	1.06	1.05

Note: $K_{e,t}$ and $K_{e,s}$ are the elastic stiffness of tested and simulated results, respectively; $P_{m,t}$ and $P_{m,s}$ are the maximum load of tested and simulated results, respectively.

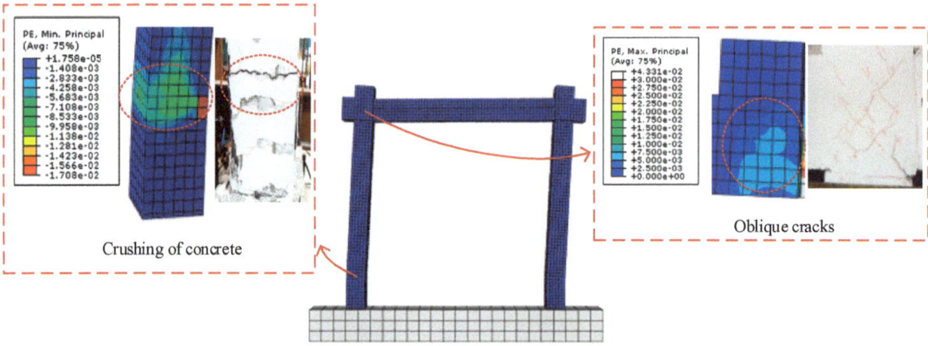

(**a**)

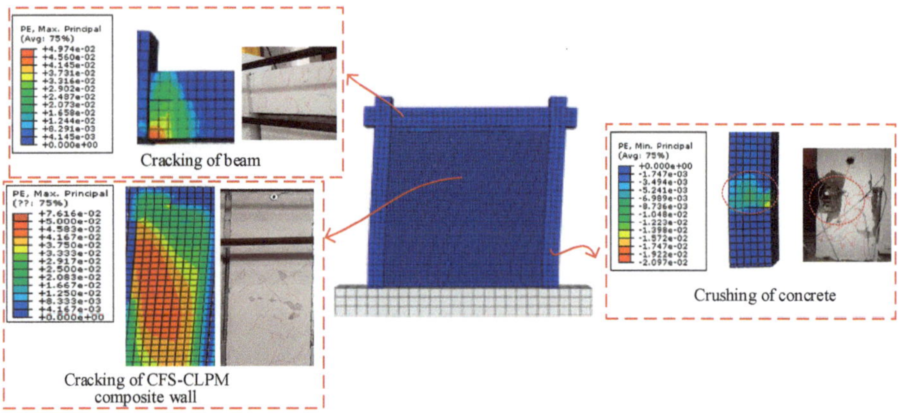

(**b**)

Figure 17. Comparison of failure characteristics: (**a**) PCF; (**b**) PCFW.

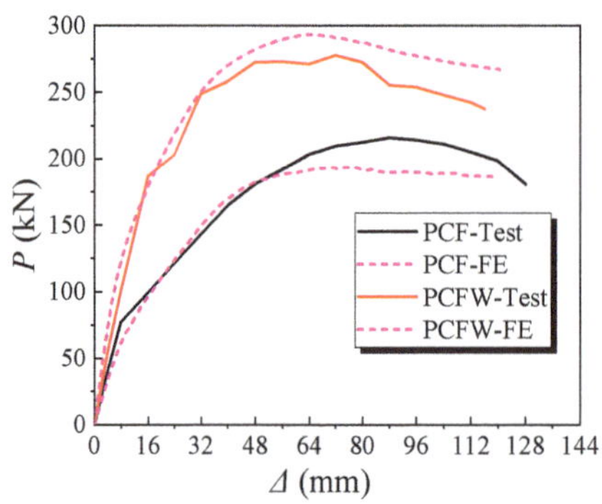

Figure 18. Comparison between experimental and numerical curves.

4.2.2. Contact Stress Distribution

Under horizontal loads, the CFS-CLPM composite walls would be compressed by the PC frame due to the column-wall interaction and wall-frame connections. Figure 19 shows the contact stress distributions on columns for specimen PCFW at various inter-story drifts. It can be seen that the initial contact stress of the left column was primarily distributed at the top of the column, and then the distribution gradually moved downward along the column. In contrast, the initial contact pressure of the right column occurred mainly at the bottom of the column, followed by two ends and middle regions of the column appeared. These differences in the contact stress distributions are due to the deformation compatibility between the shear deformation of the CFS-CLPM composite walls and the flexural deformation of the PC frame. As the inter-story drifts increased, the contact stress was larger owing to the intensely interactive compression between the columns and CFS-CLPM composite walls.

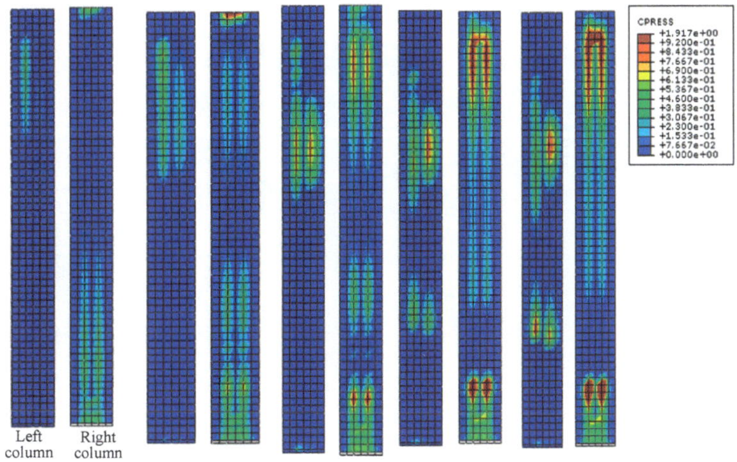

(a) (b) (c) (d) (e)

Figure 19. Contact stress distribution on columns for specimen PCFW: (**a**) drift of 0.5%; (**b**) drift of 1.0%; (**c**) peak drift of 2.0%; (**d**) drift of 3.0%; (**e**) drift of 3.65%.

4.2.3. Analyses of the CFS-CLPM Composite Walls

The minimum principal compressive stresses of CLPM are depicted in Figure 20. The compressive stress in Figure 20 was limited to a minimum of −0.92 MPa so that the stress distribution of CLPM was visible. Combined with the contact stress distribution (Figure 19), the compressive zones of the CLPM provided resistance to horizontal loads were marked out in Figure 20a–e with red dashed lines.

It seems that most of the CLPM was compressed at the drift of 0.5%, and the compressive stress was transferred through the diagonal zone of the CLPM. As the inter-story drift increased, horizontal loads transferred to the CFS-CLPM composite walls through the frame-wall joints increased significantly. Consequently, a compressive zone at the top corner of the CLPM was developed at the drift ratio of 1.0%. Compared to the CLPM stress distribution at drift of 0.5%, the diagonal compressive zone of CLPM in Figure 20b enlarged. At the peak drift of 2.0%, there were three compressive zones of CLPM, which were a big compressive zone in the diagonal direction and two small compressive zones near the frame-wall joints. In Figure 20c–e, corresponding to the drift ratio of 2.0–3.65%, the compressive zones of CLPM in the diagonal direction gradually shrunk due to the damage, while two compressive zones of CLPM near the frame-wall joints gradually extended. Eventually, the diagonal compression zone and the compression zone near the lower frame-wall joints merged. The development of the compressive zones indicated that the CFS-CLPM composite walls are considerably involved in resisting the horizontal loads and were damaged eventually. Figure 20f shows contour plots of von Mises stress in the CFS frame at the drift of 3.65%. It was found that the stress of the tracks and studs did not reach its yield strength except in small regions near frame–wall joints.

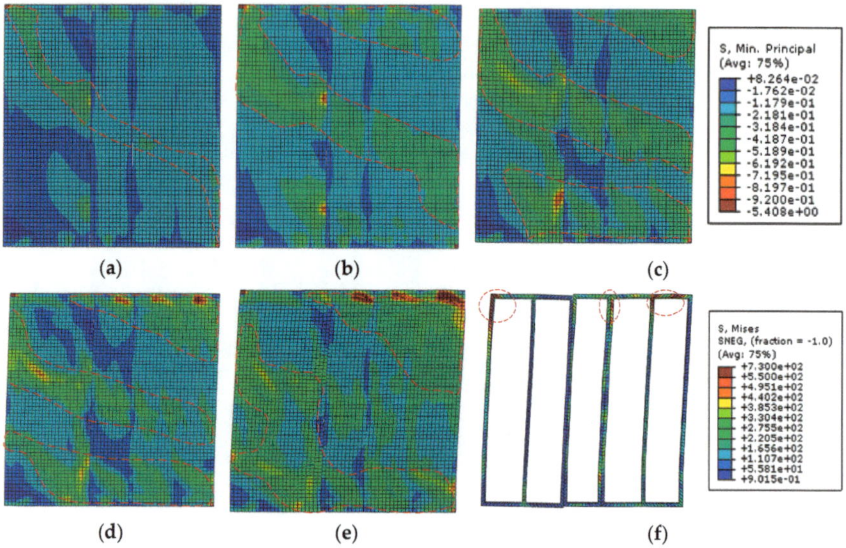

Figure 20. Stress distributions of CFS-CLPM composite walls: (**a**) drift of 0.5%; (**b**) drift of 1.0%; (**c**) peak drift of 2.0%; (**d**) drift of 3.0%; (**e**) drift of 3.65%; (**f**) drift of 3.65%.

4.3. Parametric Analyses

The experimental and numerical analyses above showed that the load capacity and elastic stiffness of the infilled PC frame was improved significantly by the CFS-CLPM composite walls. To explore the influential factors on the lateral response of the PC frame infilled with CFS-CLPM composite walls, parametric analyses using ABAQUS were conducted on material, geometry, and load ratio. The configurations of specimen PCFW were taken as a reference for the simulation sample (FE-SS). The detailed parameters of FE models are listed in Table 5. In order to analyze the influence of the various parameters, the lateral response of this structure was evaluated in terms of the maximum lateral load P_m and elastic stiffness K_e.

Table 5. Parameters of PC frame with CFS-CLPM composite walls.

FE-Models	Material			Geometry		Load Ratio
	f_c/MPa	f_p/MPa	f_y/MPa	L/H	t_w/mm	n
FE-SS	26.8	0.92	550	1.1	130	0.3
FE-11	20.1	0.92	550	1.1	130	0.3
FE-12	33.5	0.92	550	1.1	130	0.3
FE-13	40.2	0.92	550	1.1	130	0.3
FE-21	26.8	1.42	550	1.1	130	0.3
FE-22	26.8	1.92	550	1.1	130	0.3
FE-23	26.8	2.42	550	1.1	130	0.3
FE-31	26.8	0.92	235	1.1	130	0.3
FE-32	26.8	0.92	390	1.1	130	0.3
FE-41	26.8	0.92	550	0.9	130	0.3
FE-42	26.8	0.92	550	1.3	130	0.3
FE-43	26.8	0.92	550	1.7	130	0.3
FE-51	26.8	0.92	550	1.1	160	0.3
FE-52	26.8	0.92	550	1.1	190	0.3
FE-53	26.8	0.92	550	1.1	220	0.3
FE-61	26.8	0.92	550	1.1	130	0.5
FE-62	26.8	0.92	550	1.1	130	0.7
FE-63	26.8	0.92	550	1.1	130	0.85

4.3.1. Compressive Strength of Concrete (f_c)

Figure 21a shows the effect of f_c on the load-displacement curves of the PC frame infilled with CFS-CLPM composite walls. It was seen that, when f_c increased from 20.1 MPa to 26.8 MPa, 33.5 MPa and 40.2 MPa, the P_m increased by 3.11%, 3.36% and 3.42%, respectively, and the K_e increased by 3.35%, 5.7% and 7.30%. This improvement of the P_m and K_e are attributed to the increase of lateral load capacity and initial stiffness of the PC'frame.

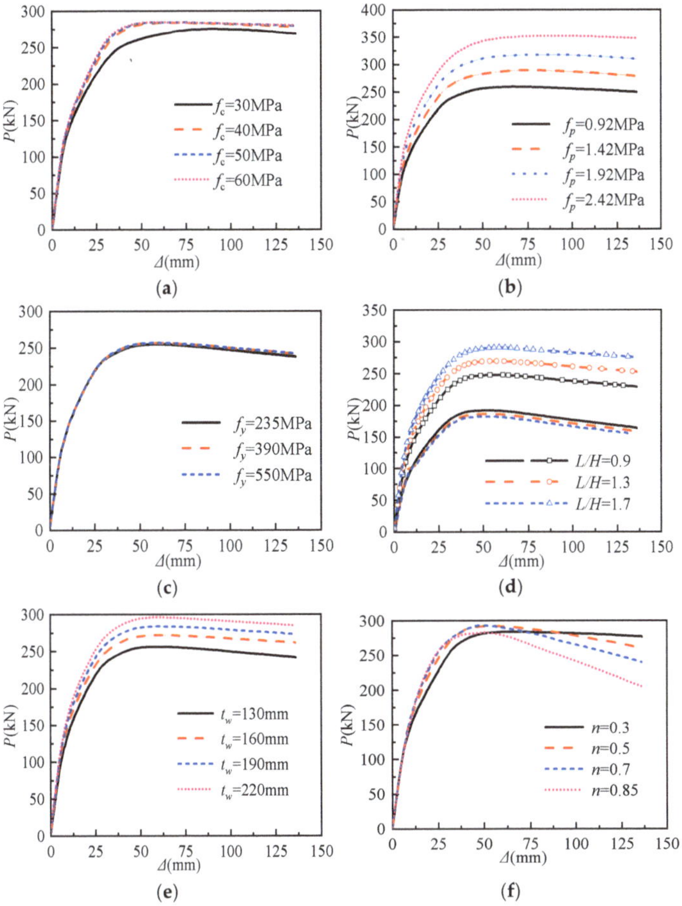

Figure 21. Influence of different parameters on lateral response of PC frame infilled with CFS-CLPM composite walls: (**a**) Compressive strength of concrete; (**b**) Compressive strength of CLPM; (**c**) Strength of cold-formed steel; (**d**) Span to height ratio; (**e**) Thickness of CFS composite walls; (**f**) Axial load ratio.

4.3.2. Compressive Strength of CLPM (f_p)

Figure 21b shows the load–displacement curves of FE models with different compressive strengths of CLPM. As the f_p increased from 0.92 MPa to 1.42 MPa, 1.92 MPa and 2.42 MPa, the P_m improved by 11.44%, 22.26% and 35.56%, respectively. Meanwhile, the K_e improved by 10.21%, 21.44% and 33.31%, respectively. It could be found that both P_m and K_e are almost improved linearly with the increase of the f_p. This phenomenon confirmed further that the parameter of f_p has a significant influence on the lateral response of the PC frame infilled with CFS-CLPM composite walls, and the CLPM fillers are important lateral resistance elements of the CFS-CLPM composite walls.

4.3.3. Strength of Cold-Formed Steel (f_y)

In the numerical analyses of the CFS-CLPM composite walls, Figure 20f shows that most of the CFS frame does not reach its yield strength. Therefore, a lower f_y was selected in parametric analyses. Figure 21c illustrates the numerical results of the structure with f_y of 235 MPa, 390 MPa, and 550 MPa. The load-displacement curves of the structure with f_y of 235 MPa and 345 MPa almost coincided with that of the structure with f_y of 550 MPa. This demonstrates that f_y has little effect on the lateral performance of the structure and it is suggested to reduce the strength of the cold-formed steel appropriately to lower the overall cost of the CFS-CLPM composite walls in engineering projects.

4.3.4. Span to Height Ratio (L/H)

The span and height of the structure were defined as depicted in Figure 3a. The FE models with different ratios (L/H) were featured with a constant height (H = 3030 mm), while the span (L) was taken as 2700 mm, 3900 mm and 5100 mm, corresponding to L/H of 0.9, 1.3 and 1.7, respectively. As L increased from 2700 to 5100 mm, the number of the separated CFS-CLPM composite walls with width of 1200 mm rose from 2 to 4.

To highlight the mechanical contribution of CFS-CLPM composite walls to the PC frame structures, the parametric analyses of L/H were conducted not only on the infilled PC frames with CFS-CLPM composite walls but also on the corresponding bare PC frames. Figure 21d shows the load-displacement curves, in which the curves of the infilled PC frame were lines with symbols. Under L/H of 0.9, P_m and K_e of the infilled PC frame were respectively 29.13% and 18.06% higher than that of the bare PC frame. Similarly, setting the bare PC frames as a benchmark at the same ratio of L/H, P_m and K_e of the infilled PC frame with L/H of 1.3 were increased by 44.83% and 26.85%, respectively, and those of the infilled PC frame with L/H of 1.7 were increased by 59.58% and 50%, respectively. It is seen that the CFS-CLPM composite walls have a notable improvement on the P_m of the PC frame structure, especially on K_e. Hence, it is suggested that the lateral load capacity and stiffness provided by the CFS-CLPM composite infill walls should be included in the lateral resistance of the infilled PC frame structure.

On the other hand, it was noted that the P-Δ curves of the bare PC frames with different L/H were almost same. On this basis, it was assumed herein that the lateral resistance contributed by the PC frame of the infilled PC frame structure was nearly identical under different L/H. Thus, compared to the P_m and K_e of the infilled PC frame structure with L/H of 0.9, P_m of the structure with L/H of 1.3 and 1.7 respectively increased by 7.99% and 16.70%, and K_e respectively increased by 7.54% and 16.63%. Generally, the span-to-height ratio of the CFS-CLPM composite walls can significantly influence the lateral response of PC frame structures.

4.3.5. Thickness of CFS-CLPM Composite Walls (t_w)

Various t_w of 130 mm, 160 mm, 190 mm and 220 mm were investigated, as depicted in Figure 21e. Compared with the numerical results with t_w of 130 mm, P_m of the structure with t_w of 160 mm, 190 mm and 220 mm increased by 6.04%, 10.74% and 15.57%, respectively. Meanwhile, K_e of the structure with t_w of 160 mm, 190 mm and 220 mm were 8.61%, 13.11% and 17.72% higher, respectively. Although the P_m and K_e of the infilled PC frame structures are enhanced to some extent by increasing of t_w, there are some disadvantages such as more expensive cost, heavier weight for the walls, and less available space of the buildings. Therefore, the thickness of the CFS-CLPM composite walls should be selected after comprehensive consideration.

4.3.6. Axial Load Ratio (n)

In view of the variation of axial loads imposed on columns in engineering practices, n varied from 0.3 to 0.85 were considered, as shown in Figure 21f. Compared to the numerical results with n of 0.3, P_m of the structure with n of 0.5, 0.7 and 0.85 were 3.87%, 4.23% and 0.66% higher, respectively, and K_e of the structure with n of 0.5, 0.7 and 0.85 were

1.01%, 0.16% higher and 3.64% lower, respectively. It can be concluded that axial loads have a minor effect on the P_m and K_e of the infilled PC frame structures, but they have a measurably adverse influence on the ductility of the structure due to the premature failure of concrete and the intensification of the second-order effects.

5. Prediction of the Elastic Stiffness

Elastic stiffness is decisive indexes for the serviceability limit state of structures and a number of researches have been conducted on the prediction method of elastic stiffness of members and structures [28]. Based on the experimental and parametric results, the elastic lateral stiffness of the infilled PC frame structure can be greatly enhanced by the CFS-CLPM composite walls. However, the lateral stiffness of the infill walls was always ignored in the previous design model of the PC frame structure, which would not effectively utilize the structural performance of the PC frame infilled with CFS-CLPM composite walls. Hence, the contribution of the CFS-CLPM composite walls was considered to accurately estimate the elastic stiffness of the infilled PC frame structure under horizontal loads in this study. The elastic lateral stiffness K of the infilled PC frame structure is provided by the PC frame and CFS-CLPM composite walls, and defined as Equation (2):

$$K = K_f + K_W \tag{2}$$

where K_f is the elastic lateral stiffness of the PC frame, and K_w is the elastic lateral stiffness of the CFS-CLPM composite walls.

Under horizontal loads, the Muto's method (modified D-method) was adopted to calculate the K_f provided by two columns of the PC frame [29]. Thereby, K_f can be obtained by Equation (3), and the relevant variables are expressed in Equations (4)–(6):

$$K_f = 2(\zeta \frac{12 i_c}{{h_c}^2}) \tag{3}$$

$$i_m = \frac{E_m I_m}{l_m} \tag{4}$$

$$\zeta = \begin{cases} \frac{k}{2+k}, \text{ columns of typical floor} \\ \frac{0.5+k}{2+k}, \text{ columns of ground floor} \end{cases} \tag{5}$$

$$k = \begin{cases} \frac{i_1+i_2+i_3+i_4}{2i_c}, \text{ columns of typical floor} \\ \frac{i_1+i_2}{i_c}, \text{ columns of ground floor} \end{cases} \tag{6}$$

where E_m is the elastic modulus of members; I_m is the inertia moment of member sections; l_m is the length of members; i_m is the line stiffness for members; h_c is the height of columns; i_c is the line stiffness for columns, and i_1~i_4 are the line stiffness of beams surrounding the column as shown in Figure 22.

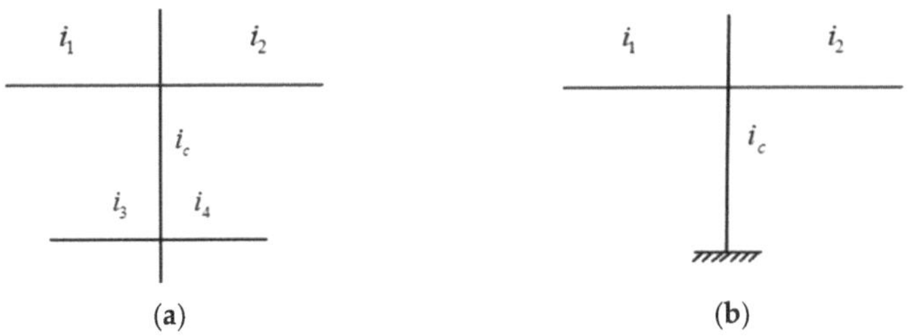

Figure 22. Line stiffness of components for a typical floor or ground floor: (**a**) typical floor; (**b**) ground floor.

Under horizontal loads, the horizontal displacements of the CFS-CLPM composite walls are induced by the shear and flexural deformation derived from a combination of the CFS frame and the CLPM fillers. In the assembly of the CFS frame, the studs were connected to the tracks using self-drilling screws that were assumed to be hinge joints in mechanical behavior. On the basis of the above assumptions, the horizontal displacement of the CFS-CLPM composite walls was determined by the shear and flexural deformation of the CLPM fillers, and was expressed as follows [11]:

$$\Delta = \frac{1.2PH}{GtL} + \frac{PH^3}{3EI} \tag{7}$$

where P is the horizontal load carried by CFS-CLPM composite walls; H, L, and t are the height, width, and thickness of the CLPM fillers, respectively; G and E are the shear and elastic modulus of the CLPM, respectively. I is the inertia moment of the CLPM fillers section. Thereby, the calculation formula of K_w is described as Equation (8):

$$K_w = \frac{P}{\Delta} = \frac{1}{\frac{1.2H}{GtL} + \frac{H^3}{3EI}} \tag{8}$$

Therefore, the elastic stiffness of the infilled PC frame structure can be calculated by substituting Equations (3) and (8) into Equation (2). The deduced formula is expressed as follows:

$$K = 2(\zeta\frac{12i_c}{h^2}) + \frac{1}{\frac{1.2H}{GtL} + \frac{H^3}{3EI}} \tag{9}$$

To verify the reliability, Equation (9) was adopted to predict the initial stiffness of specimens in Sections 3 and 4. The predictions were compared to the simulations and test results as listed in Table 6. The average ratio of K_p/K_t is 1.005 with a variation of 0.0021, which indicates that the proposed calculation method can work well for predicting the elastic lateral stiffness of the PC frame infilled with CFS-CLPM composite walls.

Table 6. Comparison of predicted lateral stiffness results.

Samples	K_t (kN/mm)	K_p (kN/mm)	K_p $/K_t$
PCF	10.22	10.347	1.012
PCFW	14.53	14.878	1.023
PE-SS	14.967	14.907	0.996
FE-21	16.604	16.375	0.986
FE-22	18.358	17.836	0.972
FE-23	20.14	19.297	0.958
FE-41	14.21	13.192	0.921
FE-42	15.67	15.73	1.004
FE-43	17.54	19.482	1.111
FE-51	15.867	15.967	1.006
FE-52	16.565	17.021	1.028
FE-53	17.269	18.075	1.047
Average	-	-	1.005
Coefficient of variation	-	-	0.0021

Note: K_t is the tested or simulated elastic lateral stiffness; K_p is the predicted elastic lateral stiffness.

6. Conclusions

In this paper, the prefabricated CFS-CLPM composite walls were adopted into the PC frame. The cyclic tests were performed to investigate the influence of the CFS-CLPM composite walls on the seismic response of the PC frame structure. Numerical simulations and analytical prediction of elastic stiffness were conducted further on the PC frame infilled with CFS-CLPM composite walls to systematically study the lateral behavior. The conclusions can be summarized as follows:

1. The failure modes of the PC frame infilled with CFS-CLPM composite walls were characterized by cracks on the PC frame, diagonal cracks on the CFS-CLPM composite walls, titling of self-drilling screws, and crushing of the columns. The CFS-CLPM composite walls remained quite intact and the proposed wall-frame joints could restrain out-of-plane movements of the walls even at failure.
2. The CFS-CLPM composite walls can significantly improve the lateral behavior of the PC frame structure. Compared with the bare frame, the lateral load capacity and elastic stiffness of the infilled frame were respectively 24.9~31.6% and 34.1~38.4% higher, respectively. Moreover, the energy dissipation capacity of the infilled frame structure increased by 14%. Despite a slight reduction in the ductility of the infilled PC frame structure owing to the infill-frame interaction, its failure drift can meet the elastic-plastic drift requirement of 2%.
3. The numerical analyses of the infilled PC frame structure revealed that CLPM fillers were the significant lateral resistance elements of the CFS-CLPM composite walls. Three compressive zones were formed on the CLPM fillers, because the horizontal shear force was transferred from the PC frame to CFS-CLPM composite walls through the frame-wall joints between the composite wall and PC beam, and the compression between the composite wall and the PC column.
4. Parametric analyses of the PC frame infill with CFS-CLPM composite walls indicated that the strength of CLPM, the span-to-height ratio, and the thickness of CFS-CLPM composite walls significantly affected the lateral capacity of the structure, while the strength of concrete, the strength of cold-formed steel and the axial load ratio affected slightly.
5. A formula considering the lateral resistance of the CFS-CLPM composite walls was proposed to predict the elastic lateral stiffness of the PC frame infilled with CFS-CLPM composite walls and the comparisons among prediction and test as well as simulation results demonstrated that the formula was reliable.

Author Contributions: Software, data curation, visualization, writing—original draft preparation, P.H.; writing—review and editing, methodology, formal analysis, Y.L.; funding acquisition, supervision, resource, J.W.; conceptualization, validation, W.W.; formal analysis, investigation, data curation G.P. All authors have read and agreed to the published version of the manuscript.

Funding: This research was funded by the Special Funds for the Central Universities of China (No. PA2021KCPYO031), the Fundamental Research Funds for the Central Universities of China (No. PA2021GDSK0078, JZ2021HGTA0158), the Natural Science Research Project of China (No. 52278170), and the Natural Science Research Project of Anhui Province (No. 2108085QE248).

Institutional Review Board Statement: Not applicable.

Informed Consent Statement: Not applicable.

Data Availability Statement: Not applicable.

Conflicts of Interest: The authors declare no conflict of interest.

References

1. Badr, A.R.; Elanwar, H.H.; Mourad, S.A. Numerical and experimental investigation on cold-formed walls sheathed by fiber cement board. *J. Constr. Steel Res.* **2019**, *158*, 366–380. [CrossRef]
2. Macillo, V.; Fiorino, L.; Landolfo, R. Seismic response of CFS shear walls sheathed with nailed gypsum panels: Experimental tests. *Thin-Walled Struct.* **2017**, *120*, 161–171. [CrossRef]
3. Ali, E.; Althoey, F. Numerical investigation on blast response of cold-formed steel framing protected with functionally graded composite material. *Buildings* **2022**, *12*, 118. [CrossRef]
4. Xu, Z.F.; Chen, Z.F.; Osman, B.H.; Yang, S.H. Seismic performance of high-strength lightweight foamed concrete-filled cold-formed steel shear walls. *J. Constr. Steel Res.* **2018**, *143*, 148–161. [CrossRef]

5. Ananthi, G.B.G.; Roy, K.; Lim, J.B.P. Experimental and numerical study of an innovative 4-channels cold-formed steel built-up column under axial compression. *Steel Compos. Struct.* **2022**, *42*, 513–538.
6. Ananthi, G.B.G. A study on cold-formed steel compound angle section subjected to axial compression. *KSCE J. Civ. Eng.* **2018**, *22*, 1803–1815. [CrossRef]
7. Mousavi, S.A.; Zahrai, S.M.; Bahrami-Rad, A. Quasi-static cyclic tests on super-lightweight EPS concrete shear walls. *Eng. Struct.* **2014**, *65*, 62–75. [CrossRef]
8. Shannag, M.J. Characteristics of lightweight concrete containing mineral admixtures. *Constr. Build. Mater.* **2011**, *25*, 658–662. [CrossRef]
9. Mydin, M.A.O.; Wang, Y.C. Structural performance of lightweight steel-foamed concrete-steel composite walling system under compression. *Thin-Walled Struct.* **2011**, *49*, 66–76. [CrossRef]
10. Yin, C.; Zhou, L.; Zou, Q.Y.; Xu, Y.F. Effect of filling phosphorgypsum on the axial compression behavior of cold-formed thin-walled steel walls. *Buildings* **2022**, *12*, 1325. [CrossRef]
11. Wu, H.H.; Sui, L.; Wang, J.Q.; Zhou, T.H. Cycle performance tests and numerical modeling of infilled CFS shear wall. *J. Constr. Steel Res.* **2020**, *168*, 106010. [CrossRef]
12. Wang, W.Q.; Wang, J.F.; Zhao, P.; Ja, L.L.; Pan, G.D. Axial compressive experiments and structural behaviour estimation of CFS composite walls sprayed with LPM. *J. Build. Eng.* **2020**, *30*, 101305. [CrossRef]
13. Wang, W.Q.; Wang, J.F.; Guo, L.; Liu, Y.; Zhang, R. Cyclic performance tests and numerical analysis of prefabricated CFS shear walls filled with lightweight EPS mortars. *Eng. Struct.* **2021**, *243*, 112554. [CrossRef]
14. Wang, J.Q.; Zhou, T.H.; Wu, H.H.; Guan, Y.; Zhang, L. Cyclic performance of steel frame fabricated with cold-formed steel composite wall structure. *Eng. Struct.* **2022**, *270*, 114892. [CrossRef]
15. Wang, F.L.; Yang, J.; Wang, X.; Azim, I. Study on progressive collapse behaviour of steel-framed substructures with sheathed CFS stud infill walls. *J. Build. Eng.* **2021**, *42*, 102720. [CrossRef]
16. Kildashti, K.; Samali, B.J.; Mortazavi, M.; Ronagh, H.; Sharafi, P. Seismic collapse assessment of a hybrid cold-formed hot-rolled steel building. *J. Constr. Steel Res.* **2019**, *155*, 504–516. [CrossRef]
17. *JGJ 355-2015*; Technical Specification for Grout Sleeve Splicing of Rebars. China Architecture & Building Press: Beijing, China, 2015.
18. *GB/T 51231-2016*; Technical Standard for Assembled Buildings with Concrete Structure. China Architecture & Building Press: Beijing, China, 2017.
19. 19. *JGJ 1-2014*; Technical Specification for Precast Concrete Structures. China Architecture & Building Press: Beijing, China, 2014.
20. *GB/T 50081-2002*; Standard for Test Method of Mechanical Properties on Ordinary Concrete. China Architecture & Building Press: Beijing, China, 2003.
21. *JG/T 266-2011*; Foamed Concrete. Architecture & Industry Press of China: Beijing, China, 2011.
22. *GB/T 228.1-2010*; Metallic Materials-Tensile Testing-Part 1: Method of Test at Room Temperature. General Administration of Quality Supervision, Inspection of PR China: Beijing, China, 2012.
23. Dautaj, A.D.; Kadiri, Q.; Kabash, N. Experimental study on the contribution of masonry infill in the behavior of RC frame under seismic loading. *Eng. Struct.* **2018**, *165*, 27–37. [CrossRef]
24. Dabiri, H.; Kaviani, A.; Kheyroddin, A. Influence of reinforcement on the performance of non-seismically detailed RC beam-column joints. *J. Build. Eng.* **2020**, *31*, 101333. [CrossRef]
25. *GB 50011-2010*; Code for Seismic Design of Buildings. China Architecture & Building Press: Beijing, China, 2010.
26. Popovics, S. A numerical approach to the complete stress-strain curve of concrete. *Cem. Concr. Res.* **1973**, *3*, 583–599. [CrossRef]
27. Cui, C.C.; Huang, Q.; Li, D.B.; Quan, C.R.; Li, H.C. Stress–strain relationship in axial compression for EPS concrete. *Constr. Build. Mater.* **2016**, *105*, 377–383. [CrossRef]
28. Lai, B.L.; Yang, L.F.; Xiong, M.X. Numerical simulation and data-driven analysis on the flexural performance of steel reinforced concrete composite members. *Eng. Struct.* **2021**, *247*, 113200. [CrossRef]
29. Mutō, K. *Aseismic Design Analysis of Buildings*; Maruzen Company, Ltd.: Tokyo, Japan, 1974.

buildings

MDPI

Article

Evaluation of Residual Lateral Capacities of Impact-Damaged Reinforced Concrete Members

Jiabin Ye [1,2], Yingtao Wang [3], Jian Cai [1,4], Qingjun Chen [1,4] and An He [1,*]

1 School of Civil Engineering & Transportation, South China University of Technology, Guangzhou 510641, China; jb_ye@foxmail.com (J.Y.); cvjcai@scut.edu.cn (J.C.); qjchen@scut.edu.cn (Q.C.)
2 Guangzhou Jishi Construction Group Co., Ltd., Guangzhou 510115, China
3 Department of Civil Engineering, Foshan University, Foshan 528000, China; scutwyt@163.com
4 State Key Laboratory of Subtropical Building Science, Guangzhou 510641, China
* Correspondence: hean@scut.edu.cn

Abstract: To study the residual lateral capacity of reinforced concrete (RC) columns after being subjected to static and horizontal impact action, static and horizontal impact tests of a total of sixteen RC columns were conducted. The variables considered in the tests included the shear-span ratio, the impact weight and the velocity. The experimental results, including the impact force, the deflection and the strain, as well as the cracking pattern and the failure mode, were discussed. Compared with the load–deflection curves under static and impact loading, it can be found that the inertial effect plays a significant part in the dynamic behaviour of the RC columns. Subsequently, static tests of six specimens with slight impact damage were carried out to obtain their residual performance. Based on the Park–Ang damage model that is widely used for assessing the post-seismic performance of RC members, an evaluation method for the structural residual capacity of RC columns after being subjected to impact loading was developed, with its accuracy confirmed by the experimental results.

Keywords: RC; horizontal impact; inertial effect; residual capacity; damage coefficient

Citation: Ye, J.; Wang, Y.; Cai, J.; Chen, Q.; He, A. Evaluation of Residual Lateral Capacities of Impact-Damaged Reinforced Concrete Members. *Buildings* **2022**, *12*, 669. https://doi.org/10.3390/buildings12050669

Academic Editors: Shan Gao, Jingxuan Wang, Dewen Kong and Yong Liu

Received: 23 April 2022
Accepted: 14 May 2022
Published: 17 May 2022

1. Introduction

With the development of international trade, marine structures are at a high risk of collision events from vessels during their life service. In the current design code of RC piers or columns, the equivalent static method is mainly adopted to consider the impact effect [1–3]. Namely, the equivalent impact force calculated by empirical formulae is regarded as the static loads acting on the structures, while the inertia effect and the strain rate effect of material induced by the impact process are neglected. Previous studies have indicated that the influence of the inertia effect and the strain rate effect on structural failure modes is significant. For example, simply supported RC beams with flexural modes under static loads may change to local shear failure when subjected to impact loads, with a shear pug developed in the mid-span impact area, especially for those without web reinforcement [4,5]. Therefore, the classical "shear failure valley" model for RC beams under static loads [6] is not applicable for those under impact loads, in which the effect of impact mass and speed should be considered. Regarding the existing design codes for equivalent impact forces of RC members, the calculation method in the Load Code of Port Engineering JTS 144-1-2010 [7] was established by a limited number of scaled physical model tests and numerical models, with only the impact mass considered; this may fail to reflect the actual response and the damage degree of RC members under impact loads. Therefore, more explicit design methods for assessing the dynamic behaviour of RC members under impact loads are required.

For RC members with damage caused by collision actions, their residual bearing capacities can be a significant index to reflect their damage degree and use as a reference for repair design [8,9]. Considering various reinforcement ratios, stirrup ratios and slenderness ratios, drop weight impact tests on a total of thirty RC beams were carried out by Adhikary et al. [10], and finite element software LS-DYNA was used for parametric analysis. By comparing the bearing capacity between the undamaged and impact-damaged specimens, the residual bearing capacity and the residual stiffness for RC beams after being subjected to impact loads were investigated. It was found that the reduction coefficient of bearing capacity for RC beams ranged from 0.8 to 1.0 for damaged specimens with obvious oblique cracks but not complete fail, while the residual bearing capacity coefficient may be decreased to 0.2 for those severely damaged specimens. Fujikake [11] conducted flexural tests on ultra-high performance concrete beams after impact, to assess their residual bearing capacity. It was found that the static load–deflection curves of the impact-damaged beams were generally consistent with the undamaged ones, but with the curves starting from the residual deflection after impact, indicating that the residual bearing capacity of RC beams could be determined according to their maximum displacements during impact loading. Based on six half-scale rectangular RC columns with different volumetric reinforcement ratios, Peng et al. [12] proposed a method for estimating the static shear performance and the residual axial capacity of RC bridge piers under near-ground lateral loads. A damage criterion relating to the reduction degree of bearing capacity was proposed by Tian et al. [13], aiming at investigating the effect of different protective measures, namely, externally bonded steel plates and external application of aluminium foam. Pei [14] conducted drop weight impact tests on twelve H-shaped steel members and residual bearing capacity static tests on the impact-damaged specimens; it was found that there is a linear relationship between the residual bearing capacity and the initial impact energy. In general, the drop-hammer impact device has been widely adopted for impact tests of structural components [15–17]. However, the horizontal impact test is rarely reported in the open literature [18,19], which has a higher capacity for applying impact loads with large mass.

In order to study the response of marine structures under vessel collision loads characterized by large mass and low speed, horizontal impact tests on a total of 13 square RC columns have been conducted by the authors [20,21], to study their dynamic behaviour under the effect of different column slenderness, impact masses and velocities. Besides, static load tests on three RC column specimens with different shear-span ratios have also been carried out. Through the comparisons of the load–displacement curves derived from the static load tests and impact tests, the difference in structural response was discussed. After the impact tests, six damaged specimens were selected for static tests, aiming at their residual capacities. Based on the experimental results from impact tests and static tests, this paper mainly focuses on the residual lateral capacities of the impact-damaged RC members, and further develops the predictive calculation model. The Park–Ang damage evolution model [22] that has been widely used for assessing the post-seismic resistances of RC members is modified, to propose an evaluation method for the post-impact damage coefficient. The accuracy of the modified Park–Ang model was assessed through the comparisons between the predicted results against the experimental ones, as well as those from RC beams subjected to drop-weight impact.

2. Previous Experimental Work

2.1. Tested Specimens

A total of 16 square cross-section RC columns were designed with various cross-sectional dimensions and diameters of the longitudinal rebars, as shown in Figure 1. For all the examined specimens, the effective height of the columns, defined as the distance from the horizontal loading point to the bottom of the column, was 1200 mm. The longitudinal steel bars were symmetrically reinforced with the reinforcement ratio being 1.4%, while the stirrups with a diameter equal to 8 mm and a spacing of 70 mm were adopted. The thickness of the concrete cover was 25 mm. All the specimens were divided into three groups according to their cross-section width (i.e., 120 mm, 150 mm and 180 mm), and the corresponding shear-span ratios were equal to 6.67, 8 and 10, with the diameters of the longitudinal bars being 12 mm, 10 mm and 8 mm, respectively. The material properties of the steel bar were measured and are listed in Table 1, in which Φ15 is the pre-stressed tendons used to apply axial load before the test. The measured average cubic compressive strength of concrete was 60.39 MPa.

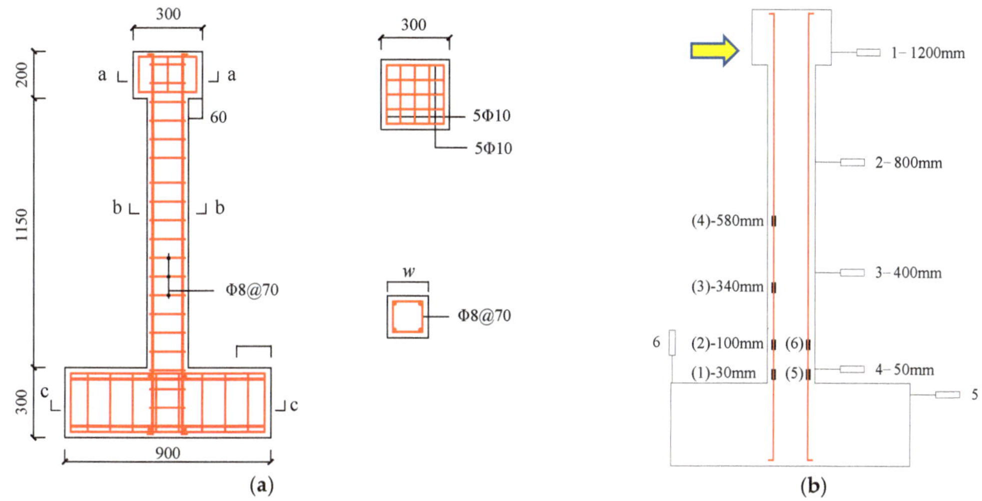

Figure 1. Details of specimen and measurements. (**a**) Dimensions of the specimens; (**b**) Measurements.

Table 1. Material properties of reinforcements.

Reinforcement	Φ8	Φ10	Φ12	Φ15
f_y (MPa)	408.8	424.6	470.4	886.2
f_u (MPa)	541.1	642.3	685.7	1135.7

Note: f_y and f_u represent the tensile yield strength and ultimate strength of rebar, respectively.

The column specimens were tested under constant vertical compression loads with either static or dynamic horizontal loads. During the tests, four linear voltage displacement transducers (LVDTs) were arranged at the column side to monitor the displacement of specimens, as shown in Figure 1b. Strain gauges were arranged on the longitudinal steel bar at the bottom area of the column, and the prestressed tendons.

2.2. Static Test

Three RC columns (i.e., SL120, SL150 and SL180) with different cross-section sizes were tested under constant axial loads and monotonic static horizontal loadings, as Figure 2 shows. The labels of the specimens start with the letters "SL" (representing static loads), followed by their cross-section width. Axial loads were first applied to the target axial compression ratio of 0.1 by the jack located at the top of the column. Then, monotonic horizontal loads were applied by an MTS actuator until the failure of the RC columns. During the testing, the horizontal loads were recorded by the MST system, while the corresponding horizontal displacement was measured by the LVDT located at the top of the column, as displayed in Figure 1b.

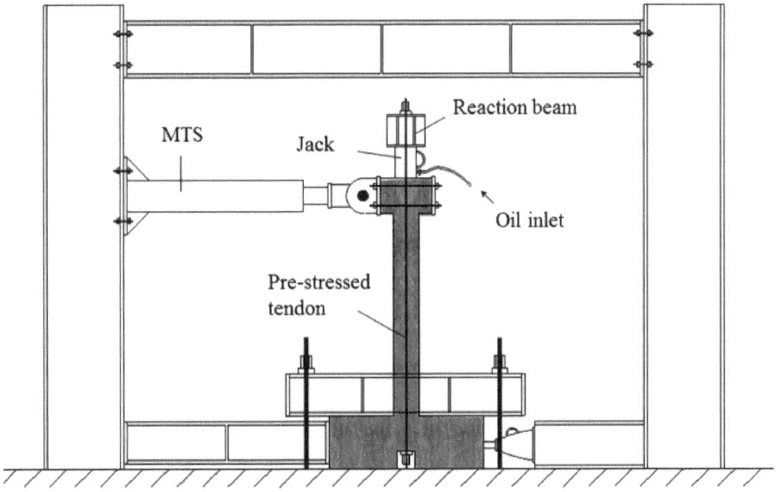

Figure 2. Static test set-up.

2.3. Impact Test

Thirteen RC column specimens were utilized for the impact tests. The details of the specimens are listed in Table 2. The labels of the specimens begin with the letters "DL" (representing dynamic loads), followed by their cross-section width (i.e., 120, 150 and 180 mm), impact mass (i.e., 1.2, 1.5 and 1.8 t) and design impact velocity (i.e., 0.4, 0.8 and 1.2 m/s), respectively.

Table 2. Details of the specimens and impact test results.

Specimen	v_0 (m/s)	F_p (kN)	t_d (s)	u_{max} (mm)	u_{res} (mm)	$E_{ab,d}$ (kJ)	Damage State
DL150-1.5-0.4	0.329	8.899	0.142	7.803	0.53	0.035	○
DL150-1.5-0.8	0.767	44.463	0.211	40.12	7.51	0.307	○
DL150-1.5-1.2	1.191	59.003	0.238	56.01	5.16	0.872	•
DL150-1.2-0.8	1.083	63.943	0.149	43.16	-	0.638	○
DL150-1.8-0.8	0.891	56.878	0.214	58.08	-	0.696	○
DL120-1.2-0.8	0.802	37.139	0.221	54.37	0.17	0.364	○
DL120-1.5-0.8	0.794	34.978	0.228	67.3	1.81	0.402	○
DL120-1.8-0.8	0.869	50.419	0.385	77.25	-	0.645	•
DL180-1.5-0.4	0.503	24.325	0.095	11.24	-	0.153	○
DL180-1.5-0.8	0.743	36.025	0.124	20.1	0.65	0.335	○
DL180-1.5-1.2	1.291	80.309	0.124	39.26	-	0.925	○
DL180-1.2-0.8	0.791	30.131	0.122	15.84	-	-	○
DL180-1.8-0.8	0.907	71.901	0.114	27.64	1.29	0.708	○

Note: v_0 is the measured impact velocity of test truck, F_p is the peak impact force, t_d is the impact duration time, u_{max} and u_{res} are the maximum and the residual displacement, respectively, $E_{ab,d}$ is the deformation energy at the maximum displacement. Damage state can be divided into two types: slight damage (○) and complete failure (•).

The setup for the impact test is indicated in Figure 3, which included a horizontal impact device and a vertical drop-hammer impact device. Both devices are connected by four fixed pulleys and steel strands. With the drop hammer falling from a certain height, the test truck accelerates along the horizontal guide rail under the traction of the steel strands until the collision. The initial self-weight of the test truck was 1.2 t, and the maximum weight can be up to 2.4 t by adding steel plates to the truck.

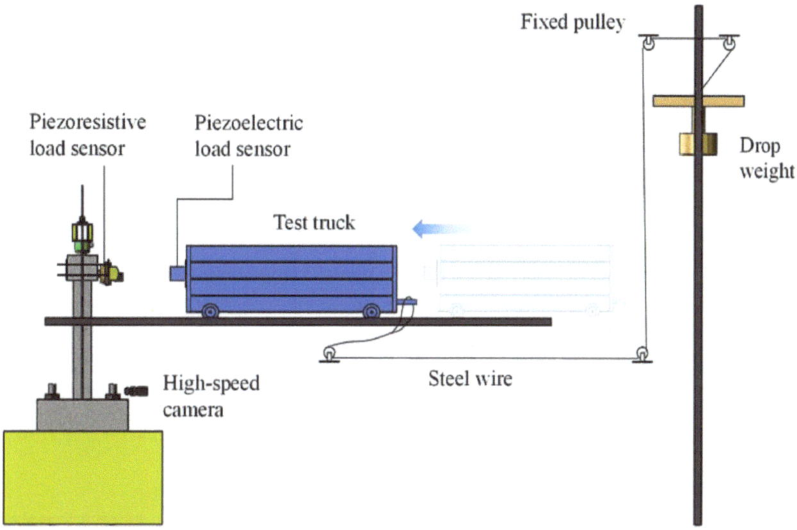

Figure 3. Impact test set-up.

In order to accurately capture the time history data of the impact force, both piezoelectric and piezoresistive load sensors were respectively installed at the front of the test truck and the column cap. The impact velocity was recorded by infrared technology at the position of 10 mm in front of the specimen. Donghua DH5922N signal acquisition instrument was used to collect the data of each sensor (including the load sensors, LVDT and strain gauges), with the sampling frequency set equal to 100 kHz. During the impact tests, a high-speed camera (OS9, Integrated Design Tools Inc.) with a frame rate of 200 fps was applied to record the crack development at the bottom of the column.

Upon completion of impact tests, residual static load tests were conducted, with the test procedure consistent with the static tests, as introduced in Section 2.2.

3. Overview of Experimental Results

3.1. Static Test Results

All the examined specimens exhibited similar experimental phenomena during monotonic static horizontal loading. With the applied static loads increased, it can be observed that bending cracks developed in the tension zone, followed by the yielding of the longitudinal steel bars. Finally, the concrete cover in the compression area was crushed and spalled. The failure mode of RC columns was shown to be bending type with good ductility, as shown in Figure 4a. The load–displacement (F–u) curves of the three static tested specimens are shown in Figure 4b, and the ultimate strengths and the corresponding displacements are listed in Table 3. Compared with the specimens SL150 and SL180, the load–displacement curve of the specimen SL120 rises slowly before reaching the ultimate strength, indicating more ductile behaviour of the specimen when subjected to static loads.

Table 3. Results of static tests.

Static Tested Specimen	F_u (kN)	u_u (mm)	$E_{ab,s}$ (kJ)
SL120	9.92	53	0.708
SL150	15.05	27.79	0.321
SL180	28.5	25.23	0.406

Note: F_u is the static flexural strength; u_u is the displacement at the ultimate strength; $E_{ab,s}$ is the absorbed energy in the static ultimate state, calculated by the surrounding area of the F-u curve and x-axis.

(a)

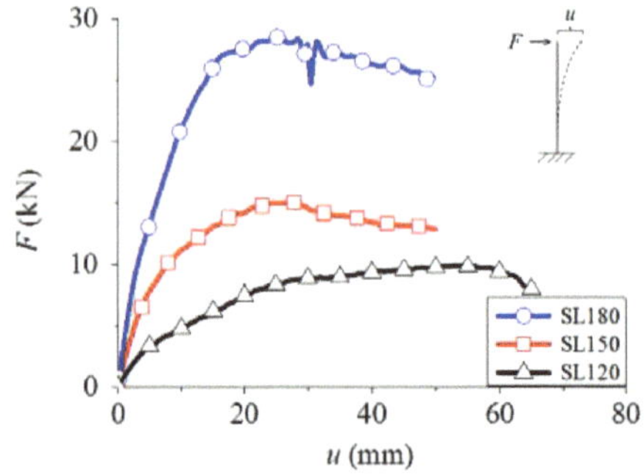

(b)

Figure 4. Static test results. (**a**) Flexural failure mode; (**b**) Load–displacement curves.

3.2. Impact Test Results

3.2.1. Crack Development

The column specimens DL150-1.5-1.2 and DL120-1.8-0.8 were severely damaged after subjected to large impact energy, with the typical flexural failure mode indicated in Figure 5a. As Figure 5a shows, the compressive concrete cover of these specimens was crushed. The rest of the column specimens did not reach the failure level in the impact tests, with the typical flexural failure mode displayed in Figure 5b. For these specimens, dense bending cracks gradually appeared on the concrete surface during the impact action, with obvious overall deformation of the columns. Until the maximum displacement was reached, the main bending cracks developed to be wide and obvious. After that, the column rebounded due to the elastic deformation energy saved during the impact process.

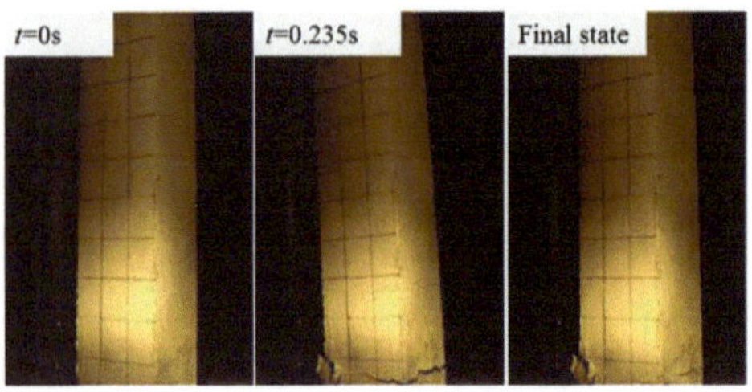

(a)

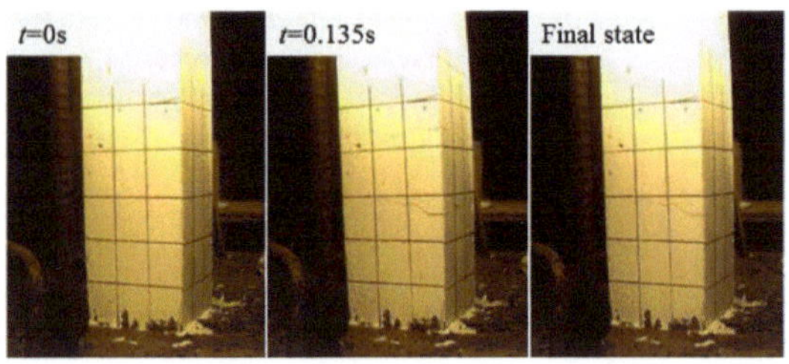

(b)

Figure 5. Process of cracking propagation. (**a**) DL120-1.8-0.8; (**b**) DL120-1.2-0.8.

3.2.2. Load–Displacement Relationship

The load–displacement (F–u) curves of all specimens under static and dynamic loads are shown in Figure 6, in which the black solid line and the red dotted line represent the impact and static loaded specimens, respectively. Note that the displacement time history data of the specimen DL180-1.2-0.8 was not successfully recorded. It can be found from Figure 6 that the existence of the inertia effect leads to a significant difference in structural response under impact and static loads. The peak impact force, the impact duration time, the maximum and residual displacement and the deformation energy at the maximum displacement of each specimen can be obtained from the load–displacement curves and are summarized in Table 2.

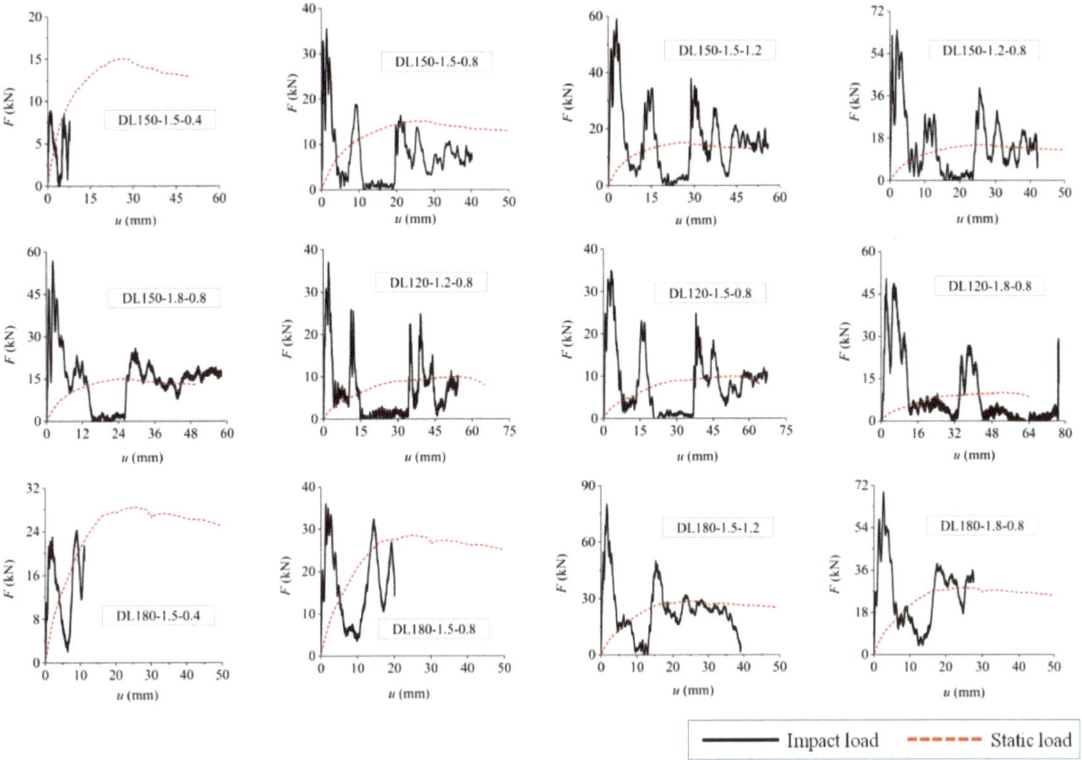

Figure 6. Load–displacement curves under static and impact loading.

According to the feature of the impact load–displacement curves, three stages of the impact action can be defined [21], namely, initial impact stage, separated stage and second peak impact stage.

With the initial impact, the RC column will process a larger velocity than that of the test truck, leading to its acceleration increasing rapidly in a short time. The impact force rose to a peak value much higher than that of the static load. In fact, the peak impact force is usually regarded as a virtual resistance [5], since it is mainly induced by the inertia effect, while the proportion that really acts on the member and results in the overall deformation is small. Therefore, the traditional ultimate capacity design method under the static condition may not be suitable for impact conditions, which will lead to a conservative design.

After the initial impact, the test truck was temporarily separated from the RC column, and the load–displacement curves showed an "unloading valley". The separated distances for the specimen series DL150 are about 10 mm. For the specimen series DL120 with large slenderness ratios and small bending stiffnesses, the separated distances may be relatively larger and up to about 16 mm.

Due to the existence of structural stiffness, the velocity of bending deformation of the specimen decreases gradually, and the test truck catches up with it again. Because the acceleration direction of the column turns to be opposite to that of the test truck, the direction of inertia force changes to be consistent with the impact force, so the impact force–time history curve develops in a fluctuated state that is slightly lower than the static load.

For the specimens (e.g., specimens DL150-1.5-0.4 and DL180-1.5-0.4) impacted with low initial energy, the longitudinal bars did not reach their yield strains, and the load–displacement curves fluctuate along the static load curve. From the residual deflection data listed in Table 2, it can be seen that most of the slightly damaged specimens were able to bounce back to the initial position eventually. Then, the free vibration with damping was performed, and the bending cracks were closed when the impact tests ended. However, for specimens DL150-1.5-1.2 and DL120-1.8-0.8 with larger impact energy, their bearing capacities have a significant decrease after the "unloading valley"; it can be concluded that these specimens have completely lost their impact-resistance capacity after the second peak impact force.

3.3. Residual Strength Test Results

After the horizontal impact tests, six RC columns with slight damage were selected for the subsequent residual bearing capacity tests, as listed in Table 4. The letters "DL" on the labels of these specimens were replaced by the letters "SDL". It can be seen from Table 2 that the residual deformations of these six specimens are small, which are in the range of 0.17~7.51 mm. Prior to the residual static tests, the residual deflection of each specimen after the impact test is reset to zero. Figure 7 shows the load–displacement curves of the specimens, while the maximum strengths and the corresponding displacement are listed in Table 4. It can be concluded that the stiffness and ultimate bearing capacity of the impact-damaged specimens decreased when compared with their undamaged counterpart, in which the ultimate bearing capacity decreased by 86% to 93%. Moreover, the initial impact energy had a slight effect on the residual bearing capacity, as the shapes of the curves were similar.

Table 4. Results of residual capacity test.

Specimen	u_u (mm)	$F_{res,ex}$ (kN)	$F_{res,eq}$ (kN)
SDL180-1.5-0.8	27.24	25.65 (0.90)	26.38 (0.93)
SDL180-1.8-0.8	30.90	25.07 (0.88)	25.10 (0.88)
SDL150-1.5-0.8	36.61	13.25 (0.88)	13.11 (0.87)
SDL150-1.8-0.8	30.91	13.94 (0.93)	11.78 (0.78)
SDL120-1.2-0.8	50.82	8.92 (0.96)	8.93 (0.90)
SDL120-1.5-0.8	49.94	8.52 (0.86)	8.73 (0.88)

Note: SDL represent static residual loaded specimen after dynamic test; $F_{res,ex}$ and $F_{res,ep}$ are the measured residual bearing capacity and the result calculated by Formula (5), respectively. The value in parentheses is the reduction factor of residual bearing capacity.

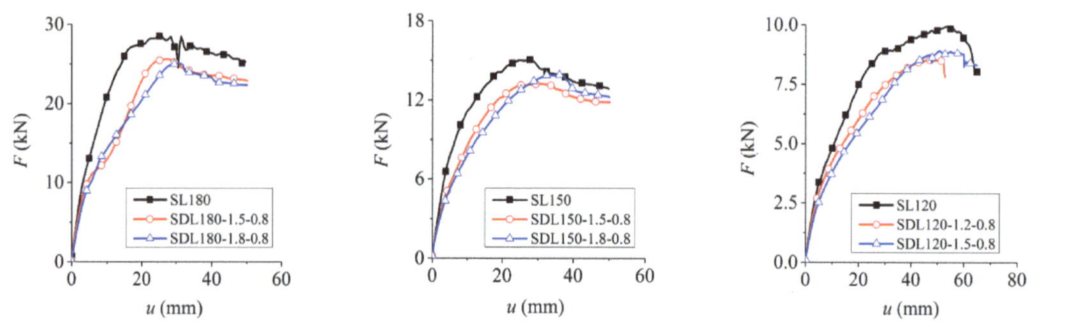

Figure 7. Load–displacement curves of residual capacity tests.

4. Damage Assessment and Residual Bearing Capacity Evaluation

4.1. Modified Park–Ang Damage Model

In order to evaluate the damage degree of RC members under seismic loading, Park and Ang (1987) [22] proposed a damage calculation model based on the structural deformation and energy absorption, as given in Equation (1).

$$D = \frac{u_m}{u_f} + \beta \frac{\int dE}{F_y u_f} \tag{1}$$

in which u_m is the maximum deformation under seismic loading, u_f is the failure deformation when the post-peak resistance of the member has decreased to 85% of its maximum strength, as indicated in Figure 8, $\int dE$ is the cumulative hysteretic energy dissipation, F_y is the yield strength and β is the energy dissipation factor and can be calculated by:

$$\beta = (-0.447 + 0.073\lambda + 0.4n_0 + 0.314\rho_t) \times 0.7^{100\rho_w} \tag{2}$$

in which λ is the shear-span ratio and is equal to 1.7 when $\lambda < 1.7$, n_0 is the axial compression ratio and is equal to 0.2 when $n_0 < 0.2$, ρ_t is the longitudinal reinforcement ratio and is equal to 0.75% when $\rho_t < 0.75\%$, and ρ_w is the transverse reinforcement ratio. In general, β is in the range between 0 and 0.85, and can be taken as 0.05 for the ductile components.

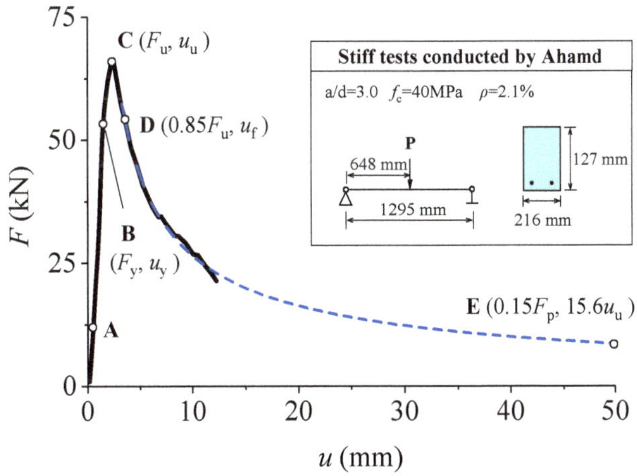

Figure 8. Typical load–deflection curve of RC members [22].

However, when predicted by the Park and Ang model [22], the damage coefficients for RC members at the condition of initial undamaged and complete failure under monotonic loading are not equal to 0 and 1, respectively. Thus, a modified formula was proposed by Chen et al. (2010) [23] based on the Park and Ang model [22], as given in Equation (3), which limited the damage coefficient to the range from 0 to 1.

$$D = (1-\beta)\frac{u_m}{u_f} + \beta \frac{\int dE}{F_y(u_f - u_y)} \tag{3}$$

in which u_y is the deformation at the yield strength. The damage coefficient is equal to 1 when the post-peak resistance of the member has decreased to 85% of its maximum strength, representing the complete failure of RC elements.

In this study, in order to evaluate the decreasing degree of ultimate bearing capacity of impact-damaged members, and satisfy the damage coefficient being equal to 1 when the member is severely damaged (i.e., its bearing capacity is equal to 0), an attenuation coefficient α_0 is introduced in Equation (3), which can be used to describe the post-peak branch of the load–displacement curve. The formula for calculating the reduction coefficient of residual bearing capacity is defined in Equation (4),

$$D_{\text{res}} = (1-\beta)\cdot\alpha_0\frac{u_{\max}}{u_{\text{u}}} + \beta\frac{E_{\text{ab,d}}}{F_{\text{y}}(u_{\text{u}} - u_{\text{y}})} \tag{4}$$

in which $\alpha_0 = 1/\mu_0$, and μ_0 is the ductility coefficient of the post-peak branch of the load–displacement curve, u_{max} is the maximum displacement during the impact process, u_u is the displacement corresponding to the static ultimate capacity, $E_{ab,d}$ is the cumulative energy dissipation, namely, the deformation energy in the impact test, and can be determined by $E_{ab,d} = 0.6E_k$ [24], in which E_k is the initial impact kinetic energy and can be calculated by $0.5mv_0^2$.

In order to obtain the whole post-peak branch of load–deflection curves, Ahamd et al. [25] designed a rigid test device for three-point bending tests of simply supported RC beams with their shear-span ratios being 2.0 and 3.0. Through these tests, the load–deflection curves with the post-peak loads decreased to 30% of the corresponding peak loads were obtained, as shown in Figure 8. On this basis, the post-peak branch of the curve can be regressed by a power function, as depicted in Figure 8 by the blue dotted line. However, the fitting curve fails to capture the bearing capacity of the member equal to 0. Therefore, the point E with its post-peak strength equal to 15% of the maximum strength is intercepted as the zero point of the bearing capacity, with the corresponding damage coefficient D_{res} taken as 1; this leads to its corresponding displacement equal to about $15.6u_u$. The ductility coefficient μ_0 of the post-peak branch can be defined as the ratio of the displacement at point E and the displacement at the maximum strength (u_E/u_u). The ductility coefficient μ_0 thus is equal to 15.6 and α_0 = 0.064.

From the above discussion, it can be found that, for general ductile members, the value of α_0 is equal to 0.064. For the members with complete brittleness, α_0 is equal to 0, while it is 0.2 for those with significant flexural failure after impact. To sum up, the residual bearing capacity of RC members after being subjected to impact can be determined by Equation (5), in which F_{res} is the residual strength and F_u is the static strength.

$$F_{\text{res}} = (1 - D_{\text{res}}) \times F_{\text{u}} \tag{5}$$

4.2. Calculation Method of Residual Bearing Capacity

The residual bearing capacity of each of the impact-damaged specimens was determined by Equations (4) and (5) and listed in Table 4. It was found that the predicted results agreed well with the experimental ones. With the objective to verify the reliability of the calculation method of residual bearing capacity after impact, the residual bearing capacity results of 29 RC beams after drop-hammer impact in References [10,23] were also collected. The details of the specimens in References [10,26] are listed in Table 5, in which the label of each specimen starts with the letters "SR" (indicating singly reinforced) or "DR" (indicating doubly reinforced), followed by the shear-span ratio, longitudinal reinforcement ratio and stirrup ratio. These specimens were subjected to an identical impact weight of 300 kg. From the load–displacement curve given in Reference [10], Specimen SR3.8-0.8 failed in complete brittle shear-type under static load, and α_0 is equal to 0. For the specimens subjected to impact loads with impact heights of 1.2 m and 1.6 m, wide oblique cracks and concrete crushing were observed after impact, thus α_0 is taken as 0.2. The residual strengths of the beam specimens are determined by Equations (4) and (5), as listed in Table 5. It was found that the predictions are in good agreement with the measured residual bearing capacity of flexural failure-oriented members after impact. The mean ratio of the measured to calculated values is 1.01, with the standard deviation being 0.11. It can be concluded that the proposed damage model can reflect the damage level of RC members after being subjected to impact and predicting the residual strength of impact-damaged RC members.

Table 5. Results of residual capacity tests in literature [10,26].

Specimen	$b \times h$ (mm)	F_u (kN)	F_y (kN)	u_u (mm)	h_0 (m)	u_{max} (mm)	$E_{ab,d}$ (J)	$F_{res,ex}$ (kN)	$D_{res,eq}$	$F_{res,eq}$ (kN)	$F_{res,ex}/F_{res,p}$	Post-Impact Failure Mode	
					0.15	3.8	264.6	116.74	0.13	104.44	1.13		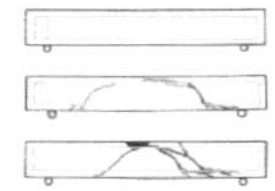
DR3.3-2.4	150 × 250	120.35	108.8	2.76	0.3	6.9	529.2	78.23	0.25	90.49	0.88	Shear-compression	
					0.6	12.7	1058.4	63.79	0.47	63.70	1.04		
					0.3	6.4	529.2	146.46	0.08	125.41	1.19		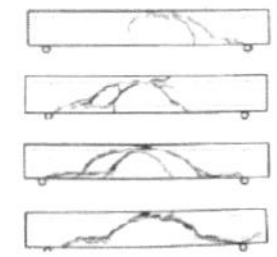
DR3.3-2.4-0.12	150 × 250	136.88	130.49	7.31	0.6	11.5	1058.4	102.66	0.16	115.50	0.91	Shear-compression	
					0.9	18.2	1587.6	106.77	0.24	103.68	1.08		
					1.2	21.6	2116.8	102.66	0.30	95.81	1.14		
					0.6	9.0	1058.4	231.54	0.04	233.45	0.97		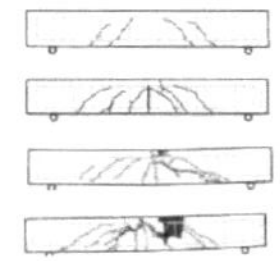
DR3.3-2.4-0.56	150 × 250	243.73	233.26	19.04	0.9	13.4	1587.6	212.04	0.06	228.40	0.90	Flexure	
					1.2	15.7	2116.8	194.98	0.19	197.73	0.90		
					1.6	19.2	2822.4	182.80	0.23	186.83	0.88		
					0.3	10.8	529.2	81.34	0.06	68.61	1.19		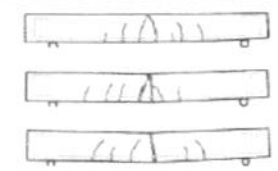
SR3.8-0.8	160 × 240	73.28	73.28	5.67	0.6	20.2	1058.4	76.21	0.13	63.95	1.19	Shear	
					0.9	29.5	1587.6	58.63	0.19	59.28	0.99		
					0.6	18.6	1058.4	86.67	0.07	74.54	1.19		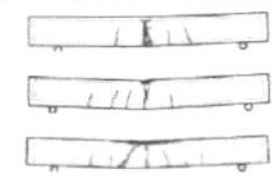
DR3.8-0.8-0.11	160 × 240	80.25	67.35	27.78	0.9	34.5	1587.6	72.22	0.12	70.47	1.07	Flexure	
					1.2	41.0	2116.8	41.73	0.35	52.02	1.01		
					0.6	19.6	1058.4	84.51	0.07	85.00	1.02		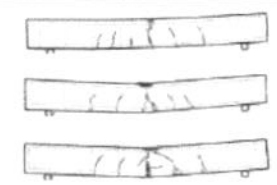
DR3.8-0.8-0.15	160 × 240	91.86	68.75	27.09	0.9	28.8	1587.6	81.76	0.11	81.69	1.05	Flexure	
					1.2	39.2	2116.8	74.41	0.15	78.13	1.01		

Table 5. *Cont.*

Specimen	$b \times h$ (mm)	F_u (kN)	F_y (kN)	u_u (mm)	h_0 (m)	u_{max} (mm)	$E_{ab,d}$ (J)	$F_{res,ex}$ (kN)	$D_{res,eq}$	$F_{res,eq}$ (kN)	$F_{res,ex}/F_{res,p}$	Post-Impact Failure Mode	
SR5.7-1.6	120 × 170	52.11	47.51	22.57	0.3	20.0	529.2	47.32	0.08	47.87	0.96	Flexure	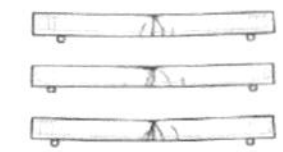
					0.45	28.8	793.8	35.36	0.29	36.88	0.86		
					0.6	37.1	1058.4	24.96	0.38	32.41	0.66		
DR5.7-1.6-0.15	120 × 170	49.43	45.98	12.09	0.3	20.0	529.2	43.50	0.15	41.84	0.96	Flexure	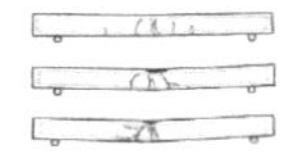
					0.45	30.0	793.8	39.54	0.23	38.05	0.92		
					0.6	39.1	1058.4	30.65	0.30	34.49	1.08		
DR5.7-1.6-0.2	120 × 170	48.93	41.78	19.41	0.3	19.1	529.2	46.46	0.10	44.25	1.00	Flexure	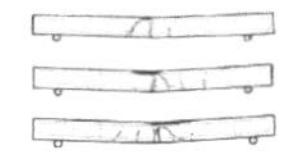
					0.45	28.8	793.8	41.57	0.14	41.89	0.93		
					0.6	37.9	1058.4	34.23	0.46	26.63	0.96		
Mean											1.01		
COV											0.11		

Note: b and h are the width and height of RC beam section, respectively; h_0 is impact height of drop hammer.

5. Conclusions

Through the static and horizontal impact tests of RC columns and the subsequent static residual bearing capacity tests, the following conclusions can be drawn:

(1) With the increase of impact energy, the RC columns finally failed in flexure-dominant type, characterized by the yielding of tensile longitudinal rebar and the crushing of concrete cover.

(2) The inertia effect had a significant influence on the impact behaviour of RC members. When subjected to impact loads with relatively low energy, the impact response of the member fluctuates along the static load–displacement curve. However, the bearing capacity shows an obvious unloading trend when subjected to a relatively high energy impact load.

(3) The Park–Ang damage evolution model for RC members under seismic loading is modified, to propose a method for evaluating the residual bearing capacity of impact-damaged members. Through the comparisons of the residual bearing capacity of six RC columns in the current study and twenty-nine RC beams subjected to drop-hammer impact in the existing literature, the accuracy of the proposed evaluation method is verified. Therefore, the proposed method can be applied to determine the damage coefficient after impact and evaluate the residual strength of RC elements in practical engineering.

Author Contributions: Conceptualization, J.Y., J.C. and A.H.; methodology, Q.C. and A.H.; software, Y.W.; validation, Y.W., Q.C. and A.H.; formal analysis, Y.W.; investigation, Y.W. and A.H.; resources, J.C.; data curation, Q.C.; writing—original draft preparation, J.Y.; writing—review and editing, A.H.; visualization, J.Y.; supervision, J.C.; project administration, A.H.; funding acquisition, J.Y. All authors have read and agreed to the published version of the manuscript.

Funding: Funding for this research project was provided by the National Natural Science Foundation of China (52108143), GuangDong Basic and Applied Basic Research Foundation (2021A1515110109), Science and Technology Planning Project of Guangzhou Municipal Construction ([2021]-KJ004; [2021]-KJ021; [2021]-KJ036; [2021]-KJ037). The support for the research is acknowledged with thanks.

Data Availability Statement: Not applicable.

Conflicts of Interest: The authors declare no conflict of interest.

References

1. Fan, W.; Sun, Y.; Yang, C.; Sun, W.; He, Y. Assessing the response and fragility of concrete bridges under multi-hazard effect of vessel impact and corrosion. *Eng. Struct.* **2020**, *225*, 111279. [CrossRef]
2. *AASHTO-LRF*; LRFD Bridge Design Specifications. American Association of State Highway and Transportation Official: Washington, DC, USA, 2007.
3. *JTG D60-2015*; General Specification for Design of Highway Bridges and Culverts. China Communication Press: Beijing, China, 2015. (In Chinese)
4. Saatci, S.; Vecchio, F.J. Effects of shear mechanisms on impact behavior of reinforced concrete beams. *ACI Struct. J.* **2009**, *106*, 78–86.
5. Fu, Y.; Yu, X.; Dong, X.; Zhou, F.; Ning, J.; Li, P.; Zheng, Y. Investigating the failure behaviors of RC beams without stirrups under impact loading. *Int. J. Impact Eng.* **2020**, *137*, 103432. [CrossRef]
6. Kani, G.N.J. Basic facts concerning shear failure. *ACI Struct. J.* **1966**, *63*, 675–692.
7. *JTS 144-1-2010*; Load Code for Harbour Engineering. China Communication Press: Beijing, China, 2010. (In Chinese)
8. Dok, G.; Caglar, N.; Ilki, A.; Yilmaz, C. Effect of impact loading on residual flexural capacity of high-strength reinforced concrete beams. *Structures* **2020**, *27*, 2466–2480. [CrossRef]
9. Esfahani, M.; Hoseinzade, M.; Shakiba, M.; Arbab, F.; Yekrangnia, M.; Pachideh, G. Experimental investigation of residual flexural capacity of damaged reinforced concrete beams exposed to elevated temperatures. *Eng. Struct.* **2021**, *240*, 112388. [CrossRef]
10. Adhikary, S.D.; Li, B.; Fujikake, K. Residual resistance of impact damaged reinforced concrete beams. *Mag. Concr. Res.* **2014**, *67*, 364–378. [CrossRef]
11. Fujikake, K. Impact Performance of ultra-high performance fiber reinforced concrete beam and its analytical evaluation. *Int. J. Prot. Struct.* **2014**, *5*, 167–186. [CrossRef]
12. Peng, T.; Chen, L.; Liu, T.; Demartino, C.; Xu, J. Static shear performance and residual axial capacity of rectangular RC bridge piers under near-ground lateral loads. *J. Struct. Eng.* **2021**, *147*, 04021219. [CrossRef]

13. Tian, L.; Zhu, C. Damage evaluation and protection technique of RC columns under impulsive load. *Eng. Mech.* **2013**, *30*, 144–150. [CrossRef]
14. Pei, C. Experimental Studies and Simulation Analysis on Dynamic Response and Residual Strength of H-Shaped Steel Member under Lateral Impact. Master's Thesis, Taiyuan University of Technology, Taiyuan, China, 2013. (In Chinese)
15. Tran, T.T.; Pham, T.M.; Huang, Z.; Chen, W.; Hao, H.; Elchalakani, M. Impact response of fibre reinforced geopolymer concrete beams with BFRP bars and stirrups. *Eng. Struct.* **2021**, *231*, 111785. [CrossRef]
16. Hao, H.; Tran, T.T.; Li, H.; Pham, T.M.; Chen, W. On the Accuracy, Reliability and Controllability of Impact Tests of RC Beams. *Int. J. Impact Eng.* **2021**, *157*, 103979. [CrossRef]
17. Tran, T.T.; Pham, T.M.; Huang, Z.; Chen, W.; Ngo, T.T.; Hao, H.; Elchalakani, M. Effect of fibre reinforcements on shear capacity of geopolymer concrete beams subjected to impact load. *Int. J. Impact Eng.* **2022**, *159*, 104056. [CrossRef]
18. Demartino, C.; Wu, J.G.; Xiao, Y. Response of shear-deficient reinforced circular RC columns under lateral impact loading. *Int. J. Impact Eng.* **2017**, *109*, 196–213. [CrossRef]
19. Xiao, Y. Development of structural testing equipment for impact and complex loading. *J. Struct. Integr. Maint.* **2021**, *6*, 1–15. [CrossRef]
20. Cai, J.; Ye, J.-B.; Chen, Q.-J.; Liu, X.; Wang, Y. Dynamic behaviour of axially-loaded RC columns under horizontal impact loading. *Eng. Struct.* **2018**, *168*, 684–697. [CrossRef]
21. Ye, J.-B.; Cai, J.; Chen, Q.-J.; Liu, X.; Tang, X.-L.; Zuo, Z.-L. Experimental investigation of slender RC columns under horizontal static and impact loads. *Structures* **2020**, *24*, 499–513. [CrossRef]
22. Park, Y.J.; Ang, A.H.S.; Wen, Y.K. Seismic Damage Analysis of Reinforced Concrete Buildings. *J. Struct. Eng.* **1987**, *111*, 740–757. [CrossRef]
23. Chen, L.; Jiang, H.; Lv, X. Modified Park-Ang damage model for reinforced concrete structures. *J. Tongji Univ. (Nat. Sci.)* **2010**, *38*, 1103–1107. (In Chinese)
24. Kishi, N.; Mikami, H.; Matsuoka, K.; Ando, T. Impact behavior of shear-failure-type RC beams without shear rebar. *Int. J. Impact Eng.* **2002**, *27*, 955–968. [CrossRef]
25. Ahmad, S.H.; Hino, S.; Chung, W.; Xie, Y. An experimental technique for obtaining controlled diagonal tension failure of shear critical reinforced concrete beams. *Mater. Struct.* **1995**, *28*, 8–15. [CrossRef]
26. Adhikary, S.D.; Li, B.; Fujikake, K. Low velocity impact response of reinforced concrete beams: Experimental and numerical investigation. *Int. J. Prot. Struct.* **2015**, *6*, 81–111. [CrossRef]

buildings

Article

Experimental Application of Cement-Stabilized Pavement Base with Low-Grade Metamorphic Rock Aggregates

Qian Yang [1], Yi Liu [2,*], Haotian Zou [2], Xiaoxiong Wang [2], Guohuan Gong [2], Yinnan Cheng [2], Liang Zhang [3] and Zhengwu Jiang [1]

1 Key Laboratory of Advanced Civil Engineering Materials of Ministry of Education, School of Materials Science and Engineering, Tongji University, Shanghai 201804, China; yqtumutj@163.com (Q.Y.); jzhw@tongji.edu.cn (Z.J.)

2 Guizhou Hongxin Chuangda Engineering Detection & Consultation Co., Ltd., Guiyang 550014, China; an_kong22@163.com (H.Z.); sxdthywxx@126.com (X.W.); guohuan_gong@163.com (G.G.); yn_cheng97@163.com (Y.C.)

3 Guizhou Transportation Planning Survey & Design Academe Co., Ltd., Guiyang 550081, China; zhang123liang2022@163.com

* Correspondence: yliuzm@163.com

Abstract: Low-grade metamorphic rock (LMR) is a kind of stone that is widely distributed in China. The alkali activity strictly prevents its application in conventional concrete. This paper evaluates the possibility of using LMR aggregate in cement-stabilized pavement base (CSPB). The compressive strength of CSPB prepared with LMR and limestone aggregates at various curing conditions was measured. Expansion rates were determined via accelerated simulation tests to assess the alkali reactivity of LMR, followed by microscopic analysis. Finally, the possibility of using LMR in CSPB was evaluated from the economic viewpoint. Results indicate that CSPB specimens prepared with LMR have similar compressive strength at each content of cement, regardless of curing conditions. The expansion rates of all CSPB specimens with LMR were lower than 0.1%, indicating the absence of an AAR, which was further validated by the absence of the AAR product in microscopic observations. It is inferred from the economic analysis that 70.9% lower cost can be achieved by the replacement of limestone aggregate with LMR aggregate. This demonstrates that technical, economic and environmental benefits endow LMR with wide market potential as the aggregate of CSPB.

Keywords: alkali–aggregate reaction; low-grade metamorphic rock; cement-stabilized pavement base; strength; microstructure

Citation: Yang, Q.; Liu, Y.; Zou, H.; Wang, X.; Gong, G.; Cheng, Y.; Zhang, L.; Jiang, Z. Experimental Application of Cement-Stabilized Pavement Base with Low-Grade Metamorphic Rock Aggregates. *Buildings* **2022**, *12*, 589. https://doi.org/10.3390/buildings12050589

Academic Editor: Jian-Guo Dai

Received: 10 March 2022
Accepted: 27 April 2022
Published: 2 May 2022

1. Introduction

Alkali–aggregate reaction (AAR) is an expansive reaction between alkali-active mineral components and the soluble alkaline solution from cement and admixtures in concrete pores [1–4]. Since AAR products have a strong water swelling capability, the expansion stress generated in the swelling process may exceed concrete strength, which, therefore, can further cause cracking damage to the concrete structures [5]. AAR is a global problem, and there are varying degrees of damage caused by AAR in all countries in the world. In 1940, the United States first discovered an AAR damage to a road in Bradley, California [6]. In the 1850s, Denmark surveyed 431 concrete buildings across the country, of which, 34% had suffered varying degrees of AAR damage, and 15% were completely destroyed. Since then, the United Kingdom, France, Australia, Spain, Switzerland, and Canada have all discovered cases of AAR causing concrete structure damage [7]. In 1990, there were many reports of AAR damage in China. AAR accidents occurred in Beijing, Tianjin, Shandong, Shaanxi, Inner Mongolia, Henan and other places. The damaged projects involved overpasses, airports, concrete bridges, railway sleepers, and civil buildings. Relevant investigations and studies have shown that the alkali content of cement is generally high in China, especially in the northern regions, and alkali-reactive aggregates are widely distributed [8–13].

Low-grade metamorphic rock (LMR) is a resource of manufactured sand, widely distributed in China with a large reserve. Using these LMRs as aggregates for concretes can reduce costs and alleviate the crisis of river sand resource shortages. However, the potential for alkali reactivity has been found in LMR aggregates [1,14–17]. If LMR is directly used without taking any action, the reactivity will cause concrete to expand and crack, posing a serious threat to the safe operation of the infrastructure. Consequently, it is very important and crucial to prevent and detect the alkali–aggregate reactivity of LMR aggregates [18,19]. Currently, there are severe application problems for the use of LMRs and a need to quickly evaluate their alkali reactivity and effectively prevent and control these harmful reactions when they are used as concrete aggregates.

Three well-known basic conditions are necessary to trigger the alkali–aggregate reaction in concrete, including a humid environment, and the presence of active aggregates and free alkali [20]. This means AAR can be avoided if the free alkali content is controlled. From this viewpoint, LRM has the potential to be used in cement-stabilized pavement base (CSPB), given that the free alkali is mainly from cement, and the amount of cement in CSPB is quite low.

Therefore, this paper evaluates the possibility of LMR aggregates for the preparation of CSPB. The strength, expansion behavior and micromorphology of cement-stabilized pavement base with low-grade metamorphic rock (LMR-CSPB) are tested. The economic benefit of using LMR in CSPB is also analyzed.

2. Materials and Methods

2.1. Materials

The 42.5 grade Ordinary Portland cement (P.O 42.5 Retarded cement) from Southwest Cement Co., Ltd. (Kaili, China), was used in this investigation. The properties of the cement are listed in Table 1, and its chemical composition is shown in Table 2.

Table 1. Physical properties of cement.

Normal Consistency (%)	Alkali Content (%)	Specific Surface Area (m^2/kg)	Setting Time (min)		Compressive/Flexural Strength (MPa)	
			Initial	Final	3d	28d
27.1	0.56	330	206	421	19.1/4.3	46.6/10.8

Table 2. The chemical composition of cement (wt%).

SiO_2	Al_2O_3	Fe_2O_3	CaO	MgO	SO_3	Loss
23.10	6.21	4.05	59.60	3.18	1.82	2.03

LMR aggregates were provided by Shi Jiangchong quarry (Kaili, China) in the southeast area of Guizhou Province, China. Crushed low-grade metamorphic rock with four particle size grades, i.e., 20–30 mm, 10–20 mm, 5–10 mm and 0–5 mm are used, the particle size distribution is listed in Table 3. The alkali activity of the LMR in the Shi Jiang Chong quarry was judged by the rapid mortar rod method, and a total of 137 groups of LMR rocks were sampled and tested for the alkali activity during different time periods between May 2017 and July 2019, which can indicate the fluctuation of the alkali activity of the LMR rock. The monitoring results of the alkali activity of the LMR in the Shi Jiang Chong quarry are illustrated in Figure 1. It can be seen that the 14 days-expansion rate of LMR is basically between 0.1% and 0.2%, and the expansion rate of almost all specimens is less than 0.3%, which meets the requirement in TB/T 3275 [21].

Table 3. The size distribution of LMR aggregates.

Type of Aggregate	Particle Size /mm	Sieve Size (mm)/Accumulated Screening Rate (%)												
		31.5	26.5	19	16	13.2	9.5	4.75	2.36	1.18	0.6	0.3	0.15	0.075
LMR	20~30	2.5	60.8	97.4	99.5	100								
	10~20		1.6	9.4	56.6	79.2	98.5	99.9	100					
	5~10				0.8	5.3	51.6	99.1	99.8	100				
	0~5							1.6	33.5	47.1	70.5	81.8	87.6	92.5

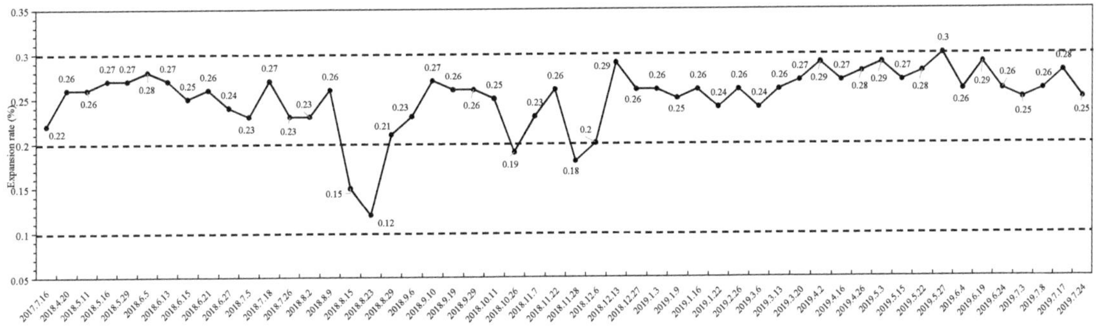

Figure 1. The monitoring results of the alkali activity of LMR.

Crushed limestone rock is used for comparison in this investigation. The gradation of limestone coarse aggregate and fine aggregate was adjusted to be the same as that of LMR rock aggregates to avoid the influence of different gradations on the test results when designing the experiment. The physical properties of LMR aggregates and limestone aggregates are shown in Tables 4 and 5. The basic performance indexes of coarse aggregate and fine aggregate of shallow metamorphic rock meet the requirements in the JTG/T F20 [22], and these can therefore be used as aggregate to prepare CSPB material.

Table 4. The physical properties of LMR coarse aggregate.

Type of Aggregate	Particle Size /mm	Apparent Density (kg/m^3)	Crushing Value (%)	Elongated or Flat Particle (%)	Water Absorption (%)	Soft Stone Content (%)	Less than 0.075 mm Particles (%)
LMR	20~30	2.712	15.3	16.4	0.2	1.8	0.6
	10~20	2.741		17.4	0.3	2.3	0.9
	5~10	2.687		19.8	0.6	2.8	1.2
Limestone	20~30	2.763	12.2	11.5	0.2	1.2	0.4
	10~20	2.801		11.8	0.2	1.6	0.7
	5~10	2.722		18.8	0.4	1.8	0.9
Requirements in JTG/T F20-2015		/	≤26	≤20	/	≤3	≤2

Table 5. The physical properties of LMR fine aggregate.

Type	Requirements in JTG/T F20-2015	LMR	Limestone
MB value (g/kg)	/	0.75	0.75
Powder content (%)	≤15	14.2	9.8
Angularity test (s)	≥30	35.9	38.3
Apparent density (kg/m^3)	/	2.704	2.731
Water absorption (%)	/	2.1	1.5
Less than 0.075 mm particles (%)	≤15	8.6	8.2
Organic matter content	/	Qualified	Qualified
Trioxide content (%)	≤0.25	0.079	0.061

2.2. Mixing Proportion Trial of the LMR-CSPB

To reduce the test deviation, the proportion of each raw material was determined based on its screening test. Then the standard sieve (pore size: 19 mm, 9.5 mm, 4.75 mm, 2.36 mm) was used to weigh each raw material in proportion to form a mixed aggregate of LMR-CSPB.

The aggregate combination of LMR-CSPB material is listed in Table 6. A unified grading was used in the mixing proportion trial when the aggregate was selected to avoid the difference in gradation from affecting the test results and to ensure the validity of the test data. The requirements of aggregate gradation for LMR-CSPB materials are shown in Table 7.

Table 6. Aggregate combination mode in LMR-CSPB.

Aggregate Combination Mode	Coarse Aggregate	Fine Aggregate
A	Low-grade metamorphic rock	Low-grade metamorphic rock
B	Limestone	Limestone
C	Low-grade metamorphic rock	Limestone

Note: The gradations of limestone coarse aggregate and fine aggregate were adjusted to be the same as those of LMR rock aggregates to avoid the influence of gradations.

Table 7. Uniform grading requirements for LMR-CSPB.

Key Sieve (mm)	19	9.5	4.75	2.36	0.075
Passing ratio (%)	80	50	30	20	0~6

The content of cement in CSPB varies from 3% to 6% in mass. Detailed mix proportions are shown in Table 8.

Table 8. Mix proportions of CSPB.

Test Plan	Cement Dosage							
	3%		4%		5%		6%	
	M Value	O Value	M Value	O Value	M Value	O Value	M Value	O Value
A	2.34	5.2	2.34	5.2	2.33	5.3	2.34	5.3
B	2.26	5.0	2.27	5.0	2.28	5.1	2.29	5.1
C	2.32	5.1	2.32	5.1	2.33	5.2	2.34	5.2

Note: In the above table, the "M value" refers to maximum dry density, and its unit is g/cm^3; the "O value" refers to optimal water content, and its unit is %.

2.3. Methods

2.3.1. Compressive Strength

Strength is one of the most important performance indicators for the pavement performance of CSPB materials. To verify the strength properties of Base metamorphic rocks as road pavement material, LCR-CSPB specimens with the dimension of φ150 mm × 150 mm were prepared and cured under standard and dry–wet cycle curing conditions, respectively. The strength was measured at 7d, 14d, and 28d, respectively, according to JTG E51 [23].

2.3.2. Accelerated Simulation Experiment

The two ends of the CSPB specimen were cut flat and then dried in an oven. After the epoxy resin was completely cured, the initial test length of the CSPB specimen was measured and placed in a constant temperature water bath at 80 °C for curing. After 24 h, the specimen was taken out, and its reference length was quickly measured. Then specimens were cured in an 80 °C, 1 mol/L NaOH solution. Finally, the expansion rate of the CSPB specimen was tested at 3d, 7d, 10d, 14d, and 28d, respectively.

2.3.3. Microscopic Observation

A field emission scanning electron microscope (FESEM, Zeiss Ultra Plus) was employed to observe the microstructure of hydration products and cracks in the interface transition zone (ITZ) of concretes.

3. Results and Discussion

3.1. Trial Calculation of Total Alkali Content in CSPB

There is a specialized calculation method for the alkali content in CSPB using the relevant standards in China and abroad. The calculation method of the total alkali content of concrete is the sum of the alkali content of each constituent material (cement, admixtures, mineral admixtures, etc.) in the concrete. Therefore, the alkali content of CSPB is calculated by referring to Formula (1), which is consistent with the formula of concrete. The results for the total alkali content of different structures are shown in Table 9.

$$m_a = (m_c + 10) \times (c_c + 0.1\%) + m_{ca} \times c_{ca} + m_{ma} \times c_{ma} \quad (1)$$

where m_a is the total alkali in concrete (kg/m^3); m_c is the amount of cement in concrete (kg/m^3); m_{ca} is the amount of admixture in concrete (kg/m^3); m_{ma} is the amount of mineral admixture in concrete (kg/m^3); C_c is the effective alkali content of cement, and it is calculated as 100% of the alkali content of the cement (%); C_{ca} is the effective alkali content of admixture, and it is calculated as 100% of the alkali content of the admixture (%); and C_{ma} is the effective alkali content of the mineral admixture (%).

Table 9. Total alkali content of various structures.

Type	C30 Concrete	Concrete Pavement	Rolling Poor Concrete Base	Cement-stabilized Pavement Base
Cement (kg/m^3)	360	400	170	80
Cement ratio (%)	16.2	23.5	7.7	3.5
Total alkali content (kg/m^3)	1.94	2.00	0.85	0.45

Note: The cement ratio refers to the ratio of cement mass to the total mass of cements and aggregates. The alkali content of the cement in cement-stabilized pavement base is calculated as 0.56%.

Using the calculation, the 20% alkali content of fly ash is considered its effective alkali content, and the effective alkali content of silica fume and granulated blast furnace slag powder can be calculated as 50% of their alkali content. However, the effective alkali content of other mineral admixtures should be determined through experiments. The value of 0.1% is the supplementary amount for the effective alkali content to compensate for the fluctuations in the alkali content of the cement.

From Table 9, the total alkali content of CSPB is lower than other groups of concrete structures. At 0.45 kg/m^3, this is much less than the level (not more than 3 kg/m^3) that induces AAR [24]. The main reason is that the cement content of CSPB is generally between 3–6% by mass, which is much less than that of other concrete structures. Therefore, its total alkali content is lower than other concrete structures, and it does not meet a necessary condition for the alkali–aggregate reaction of "sufficient soluble alkali in the concrete pore solution".

3.2. Strength of the LMR-CSPB

The strength of CSPB specimens under different curing environments and cement dosages is illustrated in Figure 2. It is clear that under different curing ages, the strength of CSPB specimens in standard maintenance is better than that in a dry–wet cycle curing environment. Under the same curing environment, strength increases with the extension of the curing age and with the increase in cement dosages. At the same time, the strength of all CSPB specimens is greater than the design value of 4.0 MPa under different curing environments, cement contents, and curing ages, which meets the design requirements in JTG/T F20.

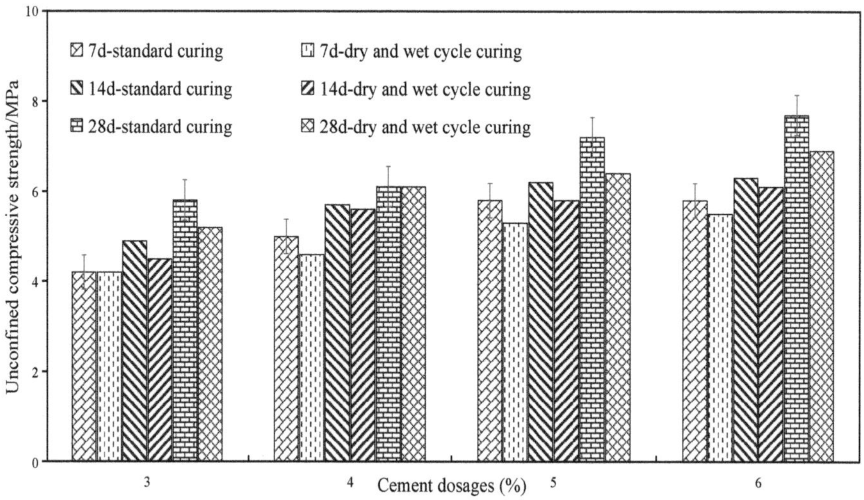

Figure 2. The strength of CSPB specimens under different curing environments and cement dosage.

To analyze the difference in strength performance of CSPB material prepared from LMR as aggregate and other commonly used rock aggregates, limestone was used as a control group for testing, and the 4% cement dosage was selected in the experiment. Figure 3 illustrates that the strength of CSPB specimens with LMR is similar to that of limestone CSPB specimens in standard maintenance and dry–wet cycle curing environments. This shows that LMR can replace general stone as aggregate to prepare CSPB material, and it meets the strength requirements. Therefore, it is feasible to use LMR as aggregate to prepare CSPB.

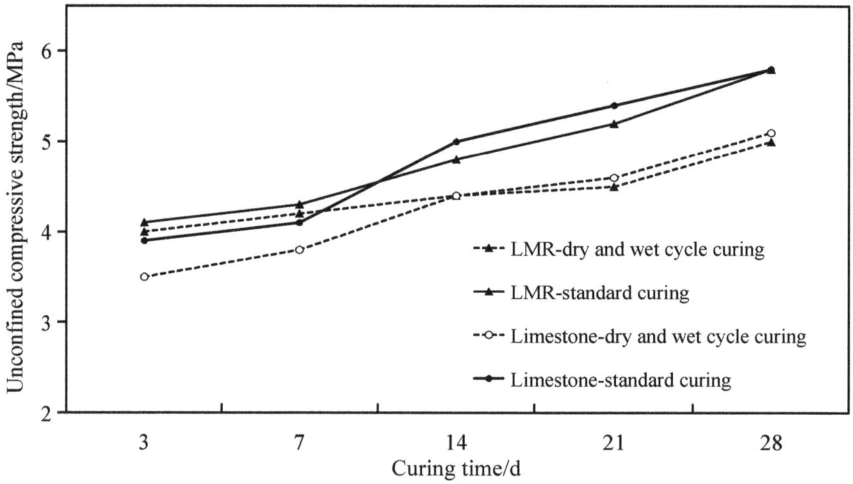

Figure 3. The strength of CSPB specimens with LMR and limestone aggregates.

3.3. Accelerated Simulation Experiment

Expansion rates of LMR-CSPB from the mortar bar method are shown in Figure 4. It is clear that the expansion of LMR-CSPB specimens increases with the curing time regardless of the cement dosage. However, the 14-d expansion rate of all specimens is less than 0.1%, which indicates that no AAR occurred in the LMR-CSPB specimens. The main reason is that the amount of cement used is very small, and its total alkali content is low. A sufficient alkaline solution environment for AAR is not provided in LMR-CSPB. Therefore, it will not cause the occurrence of AAR when the LMR is used in the CSPB within the cement dosage of 6%.

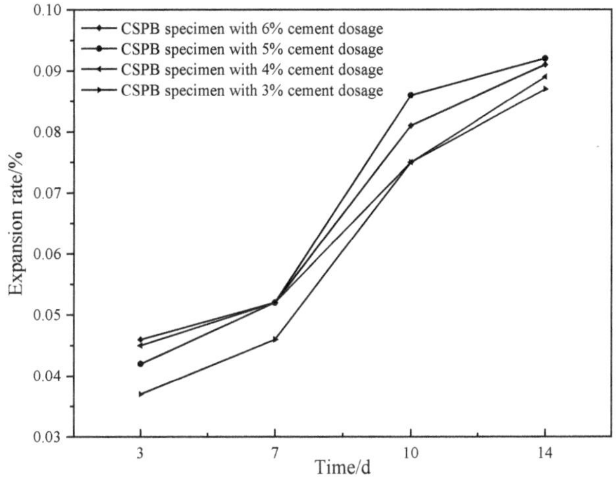

Figure 4. Accelerated simulation test results of the expansion rate of LMR-CSPB specimens.

3.4. Microscopic Observation

The AAR can be divided into alkali-silicic acid reaction (ASR) and alkali-carbonate reaction (ACR) according to the different types of alkali active minerals in the aggregate [25]. For shallow metamorphic rock aggregates, if alkali–aggregate reaction occurs, it is mostly ASR because of the high content of SiO_2. The principle of ASR is that the active SiO_2 in the aggregate reacts with the alkali in the pore solution, forming an alkali-silicic acid gel in the transition zone of the interface between the aggregate and the cement paste. The ASR product is mostly honeycomb-shaped gel with high water absorption. When the stress caused by swelling due to water absorption exceeds the strength of the concrete, it will cause the concrete to crack [26,27]. Therefore, whether there is an alkali aggregate reaction can be judged by observing the morphology around the aggregate and analyzing the product composition in the transition zone between the aggregate and the cement paste.

The morphologies of LMR-CSPB are shown in Figure 5. Figure 5a,b present the bonding state of the ITZ in the LRM-CSPB and L-CSPB specimens. It is obvious that the bonding between the aggregate and the cement paste is relatively close, and the structure is compact for both LMR-CSPB and L-CSPB. In L-CSPB specimens, an obvious crack is found at the edge of the aggregate. However, there is no crack in the interface of LRM-CSPB specimens. Magnified observation and analysis of the transition zone between the aggregate and the cement paste were carried out to clarify whether the crack is caused by AAR (see Figure 5c,d). The distribution of hydration products in the ITZ of LRM-CSPB and L-CSPB specimens is illustrated in Figure 5e,f. From Figure 5d,f, it can be seen that there is calcium hydroxide and ettringite hydrate in the crack, and there is no honeycomb alkali-silicate gel. At the same time, there are obvious rod-shaped ettringite crystals (AFt), hexagonal calcium hydroxide, and amorphous or agglomerated C-S-H gel products in the L-CSPB specimens. In addition, no AAR product can be observed.

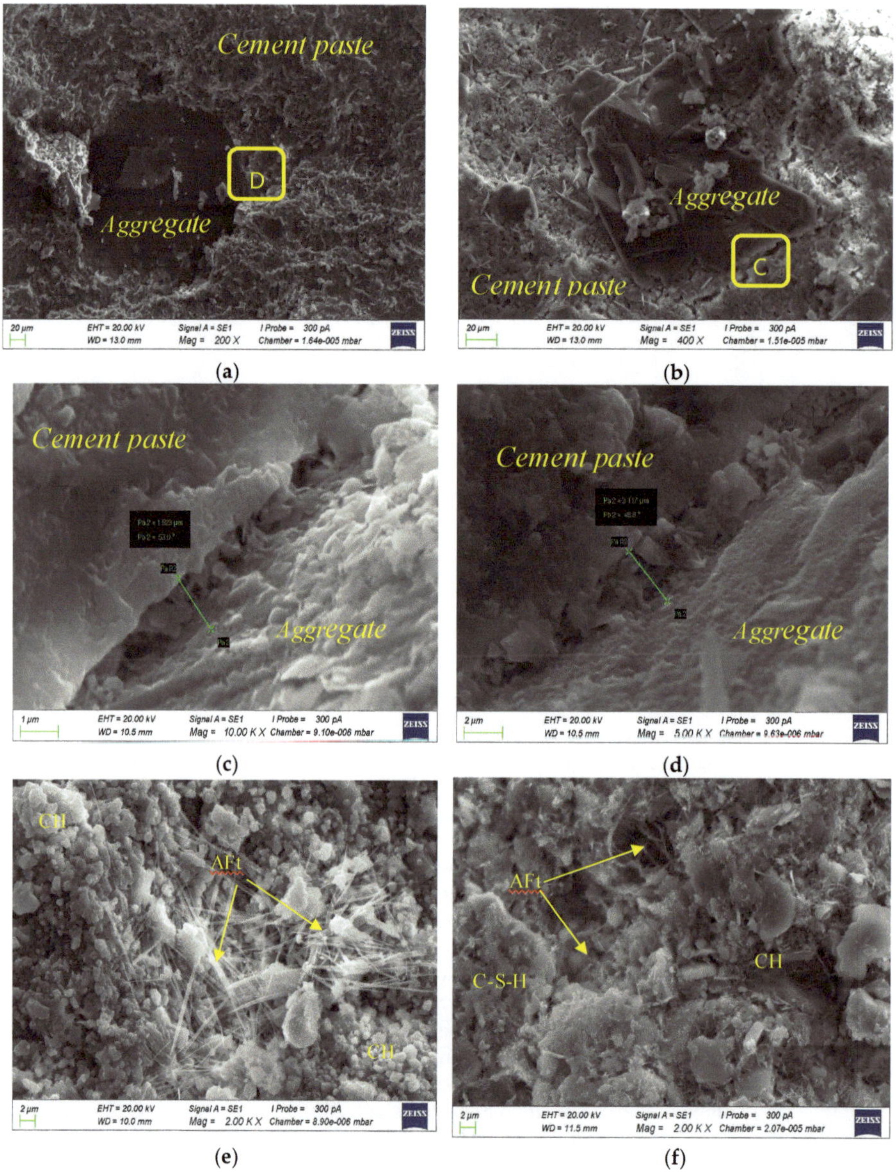

Figure 5. Sample microscopic analysis: (**a**) bonding of aggregate and cement in LMR-CSPB; (**b**) bonding of aggregate and cement in L-CSPB; (**c**) the ITZ of LMR-CSPB; (**d**) the ITZ of L-CSPB; (**e**) the hydration products in the LMR-CSPB; (**f**) the hydration products in the L-CSPB.

Similarly, the honeycomb-shaped alkali-silicate gel was not found in the ITZ of LMR-SCPB specimens, as shown in Figure 5e. There are obvious ettringite crystals, calcium hydroxide, and C-S-H gel in the LMR-CSPB specimens, and the adhesion is tight (Figure 5f). Comparison with the L-CSPB specimens shows that the LMR can be used as the raw material for the preparation of CSPB, and its strength meets the requirements of engineering design.

3.5. Economic and Environmental Analyses

Compared with the ordinary cement-stabilized pavement base, the technical and economic analysis of preparing one cubic meter of LMR-CSPB is shown in Table 10. As can be seen, the cost of one cubic meter of ordinary cement-stabilized pavement base and LMR-CSPB are about CNY 2705.18 and CNY 787.20, respectively. The LMR-CSPB can save 70.9% of the cost of an ordinary cement-stabilized pavement base.

Table 10. The technical and economic analysis of preparing LMR-CSPB.

Raw Materials	L-CSPB/m^3		LMR-CSPB/m^3		LMR-CSPB/m^3			L-CSPB/m^3		
	Dosage/t	Unit Price CNY(USD)/t	Dosage/t	Unit Price CNY(USD)/t	Dosage/t	Shipping Cost CNY(USD)/t	Unit Price CNY(USD)/t	Dosage/t	Shipping Cost CNY(USD)/t	Unit Price CNY(USD)/t
Manufactured sand	1	33(5.10)	1	40(6.18)	5.33	0	40(6.18)	5.33	103 (15.92)	136 (21.03)
Gravel	1	35(5.41)	1	40(6.18)	14.35	0	40(6.18)	14.35	103 (15.92)	138 (21.33)
Total cost CNY(USD)/m^3		-		-		787.2 (121.70)			2705.12 (418.21)	
Cost saving/%						70.9				

Note: CNY is a unit of RMB, which is converted to USD at 0.1546 rates. Economic benefit of the cement-stabilized pavement base with low-grade metamorphic rock is relative to ordinary cement-stabilized pavement.

In addition, the preparation of LMR-CSPB can involve the recycling of waste resources, such as shallow metamorphic rocks. It can also promote the scientific utilization of shallow metamorphic rock in the pavement base layer, which can improve the quality of construction and operation of the project and make the application of LMR-CSPB material on the pavement in areas enriched with shallow metamorphic rocks reach a new level. This promotes the sustainable development of engineering construction in areas enriched with shallow metamorphic rocks. Thus, it is summarized that LMR-CSPB has a great engineering significance and economic value, and its application prospects are very broad.

From the environmental viewpoint, the replacement of river sand with manufactured sand can eliminate severe impacts in river basin environments, such as decreased species diversity and biomass of aquatic organisms [28,29]. In addition, the replacement of locally-produced LMR aggregate for far-shipped river or manufactured limestone sand not only makes use of local resources but also reduces CO_2 emissions due to transportation.

4. Conclusions

This paper investigates the possibility of using LMR in CSPB, given that the alkali content of CSPB is low compared to traditional concrete due to the limited usage of concrete. Based on the findings, the main conclusions can be drawn as follows:

1. The total alkali content of CSPB is 0.45 kg/m^3, which is lower than the required value (normally higher than 3.0 kg/m^3), for triggering AAR.
2. CSPB has a higher compressive strength with a higher content of cement. In addition, the compressive strength of CSPB prepared with LMR is similar to that of limestone aggregate regardless of standard curing or wet–dry curing.
3. An accelerated simulation test demonstrated that there is no AAR in LMR-CSPB, given that the 14-d expansion rates are lower than 0.1% for all mixtures. This was further validated by microscopic analysis showing that no AAR product was found in any area of LMR-CSPB.
4. LMR-CSPB has a cost 70.9% lower than ordinary cement-stabilized pavement base, demonstrating the significant benefit of using LMR for the preparation of CSPB. Destruction of river basin environments can be avoided, and CO_2 emission caused by sand transportation can be reduced by using locally-produced LMR aggregate.

Therefore, it is concluded that LMR has a high potential to be used for preparing CSPB from technical, economic and environmental viewpoints.

Author Contributions: Conceptualization, Q.Y. and Z.J.; methodology, Y.L., H.Z., G.G. and Y.C.; software, H.Z.; formal analysis, Y.L., X.W. and L.Z.; investigation, Y.L., Y.C. and L.Z.; writing—original draft preparation, Y.L.; writing—review and editing, Z.J.; supervision, Q.Y.; project administration, Q.Y. All authors have read and agreed to the published version of the manuscript.

Funding: This project was financially supported by the Scientific Project Guizhou Province Transportation Office (2019-122-003), (2022-323-005), and the China Construction West Project (ZJXJ-2020-01). We also thank those factories for supplying raw materials for our research work.

Institutional Review Board Statement: Not applicable.

Informed Consent Statement: Not applicable.

Data Availability Statement: Not applicable.

Conflicts of Interest: The authors declare no conflict of interest.

References

1. Šťastná, A.; Šachlová, Š.; Pertold, Z.; Přikryl, R. Factors affecting alkali-reactivity of quartz-rich metamorphic rocks: Qualitative vs. quantitative microscopy. *Eng. Geol.* **2015**, *187*, 1–9.
2. Fournier, B.; Bérubé, M.A. Alkali–aggregate reaction in concrete: A review of basic concepts and engineering implications. *Can. J. Civ. Eng.* **2000**, *27*, 167–191. [CrossRef]
3. Bulteel, D.; Rafaï, N.; Degrugilliers, P.; Garcia-Diaz, E. Petrography study on altered flint aggregate by alkali–silica reaction. *Mater. Charact.* **2004**, *53*, 141–154. [CrossRef]
4. Broekmans, M.A.T.M. Deleterious reactions of aggregate with alkalis in concrete. *Rev. Mineral. Geochem.* **2012**, *74*, 279–364. [CrossRef]
5. Garcia-Diaz, E.; Riche, J.; Bulteel, D.; Vernet, C. Mechanism of damage for the alkali–silica reaction. *Cem. Concr. Res.* **2006**, *36*, 395–400. [CrossRef]
6. Stanton, T.E. Expansion of concrete through reaction between cement and aggregate. In Proceedings of the ASCE, Los Angeles, CA, USA, December 1940; Volume 66, pp. 1781–1811.
7. Tang, M.S. General situation of alkali-aggregate reaction in various countries in the world. *Cem. Eng.* **1999**, *4*, 1–6. (In Chinese)
8. Xu, H.; Chen, M. Alkali-Aggregate Reaction in Chinese Engineering Practices. *J. Yangtze River Sci. Res. Inst.* **1989**, *10*, 28–35.
9. Lin, L.; You, Y. The investigation and research of Alkali-Aggregaion reaction for Tianjin's concrete engineering. *J. Tianjin Urban Constr. Inst.* **2001**, *31*, 1015–1022. (In Chinese)
10. Qin, Y.; Xu, H.; Zhang, D. The research of restrain concrete alkali-aggregate reaction in Xinjiang area's buildings. *Sichuan Build. Sci.* **2009**, *22*, 941–948. (In Chinese)
11. Li, J.Y. Alkali aggregate reaction in dam concrete in China. *Hydroelectr. Power* **2005**, *31*, 34–37. (In Chinese)
12. Li, G.W.; Zhou, Q.W. Alkali-aggregate reaction of dam concrete of Jinping I Hydropower Station. In Proceedings of the Hydraulic Dam Concrete Materials and Temperature Control Academic Exchange Meeting, Chengdu, China, 6–9 July 2009. (In Chinese).
13. Deng, M.; Lan, X.; Xu, Z. Petrographic characteristics and distributions of reactive aggregates in China. In Proceedings of the 12th International Conference on Alkali-Aggregate Reaction in Concrete, Nanjing, China, 15–19 October 2004; Volume 1.
14. *ASTM C1260*; American Society for Testing and Materials. Philadelphia, American Society for Testing and Materials: West Conshohocken, PA, USA, 2011.
15. Bragg, D. Alkali-aggregate reactivity in Newfoundland, Canada. *Can. J. Civil. Eng.* **2000**, *27*, 192–203. [CrossRef]
16. Jiang, Z.W.; Li, X. Alkali reactivity of metamorphic rock aggregate and its prevention measures in the southeast area of Guizhou Province. *J. Build. Mater.* **2010**, *13*, 22–26. (In Chinese)
17. Wei, B. The engineering application and restraining technology for the alkali aggregate reaction of concretes with metamorphic rock aggregate. *J. China Foreign Highw.* **2014**, *51*, 761–769. (In Chinese)
18. Lu, D.; Fournier, B.; Grattan-Bellew, P. Evaluation of accelerated test methods for determining alkali-silica reactivity of concrete aggregates. *Cem. Concr. Comp.* **2006**, *28*, 546–554. [CrossRef]
19. Tayfur, S.; Yüksel, C.; Alver, N.; Akar, O.; Andiç-Çakır, Ö. Evaluation of alkali–silica reaction damage in concrete by using acoustic emission signal features and damage rating index: Damage monitoring on concrete prisms. *Mater. Struct.* **2021**, *54*, 1–17. [CrossRef]
20. Lee, W.E.; Gadow, R.; Mitic, V. Alkali-Aggregate Reactions in Concrete. In *Proceedings of the III Advanced Ceramics and Applications Conferenc*; Atlantis Press: Paris, France, 2016; pp. 221–240.
21. *TB/T 3275*; Concrete for Railway Construction. Industry Standards of People's Republic of China: Beijing, China, 2018.
22. *JTG/T F20*; Technical Guidelines for Construction of Highway Roadbases. Industry Standards of People's Republic of China: Beijing, China, 2015.
23. *JTG E51*; Test Methods of Materials Stabilized with Inorganic Binders for Highway Engineering. Industry Standards of People's Republic of China: Beijing, China, 2009.
24. *GB/T 50733*; Technical Specification for the Prevention of Alkali-Aggregate Reaction in Concrete. National Standards of People's Republic of China: Beijing, China, 2011.
25. García-Lodeiro, I.; Palomo, A.; Fernández-Jiménez, A. Alkali–aggregate reaction in activated fly ash systems. *Cem. Concr. Res.* **2007**, *37*, 175–183. [CrossRef]
26. Zhang, X.Y.; Gallucci, E.; Scrivener, K. Prognosis of Alkali Aggregate Reaction with SEM. *Adv. Mater. Res.* **2011**, *194*, 1012–1016. [CrossRef]

27. Grimal, E.; Sellier, A.; Pape, Y.L. Creep, Shrinkage, and Anisotropic Damage in Alkali-Aggregate Reaction Swelling Mechanism-Part I: A Constitutive Model. *Aci. Mater. J.* **2008**, *105*, 227–235.
28. Xu, F.; Jia, Y.; Wang, Y.; Zhang, F.; Li, L.; Li, Y.; Ren, L.; Wang, D.; Zhang, T. Does sand mining affect the remobilization of copper and zinc in sediments?—A case study of the Jialing River (China). *Environ. Res.* **2021**, *200*, 111416. [CrossRef] [PubMed]
29. Meng, X.; Jiang, X.; Li, Z.; Wang, J.; Cooper, K.M.; Xie, Z. Responses of macroinvertebrates and local environment to short-term commercial sand dredging practices in a flood-plain lake. *Sci. Total Environ.* **2018**, *631*, 1350–1359. [CrossRef] [PubMed]

buildings

MDPI

Article

Study on Erosion Characteristics and Mechanisms of Recycled Concrete with Tailings in Salt Spray Environments

Jin Xu [1,2,3], Tao Li [4,5,*], Meng Zhan [4], Xiuyun Chen [4], Fan Xu [6] and Sheliang Wang [3]

1 School of Civil Engineering, Xijing University, Xi'an 710123, China; 20160030@xijing.edu.cn
2 Shaanxi Key Laboratory of Safety and Durability of Concrete Structure, Xi'an 710123, China
3 School of Civil Engineering, Xi'an University of Architecture and Technology, Xi'an 710055, China; sheliangw@163.com
4 College of Architecture Engineering, Huanghuai University, Zhumadian 463000, China; zhanyi313@163.com (M.Z.); chenxiuyun@huanghuai.edu.cn (X.C.)
5 College of Urban, Rural Planning and Architectural Engineering, Shangluo University, Shangluo 726000, China
6 School of Civil and Architecture Engineering, Xi'an Technological University, Xi'an 710021, China; xufanxf1205@163.com
* Correspondence: litao623114@126.com

Citation: Xu, J.; Li, T.; Zhan, M.; Chen, X.; Xu, F.; Wang, S. Study on Erosion Characteristics and Mechanisms of Recycled Concrete with Tailings in Salt Spray Environments. *Buildings* **2022**, *12*, 446. https://doi.org/10.3390/buildings12040446

Academic Editor: Yann Malecot

Received: 11 February 2022
Accepted: 1 April 2022
Published: 5 April 2022

Abstract: To improve the utilization efficiency of iron tailings (IOT) and recycled coarse aggregate (RCA), the mechanical properties, erosion depth and other erosion characteristics of recycled aggregate concrete (RAC) with different IOT amounts were studied in salt spray erosion environments and the erosion mechanisms were analyzed by SEM technology. The results showed that at the same erosion age, IOT caused the compressive strength and splitting tensile strength of RAC to tend to first increase and then decrease, with the optimum mixing amount being approximately 40%. Under the same conditions, the erosion depth of RAC was much higher than that of ordinary concrete. The erosion depth first decreased and then increased with an increasing amount of IOT. When the IOT content was 30–40%, the salt spray erosion depth reached its minimum. The solidification coefficient K_1 first decreased and then increased with the increase in iron tailings content. At its lowest point, the iron tailings content was approximately between 30% and 50%, which demonstrated that the higher the salt spray erosion age, the larger the solidification coefficient. Through SEM microscopic images, it could be seen that the appropriate amount of iron tailings caused the formation of salt spray erosion crystals and that the effect of physical expansion pressure caused a reduction in the porosity of RAC and a slight increase in its mechanical properties and salt spray erosion resistance. When the iron tailings content was large, the optimal mix ratio of the concrete also changed and then harmful pores and cracks were regenerated. Therefore, resistance to salt spray erosion was weakened. The research in this paper provides a theoretical basis for the engineering application of recycled concrete with tailings in salt spray environments.

Keywords: iron tailings; recycled concrete; salt spray environment; erosion characteristics; erosion mechanisms

1. Introduction

As granular waste from iron ore smelting, IOT have the obvious characteristics of fine powder and pozzolanic activation [1]. The porosity and related mechanical properties of recycled aggregate concrete are lower than those of ordinary concrete due to the existence of adhesive mortar [2,3]. At the same time, the utilization rate of IOT in China is less than 30%, which is far lower than the 90% rates in other developed countries [4,5]. Finally, a large amount of funds needs to be invested every year to deal with this problem, which has become the main factor in the restriction of the economic and ecological development of our country. When we are able to give full play to the respective performance of IOT and combine them with RAC to make recycled concrete with tailings, it improves their performance and provides a strong push to respond to the national policy of "clear waters and green mountains are as good as mountains of gold and silver".

Under the action of salt spray, chloride ions that attach to the concrete surface gradually erode and soak into the concrete, resulting in the destruction of the passive film of the reinforcement, the erosion of the reinforcement and even the durability failure of the structure [6]. Kamiharako [7] analyzed the relationship between the number of chloride ions attached to a structure's surface and the shape of the structure through numerical simulation and software analysis. Then, a set of techniques were proposed for applying these results to the prediction of the amount of chloride ingress in the concrete members of a bridge superstructure. Chendra [8] analyzed the main influencing factor of flow field on the chloride ion adsorption capacity of a concrete surface by using the wind tunnel test. Then, the amount of sea salt in the air was estimated through numerical simulation (computational fluid dynamics). Mohamed [9] developed a new artificial neural network (ANN) to predict the chloride ion penetration level and compressive strength of a self-compacting concrete mixture, which was in good agreement with our test results. Ariyachandra [10] discussed the effect of NO_2 isolated recycled concrete aggregate (NRCA) on the chloride diffusion and chloride binding capacity of concrete, the results of which showed that the addition of NRCA increased the formation of Friedel's salt and Kuzel's salt in concrete and enhanced the chloride ion binding ability of concrete. The results also identified the mechanisms for enhancing the chloride ion erosion resistance of NRCA mixed concrete. Rajamallu [11] studied the chloride ion permeability and erosion characteristics of cement concrete that was partially replaced by slag and silica fume and showed that it could strengthen erosion resistance. Su [12] simulated the combined effects of marine salt spray environments and fatigue alternating loads. The results showed that the greater the stress level and the greater the loading frequency, the greater the chloride ion content at the same depth.

At present, experts and scholars have systematically studied the erosion characteristics and erosion mechanisms of ordinary concrete in salt spray environments, but research on recycled concrete with iron tailings has been relatively less extensive. Oritola [13] used microscopic and physical examination techniques to test and evaluate five types of iron tailings that were obtained from different places and compared them to ordinary natural aggregate. The results showed that iron tailings could be used well in concrete. Protasio [14] studied the pozzolanic activity and comprehensive pore size distribution of concrete with iron tailings using thermogravimetric analysis, XRD and uniaxial compression tests. Then, the influence laws of different replacement proportions of iron tailings on the compressive strength and workability of concrete were obtained. Wang [15] studied the strength of recycled concrete that was mixed with construction waste and iron tailings. When the additional water content of the recycled aggregate was 0% and the content of iron tailings was 60%, the compressive strength of the recycled concrete was at its highest. Wei [16] conducted experimental research on 12 groups of high-ductility recycled concrete with iron tailings and systematically studied the main mechanical properties, such as compressive strength, flexural strength and splitting tensile strength. The results showed that the mechanical properties first increased and then decreased with the increase in iron tailings content and reached a peak when the iron tailings content was 30%. Cui [17] used the variance analysis method to analyze the compressive strength of recycled concrete with iron tailings under different working conditions. Professor Wang of the Xi'an University of Architecture and Technology [18–22] conducted a systematic macro- and micro-analysis on the strength, mechanical properties, deformation properties, carbonation, freeze–thaw cycles and other durability factors of recycled concrete with tailings, which provided a valuable basis for the development of recycled concrete with tailings.

Therefore, this paper studied the erosion characteristics and mechanisms of recycled concrete with different amounts of iron tailings under different periods of salt spray erosion in order to find the optimal amount of iron tailings for salt spray environments and provide a theoretical basis for engineering applications.

2. Test Materials

Qinling ordinary portland cement (p.o.42.5) and Bahe natural river sand were used in our tests. Natural coarse aggregate (NCA) was made from artificial crushed stone with particle size range of 5–20 mm and continuous grading. Recycled coarse aggregate (RCA) was a common material found in Xi'an market, which has been in service for 30 years. The strength grade of the original concrete was C30. After screening, cleaning, drying, bagging and the other procedures, it could be used for later tests.

The iron tailings (IOT) came from the YAOGOU tailings reservoir in Shangluo, Shaanxi Province, which was accumulated by the waste slag that was generated after steelmaking and covered an area of nearly 1000 square meters up to 6.8 m high. According to incomplete statistics, the accumulation of tailings accounted for 45.67% of industrial waste in China. In 2020–2021, the comprehensive utilization of industrial solid waste was 2.059 billion tons with a utilization rate of 62.3%, but this rate for tailings was only 18.9%, which still leaves a lot of room for development. In this paper, the constituent elements of iron tailings were determined using energy spectrum analysis with the Vario EL cube and the particle gradation was determined using a negative pressure grading screen, as shown in Figure 1, Tables 1–3. The performances of the main test materials met the requirements for the relevant specifications [23,24].

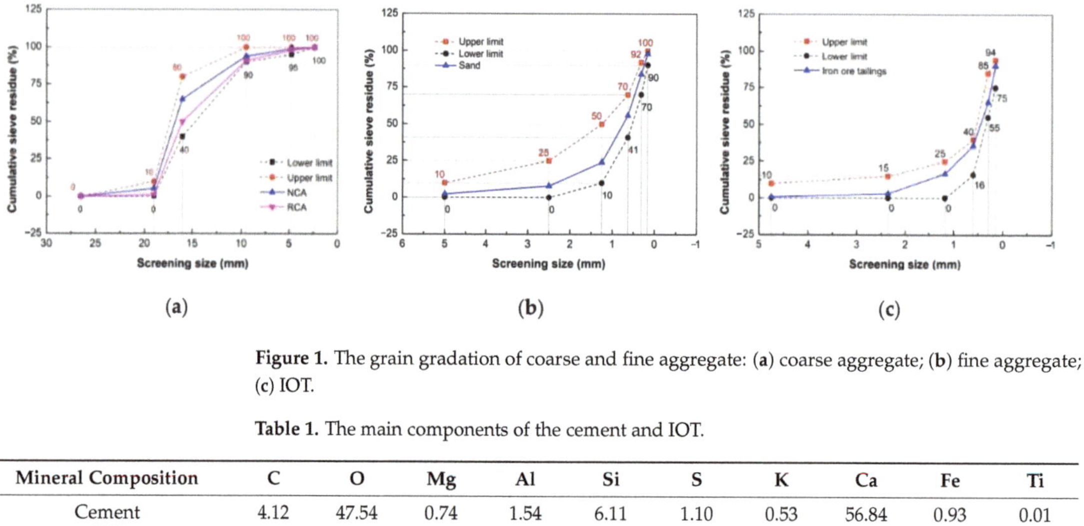

(a) (b) (c)

Figure 1. The grain gradation of coarse and fine aggregate: (**a**) coarse aggregate; (**b**) fine aggregate; (**c**) IOT.

Table 1. The main components of the cement and IOT.

Mineral Composition	C	O	Mg	Al	Si	S	K	Ca	Fe	Ti
Cement	4.12	47.54	0.74	1.54	6.11	1.10	0.53	56.84	0.93	0.01
IOT	0	56.49	6.95	8.38	17.46	-	3.02	1.68	5.28	0.74

Table 2. The main performance index of the cement.

Water Requirement of Normal Consistency (%)	Initial Setting Time (min)	Final Setting Time (min)	Fineness (45 μm)	Stability	Flexural Strength (MPa)		Compressive Strength (MPa)	
					3 d	28 d	3 d	28 d
28	**160**	**280**	**2.8**	Qualified	5.2	6.8	19.5	42.2

At the same time, based on previous results obtained by the research group, the replacement rate of recycled aggregate was 30% [18–22]. For the convenience of comparative analysis, the tailings dosage was 0%, 10%, 20%, 30%, 40%, 50%, 70% and 100% and the water–cement ratio and the sand ratio were 0.4 and 0.35, respectively. The mix proportion was also carried out according to JGJ/T443-2018 [25]. The mix proportion of each test is shown in Table 4.

Table 3. The performance index of the main materials.

Performance Index	Apparent Density (kg/m^3)	Bulk Density (kg/m^3)	Crushing Value Index (%)	Water Absorption (%)	Sediment Percentage (%)	Moisture Content (%)	Organic Matter Content	Alkali Aggregate Reaction
NCA	2941	1749	10.3	1.33	0.72	0.8	Qualified	Qualified
RCA	2536	1467	14.8	7	1.86	3.02	Qualified	Qualified
Norm value [23]	≥2500	≥1300	≤16	-	≤1.0	-	Qualified	Qualified
Sand	2764	1830	12	2.12	1.2	4.1	Qualified	Qualified
IOT	2745	1824	19.53	8.7	2.9	1.45	Qualified	Qualified
Norm value [23]	-	-	≤10	-	≤3.0	-	Qualified	Qualified

Table 4. The mix design of the recycled concrete under different working conditions (kg/m^3).

Serial Number	No	Cementitious Materials	Coarse Aggregate		Fine Aggregate		Water
		Cement	NCA	RCA	Sand	IOT	
1	NAC	538	1063	0	572	0	215
2	RAC-1	538	735	315	566	0	215
3	RAC-2	538	739	317	512	57	215
4	RAC-3	538	743	319	458	114	215
5	RAC-4	538	744	319	402	172	215
6	RAC-5	538	751	322	343	229	215
7	RAC-6	538	755	324	290	290	215
8	RAC-7	538	763	327	176	410	215
9	RAC-8	538	773	331	0	594	215

3. Test Setup

To simulate the chloride ion erosion characteristics of recycled concrete with tailings in salt spray environments, cube test blocks (100 mm × 100 mm × 100 mm) were used to test the salt spray erosion depth and change in ion content characteristics. To avoid the influence of multi-dimensional salt spray erosion on the erosion depth and ion concentration of the test block, a non-pouring surface was reserved as the salt spray erosion surface after the maintenance. The test process is shown in Figure 2.

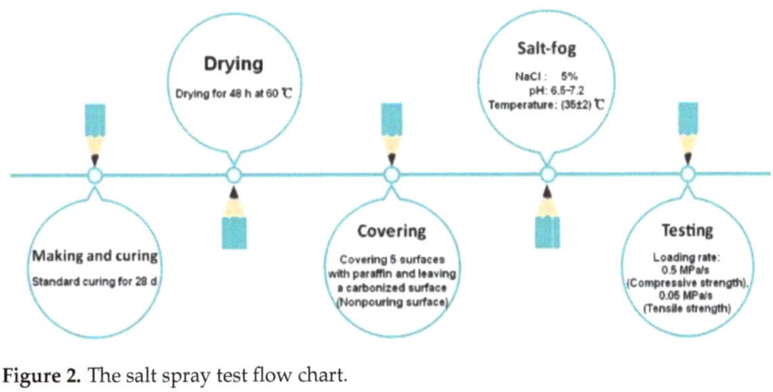

Figure 2. The salt spray test flow chart.

The test used a large salt spray erosion test box that was produced by Yingbai Technology Co., Ltd. in Wuxi Wuxi City, Jiangsu Province, China. When the test block had eroded to the specified age (7 d, 14 d, 28 d and 90 d), the salt spray erosion depth and ion content were measured using the following procedures:

(1) For the erosion depth measurement, a split along the erosion surface was sprayed with a 0.1 mol/L $AgNO_3$ solution and after 15 min, 10 points on the cleavage surface were measured with a digital depth meter and the average value was calculated;
(2) For the ion concentration value, the single side grinding method was used to collect powder from the eroded surface every 2 mm, layer by layer up to 10 mm. Then, holes were drilled every 5 mm to collect the powder and the powder was then filtered through a 0.16 mm sieve. According to the specifications [26], the total chloride ion concentration C_t and the free chloride ion concentration C_f were then extracted with dilute nitric acid mixed with distilled water and titrated potassium thiocyanate mixed with potassium chromate, respectively. The solid–liquid extraction method [27] was used to determine the mass fraction.

4. Test Results

4.1. Slump Values

Table 5 shows the slump values of the RAC with different IOT contents. It can be seen that the values decreased significantly with the increase in IOT content. When the content reached 100%, its value was 48.2% lower than that of NAC. The main reason for this is that the IOT had fine particles and a large specific surface area, which meant that its active components could promote the hydration of cement and then reduce the amount of free water and slump.

Table 5. The slump values of RAC with IOT.

No.	NAC	RAC-1	RAC-2	RAC-3	RAC-4	RAC-5	RAC-6	RAC-7	RAC-8
Cement	141	134	122	108	103	95	81	76	73

4.2. Compressive Strength

Figure 3 shows the change curve of the compressive strength of the cubes and its growth coefficient after salt spray erosion. The growth coefficient was defined as the ratio between the actual strength and the strength for the same age of salt spray erosion at 0 days (natural curing for 28 days). It can be seen from Figure 3a that the influence trends of the different tailings dosage conditions for the same erosion age were the same. Except for ordinary concrete, the compressive strength increased when the tailings content was lower ($\leq$40%) for the same salt spray erosion age and the higher the tailings content, the greater the compressive strength. For example, when the content of IOT increased from 0% to 40%, the compressive strength increased by 24.52% (0 d), 21.72% (7 d), 18.86% (14 d), 18.82% (28 d) and 19.60% (90 d). The main reason for this is that the tailings presented a certain fine powder and activity. Under the same conditions, the concrete became denser, the porosity decreased and the compressive strength increased with the addition of a 40% content of IOT. At the same time, when the IOT content was higher (>40%), the compressive strength decreased. This change trend was similar to that found in the results of other studies [27–29] on ordinary concrete.

Figure 3b shows the change rule for the compressive strength growth coefficient to salt spray erosion age under the same iron tailings mixture ratio. It can be seen that the change trends of the compressive strength growth coefficient caused by salt spray erosion were similar. With the same tailings content, the increase in erosion time generally demonstrated the trend of the growth coefficient of the compressive strength first increasing and then decreasing. Most of the concrete that was mixed with tailings had a peak point at 14 days, but the increased value was relatively limited with most being below 10% and some being less than 4%. In addition, there were a few concrete mixtures with peak points at 28 d, such as RAC-7 (70%) and RAC-8 (100%). The main reason for this is that when the tailings content was high, the design mix ratio of the concrete changed, the porosity increased and the peak point was pushed back. At the same time, when the salt spray erosion age was 90 d, the tailings content was 100%. In this case, the compressive strength was also increased by 0.78% compared to 0 d.

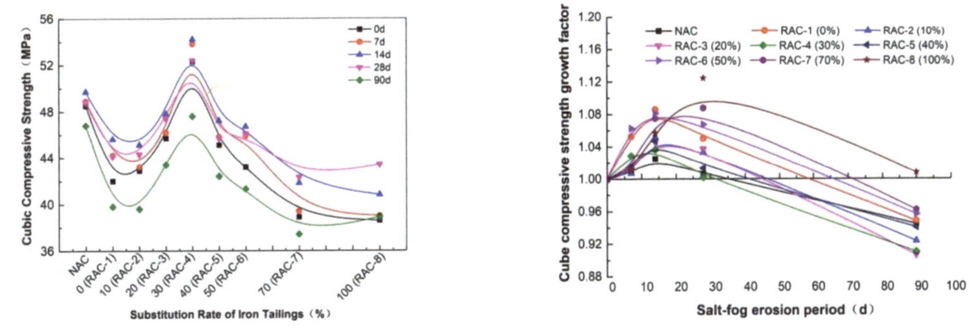

(a) (b)

Figure 3. The influence of IOT content on the compressive strength of the cube: (**a**) the compressive strength value of the cubes; (**b**) the growth factor of the compressive strength of the cubes.

The above analysis showed that porosity was very significant in the salt spray erosion of concrete. Analyzing the reasons for this: on the one hand, the salt spray erosion products filled the pore structures and on the other hand, the salt crystals also had a swelling and densification effect on the pore structures [30]. Therefore, the greater the porosity, the more erosion products could accumulate and the less destructive the effect of erosion on the concrete.

4.3. Splitting Tensile Strength

Figure 4 shows the change curve between the splitting tensile strength of the recycled concrete cubes with different tailings contents and their corresponding growth coefficients under salt spray erosion.

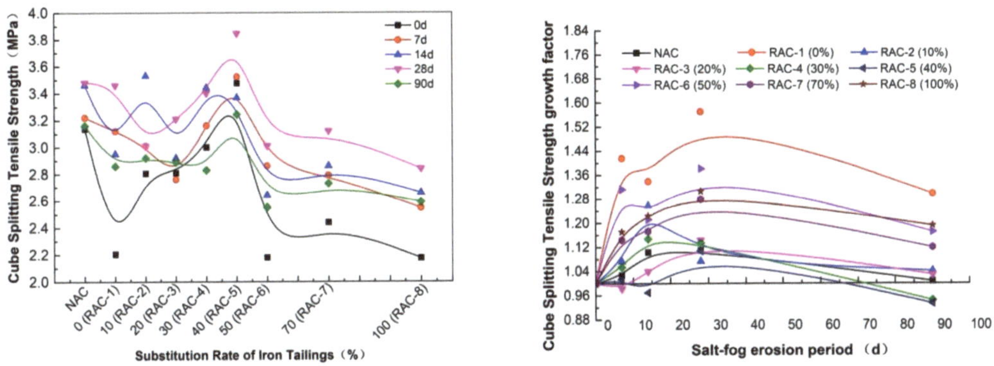

(a) (b)

Figure 4. The influence of IOT content on the splitting tensile strength of the cubes: (**a**) the splitting tensile strength of the cubes; (**b**) the growth coefficient of the splitting tensile strength of the cubes.

Due to the low splitting tensile strength of the cubes, which was greatly influenced by human, mechanical and material factors, the value of the splitting tensile strength fluctuated to varying degrees; however, on the whole, when the content of tailings was low (≤40%), the splitting tensile strength values for different erosion ages showed an undulating trend with the peak point appearing when the content of tailings was 30–40%.

When the peak point was passed, the IOT content increased rapidly and then tended to level off. Most results showed that before 28 days of erosion, the value of splitting tensile strength gradually increased as the salt crystallization effect of the erosion increased, but then gradually decreased. The splitting tensile strength values at 90 days were even lower than those at 7 days, which showed that with the gradual increase in crystalline salt during those periods, a large expansion stress was generated in the cracks and pores and the tensile strength values were reduced accordingly.

Figure 4b shows the increasing growth coefficient of splitting tensile strength with the increase in erosion age with the same tailings content. Compared to the growth coefficient of salt spray of the compressive strength of the cubes, erosion age had less of an effect on splitting tensile strength and the intuitive phenomenon was that the slope of the curve was relatively small. On the whole, the growth coefficient for the same tailings content occurred after salt spray erosion for 14 to 28 days. Similar to the growth coefficient of compressive strength, the maximum value of the growth coefficient for splitting tensile strength occurred at 28 d for the condition of increased porosity, such as with RAC-1, RAC-7 (70%) and RAC-8 (100%). The growth coefficient at the peak point was also greater than that of compressive strength. For example, the splitting tensile strength growth coefficient of RAC-21, RAC-6 (60%) and RAC-8 (100%) increased by 56.9%, 38.1% and 22.3%, respectively, at the maximum point.

5. Erosion Depth

Figure 5 shows the erosion depth of the recycled concrete at different ages. It can be seen that the salt spray resistance of recycled concrete (RAC-1) was much lower than that of ordinary concrete (NAC) at the same erosion age. With the increase in IOT content, the salt spray erosion depth presented an approximate trend of first decreasing and then increasing, but the change trends were different. Approximately, when the tailings content was 30–40%, the erosion depth reached its minimum value. For example, after 90 days of salt spray erosion, the erosion depth of RAC-1 was 10.07% higher than that of NAC and the erosion depth of RAC-5 (u_{IOT} = 40%) was 14.29% lower than that of RAC-1 (u_{IOT} = 0%), indicating that the effect of an appropriate IOT amount on salt spray erosion was relatively obvious. The main reason for this is that the porosity of RAC was high and there were many interfacial transition zones, so salt spray could penetrate into the concrete more easily. As mineral admixtures, IOT particles are fine and have an obvious filling effect and activity, meaning that an appropriate amount could effectively promote the secondary hydration of cement particles and improve the pore structures and the compactness of the concrete matrix structures. As IOT replaced the fine aggregate in concrete, when the tailings content was too high, the optimal mix proportion of the concrete aggregate was, which led to an increase in the porosity of the concrete and a reduction in its anti-erosion ability.

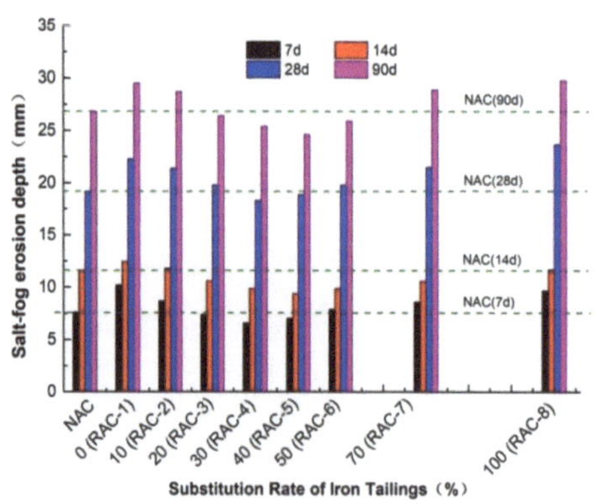

Figure 5. The effect of different IOT contents on the salt spray erosion depth of RAC.

6. Curing Ability of Chloride Ions

The absorption and curing effect of the chloride ions on concrete can greatly reduce the concentrations of free chloride ions and total chloride ions and the transmission rate of chloride ions, thereby reducing the risk of steel corrosion. Therefore, it was necessary to study the relationship between these three factors.

Mohammed [31] conducted a chloride ion erosion test on different cement types, cement components and concrete mixed with mineral powder, steel slag and other admixtures. It was found that the free chloride ion and total chloride ion contents conformed to the following formula:

$$C_t = K \cdot C_f \tag{1}$$

where K is the curing coefficient of concrete with different material types. The larger the K value, the stronger the curing ability of the concrete. In view of the diversity of cement varieties and cement components, this value is usually obtained by fitting the test results and reference [32] also used this model. However, some scholars [33] found that this model was not comprehensive and that it had a higher correlation when expressed as a complete linear model, i.e.:

$$C_t = K_1 \cdot C_f + K_2 \tag{2}$$

where K_1 is the influence coefficient and K_2 is a constant for the amount of total chloride ions that can be solidified and absorbed per unit mass of concrete, i.e., the amount of combined chloride ions when the free chloride ion content is 0. In our tests, Equation (2) was used to analyze the test results. In view of the small erosion depth and relatively few test data points in the early stages of the test (7 d and 14 d), the correlation between the free chloride ion and total chloride ion contents at 28 d and 90 d was studied in order to improve the accuracy of the study, as shown in Figure 6.

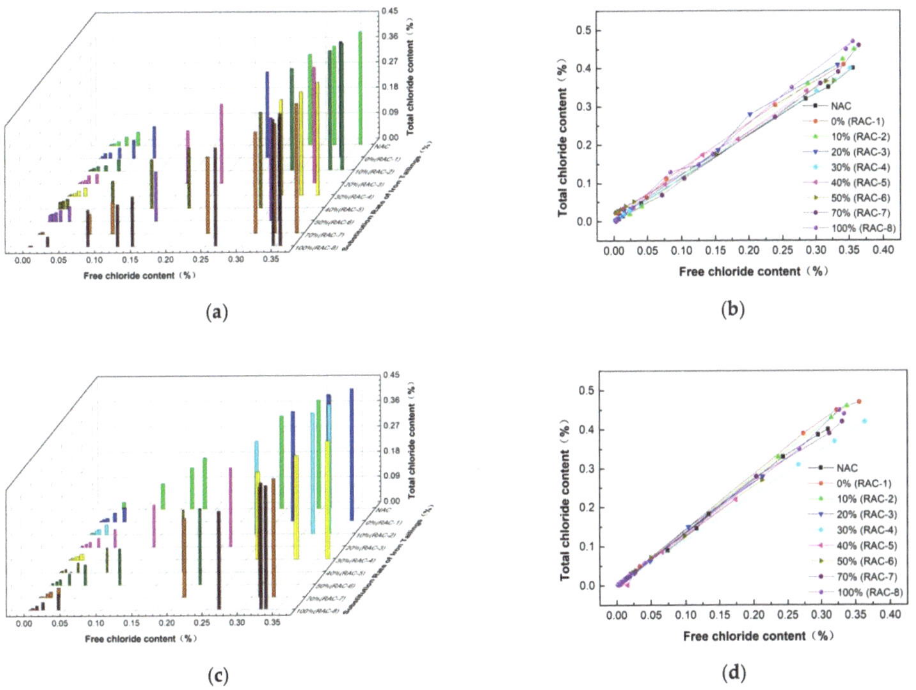

Figure 6. The relationship between C_t and C_f with different tailings dosages: (**a**) three-dimensional drawing (28 d); (**b**) plane figure (28 d); (**c**) three-dimensional drawing (90 d); (**d**) plane figure (90 d).

Figure 6a,c shows the location and size relationships of numerical points at the relevant erosion ages (28 d and 90 d), while Figure 6b,d represents the direct ratio relationships. Due to the low content of chloride ions, the position of powder collection and the density and permeability of the concrete had a great influence on the calculated chloride ion content. Compared to Figure 6a,c, the numerical points at the two erosion ages were in an approximately random distribution; however, it can be seen that the value of $C_f > 0.35\%$ when the figure was decreasing, indicating that the free chloride ion concentration (C_f) of concrete with the same tailings content gradually decreased with the increase in erosion time. As can be seen from Figure 6b,d, the ratio of free chloride ions to total chloride ions changed linearly (approximately and except for some scattered numerical points) and the curing coefficient K_1 changed very little. Figure 7 shows the change laws of the curing coefficients of different tailings content conditions at different erosion ages.

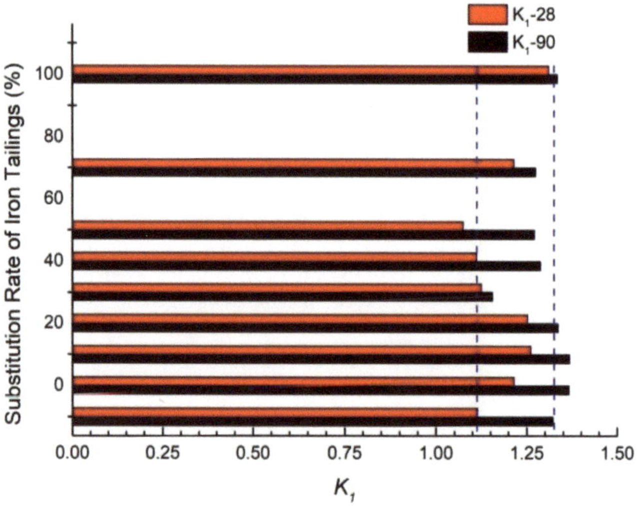

Figure 7. The comparative analysis of curing coefficients at 28 d and 90 d of salt spray erosion.

It can be seen that the curing coefficient increased with the increase in erosion age, which indicated that the curing ability of concrete gradually became stronger, i.e., more free chloride ions were converted into bound chloride ions. This was mainly because the longer the erosion time, the more calcium aluminate hydrate took part in the reaction and converted chloride ions into Friedel's salt and then, the corresponding bound chloride ion content increased. However, the increase degree was different for different IOT contents. When IOT content was 30%, 40% and 50%, the curing coefficient increased by 2.75%, 15.79% and 18.29%, respectively. The blue dotted line in the figure indicates the ratio of NAC at the corresponding erosion age. It can be seen that when the erosion age was small (28 d), the curing ability of the chloride ions was enhanced by high tailings content. When the erosion age higher (90 d), the alkali content of the concrete was reduced due to the tailings promoting the secondary hydration of concrete, which had an impact on the alkaline environment of Friedel's salt. Therefore, the curing performance of chloride ions decreased in the long term. However, due to the low activity of tailings, their impact on the alkaline environment of concrete was relatively limited. At the same time, the alkaline environment of RAC was stronger than that of NAC and the porosity of RAC was also higher, which caused the contact area with free chloride ions to be larger. Therefore, the improvement was relatively large in the short term, so the long-term curing ability would continue to increase.

7. Erosion Mechanisms

To explore the influence of IOT on the salt spray erosion resistance of RAC, an SEM analysis using the Hitachi S-4800 model at Chang'an University was carried out before and after the erosion took place and the degradation mechanisms were also studied.

Clusters of crystal products were formed at both 28 and 90 days of salt spray erosion. Under the same working conditions, the longer the erosion time, the more crystalline products were produced and the more obvious the cluster-like products were. Compared to the working conditions before erosion, the addition of RAC made the matrix structure relatively loose and also provided space for the formation, growth and development of chloride crystals. It was intuitively shown that the matrix structure of the concrete was covered by salt crystals due to the formation and physical expansion pressure of those salt crystals, meaning that the micro-cracks and harmful pores of the matrix structure were covered by salt crystals. At the same time, some connected fractures were gradually transformed into closed pores and harmful pores were further transformed into harmless pores. Meanwhile, when the erosion age was small, the mechanical properties and resistance to chlorine–salt erosion were slightly increased.

As shown in Figure 8b,d,f, when the erosion age was higher (90 d), it could be seen that the pores and micro-cracks of the concrete matrix structure were significantly higher than those at lower erosion ages (28 d), which showed that when the erosion time was longer, the concrete matrix structure was destroyed by the accumulation of salt crystalline substances. The reason for this is that, on the one hand, the physical expansion of the chloride crystals that accumulated in the pores caused tensile stress in the matrix structure and made it easy to crack. On the other hand, both cement hydration and the generation of chloride crystals needed to consume free water, which resulted in adverse effects on the matrix structure due to water migration as the amount of free water in the concrete was relatively small. To sum up, the mechanical properties and durability of almost all working conditions with high erosion ages decreased to varying degrees, which was similar to the results of Xue [34] and Yue [35].

Figure 8. The scanning microscopic morphology of SEM images after different erosion cycles: (**a**) NAC (28 d); (**b**) NAC (90 d); (**c**) RAC (28 d); (**d**) RAC (90 d); (**e**) RAC-4 (28 d); (**f**) RAC-4 (90 d).

Combined with Figure 8d,f, Figure 9 shows the scanning microscopic morphology of SEM images of salt spray erosion on RAC with different tailings contents at 90 d.

(**a**) (**b**) (**c**)

Figure 9. The scanning microscopic morphology of SEM images for different tailings contents after 90 d: (**a**) RAC-1; (**b**) RAC-5; (**c**) RAC-8.

Due to the introduction of recycled aggregate, when the IOT amount was small, a large number of pores appeared, as shown in Figures 8c and 9a. On the periphery of the pores, cluster-like crystal products, flocculent CSH and needle-like ettringite could be found. These three features intersected with each other within the pores and were embedded around the micro-pores. The concrete was refined, which effectively improved its porosity and pore structure and enhanced its resistance to chloride ion erosion. It can be seen from Figure 8b,e that there was still a small number of pores in the matrix structure, but they were relatively complete. As shown in Figure 9c, when the IOT content was high, a large number of micro-cracks and micro-pores were generated in the matrix structure, which greatly affected the integrity of the matrix structure and reduced its mechanical properties and erosion resistance, as shown in Figures 3a, 4a and 5. The main reason for this is that a small IOT content could promote hydration and fill pores, while excess additions changed the optimal mix proportion, increased porosity and caused the deterioration of each index.

8. Conclusions

In this paper, the mechanical properties, chloride ion concentration and chloride ion curing properties of recycled concrete with different IOT contents are analyzed at different salt spray ages (7 d, 14 d, 28 d and 90 d). The conclusions were as follows:

(1) At the same erosion age, when the content of iron tailings increased from 0% to 40%, the compressive strength of the concrete increased by 24.52% (0 d), 21.72% (7 d), 18.86% (14 d), 18.82% (28 d) and 19.60% (90 d), indicating that the higher the tailings content, the greater the compressive strength. However, when the tailings content was too high (>40%), it had the opposite effect. With the increase in erosion age, the compressive strength first increased and then decreased with a peak at 14–28 d. When the content of tailings was lower ($\leq$40%), the splitting tensile strength increased and then decreased rapidly. The effect of erosion age on splitting tensile strength was less than that on compressive strength;

(2) The salt spray erosion resistance of RAC was much lower than that of NAC and when the IOT content was 30–40%, the salt spray erosion depth reached its lowest point;

(3) The higher the salt spray erosion age, the greater the corresponding growth coefficient. When the dosage of IOT was 30%, 40% and 50%, the curing coefficient at 90 d was 2.75%, 15.79% and 18.29% higher than that at 28 d, respectively. At the same erosion age, the ratio of free chloride ions to total chloride ions changed linearly and its curing coefficient K_1 first decreased and then increased, with the optimum IOT content being 30–50%;

(4) Through SEM microscopic morphology, it could be concluded that when the erosion age was low, the formation of salt crystals and the effect of physical expansion pres-

sure caused porosity to decrease. When the erosion age was higher, the cumulative expansion of crystals made them decrease in size. When added together, the crystallization and hydration products grew and developed in the micro-pores. When added excessively, the optimal mix ratio of the concrete changed, harmful pores and cracks were generated and resistance to salt spray erosion was reduced.

Author Contributions: Conceptualization, J.X. and S.W.; Data curation, T.L. and M.Z.; Formal analysis, J.X.; Investigation, M.Z.; X.C. and F.X.; Methodology, J.X.; Project administration, J.X.; Resources, J.X.; Software, T.L.; Supervision, J.X.; Validation, J.X.; Visualization, T.L., M.Z., X.C. and F.X.; Writing—original draft, J.X.; Writing—review & editing, T.L. All authors have read and agreed to the published version of the manuscript.

Funding: This research was funded by the Natural Science Foundation of China (grant number 51678480), the Ministry of Education Cooperative Education Project (grant number 201802308007), the Key Research and Development Program of Shaanxi (grant number 2021SF-521), the Natural Science Foundation of Shaanxi Province (grant number 2021JQ-844) and the Scientific and Technological Project in Henan Province (grant numbers 222102320311 and 222102320199).

Institutional Review Board Statement: Not applicable.

Informed Consent Statement: Not applicable.

Conflicts of Interest: The authors declare no conflict of interest.

References

1. Figueiredo, R.; Brando, P.; Soutsos, M.; Henriques, A.B.; Mazzinghy, D.B. Producing sodium silicate powder from iron ore tailings for use as an activator in one-part geopolymer binders. *Mater. Lett.* **2021**, *288*, 129333. [CrossRef]
2. Kaplan, G.; Bayraktar, O.Y.; Gholampour, A.; Gencel, O.; Koksal, F.; Ozbakkaloglu, T. Mechanical and durability properties of steel fiber-reinforced concrete containing coarse recycled concrete aggregate. *Struct. Concr.* **2021**, *22*, 2791–2812. [CrossRef]
3. Nazarpour, H.; Jamali, M. Mechanical and freezing cycles properties of geopolymer concrete with recycled aggregate. *Struct. Concr.* **2020**, *21*, 1004–1012. [CrossRef]
4. Lv, X.D.; Liu, Z.A.; Zhu, Z.G.; Li, B.X. Study of the progress of tailings comprehensive utilization of raw materials in cement and concrete. *Mater. Rev.* **2018**, *32*, 452–456.
5. Liu, K.; Chen, Y.D.; Huang, D. Analysis on the research current situation and ruture trends of recycled concrete and analysis of future research trends. *Concrete* **2020**, *10*, 47–50. [CrossRef]
6. Jin, W.L.; Zhao, Y.X. *Durability of Concrete Structure*; Science Press: Beijing, China, 2014.
7. Kamiharako, A.; Tomiyama, J.; Arai, K. Estimation of the Amount of Chloride Ingress in Concrete Taking into Account Differences in the Amount of Adhering Salt on Bridge Members. *Concr. J.* **2016**, *54*, 164–169. [CrossRef]
8. Ronald, C.; Katsuchi, H.; Yamada, H.; Sasaki, H. Study on estimation of amount of airborne sea salt around bridge deck. *Kozo Kogaku Ronbunshu A* **2012**, *58*, 528–541. [CrossRef]
9. Mohamed, O.; Kewalramani, M.; Ati, M.; Hawat, W.A. Application of ann for prediction of chloride penetration resistance and concrete compressive strength. *Materialia* **2021**, *17*, 101123. [CrossRef]
10. Ariyachandra, E.; Peethamparan, S.; Patel, S.; Orlov, A. Chloride diffusion and binding in concrete containing NO_2 sequestered recycled concrete aggregates (nrcas). *Constr. Build. Mater.* **2021**, *291*, 123328. [CrossRef]
11. Rajamallu, C.; Reddy, T.C.; Arunakanthi, E. Service life prediction of self-compacted concretes with respect to chloride ion penetration. *Mater. Today Proc.* **2021**, *46*, 677–681. [CrossRef]
12. Su, L.W.; Cai, J.; Liu, P.H. Experimental Investigation in to RC Beam under the Action of Alternating Load in Salt-Spray Environment. *J. South China Univ. Technol.* **2017**, *45*, 97–104. [CrossRef]
13. Oritola, S.F.; Saleh, A.L.; Sam, A. Characterization of iron ore tailings as fine aggregate. *ACI Mater. J.* **2020**, *117*, 125–134. [CrossRef]
14. Protasio, F.; Avillez, R.; Letichevsky, S. The use of iron ore tailings obtained from the germano dam in the production of a sustainable concrete. *J. Clean. Prod.* **2020**, *278*, 123929. [CrossRef]
15. Wang, P.P.; Hu, J.L.; Yang, X. Study on the Strength of Recycled Concrete with Building Waste Mixed with Iron Tailings. *J. Hebei Inst. Archit. Civ. Eng.* **2019**, *37*, 38–42. [CrossRef]
16. Wei, T.; Quan, X.Y.; Yan, Q.Q. Experimental study on mechanical properties of recycled concrete with high ductility iron tailings. *China Concr. Cem. Prod.* **2019**, *8*, 93–96. [CrossRef]
17. Cui, H.H.; Yang, X.; Hu, J.L. Study on the matching and mechanical properties of regenerated concrete with iron tailings. *Sichuan Build. Sci.* **2018**, *44*, 100–105. [CrossRef]
18. Li, T.; Wang, S.; Xu, F.; Meng, X.; Li, B.; Zhan, M. Study of the basic mechanical properties and degradation mechanism of recycled concrete with tailings before and after carbonation. *J. Clean. Prod.* **2020**, *259*, 120923. [CrossRef]

19. Xu, F.; Wang, S.; Li, T.; Liu, B.; Li, B.; Zhou, Y. Mechanical properties and pore structure of recycled aggregate concrete made with iron ore tailings and polypropylene fibers. *J. Build. Eng.* **2021**, *33*, 101572. [CrossRef]
20. Tong, X.; Wang, S.L. Mechanical properties and microstructure analysis of recycled aggregate concrete with iron tailings. *Concrete* **2021**, *1*, 91–97. [CrossRef]
21. Xu, F.; Wang, S.; Li, T.; Liu, B.; Zhao, N.; Liu, K. The mechanical properties and resistance against the coupled deterioration of sulfate attack and freeze-thaw cycles of tailing recycled aggregate concrete. *Constr. Build. Mater.* **2021**, *269*, 121273. [CrossRef]
22. Li, T.; Wang, S.; Xu, F.; Li, B.; Dang, B.; Zhan, M.; Wang, Z. Study on Carbonation Damage Constitutive Curve and Mircroscopic Damage Mechanism of Tailing Recycled Concrete. *J. Renew. Mater.* **2021**, *9*, 1413–1432. [CrossRef]
23. Ministry of Housing and Urban-Rural Construction of the People's Republic of China. *JGJ52-2006*; Standard for Technical Requirements and Test Method of Sand and Crushed Stone (or Gravel) for Ordinary Concrete. China Architecture & Building Press: Beijing, China, 2006.
24. Ministry of Housing and Urban-Rural Construction of the People's Republic of China. *JGJ63-2006*; Standard of Water for Concrete. China Architecture & Building Press: Beijing, China, 2006.
25. Ministry of Housing and Urban-Rural Construction of the People's Republic of China. *JGJ/T 443-2018*; Technical Standard for Recycled Concrete Structure. China Architecture & Building Press: Beijing, China, 2018.
26. Ministry of Transport of the People's Republic of China. *JTS/T236-2019*; Technical Specification for Concrete Testing of Port and Waterway Engineering. China Communications Press: Beijing, China, 2019.
27. Choi, S.J.; Kim, Y.U.; Oh, T.G.; Cho, B.S. Compressive strength, chloride ion penetrability, and carbonation characteristic of concrete with mixed slag aggregate. *Materials* **2020**, *13*, 940. [CrossRef] [PubMed]
28. Gehlot, T.; Sankhla, S.S.; Parihar, S. Compressive strength and chloride ion permeability test of concrete incorporating fly ash and ggbs with diverse water binder ratio. *Mater. Sci. Forum* **2021**, *1*, 789. [CrossRef]
29. Ge, Y.F. *Fine Numerical Simulation and Experimental Study of Chloride Corrosion of Self-Compacting Concrete*; Southeast University: Nanjing, China, 2017.
30. Mousavinejad, S.; Sammak, M. Strength and chloride ion penetration resistance of ultra-high-performance fiber reinforced geopolymer concrete. *Structures* **2021**, *32*, 1420–1427. [CrossRef]
31. Mohammed, T.U.; Hamada, H. Relationship between free chloride and total chloride contents in concrete. *Cem. Concr. Res.* **2003**, *33*, 1487–1490. [CrossRef]
32. Wang, Y.; Liu, C.; Tan, Y.; Wang, Y.; Li, Q. Chloride binding capacity of green concrete mixed with fly ash or coal gangue in the marine environment. *Constr. Build. Mater.* **2020**, *242*, 118006. [CrossRef]
33. Zeng, J.J.; Fan, Z.H.; Xiong, J.B. Research on Durability of Metakaolin Concrete Based on Long-Term Exposure Experiment at In-Situ Marine Splash Zone. *J. South. China Univ. Technol.* **2018**, *46*, 53–60. [CrossRef]
34. Pillay, D.L.; Olalusi, O.B.; Awoyera, P.O.; Rondon, C.; Kolawole, J.T. A review of the engineering properties of metakaolin based concrete: Towards combatting chloride attack in coastal/marine structures. *Adv. Civ. Eng.* **2020**, *2020*, 8880974. [CrossRef]
35. Yue, J. *Study on Carbonation and Chloride Resistance of Autoclaved Nano-Modified Concrete*; China Mining University: Xuzhou, China, 2019.

buildings

MDPI

Article

Proportion and Performance Optimization of Lightweight Foamed Phosphogypsum Material Based on an Orthogonal Experiment

Tao Zheng [1], Xun Miao [1], Dewen Kong [1,*], Lin Wang [1], Lili Cheng [2] and Ke Yu [3]

1 College of Civil Engineering, Guizhou University, Guiyang 550025, China; tzheng@gzu.edu.cn (T.Z.); miaoxun2022@163.com (X.M.); lwang@gzu.edu.cn (L.W.)
2 Department of Civil Engineering, Guiyang College of Information Science and Technology, Guiyang 550025, China; llcheng@gzu.edu.cn
3 Guizhou LianJian Civil Engineering Quality Inspection and Monitoring Center Co., Ltd., Guiyang 550025, China; huzhou_1994@126.com
* Correspondence: dwkong@gzu.edu.cn

Abstract: A lightweight foam phosphogypsum material (LFPM) was prepared by multi-factor orthogonal and optimization experiments. The effects of foam, quicklime, silica fume and cement on the mechanical and physical properties of this LFPM were studied. The orthogonal experimental results showed that the silica fume content exhibited the most significant effect on the strength of this material, and the cement content exhibited the most obvious influence on the softening coefficient. The comprehensive index analysis indicated that the LFPM with 8% foam, 3.5% quicklime, 3% silica fume and 15% cement was selected as the optimal proportion. The 28 d compressive strength and flexural strength were 3.15 and 0.97 MPa, respectively. The dry density was 809.1 kg/m^3, and the 28 d softening coefficient was 0.628. The optimization experimental results showed that the strength and dry density of the sample increased first and then decreased with an increase in the foam stabilizer content. The strength and dry density increased, and water absorption decreased with increasing waterproof agent content.

Keywords: lightweight foam phosphogypsum material; orthogonal experiment; optimization experiment; mechanical properties

Citation: Zheng, T.; Miao, X.; Kong, D.; Wang, L.; Cheng, L.; Yu, K. Proportion and Performance Optimization of Lightweight Foamed Phosphogypsum Material Based on an Orthogonal Experiment. *Buildings* **2022**, *12*, 207. https://doi.org/10.3390/buildings12020207

Academic Editor: Antonio Francesco Paolo Portioli

Received: 7 January 2022
Accepted: 3 February 2022
Published: 11 February 2022

1. Introduction

Phosphogypsum (PG) is a by-product of solid waste produced from the production of wet-process phosphoric acid; for instance, one ton of phosphoric acid produces 5 tons of phosphogypsum [1,2]. PG is mainly composed of calcium sulfate dihydrate ($CaSO_4 \cdot 2H_2O$), accounting for more than 90% of its structure. In addition, it also contains a small amount of impurities such as phosphoric acid, fluorine, silicon, iron, aluminum and organic matter [2–5]. Currently, approximately 20–30 billion tons of PG are produced annually, which has created tremendous pressure for environmental protection and has created a huge challenge for its recycling [6]. Large amounts of PG not only occupy a large number of land resources, but they also cause serious pollution problems to soil, water, and the atmosphere, as well as to human settlement environments. Therefore, it is a great challenge for governments and relevant enterprises to accelerate PG consumption and to develop new uses.

Currently, PG is mainly used in the cement industry, construction road industry, agriculture, and sulfuric acid and sulfate industries [7]. PG is mainly used in cement production as a mineralizer, retarder, and activator [8–10]. In the most used industry, construction road, PG is mainly used to prepare PG bricks and blocks [11–13], paper gypsum board [2], construction gypsum powder [14], road construction [15,16], and mine filling [17]. Due to advantages such as thermal insulation, sound insulation and fire resistance, researchers at home and abroad have paid more attention to the development of new PG building materials in recent years [18]. Among them, lightweight foam phosphogypsum material (LFPM), with a large consumption of gypsum, has many advantages, such as being light weight and having thermal insulation and sound insulation as well as fire resistance [13,19]. Therefore, investigations regarding the properties of LFPMs have attracted considerable attention in the preparation of phosphogypsum materials. In order to use PG to prepare lightweight building materials, Wang et al. [20] added 25% Portland cement, 10–20% fly ash, 10% ground slag, 6% hydrated lime and 60% foam into PG to prepare lightweight building materials with a compressive strength of 1.7 MPa, bulk density of 521.7 kg/m^3 and thermal conductivity of 0.0724 w/(m·k). Feng et al. [1] studied the effects of foam volume and cement content on thermal conductivity, the water resistance coefficient and the mechanical strength of foamed phosphogypsum, and obtained the effect of the foam content for each performance. Cement can increase thermal conductivity, water resistance and mechanical strength. Additionally, hemihydrate phosphogypsum (HPG), as a base material, is used to prepare composite materials by adding mineral admixtures, alkaline substances, water reducing agents, retarders and cement (Jian Wang [21], Jun Zhou [12], Xiaoyu Guo [22], Zhu Lu [23]). On the basis of this composite material, the foam is used to prepare lightweight phosphogypsum materials. A microscopic analysis was performed to develop well-qualified products meeting the requirements of relevant reference standards. Gypsum exhibits excellent fire resistance, air permeability, sound absorption and decorative effects, which are appropriative to apply to lightweight insulation materials [24,25]. Many researchers have performed numerous studies on the waterproofness of gypsum products and have also made many meaningful achievements [26–31]. One fundamental method is to add organic additives (e.g., paraffins, stearic acid and organosilicon) into gypsum materials, which form a waterproof film on the surface of the gypsum crystals to reduce the dissolution rate of gypsum. Another method is to directly incorporate Portland cement, blast furnace slag or active minerals consisting of amorphous silicon into gypsum to produce hydraulically rigid products. These products are wrapped on the surface of the gypsum to reduce its dissolution rate. Another strategy is to directly spray an organic waterproofing agent onto the surface of the gypsum or to cover the waterproofing layer on the surface of the gypsum. However, these techniques, owing to their temporary effects, cannot fundamentally solve the long-term waterproofing and moisture-proofing problems of gypsum products. In addition, the mechanical strength of gypsum is relatively low, and various fibers are usually used as reinforcing materials in gypsum products to improve their mechanical properties [32–37]. However, most of these above-mentioned studies were based on compact gypsum. Compared with compact gypsum, foam gypsum has a high porosity and large pore size. Moreover, the effects of water repellents, fibers, and admixtures on gypsum-based foam materials have rarely been reported in the existing studies.

In this work, hemihydrate phosphogypsum (HPG) and raw phosphogypsum (RPG) (70:30) were chosen as the main raw materials. The foam, admixture and other additives were mixed to prepare light phosphogypsum materials to investigate the properties. The material proportion and property optimization were carried out by multi-factor orthogonal experiments and optimization tests. The intuitive, range, analysis of variance and comprehensive analysis were investigated in the orthogonal test to obtain the optimal mix proportion. The effects of additives on the properties of LFPMs were further optimized on the basis of the optimal mix proportion. Moreover, the internal morphology of this LFPM was analyzed by scanning electron microscope (SEM).

2. Materials and Methods

2.1. Materials

(1) Phosphogypsum (PG) was divided into raw phosphogypsum (RPG) and hemihydrate phosphogypsum (HPG). RPG: from Guizhou Kai Phosphate phosphogypsum

Comprehensive Utilization Co., Ltd., Guiyang, China, gray, moisture content 21.98%, PH value 6.8. After natural drying and passing through 0.15 mm square hole sieve for backup use; HPG: placed 0.15 mm RPG in 160 °C oven to bake for 2 h, sealed and aged for about 7 d. The raw material and XRD spectrum of RPG and HPG is given in Figure 1, and the morphology of RPG and HPG from scanning electron microscope (SEM) is listed in Figure 2.

(2) Cement: PO 42.5 cement purchased from market; silica fume: produced by Gongyi Hengnuo Filter Co., Ltd., Gongyi, China (gray powder); lime: Yibin Chuanhui Biotechnology Co., Ltd., Yibin, China, Production (white powder); water reducing agent: polycarboxylate water reducing agent (powder), Shanghai Chenqi Chemical Technology Co., Ltd., Shanghai, China, Production; foaming agent: polymer compound foaming agent, produced by Hefei Baile Energy Equipment Co., Ltd., Hefei, China, with foaming multiple > 20 times and PH value of 7.2; foam stabilizer: produced by Hengshui Zhongda New Materials Co., Ltd., Hengshui, China; waterproofing agent: redispersible powder, market. The main chemical components of the main raw materials are shown in Table 1.

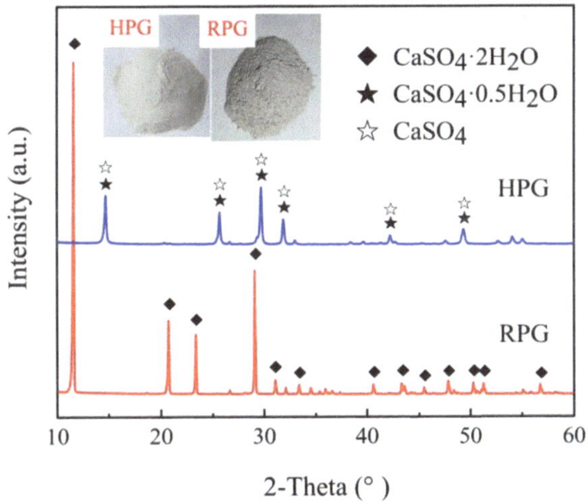

Figure 1. XRD spectra of raw materials RPG and HPG.

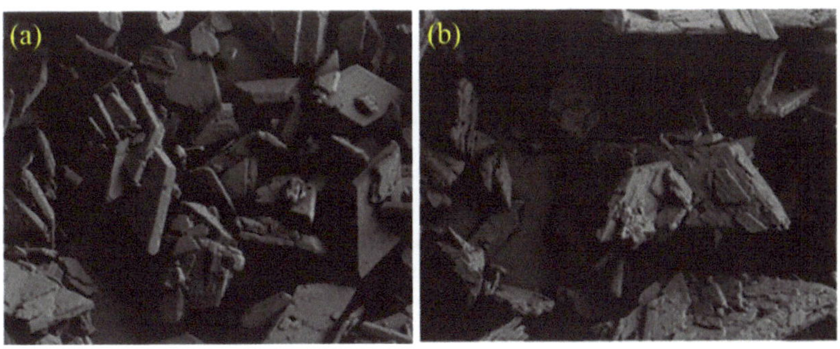

Figure 2. SEM morphology of RPG (**a**) and HPG (**b**).

Table 1. Main chemical compositions of raw materials (wt/%).

Item	SO_3	CaO	SiO_2	P_2O_5	Fe_2O_3	Al_2O_3
RPG	55.28	39.52	2.68	0.89	0.37	0.3
HPG	53.6	41.84	2.71	0.86	0.38	0.29
Silica fume	/	0.11	96.74	0.01	0.08	0.32
cement	3.96	61.71	19.9	0.17	4.46	5.16
quick lime	0.238	98.292	0.599	/	0.111	0.14

2.2. Experimental Design

RPG: HPG relative dosage ratio was set to be 3:7 in the experiments, the ratio between water and material was 0.25, and water reducing agent dosage was 0.72%. The dry mass percentage of foam, cement, silica fume and lime was calculated according to experimental requirements.

(1) Orthogonal experiment

The orthogonal design of the experiment was a method to arrange and analyze experiments with factors and levels utilizing the orthogonal table. Representative tests were selected from all combinations to analyze the comprehensive experiments and to obtain the optimum combination through these test results [38]. In order to study the effect of foam and admixture on the compressive and flexural strength, dry density, and the softening coefficient of the LFPM, a four-factor and four-level orthogonal table L_{16} (4^5) was used to design the experimental ratio. The four factors included foam (A), quicklime (B), silica fume (C) and cement (D). The specific values of each factor level are shown in Table 2, and each mass dosage was the proportion fraction of total dry mass. The orthogonal experiment results were discussed from intuitive, range, variance and comprehensive analysis to determine the optimum proportion.

Table 2. Factor levels of orthogonal experiments.

Levels	Factors			
	A (%)	B (%)	C (%)	D (%)
1	7.0	2.5	2.0	7.5
2	7.5	3.0	3.0	10.0
3	8.0	3.5	4.0	12.5
4	8.5	4.0	5.0	15.0

(2) Optimization experiment

The existing investigations [26,39] indicated that the poor water and moisture resistance of the LFPM limited the application of gypsum. In addition, a large number of bubbles with nonuniform sizes were generated during the early preparation, which had an effect on the property promotion of gypsum-based materials. The foam stabilizer can improve stability and uniformity, and the waterproof agent can reduce the water absorption of the material. Therefore, the influence of the foam stabilizer and waterproof agent on the strength, dry density and softening coefficient of the composites was analyzed. Specifically, based on the recommended optimal mix proportion from orthogonal experiments, foam stabilizer and waterproofing agent were added. According to the relevant references, the foam stabilizer content was 0.1%, 0.2%, 0.3%, 0.4%, 0.5%, and the waterproofing agent content was 1.5%, 3.0%, 4.5%, 6%, 7.5%, respectively. The effect of the dosage on the performance of LFPM was discussed.

(3) Main test instruments

The test instruments in the present study are listed in Table 3.

Table 3. The test instruments.

Serial Number	Apparatus	Model	Manufacturer
1	Electronic weight scale	ZCS	Rui 'an Hao Exhibition Scale Co., Ltd., Guiyang, China
2	Microcomputer controlled pressure testing machine	CXYAW-2000S	Zhejiang Chenxin Machinery Equipment Co., Ltd., Zhejiang, China
3	Automatic cement bending and compression integrated machine	YAW-300	Zhejiang Lixian Test Instrument Manufacturing Co., Ltd., Zhejiang, China
4	Electric drying oven	XMA-2000	Shanghai Qiuzuo Scientific Instruments Co., Ltd., Shanghai, China
5	X-ray diffraction	Empyrean	PANalytical B.V.
6	scanning electron microscope	ZEISS Gemini 300	Thermo Scientific, Germany
7	Electric agitator	OULAIDE	German Olyde Company, Germany
8	Electric vibrating screen machine	ZBSX-92A	Zhejiang Shangyu Zhangxing Yarn Screen Factory, Zhejiang, China
9	Cement mortar test mold	40 × 40 × 160 mm	Zhejiang Qishun Instrument Technology Co., Ltd., Zhejiang, China
10	Concrete test block mold	100 × 100 × 100 mm	Hebei Xinfu Zhengyuan Environmental Protection Equipment Manufacturing Co., Ltd., Hebei, China
11	Thermal conductivity instrument	CD-DR3030	Shenyang Ziweiheng Testing Equipment Co., Ltd., Shenyang, China
12	Micro-cement foaming machine	TH-29A	Zhejiang Tenghe Machinery Co., Ltd., Zhejaing, China

(4) Sample preparation and experimental method

According to the mix proportion of each group, the dry material and additives were poured into the mixing barrel, and the mixing machine was used to evenly stir. After adding water, the foam was poured into the mixing barrel and stirred evenly and then placed into the 100 × 100 × 100 mm and 40 × 40 × 160 mm triple mold, and then vibrated and scraped. After 24 h curing in the natural environment, the molds were removed and maintained for 7, 14, and 28 d and dried to constant weight. The corresponding performance indexes of the specimens at different ages were determined. Dry density and absolute dry compressive strength was measured according to Chinese standard JGT266-2011 foam concrete standard specification [40]. The absolute dry flexural strength was measured according to the Chinese standard GB/T9776-2008 building gypsum determination [41]. The water resistance index softening coefficient test referred to China standard JC/T698-2010 gypsum block [42]. The microstructure was scanned using a scanning electron microscope (SEM, ZEISS MERLIN Compact).

3. Results and Discussions

3.1. Results of Orthogonal Experiments

3.1.1. Intuitive Analysis

After the LFPM samples were cured under natural conditions for 7, 14 and 28 d, the strengths at different ages were measured by compressive and flexural testing machines. Dry density and softening coefficients were measured at 28 d. The test results of compressive strength, flexural strength, dry density and softening coefficient are shown in Table 4.

Table 4. Orthogonal experimental results.

Group Number	Dry Density (kg/m^3)	7 d Compressive Strength (MPa)	14 d Compressive Strength (MPa)	28 d Compressive Strength (MPa)	7 d Flexural Strength (MPa)	14 d Flexural Strength (MPa)	28 d Flexural Strength (MPa)	28 d Softening Coefficient
1	794.3	1.58	1.76	1.68	0.56	0.6	0.67	0.625
2	876.5	2.90	3.36	3.52	1.00	1.29	1.15	0.645
3	898.5	3.20	4.33	4.38	1.15	1.39	1.37	0.633
4	885.8	3.63	4.08	4.85	1.19	1.34	1.54	0.703
5	873.0	3.36	3.43	3.85	0.74	1.01	1.24	0.727
6	787.1	2.05	2.27	2.46	0.80	0.52	0.74	0.707
7	812.8	2.16	2.39	2.54	0.67	0.58	0.82	0.555
8	804.3	2.14	2.38	2.51	0.59	0.68	0.69	0.637
9	774.3	1.84	2.69	3.06	0.66	0.64	0.96	0.578
10	756.7	1.90	2.13	2.51	0.56	0.71	0.79	0.610
11	777.0	1.92	1.97	1.77	0.62	0.68	0.68	0.684
12	890.7	3.34	4.15	3.89	1.12	1.42	1.17	0.596
13	783.5	2.25	2.41	2.46	0.70	0.76	0.76	0.569
14	813.6	2.45	2.45	2.80	0.94	0.83	0.86	0.500
15	812.8	3.16	3.08	3.10	1.00	1.02	0.92	0.742
16	821.7	2.41	2.46	2.11	0.69	0.66	0.68	0.768

The strength of the fourth group was highest, and the dry density and softening coefficient were also relatively high. The 28 d compressive and flexural strength reached 4.85 and 1.54 MPa, respectively. The dry density was 885.8 kg/m^3, and the 28 d softening coefficient was 0.703. Figure 3 shows the comparison of 7, 14, and 28 d compressive strength, flexural strength, 28 d softening coefficient, and dry density for 16 groups of specimens. From Figure 3a,b, a small strength difference of the LFPM between 14 and 28 d was observed, indicating that most of the hydration reactions were completed after 14 d of curing. In the test, the dry density of the 10th group was 756.7 kg/m^3, which was the minimum value in the present study. For the 10th group, the 28 d compressive and flexural strengths were 2.51 and 0.79 MPa, respectively, and the softening coefficient was 0.61.

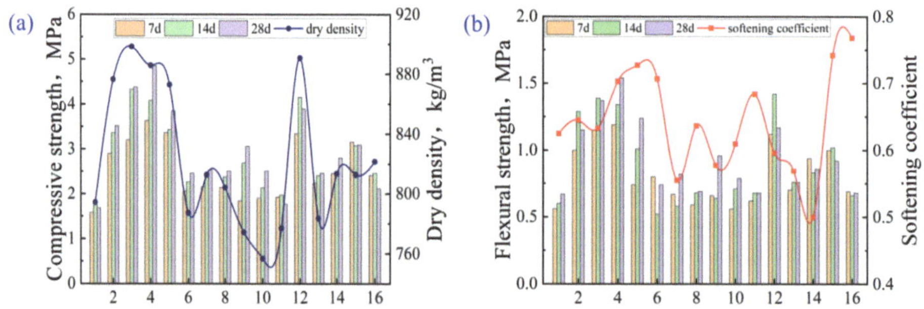

Figure 3. Test results of (**a**,**b**) orthogonal samples.

3.1.2. Range Analysis

The range analysis method refers to the R method [43]. In the analysis of the orthogonal experimental results, the larger the R value of a certain factor, the greater the influence of this factor on the test index. Therefore, the importance of each factor was determined by the R value in this work. The calculation formula of the range (R) is as follows:

$$K_{mn} = \frac{1}{N} * \sum_{i=1}^{N} p_i \tag{1}$$

$$R_m = \max(K_{m1}, K_{m2}, \ldots, K_{mn}) - \min(K_{m1}, K_{m2}, \ldots, K_{mn}) \quad (2)$$

In the formula: K_{mn} is the average value of the corresponding index at the n level of the m-th factor; p_i is the index value; R_m is the range of factor m.

The compressive strength, flexural strength, dry density and softening coefficient of the LFPM at different ages (7, 14 and 28 d) were analyzed by range analysis, and the corresponding results are summarized in Table 5.

Table 5. Results of range analysis.

Types of Range		Level	A	B	C	D	Significance	Optimal Solution
compressive strength (MPa)	R_{7d}	k1	2.83	2.26	1.99	2.38	C > B > A > D	$A_1B_4C_2D_3$
		k2	2.43	2.33	3.19	2.3		
		k3	2.25	2.61	2.41	2.72		
		k4	2.57	2.88	2.49	2.67		
		Range	0.58	0.62	1.2	0.42		
	R_{14d}	k1	3.38	2.57	2.12	2.69	C > A > B > D	$A_1B_4C_2D_3$
		k2	2.62	2.55	3.51	2.53		
		k3	2.74	2.94	2.96	3.09		
		k4	2.6	3.27	2.75	3.03		
		Range	0.78	0.72	1.39	0.56		
	R_{28d}	k1	3.61	2.76	2.01	2.73	C > A > D > B	$A_1B_4C_2D_4$
		k2	2.84	2.82	3.59	2.57		
		k3	2.81	2.95	3.19	3.21		
		k4	2.62	3.34	3.09	3.37		
		Range	0.99	0.58	1.59	0.8		
flexural strength (MPa)	R_{7d}	k1	0.98	0.67	0.67	0.82	C > A > B > D	$A_1B_4C_2D_4$
		k2	0.70	0.83	0.97	0.73		
		k3	0.74	0.86	0.84	0.79		
		k4	0.83	0.90	0.78	0.91		
		Range	0.28	0.23	0.3	0.18		
	R_{14d}	k1	1.16	0.75	0.62	0.86	C > A > B > D	$A_1B_4C_2D_3$
		k2	0.70	0.84	1.19	0.85		
		k3	0.86	0.92	0.89	0.94		
		k4	0.82	1.03	0.85	0.88		
		Range	0.46	0.28	0.57	0.09		
	R_{28d}	k1	1.18	0.91	0.69	0.88	C > A > D > B	$A_1B_4C_2D_4$
		k2	0.87	0.89	1.12	0.82		
		k3	0.90	0.95	0.97	1.02		
		k4	0.81	1.02	0.98	1.04		
		Range	0.38	0.14	0.43	0.22		
dry density (kg/m^3)	R_{28d}	k1	863.8	806.3	795	827.9	C > A > B > D	$A_3B_1C_1D_2$
		k2	819.3	808.5	863.3	810.3		
		k3	799.7	825.3	822.7	837.5		
		k4	807.9	850.6	809.7	815.0		
		Range	64.1	44.4	68.2	27.1		
softening coefficient	R_{28d}	k1	0.652	0.625	0.696	0.569	D > C > B > A	$A_2B_4C_1D_3$
		k2	0.657	0.616	0.678	0.634		
		k3	0.617	0.654	0.587	0.685		
		k4	0.645	0.676	0.609	0.683		
		Range	0.040	0.061	0.109	0.116		

It can be seen from Table 5 that the primary and secondary order of influencing factors of 28 d strength was silica fume > foam > cement > quick lime (C > A > D > B); the primary and secondary order of influencing factors of 14 d strength was silica fume > foam > quicklime > cement (C > A > B > D); the order of influencing factors of 7 d compressive and flexural strength was silica fume > quicklime > foam > cement (C > B > A > D), and silica fume > foam > quicklime > cement (C > A > B > D). Therefore, the silica fume content was the first major factor affecting the early and late strength of LFPM. Foam content was the second major factor on 7 d flexural strength and 14 and 28 d compressive and flexural strengths. Compared with the cement content, the lime content occupied the main position in the early stage, and the influence of cement was more obvious in the later stage. The order of dry density influence factor was silica fume > foam > lime > cement (C > A > B > D). For LFPM, the smaller dry density represented the better property. From Table 4, the influences of various factors on dry density are explained by the range results. The range of silica fume and foam content was between 863.8 and 863.3, which stated that the silica fume and foam content had a primary influence on the dry density. The primary and secondary factors for the softening coefficient were cement > quicklime > silica fume > foam (D > C > B > A). This result can show that cement is a major factor on the softening coefficient of LFPM, followed by silica fume. It can be summarized from Table 5 that the silica fume content has the greatest influence on the compressive strength, flexural strength and dry density of each age. Regarding the above results, the optimal ratio scheme of each performance is listed in Table 5.

3.1.3. Analysis of Variance

Analysis of variance (ANOVA) is the most common statistical processing method for experimental results [38] and is used to determine a significant effect factor of LFPM. ANOVA can distinguish the reason for the result difference between each level of each factor (different factor level or experimental error) [4]. The total variation values in this experiment were composed of four parts: factors A, B, C, and D; thus, the corresponding error variation was calculated. Therefore, the decomposition formula of the square sum and degree of freedom in variance analysis is:

$$SS_T = SS_A + SS_B + SS_C + SS_D + SSe \tag{3}$$

$$df_T = df_A + df_B + df_C + df_D + df_e \tag{4}$$

n represents the number of the tests; a, b, c and d represent the level of different factors (A, B, C and D); k_a, k_b, k_c and k_d represent the level under repetition of factors A, B, C and D. In this experiment, $n = 16$, $a = b = c = d = 4$, $k_a = k_b = k_c = k_d = 4$. The equations from (3) to (9) were utilized to calculate the variation and degrees of freedom caused by factors A–D.

$$C_0 = T^2/n \tag{5}$$

$$SS_T = \Sigma x^2 - C_0 \tag{6}$$

$$SS_A = \Sigma T_A^2/k_a - C_0 \tag{7}$$

$$SS_B = \Sigma T_B^2/k_b - C_0 \tag{8}$$

$$SS_C = \Sigma T_C^2/k_c - C_0 \tag{9}$$

$$SS_D = \Sigma T_D^2/k_d - C_0 \tag{10}$$

$$df_e = df_T - df_A - df_B - df_C - df_D \tag{11}$$

C_0 is the correction number; SS_T is the total sum of squares; SS_A is the sum of squares of factor A; SS_B is the sum of squares of factor B; SS_C is the sum of squares of factor C; SS_D is the sum of squares of factor D; df_A, df_B, df_C and df_D are the degrees of freedom of factors A, B, C and D, respectively; df_T is the total degree of freedom; df_e is the degree of freedom of error.

According to the above calculation rules, ANOVA was conducted on the strength, dry density and softening coefficient at 28 d, and the experimental results are shown in Table 6. The meaning of each indicator in Table 6 is as follows: (1) The source of difference comes from the factor, interaction or error. (2) SS is the sum of squares between the factor and error. (3) DF (degree of freedom) is the degree of freedom of each factor, which is the difference between level factor number and 1. Since the degree of freedom of each factor was 4 in the orthogonal experiment, the degree of freedom of each factor was 3. (4) MS (mean square) is the mean square divided by degrees of freedom. (5) The F value is the ratio of two mean squares (effect term/error term). The larger F value (compared with the standard F value at a given significant indigenous level) indicates the more obvious effect (difference). The smaller error term represents the higher test accuracy. The F value is obtained by dividing the effect value MS by the error MS, and the ratio is compared with the critical value F in the table to determine a significant factor. In Table 6, * is indicated in F(0.1). The ANOVA results of the LFPM are shown in Table 6, indicating that the results from the significant variance analysis of different factor influences are consistent with the range analysis results.

Table 6. Results of ANOVA.

Item	Factor	SS	DF	MS (Effect)	MS9 (Error)	F	Significance
dry density (kg/m^3)	A	9792	3	3263.9	1940.0	1.68	C > A > B > D
	B	5034	3	1678.1	1940.0	0.86	
	C	10317	3	3438.9	1940.0	1.77	
	D	1829	3	609.7	1940.0	0.31	
28 d compressive strength (MPa)	A	2.30	3	0.77	0.70	1.09	C > A > D > B
	B	0.81	3	0.27	0.70	0.39	
	C	5.51	3	1.84	0.70	2.62	
	D	1.76	3	0.59	0.70	0.84	
28 d flexural strength (MPa)	A	0.33	3	0.111	0.074	1.49	C > A > D > B
	B	0.04	3	0.014	0.074	0.19	
	C	0.38	3	0.128	0.074	1.72	
	D	0.14	3	0.045	0.074	0.62	
28 d softening coefficient	A	0.0037	3	0.00124	0.00072	1.72	D* > C* > B > A
	B	0.0091	3	0.00305	0.00072	4.22	
	D	0.0331	3	0.01103	0.00072	15.26	
	C	0.0354	3	0.01179	0.00072	16.31	

The optimal mix ratio of the dry density test was $A_3B_1C_1D_2$ (foam 8%, quicklime 2.5%, silica fume 2%, cement 10%). The optimal proportion for 28 d compressive strength and flexural strength was $A_1B_4C_2D_4$ (7% foam, 4% quicklime, 3% silica fume and 15% cement). The optimal mix proportion of 28 d softening coefficient was $A_2B_4C_1D_3$, (7.5% foam, 4% quicklime, 2% silica fume and 12.5% cement). The critical value of F(0.1) was 5.39; therefore, cement and silica fume content have significant effects on the 28 d softening coefficient at F(0.1) and have no obvious effect on 28 d strength and dry density.

3.1.4. Comprehensive Analysis

The orthogonal experimental results of the LFPM were comprehensively analyzed for each performance index to obtain the optimal mix proportion. The 28 d compressive strength, flexural strength, dry density and 28 d softening coefficient were selected as performance indexes. The compressive strength and dry density were the main indexes, and the flexural strength and softening coefficient were the secondary indexes. The influences of various factors on 28 d compressive strength, flexural strength, dry density and softening coefficient of this material are shown in Figure 4.

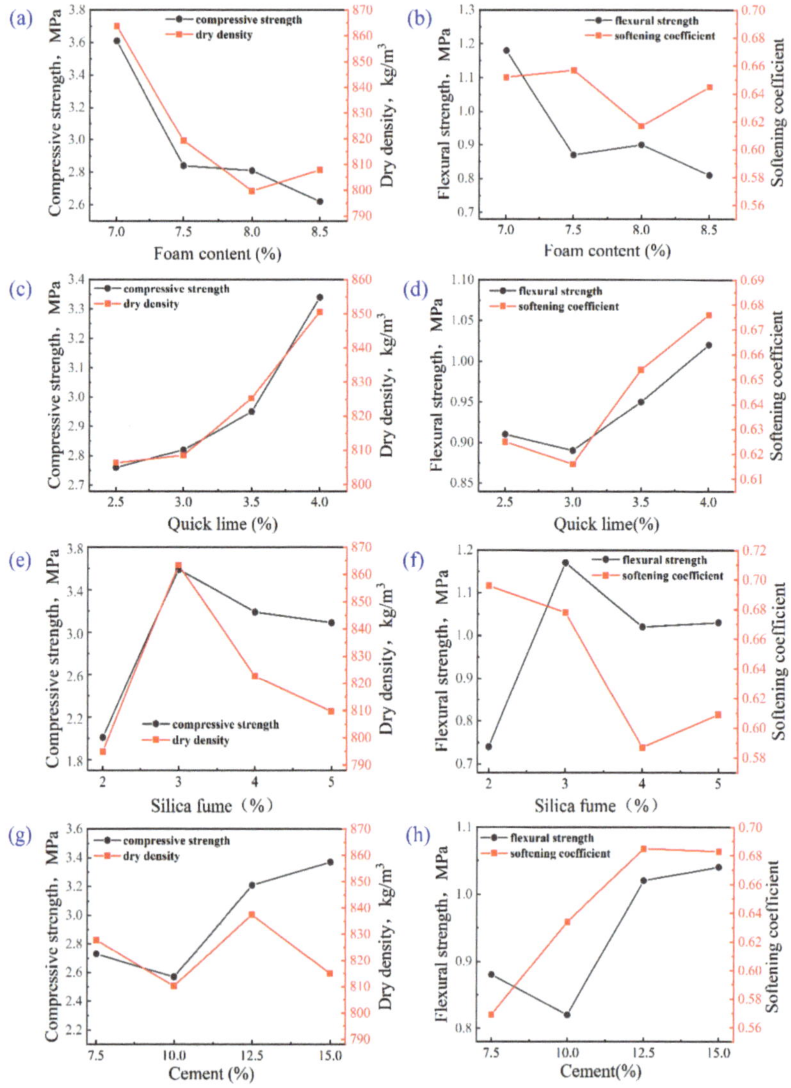

Figure 4. Relationship between strength, dry density and softening coefficient of each factor. (**a**) Effects of foam content on compressive strength and dry density;(**b**) Effects of foam content on flexural strength and softening coefficient; (**c**) Effects of quick lime on compressive strength and dry density; (**d**) Effects of quick lime on flexural strength and softening coefficient; (**e**) Effects of silica fume on compressive strength and dry density; (**f**) Effects of silica fume on flexural strength and softening coefficient; (**g**) Effects of cement on compressive strength and dry density; (**h**) Effects of cement on flexural strength and softening coefficient.

It can be seen from Figure 4a,b that with the increase in foam content, the compressive and flexural strengths decreased and increased slightly at a foam content of 8%. The softening coefficient fluctuated up and down, and the range of softening coefficient of the LFPM was 0.04 in the range analysis, illustrating a small effect. However, the minimum dry density appeared at the foam content of 8%. The dry density and compressive strength were considered first to select the optimal dosage. Therefore, the foam content was selected to be 8.0% according to the experimental results of the dry density, compressive strength and flexural strength.

Figure 4c,d shows that the compressive strength and dry density increased significantly with an increase in lime content. The flexural strength and softening coefficient decreased at first and then increased significantly with an increase in lime content. When the content of quicklime was 4%, the properties of the composites reached the maximum value, and the compressive strength, flexural strength, dry density and softening coefficient were 3.34 MPa, 1.02 MPa, 850.6 kg/m^3 and 0.676, respectively. However, when the content was 4%, some microcracks were observed on the surface, which affected the smoothness and cleanliness of the specimen. This was because excessive lime was added to generate needle-column calcium vanadate and a large number of C-S-H gel expanded, causing microcracks. However, when the lime content was 3.5%, the surface of the LFPM specimen was smooth and tidy, and fewer microcracks were observed. Moreover, at a content of 3.5%, the compressive strength, flexural strength, dry density and softening coefficient were 2.95 MPa, 0.95 MPa, 825.3 kg/m^3 and 0.654, respectively. Compared with the content of 4%, the flexural strength and softening coefficient exhibited small differences, and the dry density decreased by 25.3 kg/m^3. Therefore, the lime content was selected to be 3.5% in this study.

Figure 4e,f shows the effects of the silica fume content on the compressive strength, dry density, flexural strength and softening coefficient of the LFPM. The dry density, compressive strength and flexural strength first increased rapidly and then decreased at the transition content of 3%. Although the dry density of this material was highest at 3% content, the compressive and flexural strength reached the maximum value (3.59 and 1.12 MPa). Moreover, the softening coefficient appeared to be a rapid reduction as the content exceeded 3% from Figure 4f. When the appropriate amount of silica fume was added in the excitation of the alkaline environment, due to the good chemical activity and micro-aggregate effect of silica fume [44], the main component SiO_2 quickly reacted with Ca $(OH)_2$ to generate a large number of C-S-H gels, which compacted the internal structure of the matrix, improving the dry density and strength of the LFPM. However, excessive silica fume can result in hydration heat generated by the chemical reaction inside the sample to produce temperature stress. The higher temperature stress can lead to the microcracks at the interfaces, which was seen to reduce the strength of materials. Therefore, the optimal content of silica fume was 3% according to the effect of silica fume content on the compressive and flexural strength of the LFPM.

From Figure 4g,h, the softening coefficient of this material increased until 12.5% with an increase in cement content and exhibited a slight decrease at 15%. The compressive strength and flexural strength exhibited a decrease at 10% content and increased during the content range from 10% to 15%. Additionally, the dry density fluctuated up and down during the range from 10% to 15%. The appropriate addition of cement hydrated to form acicular and reticular ettringite crystals; therefore, the strength and softening coefficient of the LFPM increased due to the high hardness of the ettringite crystals [22]. The hydration process of the cement absorbed water; thus, the water demand of the material relatively increased. Ordinary Portland cement, as an additive, improved the water resistance of phosphogypsum in the LFPM and increased the compactness of the gypsum products [45]. The tricalcium aluminate hydrated in Portland cement to form calcium aluminate crystals, and $SO_4{}^{2-}$ ionized by $CaSO_4 \cdot 2H_2O$ in the calcium aluminate crystal binding system to form ettringite crystals. The chemical reaction formula is shown in Equation (12). Therefore, the strength and softening coefficient of this composite can improve by adding an appropriate amount of cement. At the cement content of 15%, the compressive strength, flexural strength and dry density was 3.37 MPa, 1.04 MPa and 815 kg/m^3, respectively. Therefore, the optimal dosage of 15% was determined from the softening coefficient and strength results.

$$3CaO \cdot Al_2O_3 + 3CaSO_4 \cdot 2H_2O + 26H_2O \rightarrow 3CaO \cdot Al_2O_3 \cdot 3CaSO_4 \cdot 32H_2O \quad (12)$$

Based on the effects of the above factors on the performance of the LFPM, the optimal mix proportion of various factors was determined to be $A_3B_3C_2B_4$ from the orthogonal experiments of the LFPM; namely, the foam content was 8%, the lime content was 3.5%, the silica fume content was 3%, and the cement content was 15%. In order to verify this mixture ratio, the verification experiments were conducted on the $A_3B_3C_2B_4$ specimen, and the experimental results are summarized in Table 7. These data revealed that the performance parameters of $A_3B_3C_2B_4$ can meet the standard requirements of A09- and C3-qualified products in JGT266-2011 foam concrete. The test block with the smooth surface of foamed phosphogypsum material prepared under the combination ratio of $A_3B_3C_2B_4$ is shown in Figure 5.

Table 7. Verification experiment results of the optimal mix ratio.

	28 d Compressive Strength (MPa)	Dry Density (kg/m^3)	28 d Flexural Strength (MPa)	28 d Softening Coefficient
Results	3.15	809.1	0.97	0.628

Figure 5. The surface of the $A_3B_3C_2B_4$ specimen.

3.2. Optimization Experimental Results

3.2.1. Results analysis

Tables 8 and 9 give the experimental results of $A_3B_3C_2B_4$ specimens with foam stabilizer and waterproofing agents, respectively. The effects of foam stabilizer content and waterproofing agent content on the performance parameters are summarized in Figure 6.

Table 8. Optimization results of specimens with different foam stabilizer contents.

Group Number	Foam Stabilizer (wt%)	Dry Density (kg/m^3)	7 d Compressive Strength (MPa)	28 d Compressive Strength (MPa)	7 d Flexural Strength (MPa)	28 d Flexural Strength (MPa)	28 d Softening Coefficient
1	0	809.1	2.84	3.15	0.63	1.03	0.628
2	0.1	812.4	3.10	3.55	0.56	1.09	0.635
3	0.2	830.3	3.80	4.08	0.97	1.26	0.603
4	0.3	805.6	2.97	3.01	0.56	0.98	0.636
5	0.4	806.7	2.75	3.09	0.5	0.97	0.71
6	0.5	810.1	2.80	3.10	0.85	1.02	0.822

Table 9. Optimization results of specimens with different waterproofing agent contents.

Group Number	Waterproofing Agent (wt%)	Dry Density (kg/m^3)	7 d Compressive Strength (MPa)	28 d Compressive Strength (MPa)	7 d Flexural Strength (MPa)	28 d Flexural Strength (MPa)	28 d Softening Coefficient	Water Absorption
1	0	809.1	2.84	3.15	0.63	1.03	0.628	0.542
2	1.5	812.3	2.75	3.17	0.86	1.16	0.635	0.538
3	3.0	840.6	3.05	3.89	0.98	1.49	0.603	0.426
4	4.5	896.8	5.5	6.47	1.59	2.27	0.636	0.382
5	6.0	1013.8	7.8	8.6	2.29	2.77	0.71	0.238
6	7.5	865.3	4.85	5.31	1.91	2.31	0.822	0.442

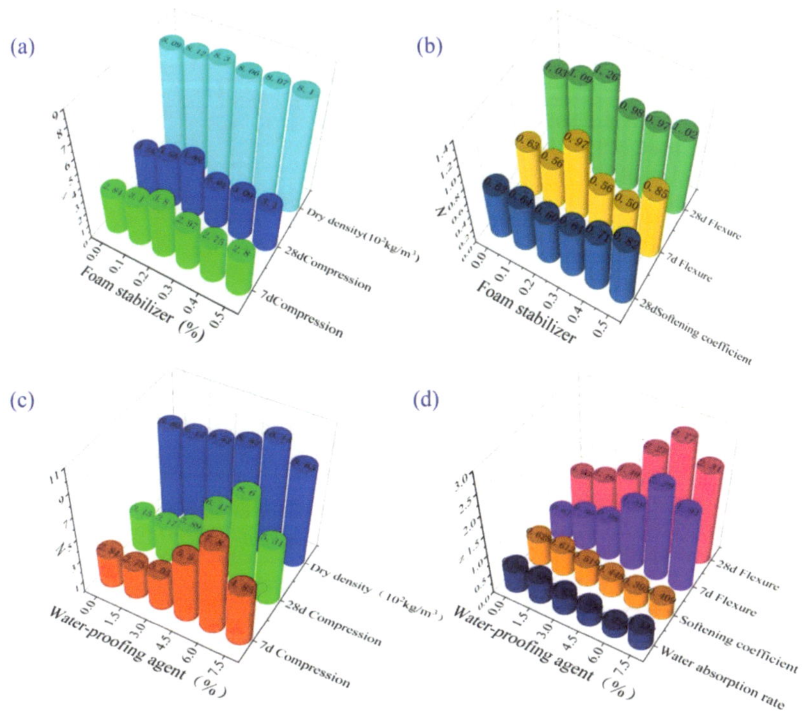

Figure 6. The effects of foam stabilizer and waterproofing agent content on the properties of the LFPM. (**a**,**b**) Effects of the foam stabilizer, (**c**,**d**) Effects of the water-proofing agent.

It can be seen from Table 8 and Figure 6a,b that with the increase in foam stabilizer, the strength and dry density first increased and then decreased, and the softening coefficient increased. The appropriate foam stabilizer made the foam more stable and uniform and made the internal structure denser. As the foam stabilizer content was 0.2%, the compressive strength, flexural strength and dry density at 7 and 28 d reached the maximum value. The compressive and flexural strengths at 28 d were 4.08 and 1.26 MPa, respectively. The dry density was 830.3 kg/m^3, and the softening coefficient was 0.6. It was suggested that the optimal content of the foam stabilizer was located in the range from 0.1% to 0.2%. From Table 9 and Figure 6c,d, the waterproof agent content exhibited a significant effect on the performance of the foamed phosphogypsum material. With an increase in the waterproof agent content, the strength, dry density and water absorption increased; however, the softening coefficient decreased. This was because the waterproof agent (latex powder) dispersed in water enhanced the flexibility and adhesion of the material, thereby increasing the strength of the material [13]. Moreover, the bubbles became smaller and denser, and

Abbreviations

PG	phosphogypsum
RPG	raw phosphogypsum
HPG	hemihydrate phosphogypsum
LFPM	lightweight foam phosphogypsum material
SEM	scanning electron microscope
XRD	X-ray diffraction
R method	range analysis method
ANOVA	analysis of variance
SS	sum of squares
DF	degree freedom
MS	mean square

References

1. Feng, L.; Jin, K.; Wang, H. Research on the Thermal Conductivity and Water Resistance of Foamed Phosphogypsum. *Coatings* **2021**, *11*, 802. [CrossRef]
2. Zhou, J.; Li, X.; Zhao, Y.; Shu, Z.; Wang, Y.; Zhang, Y.; Shen, X. Preparation of paper-free and fiber-free plasterboard with high strength using phosphogypsum. *Constr. Build. Mater.* **2020**, *243*, 118091. [CrossRef]
3. Fornés, I.V.; Vaiiukynien, D.; Nizeviien, D.; Dorosevac, V.; Michalik, B. A comparative assessment of the suitability of phosphogypsum from different origins to be utilised as the binding material of construction products. *J. Build. Eng.* **2021**, *44*, 102995. [CrossRef]
4. Zhang, L.; Zhang, A.; Li, K.; Wang, Q.; Han, Y.; Yao, B.; Gao, X.; Feng, L. Research on the pretreatment and mechanical performance of undisturbed phosphogypsum. *Case Stud. Constr. Mater.* **2020**, *13*, e00400. [CrossRef]
5. Rashad, A.M. Phosphogypsum as a construction material. *J. Clean. Prod.* **2017**, *166*, 732–743. [CrossRef]
6. Cao, Y.; Cui, Y.; Yu, X.; Li, T.; Chang, I.; Wu, J. Bibliometric analysis of phosphogypsum research from 1990 to 2020 based on literatures and patents. *Environ. Sci. Pollut. Res.* **2021**, *28*, 66845–66857. [CrossRef]
7. Liu, L.; Zuo, H.; Xu, Z. Resource utilization approach of industrial gypsum and its prospect. *Inorg. Chem. Ind.* **2021**, *53*, 1–9. (In Chinese)
8. Liu, S.; Fang, P.; Ren, J.; Li, S. Application of Lime Neutralised Phosphogypsum in Supersulfated Cement. *J. Clean. Prod.* **2020**, *272*, 122660. [CrossRef]
9. Jin, Z.; Ma, B.; Su, Y.; Lu, W.; Qi, H.; Hu, P. Effect of calcium sulphoaluminate cement on mechanical strength and waterproof properties of beta-hemihydrate phosphogypsum. *Constr. Build. Mater.* **2020**, *242*, 118198. [CrossRef]
10. Zeng, L.L.; Bian, X.; Zhao, L.; Wang, Y.J.; Hong, Z.S. Effect of phosphogypsum on physiochemical and mechanical behaviour of cement stabilized dredged soil from Fuzhou, China. *Geomech. Energy Environ.* **2020**, *25*, 100195. [CrossRef]
11. Shi, Y. Study on Preparation Technology of Cement-Based and Phosphogypsum-Based Foams. Ph.D. Thesis, Kunming University of Science and Technology, Kunming, China, 2020. (In Chinese)
12. Zhou, J.; Hui, G.; Zhu, S.; Wang, Y.; Yan, C. Utilization of waste phosphogypsum to prepare non-fired bricks by a novel Hydration–Recrystallization process. *Constr. Build. Mater.* **2012**, *34*, 114–119. [CrossRef]
13. Wang, G. Experimental Study on Fit Ratio and Basic Properties of Foamed Phosphogypsum Wall Material. Ph.D. Thesis, Zhengzhou University, Zhengzhou, China, 2019. (In Chinese)
14. Zhang, K. Study on Preparation of Building Gypsum Powder by Modifying Agent—Calcination Combined with Modified Wet Fluoro Gypsum. Ph.D. Thesis, Central South University, Changsha, China, 2013. (In Chinese)
15. Amrani, M.; Taha, Y.; Kchikach, A.; Beznaazoua, M.; Hakkou, R. Phosphogypsum recycling: New horizons for a more sustainable road material application. *J. Build. Eng.* **2020**, *30*, 101267. [CrossRef]
16. Folek, S.; Walawska, B.; Wilczek, B.; Miskiewicz, J. Use of phosphogypsum in road construction. *Pol. J. Chem. Technol.* **2011**, *13*, 18–22. [CrossRef]
17. Jiang, G.; Wu, A.; Wang, Y.; Lan, W. Low cost and high efficiency utilization of hemihydrate phosphogypsum: Used as binder to prepare filling material. *Constr. Build. Mater.* **2018**, *167*, 263–270. [CrossRef]
18. Liu, F.; Wu, J.; Wang, C. Study on properties of yellow phosphorus slag—Phosphogypsum composite cementitious material. *Mater. Rep.* **2007**, *5*, 530–532. (In Chinese)
19. Shi, C. The Research on Preparation and Performance of Gypsum-Based Foam Insulation Material. Ph.D. Thesis, Hunan University, Changsha, China, 2016. (In Chinese)
20. Wang, T.; Gao, X.J.; Wang, J. Preparation of Foamed Phosphogypsum Lightweight Materials by Incorporating Cementitious Additives. *Mater. Sci.* **2019**, *25*, 340–347. [CrossRef]
21. Wang, J. Preparation and Basic Properties of Foamed Phosphogypsum Lightweight Material. Ph.D. Thesis, Harbin Institute of Technology, Harbin, China, 2016. (In Chinese)
22. Guo, X.; Singh, S.K.; Zhou, C.; Ling, X.; Li, J.; Fan, C. Preparation and Characterization of Lightweight Wall Materials Based on a Binder Mainly Including Phosphor-gypsum. *J. Adv. Concr. Technol.* **2020**, *18*, 689–698. [CrossRef]

23. Zhu, L. Preparation and properties of phosphogypsum-based functional foam gypsum. Ph.D. Thesis, Wuhan University of Technology, Wuhan, China, 2018. (In Chinese)
24. Karni, J.; Karni, E. Gypsum in construction: Origin and properties. *Mater. Struct.* **1995**, *28*, 92–100. [CrossRef]
25. Lushnikova, N.; Dvorkin, L. Sustainability of gypsum products as a construction material. *Sustain. Constr. Mater.* **2016**, *25*, 643–681.
26. Kondratieva, N.; Barre, M.; Goutenoire, F.; Sanytsky, M. Study of modified gypsum binder. *Constr. Build. Mater.* **2017**, *149*, 535–542. [CrossRef]
27. Li, J.; Li, G.; Yu, Y. The influences of gypsum water-proofing additive on gypsum crystal growth. *Mater. Lett.* **2007**, *61*, 872–876. [CrossRef]
28. Singh, M.; Garg, M. Investigation of a durable gypsum binder for building materials. *Constr. Build. Mater.* **1992**, *6*, 52–56. [CrossRef]
29. Singh, M.; Garg, M.; Rehsi, S. Durability of phosphogypsum based water-resistant anhydrite binder. *Cem. Concr. Res.* **1990**, *20*, 271–276. [CrossRef]
30. Magallanes-Rivera, R.; Juarez-Alvarado, C.; Valdez, P.; Mendoza-Rangel, J.M. Modified gypsum compounds: An ecological-economical choice to improve traditional plasters. *Constr. Build. Mater.* **2012**, *37*, 591–596. [CrossRef]
31. Singh, M.; Garg, M. Relationship between mechanical properties and porosity of water-resistant gypsum binder. *Cem. Concr. Res.* **1996**, *26*, 449–456. [CrossRef]
32. Hua, S.; Wang, K.; Yao, X.; Xu, W.; He, Y. Effects of fibers on mechanical properties and freeze-thaw resistance of phosphogypsum-slag based cementitious materials. *Constr. Build. Mater.* **2016**, *121*, 290–299. [CrossRef]
33. Gencel, O.; Jose, D.; Sutcu, M.; Koksal, F.; Rabanal, F.P.A.; Martinez-Garrera, G. A novel lightweight gypsum composite with diatomite and polypropylene fibers. *Constr. Build. Mater.* **2016**, *113*, 732–740. [CrossRef]
34. Martias, C.; Joliff, Y.; Favotto, C. Effects of the addition of glass fibers, mica and vermiculite on the mechanical properties of a gypsum-based composite at room temperature and during a fire test. *Compos. B Eng.* **2014**, *62*, 37–53. [CrossRef]
35. Iucolan, F.; Liguori, B.; Aprea, P.; Caputo, D. Thermo-mechanical behaviour of hemp fibers-reinforced gypsum plasters. *Constr. Build. Mater.* **2018**, *185*, 256–263. [CrossRef]
36. Zhu, C.; Zhang, J.; Peng, J.; Cao, W.; Liu, J. Physical and mechanical properties of gypsum-based composites reinforced with PVA and PP fibers. *Constr. Build. Mater.* **2018**, *163*, 695–705. [CrossRef]
37. Gencel, O.; Diaz, J.; Sutcu, M.; Koksal, F.; Rabanal, F.P.A.; Martinez-Barrera, G.; Brostow, W. Properties of gypsum composites containing vermiculite and polypropylene fibers: Numerical and experimental results. *Energy Build.* **2014**, *70*, 135–144. [CrossRef]
38. Feng, J.; Yin, G.; Tuo, H.; Nu, Z. Parameter optimization and regression analysis for multi-index of hybrid fiber-reinforced recycled coarse aggregate concrete using orthogonal experimental design. *Constr. Build. Mater.* **2020**, *267*, 121013. [CrossRef]
39. Guo, L.; Jian, L.; Rui, G. Investigation of the water resistance of gypsum materials. *ZKG Int.* **2003**, *56*, 87–93.
40. JG/T 266-2011, Foamed Concrete, China Standards Press, Beijing, China. Available online: https://www.chinesestandard.net/PDF/English.aspx/JGT266-2011 (accessed on 6 January 2022).
41. GB/T 9776-2008, Building Gypsum, China Standards Press, Beijing, China. Available online: https://www.chinesestandard.net/PDF/English.aspx/GBT9776-2008 (accessed on 6 January 2022).
42. JC/T 698-2010, Gypsum Block, China Standards Press, Beijing, China. Available online: https://www.chinesestandard.net/PDF/English.aspx/JCT698-2010 (accessed on 6 January 2022).
43. Zhao, F.; Hu, J.; Zeng, P.; Wang, L.; Zhao, L. Optimization of the ratio of base phosphogypsum cemented backfill based on orthogonal experiment. *Chin. J. Nonferrous Met.* **2021**, *31*, 10. (In Chinese)
44. Nili, M.; Afroughsabet, V. Combined effect of silica fume and steel fibers on the impact resistance and mechanical properties of concrete. *Int. J. Impact Eng.* **2010**, *37*, 879–886. [CrossRef]
45. Zhang, F.; Li, G.; Li, J.; Ma, Y.; Tang, H.; Zhang, J.; Liu, L. The Experimental Research on Additive Improving Water Resistance of Gypsum. *Non-Met. Mines* **2015**, *38*, 26–28. (In Chinese)

MDPI AG
Grosspeteranlage 5
4052 Basel
Switzerland
Tel.: +41 61 683 77 34

Buildings Editorial Office
E-mail: buildings@mdpi.com
www.mdpi.com/journal/buildings